SOME MASSES

Atomic mass unit	$1\,u = 1.660 \times 10^{-27}\,kg = 931.5\,MeV/c^2$
Electron	$m_e = 9.11 \times 10^{-31}\,kg = 5.49 \times 10^{-4}\,u = 0.511\,MeV/c^2$
Proton	$m_p = 1.673 \times 10^{-27}\,kg = 1.007\,u = 938.3\,MeV\,c^2$
Neutron	$m_n = 1.675 \times 10^{-27}\,kg = 1.009\,u = 939.6\,MeV/c^2$
Hydrogen atom	$M_H = 1.674 \times 10^{-27}\,kg = 1.008\,u = 938.8\,MeV/c^2$
Helium atom	$M_{^4He} = 6.65 \times 10^{-27}\,kg = 4.003\,u = 3728\,MeV/c^2$
Alpha particle	$m_\alpha = 6.64 \times 10^{-27}\,kg = 4.002\,u = 3727\,MeV/c^2$

MISCELLANEOUS CONSTANTS

Standard acceleration of gravity	$g = 9.8\,m/s^2$
Density of water	$1000\,kg/m^3$
Density of air (0°C, 1 atm)	$1.29\,kg/m^3$
Speed of sound in air (0°C, 1 atm)	$331\,m/s$
Molecular mass of air	$28.98\,g/mole$
Mechanical equivalent of heat	$1\,cal = 4.19\,J$

CONVERSION FACTORS

year	$1\,year = 3.16 \times 10^7\,s$
light-year	$1\,light\text{-}year = 9.46 \times 10^{15}\,m$
angstrom	$1\,Å = 10^{-10}\,m$
fermi	$1\,fm = 10^{-15}\,m$
barn	$1\,b = 10^{-28}\,m^2$
gauss	$1\,gauss = 10^{-4}\,T$
Absolute zero	$0\,K = -273.15°C$
electron volt	$1\,eV = 1.60 \times 10^{-19}\,J$

MODERN PHYSICS

The sequence of pictures on the cover illustrates the spontaneous fission of a ^{8}Be nucleus into two alpha particles. The colored surfaces show contours of constant density within the nucleus, calculated from a theoretical description of the fission process. The calculations were performed on the Cray computer at Los Alamos National Laboratory. (Courtesy J. W. Negele, Massachusetts Institute of Technology.)

MODERN PHYSICS

Hans C. Ohanian

Rensselaer Polytechnic Institute

PRENTICE-HALL, INC. / Englewood Cliffs, NJ 07632

Library of Congress Cataloging-in-Publication Data

Ohanian, Hans C.
  Modern physics.

  Bibliography: p.
  Includes index.
  1. Physics.   I. Title.
QC21.2.036   1987   530   86-8163
ISBN 0-13-596123-8

Editorial/production: Nicholas C. Romanelli
Cover design: Bruce Kensellar
Manufacturing buyer: John Hall

Printed in the United States of America

10 9 8 7 6 5 4 3 2 1

ISBN 0-13-596123-8

Prentice-Hall International (UK) Limited, *London*
Prentice-Hall of Australia Pty Limited, *Sydney*
Prentice-Hall of Canada Inc., *Toronto*
Prentice-Hall Hispanoamericana, S.A., *Mexico*
Prentice-Hall of India Private Limited, *New Delhi*
Prentice-Hall of Japan, Inc., *Tokyo*
Prentice-Hall of Southeast Asia Pte., Ltd., *Singapore*
Editora Prentice-Hall do Brasil, Ltda., *Rio de Janeiro*

To T. S. S.

Contents

* This chapter is optional.

7 **Spin and the Exclusion Principle**   **256**

8 **Electrons in Solids**   **310**

9 **Nuclear Structure**   **352**

Preface

This book provides an introduction to modern physics, i.e., the physics of the twentieth century. It is intended for science and engineering students who have become acquainted with classical mechanics and electrodynamics in the usual calculus-based physics course, but who have not yet been exposed to quantum physics or relativity.

The book begins with a review of the properties of particles and waves in classical physics, and then covers the theory of special relativity, the discovery of quanta of energy, the "old" quantum theory of Bohr, wave mechanics, angular momentum and spin, atoms and molecules, lasers, solid-state physics (including superconductivity), nuclear physics (including fission and fusion), and elementary particles. The arrangement of the topics is more or less chronological; however, for the sake of preserving the continuity of quantum physics, relativity is placed before the discovery of quanta. In spite of its early placement in the book, the chapter on relativity is optional—it is not needed in later chapters. I have avoided the use of relativity in the derivation of the Compton effect, relying instead on Sommerfeld's approximate nonrelativistic derivation, which yields the same result as the somewhat more messy relativistic calculation. Likewise, I have avoided the use of relativity in the derivation of the mass–energy formula, relying instead on Einstein's simple *Gedanken-experiment* exploiting momentum conservation and the pressure of light. Although my deliberate avoidance of relativity may seem out of tune with twentieth-century physics, I believe it is consistent with the historical record: until the rise of quantum electrodynamics and the construction of high-energy accelerators in the 1930s, relativity played a rather marginal role in the development of physics.

The book grew out of my experiences in teaching a modern physics course to students at Union College. I was afraid that the mathematical level of most of the available textbooks demanded of the students an excessively large quantum jump in mathematical level. From dealing with simple derivatives and integrals in their classical physics course, they were suddenly expected to deal with eigenvalue problems involving partial differential equations, separation of variables, and a host of special mathematical functions. The emphasis on this kind of fancy mathematics tends to distract the student's attention from the underlying physical concepts, which are ultimately much more important. I decided to avoid this pitfall by restricting the full mathematical treatment to simple examples of one-dimensional piecewise constant potentials, and by handling all examples of variable potentials—such as the harmonic oscillator and the hydrogen atom—by the WKB approximation for energy eigenvalues ($\oint \sqrt{2m(E - U)}\, dx = nh$). This approximation has the advantage of placing before the student's eyes a sharp picture of the relationship between eigenvalues and standing waves. Although the WKB approximation is valid only for large quantum numbers, it usually gives good results even for low quantum numbers (sometimes perhaps excessively good results, which carry some risk of deluding the student, a risk I reckoned as tolerable).

Throughout the book I have called attention to recent practical applications, such as resonance-ionization spectroscopy (Section 4.6), Doppler-free spectroscopy (Section 7.7), Josephson-effect potentiometer (Section 8.5), NMR imaging (Section 9.6), and also diverse metrological applications which have led to the latest high-precision determinations of fundamental constants. I have taken pains to ensure that the material is up to date. For instance, the chapter on Special Relativity includes a discussion of the most recent experimental tests of the time-dilation effect, and the chapter on particles includes a discussion of electroweak theory, broken symmetry, and the discovery of the intermediate bosons.

Each chapter ends with a brief summary of important formulas for ready reference and a generous selection of problems, ranging in difficulty from simple plug-in exercises to fairly complicated derivations of supplementary results (problems of exceptional difficulty are marked with a dagger, [†]). The system of units used in the book is SI. However, I have retained the Ångstrom as the traditional unit for wavelengths.

I thank professors Betty L. Atkinson (University of Oklahoma), Robert L. Chasson (University of Denver), Edward F. Gibson (California State University), Gaylord T. Hageseth (University of North Carolina), James C. Ho (Wichita State University), and Bernard Kramer (Hunter College) for their thorough critical reviews of the earlier drafts of this book. Their valuable suggestions permitted me to make many improvements. I thank Madhusree Mukerjee (University of Chicago) for her careful scrutiny of the final version. I am indebted to Nicholas Romanelli, my editor at Prentice-Hall, for his thoughtful and remarkably efficient handling of the multitude of vexatious details involved in the production process. And I thank Judy Peck for her skillful transmutation of my well-nigh undecipherable handwritten manuscript into neat typescript.

H.C.O

MODERN PHYSICS

Particles and Waves
in Classical Physics

In his autobiographical notes, Albert Einstein recapitulated the belief held by the physicists of the late nineteenth century:

> *In the beginning (if there was such), God created Newton's laws of motion and the necessary masses and forces. That is all; everything else follows by deduction upon the development of suitable mathematical methods.* *

Classical Newtonian physics had stood unchallenged for more than two hundred years and it had achieved many spectacular successes—especially in celestial mechanics—which confirmed physicists in their dogmatic belief. But this dogma was shattered by the revolutionary theories discovered in the early years of the twentieth century: quantum theory and relativity theory. These new theories swept away Newton's concept of a particle and his concepts of space and time. A new set of laws of quantum physics and relativistic physics replaced the laws of classical Newtonian physics, which were seen to be merely approximations, valid only in the limiting case of fairly large masses, large energies, and low speeds (compared to the speed of light).

The discoveries of quantization and relativity mark the beginning of modern theoretical physics. However, modern experimental physics had its beginning somewhat earlier, in the last decade

---

* *Albert Einstein: Philosopher-Scientist*, P. A. Schilpp, ed., p. 18. Quoted with permission of The Hebrew University of Jerusalem.

of the nineteenth century, with the discoveries of electrons, ions, X rays, and radioactivity. These discoveries gave experimenters the first glimpse of the inhabitants of the subatomic world.

In this chapter we will briefly review the properties of particles and of waves in classical Newtonian physics, and discuss some experimental discoveries of the late nineteenth century and some subsequent experimental developments of the twentieth century. Throughout this chapter we will rely on Newton's Laws. Although these laws are only approximations, they are very useful approximations. They are adequate for the motion of all the macroscopic bodies of our everyday experience. And they are even adequate for the motion of electrons and other subatomic particles, provided that the orbits are large, and the speeds are neither too low nor too high. Furthermore, quantum physics and relativistic physics borrow terminology and concepts from Newtonian physics, and therefore a review of Newtonian physics will lay the groundwork for the developments in later chapters.

1.1 **Classical Particles**

In classical physics, a **particle** is a pointlike body that has no discernible size or internal structure. The only attributes of a classical particle are its mass m and its electric charge q. At any given instant of time, the state of a classical particle is completely described by its position vector $\mathbf{r}$ and its velocity vector $\mathbf{v} = d\mathbf{r}/dt$. According to classical physics, the position and velocity of a particle are measurable, in principle, with arbitrary precision. The only limitations on the precision of measurement arise from practical considerations; for instance, in a length measurement employing the standard meter bar that used to serve as official standard of length, the coarseness of the scratches marking the ends set a practical limitation of about $\pm 10^{-8}$ m on the precision of the measurement; but obviously this was not a fundamental limitation—it could have been reduced by using finer scratches.

The measurements of position and velocity require that we specify a **reference frame** relative to which these measurements are to be made. Such a reference frame is an (imagined) three-dimensional array of measuring rods and clocks that permit us to assign space and time coordinates to any spacetime point, or **event.** Unless otherwise noted, we will take it for granted that the reference frame used for the description of the motion of any particle is an

inertial reference frame, i.e., a reference frame in which the motion of a free particle proceeds with constant velocity. In such a reference frame, the motion of any particle subjected to a force obeys Newton's Second Law,

$$m\frac{d\mathbf{v}}{dt} = \mathbf{F}$$

or

$$m\mathbf{a} = \mathbf{F} \qquad (1)$$

This is called the **equation of motion.** Under the assumption that the force is a known function, this equation permits the calculation of the position and velocity of the particle at any time, starting with the position and velocity at one initial instant of time. Thus, classical mechanics is deterministic: if the initial state of the particles in a system is given, then the state at any future (or past) time is completely predictable. Of course, in practice there are limitations to the accuracy of the predictions of classical mechanics. For instance, the positions of the major planets can be predicted for twenty or thirty years with an accuracy of a few seconds of arc; but for longer time intervals, the accuracy deteriorates drastically because of propagated errors of the initial data, and because of round-off and truncation errors in the computations.

Some of the important quantities in the study of particle dynamics are the **kinetic energy:**

$$K = \tfrac{1}{2}mv^2 = \tfrac{1}{2}m(v_x^2 + v_y^2 + v_z^2) \qquad (2)$$

the **momentum:**

$$\mathbf{p} = m\mathbf{v} \qquad (3)$$

and the (translational) **angular momentum:**

$$\mathbf{L} = \mathbf{r} \times \mathbf{p} = m\mathbf{r} \times \mathbf{v} \qquad (4)$$

Under appropriate circumstances these quantities obey conservation laws. In the absence of external forces, the net kinetic energy of a system of particles is conserved in *elastic* collisions among the particles, i.e., the net kinetic energy is the same before and after the collision. The net momentum is conserved in *all* collisions. The net angular momentum is conserved in *all* collisions and in all interactions provided the particles interact via central forces (forces that act along the lines joining the particles).

In atomic physics, in nuclear physics, and in high-energy particle physics, experimenters manipulate charged particles by means of electric and magnetic fields. The force on a particle of charge q subjected to an electric field $\mathbf{E}$ and a magnetic field $\mathbf{B}$ is

$$\mathbf{F} = q\mathbf{E} + q\mathbf{v} \times \mathbf{B} \tag{5}$$

This formula is written in metric (SI) units: the charge q is measured in coulomb (C), the electric field in volt/meter (V/m), and the magnetic field in tesla (T). Let us consider two simple special cases of this general formula. For a particle in a uniform electric field ($\mathbf{E}$ = const., $\mathbf{B} = 0$), the force is $q\mathbf{E}$ and the equation of motion is

$$m\mathbf{a} = q\mathbf{E} \tag{6}$$

Assuming that the direction of the electric field is along the x-axis, this becomes

$$ma_x = qE \tag{7}$$

Thus, the acceleration is constant. For a particle that starts at $x = 0$ with $v_x = 0$, the position and velocity are then

$$v_x = a_x t = \frac{q}{m} Et \tag{8}$$

and

$$x = \frac{1}{2} a_x t^2 = \frac{1}{2} \frac{q}{m} Et^2 \tag{9}$$

The kinetic energy of the particle is

$$K = \frac{1}{2} m v_x^2 = \frac{1}{2} m \left(\frac{q}{m} Et \right)^2 = \frac{1}{2} \frac{q^2}{m} E^2 t^2 \tag{10}$$

In view of Eq. (9), this is the same as

$$K = xqE \tag{11}$$

The right side of this equation is the product of the displacement and the force, i.e., it is the work done by the force. We can also write Eq. (11) as

$$K = q\,\Delta V \tag{12}$$

where $\Delta V = xE$ is the product of displacement and electric field, i.e., it is the change of electrical potential.

In atomic, nuclear, and particle physics, energies are commonly expressed in **electron-volt** (eV), million electron-volt (MeV), or billion electron-volt (GeV). The electron volt is the energy that a particle of charge equal to the proton charge ($q = e = 1.60 \times 10^{-19}$ C) acquires when "falling" through a potential difference of 1 volt, thus

$$1\,\text{eV} = 1e \times 1V = (1 \times 1.60 \times 10^{-19}\,\text{C}) \times (1\,\text{V})$$

$$= 1.60 \times 10^{-19}\,\text{C} \cdot \text{V} = 1.60 \times 10^{-19}\,\text{J} \qquad (13)$$

Let us now consider a particle moving in a uniform magnetic field ($\mathbf{E} = 0$, $\mathbf{B} = $ constant). The force is $q\mathbf{v} \times \mathbf{B}$ and the equation of motion is

$$m\mathbf{a} = q\mathbf{v} \times \mathbf{B} \qquad (14)$$

This shows that the acceleration is always perpendicular to $\mathbf{B}$; hence the component of the velocity parallel to $\mathbf{B}$ remains constant. We will hereafter ignore this component and assume that the motion is confined to a plane perpendicular to $\mathbf{B}$ (see Fig. 1.1a). Equation (14) also shows that the acceleration is always perpendicular to the velocity, i.e., the change of the velocity is perpendicular to the velocity. Consequently, the magnitude of the velocity does not change; only its direction changes. Such behavior of the velocity

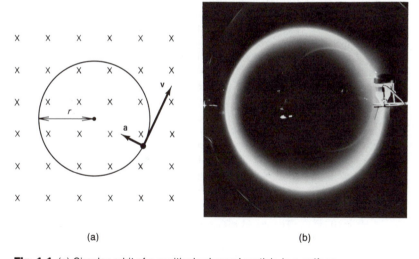

(a)   (b)

Fig. 1.1 (a) Circular orbit of a positively charged particle in a uniform magnetic field. The magnetic field is perpendicular to the plane of the page; its direction is into the page. The magnitude of the centripetal acceleration is $a = v^2/r$. (b) Electrons orbiting in a uniform magnetic field. The electrons emerge from a cathode at right center. They move in an evacuated tube containing small amounts of mercury vapor, which glows when struck by the beam of electrons. (Courtesy Leybold-Heraeus.)

is characteristic of uniform circular motion. Figure 1.1b is a photograph of the orbit of charged particles in circular motion in a uniform magnetic field. If the radius of the circle is r, the magnitude of the centripetal acceleration is v^2/r and the magnitude of the left side of Eq. (14) is mv^2/r. The magnitude of the right side is qvB, so that

$$m\frac{v^2}{r} = qvB \tag{15}$$

From this, we can calculate the radius of the circular motion:

$$r = \frac{mv}{qB} \tag{16}$$

We can also calculate the frequency of the circular motion:

$$v = \frac{v}{2\pi r} = \frac{qB}{2\pi m} \tag{17}$$

This frequency is called the **cyclotron frequency.** Note that it depends only on the magnetic field and on the ratio of charge to mass (q/m) of the particle; it does not depend on the velocity.*

Example 1   A **cyclotron** is a device for accelerating particles to high speeds. In the cyclotron a magnetic field, produced by a powerful electromagnet, holds the particles in a circular orbit within two semicircular metallic cans, or **Dees** (see Fig. 1.2a). A radio-frequency potential difference applied to the Dees generates an electric field in the gap between them. This potential difference oscillates with the frequency (17), so that whenever the particle crosses the gap, the electric field gives it a push, accelerating it to higher speeds. Thus the orbit of the particle consists of a sequence of semicircular arcs of stepwise increasing radius. The first successful cyclotron (see Fig. 1.2b), had a diameter of 11 cm and accelerated protons to an energy of 80 keV. What magnetic field was required to hold such a proton in an orbit of diameter 11 cm?

Solution   The mass of the proton is 1.67×10^{-27} kg and the charge is 1.60×10^{-19} C. The speed of a proton of kinetic energy $K = 80$ keV $= 1.3 \times 10^{-14}$ J is

$$v = \sqrt{\frac{2K}{m_p}} = \sqrt{\frac{2 \times 1.3 \times 10^{-14}\,\text{J}}{1.67 \times 10^{-27}\,\text{kg}}}$$

$$= 3.9 \times 10^6\,\text{m/s} \tag{18}$$

---

* For relativistic particles, the cyclotron frequency depends on the velocity because the equation of motion (15) is modified by an extra factor $1/\sqrt{1 - v^2/c^2}$ on the left side (see the next chapter).

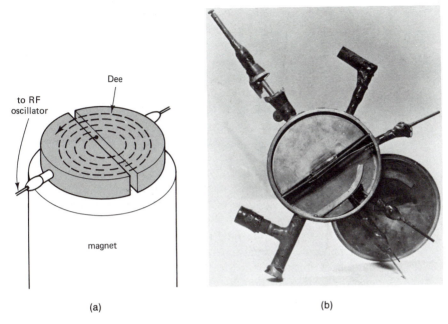

(a)   (b)

Fig. 1.2 (a) Schematic diagram of a cyclotron and of the orbit of a proton (dashed). The upper pole of the magnet has been omitted for the sake of clarity. (b) The Dees of the first successful cyclotron, built in 1930 by E. O Lawrence and M. S. Livingston. (Courtesy Lawrence Berkeley Laboratory.)

Hence, Eq. (16) gives

$$B = \frac{m_p v}{qr} = \frac{1.67 \times 10^{-27}\,\text{kg} \times 3.9 \times 10^6\,\text{m/s}}{1.60 \times 10^{-19}\,\text{C} \times (0.11/2)\,\text{m}} = 0.74\,\text{T} \quad \blacksquare \quad (19)$$

1.2 **The Discovery of the Electron**

The electron was the first subatomic and elementary particle to be discovered. Toward the end of the nineteenth century, physicists were experimenting with electrical discharges in rarefied gases. In such experiments a partially evacuated glass tube containing low-pressure gas with an electrode at each end is connected to a high-voltage generator (see Fig. 1.3). This produces an electric discharge in the gas, i.e., the gas becomes a conductor permitting the flow of electric current from one electrode to the other. The character of this discharge depends on the pressure. At a pressure of about

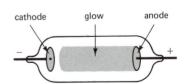

Fig. 1.3 A discharge tube.

10^{-3} atm, the discharge takes the form of a luminous glow filling the tube, as in a neon sign. At lower pressures, the luminosity disappears, even though the electric current continues to flow through the gas. In 1858 J. Plücker* discovered that at a pressure below 10^{-6} atm, a new phenomenon occurs in the tube: the negative electrode, or cathode, emits invisible rays that propagate through the nearly empty space in the tube. Although the rays by themselves are invisible, they make their presence known when they strike the walls of the glass tube: the impact makes the glass phosphoresce in greenish or bluish colors. The rays were called **cathode rays.**

Investigations of these rays by W. Crookes[†] and other experimenters showed that if an obstruction is placed near the cathode, the rays throw a shadow of this obstruction on the walls of the tube, which establishes that the rays propagate along straight lines. Also, it was soon discovered that the rays were deflected by magnetic fields, as expected for a beam of charged particles. But the early experimenters failed in their attempts to deflect the rays by electric fields, and this led to a great deal of confusion regarding the nature of the rays. The matter was finally cleared up by J. J. Thomson[‡] in 1897. Using more thoroughly evacuated tubes, he succeeded in deflecting the cathode rays by an electric field, and he established that they behave like small particles with a negative electric charge. These particles were later named **electrons.**

Figure 1.4a shows a cathode-ray tube used by Thomson to investigate the deflection of a beam of electrons by electric and magnetic fields. The beam of electrons emerges from the cathode at the left end of the tube with some (unknown) horizontal velocity v_x. In the central part of the tube, a pair of parallel capacitor plates supplies a uniform vertical electric field, while an electromagnet supplies a uniform horizontal magnetic field. For a start, suppose that only the electric field is switched on. An electron approaching the plates from the left with a horizontal velocity v_x will be deflected vertically by the electric field (see Fig. 1.4b). The electron will acquire a vertical velocity v_z, while its horizontal velocity

---

* **Julius Plücker,** 1801–1868, German mathematician and physicist, professor at Bonn.

[†] Sir **William Crookes,** 1832–1919, English physicist and chemist.

[‡] Sir **Joseph John** (J. J.) **Thomson,** 1856–1940, English physicist, Cavendish professor at Cambridge, where he succeeded Rayleigh. He was awarded the Nobel Prize in 1906 for his research on the conduction of electricity by gases at low pressure. After establishing the nature of cathode rays and canal rays by deflecting them with magnetic and electric fields, Thomson applied the same method to ions of neon and discovered different isotopes of this element.

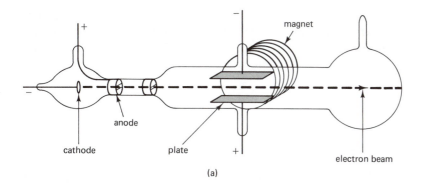

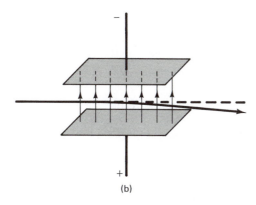

Fig. 1.4 (a) Thomson's cathode-ray tube. One pole of the magnet is behind the tube; the other pole has been omitted in the diagram. (b) Trajectory of an electron in the cathode-ray tube. The electric field points vertically up. The magnetic field is absent.

remains constant. The final vertical velocity will be

$$v_z = a_z t = -\frac{eE}{m_e} t \tag{20}$$

where t is the time spent in the electric field. Given that the length of the plates is l, this time is $t = l/v_x$ and therefore

$$v_z = -\frac{eEl}{m_e v_x} \tag{21}$$

The deflection angle of the electron beam is then given by

$$\tan \theta = \frac{v_z}{v_x} = -\frac{eEl}{m_e v_x^2} \tag{22}$$

Next, suppose that the magnetic field is switched on and its strength is adjusted so that the deflection due to the magnetic field exactly cancels the deflection due to the electric field. This means that the net vertical force on an electron in the beam is zero,

$$F_z = -eE + ev_x B = 0 \tag{23}$$

which requires that

$$v_x = \frac{E}{B} \tag{24}$$

Thus, the ratio of E to B for zero deflection equals the (originally unknown) velocity of the electrons. Combining Eqs. (22) and (24), we obtain

$$\frac{e}{m_e} = -\frac{E}{B^2 l} \tan \theta \tag{25}$$

This expresses the ratio e/m_e in terms of known and measurable quantities.

Thomson's experimental results were somewhat contaminated by systematic errors; his value for e/m_e was too small by a factor of 2. The best modern value for the charge-to-mass ratio of an electron is

$$\frac{e}{m_e} = 1.758805 \times 10^{11} \, \text{C/kg} \tag{26}$$

Deflection experiments with electrons in electric or magnetic fields can determine only the ratio e/m_e, but not e or m_e separately. Thomson and his colleagues J. S. E. Townsend and H. A. Wilson tried to determine the value of e by investigating the motion of electrically charged water droplets formed in a cloud chamber. In such a chamber, invented by C. T. R. Wilson,* a volume of humid air is suddenly cooled by expansion, causing condensation of water droplets. The droplets condense preferentially around grains of dust or, if the air is dust free, around any ions that may be present. Such ions are atoms or molecules that have lost or gained one or several electrons; consequently, the electric charge of an ion—and the electric charge of a water droplet condensed around an ion—is a multiple of the electric charge of the electron. In Wilson's version of the experiment, an abundance of ions is produced in the cloud

---

* **Charles Thomson Rees Wilson,** 1869–1959, British physicist, professor at Cambridge. For his invention of the cloud chamber, he was awarded the Nobel Prize in 1927.

chamber by irradiation of the air with X rays. The average mass of the water droplets condensed around these ions is calculated from the terminal velocity attained by a cloud of droplets when falling under the simultaneous actions of gravity and of the frictional drag of the air. The average electric charge of a droplet is then calculated from the change in terminal velocity produced by the action of an extra vertical electric field.

In 1909, this experiment was much improved by R. A. Millikan,* who noticed that it was possible to perform observations of the motion of *individual* droplets, instead of observations of entire clouds of droplets. Furthermore, Millikan made use of oil droplets, thereby avoiding uncertainties introduced by the partial evaporation of the water droplets during the measurement. Figure 1.5 shows Millikan's apparatus. The oil droplets are produced by an atomizer in the large chamber C at the top. The plates M and N are connected to a high-voltage battery, so that an electric field is generated between these plates. Occasionally, a droplet will pass through the pinhole b into the region between the plates. If the electric field is switched off, the droplet will fall under the actions of gravity and the frictional drag of the air and attain a terminal velocity. According to Stokes' law, the frictional-drag force is proportional to the velocity,

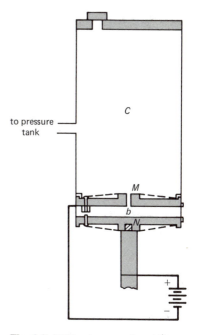

Fig. 1.5 Millikan's apparatus. (After R. A. Millikan, *The Electron*.)

$$f = 6\pi\eta Rv \tag{27}$$

where η is the viscosity of air and R is the radius of the droplet. When the droplet attains its terminal velocity v_1, this force is balanced by the weight, i.e.,

$$mg = 6\pi\eta Rv_1 \tag{28}$$

Since the radius can be expressed in terms of the mass and the density of the droplet ($4\pi R^3/3 = m/\rho$, where ρ is the density of the oil), Eq. (28) permits a calculation of the mass of the droplet in terms of known and measurable quantities. If the electric field is now switched on, the droplet will accelerate and attain a new terminal velocity, say, in the upward direction. At this new terminal velocity v_2, the frictional force is balanced by the electric force minus the

---

* **Robert Andrews Millikan,** 1868–1953, American experimental physicist, professor at Chicago and director of the N. Bridge Laboratory at the California Institute of Technology. Millikan was a very meticulous experimenter, and took several years to perfect his measurements of the electronic charge. At the same time he worked on the photoelectric effect, and confirmed Einstein's equation. For these contributions he was awarded the Nobel Prize in 1923.

weight,

$$qE - mg = 6\pi\eta Rv_2 \tag{29}$$

Dividing Eq. (29) by Eq. (28), we obtain

$$\frac{qE - mg}{mg} = \frac{v_2}{v_1}$$

and

$$q = \frac{mg}{E}\frac{v_2 + v_1}{v_1} \tag{30}$$

This expresses the charge on the droplet in terms of known and measurable quantities.

Example 2   In one of his early measurements, Millikan* found that an oil drop acquired a downward terminal velocity of 8.584×10^{-4} m/s when falling in the absence of an electric field, and it acquired an upward terminal velocity of 2.042×10^{-5} m/s when under the influence of an electric field of 3.178×10^5 V/m. What was the electric charge on this drop? The density of the oil is 919.9 kg/m^3 and the viscosity of air is 1.825×10^{-5} N·s/m^2.

Solution   With $R = (3m/4\pi\rho)^{1/3}$, Eq. (28) becomes

$$mg = 6\pi\eta\left(\frac{3m}{4\pi\rho}\right)^{1/3}v_1$$

from which

$$m = \left(\frac{6\pi\eta v_1}{g}\right)^{3/2}\left(\frac{3}{4\pi\rho}\right)^{1/2} \tag{31}$$

$$= \left(\frac{6\pi \times 1.825 \times 10^{-5}\,\text{N·s/m}^2 \times 8.584 \times 10^{-4}\,\text{m/s}}{9.81\,\text{m/s}^2}\right)^{3/2}$$

$$\times \left(\frac{3}{4\pi \times 919.9\,\text{kg/m}^3}\right)^{1/2}$$

$$= 8.413 \times 10^{-14}\,\text{kg}$$

From Eq. (30) we then find

$$q = \frac{8.413 \times 10^{-14}\,\text{kg} \times 9.81\,\text{m/s}^2}{3.178 \times 10^5\,\text{V/m}}\left(\frac{8.584 \times 10^{-4}\,\text{m/s} + 2.042 \times 10^{-5}\,\text{m/s}}{8.584 \times 10^{-4}\,\text{m/s}}\right)$$

$$= 2.66 \times 10^{-18}\,\text{C}\quad\blacksquare$$

---

* R. Millikan, *The Electron*, pp. 67–85.

In our simple calculations leading to Eqs. (30) and (31), we ignored two small corrections that Millikan included in his somewhat more careful calculations: he included the buoyant force due to the air, and he included an empirical correction factor in the Stokes formula (27) to take care of a small dependence of the effective viscosity on the droplet radius.

Upon measuring the charges q of a large number of droplets, Millikan found that all these charges could be expressed as integer multiples of a smallest amount of charge, which he identified as the magnitude of the charge of a single electron. The best result obtained by Millikan was $e = 1.592 \times 10^{-19}$ C, quite close to the best modern result,

$$e = 1.602189 \times 10^{-19} \, \text{C} \tag{32}$$

Dividing his experimental result for e by the known result for e/m_e, Millikan obtained the value of the electron mass. The best modern value for this mass is

$$m_e = 9.10953 \times 10^{-31} \, \text{kg} \tag{33}$$

Incidentally, this best modern value of the electron mass is obtained from an indirect comparison of the electron and the proton masses (via their intrinsic magnetic moments), not from a comparison of e and e/m_e.

The charge e given in Eq. (32) is the fundamental quantum of charge. The electric charges of all particles found in nature are integer multiples of this fundamental quantum of charge, i.e., the electric charges are always 0, $\pm e$, $\pm 2e$, $\pm 3e$, etc. In the case of ions, this quantization rule is self-evident: an ion is an atom or molecule that has lost or gained one or several electrons, hence its charge is necessarily a multiple of the electron charge.* However, the rule of quantization of charge also applies to all the subatomic particles and all the particles discovered by high-energy physicists. For instance, the charge of the neutron is zero, that of the proton is $+e$, that of the alpha particle (or helium nucleus) is $+2e$, that of the muon is $-e$, that of the positive pion is $+e$, etc.

Of late, much effort has been expended on the search for charges $\pm\frac{1}{3}e$ and $\pm\frac{2}{3}e$, these being the charges of the quarks that

---

* Nonionized atoms and molecules are known to be precisely neutral. High-precision experiments testing the neutrality of small levitated steel spheres [see, e.g., M. Marinelli and G. Mopurgo, *Phys. Lett.* **137B**, 439 (1984)] have established that in nonionized atoms the negative charge of the electrons balances the positive charge on the nucleus to within better than 1 part in 10^{21}.

Fig. 1.6 The apparatus of Hodges et al. (Courtesy C. L. Hodges, San Francisco State University.)

are supposed to exist inside protons and neutrons (we will discuss quarks in some detail in Chapter 11). But no such fractional charges, nor any other fractional charges, have ever been found. One method of search relies on a modified, automated version of the Millikan experiment. Figure 1.6 shows the apparatus used for this experiment by Hodges et al.* The apparatus automatically separates a sample of liquid into small drops and determines the charge of each drop. Samples of water and of mercury as large as 10^{-4} g have been examined drop by drop. The absence of fractional charges sets stringent limits on the abundance of free quarks in such samples of liquids: less than one free quark per 10^{19} protons.

1.3 Ions and Isotopes

In 1886, E. Goldstein[†] discovered that if the cathode of a discharge tube has a hole, some kind of rays emerge from this hole in the backward direction, i.e., in the direction *away* from the positive electrode, or anode. He called these rays **canal rays,** and it was soon demonstrated that they were beams of positively charged particles whose ratio of charge to mass was several thousand times smaller than that of electrons. These particles are positive **ions**— atoms or molecules of gas that lost one or two electrons when under bombardment by the cathode rays streaming through the tube, and were then attracted by the cathode, passed through it, and formed a beam in the backward direction.

The determination of the ratio of charge to mass for such ions is rather more difficult than for electrons because the ions in the beam usually have a wide range of speeds. Thomson devised an elegant method for comparing the e/m ratios for different kinds of ions. Figure 1.7 shows his apparatus, called a positive-ray tube. The round portion on the left is the discharge tube, in which cathode rays streaming from the cathode to the anode generate positive ions, which travel toward the cathode. Most ions strike the cathode and are absorbed; but some manage to pass through a fine hole in the cathode and emerge on the far side as a well-collimated beam. This beam is subjected to the simultaneous actions of horizontal electric and magnetic fields. The electric field produces a

---

* C. L. Hodges et al., *Phys. Rev. Lett.* **47**, 1651 (1981); D. C. Joyce et al., *Phys. Rev. Lett.* **51**, 731 (1983).

† **Eugen Goldstein,** 1850–1930, German physicist, professor at Berlin.

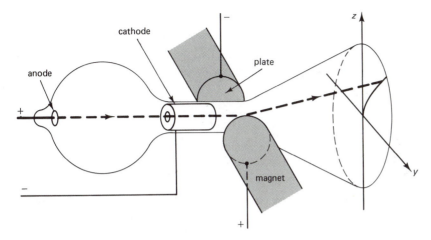

Fig. 1.7 Thomson's positive-ray tube.

horizontal deflection, and the magnetic field a vertical deflection. The magnitudes of these deflections depend on the energies of the particles: the deflections are small if the energy is high and the deflections are large if the energy is low. The locus of impact points traced out on the face of the tube as a function of energy is one-half of a parabola (see Fig. 1.8).

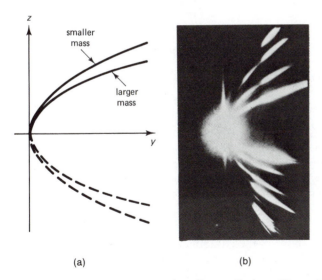

(a)   (b)

Fig. 1.8 (a) The locus of impact points of ions on the face of the positive-ray tube is one half of a parabola. Ions with different values of e/m fall on different parabolas. (b) Photograph of parabolas obtained by J. J. Thomson with ions of xenon, krypton, argon, and neon. [From J. J. Thomson, *Proc. Roy. Soc.* **A89**, 1 (1913).]

Example 3   Derive the equation for the locus of impact points for singly ionized ions of charge e, mass m.

Solution   Suppose that the ion enters the electric and magnetic fields with horizontal velocity v_x. If the transverse deflections are small, we can regard this horizontal velocity as approximately constant; thus, the ion spends a time $t = l/v_x$ in the fields. During this time the electric field gives the ion a constant acceleration in the y direction ($a_y = eE/m$), which results in a deflection in the y direction:

$$\tan\theta_y = \frac{a_y t}{v_x} = \frac{eE}{m}\frac{l}{v_x^2} \tag{34}$$

Simultaneously, the magnetic field gives the ion an acceleration perpendicular to the magnetic field, i.e., an acceleration in the z-direction. Within the approximation $v_x = $ constant, this acceleration is also constant and it results in a deflection in the z-direction:

$$\tan\theta_z = \frac{a_z t}{v_x} = \frac{eBv_x}{m}\frac{l}{v_x^2} = \frac{eB}{m}\frac{l}{v_x} \tag{35}$$

If we eliminate the velocity v_x from Eqs. (34) and (35), we obtain the relationship between the tangents of the deflection angles for an ion emerging from the field region,

$$\tan\theta_y = \frac{E}{B^2 l}\frac{m}{e}\tan^2\theta_z \tag{36}$$

Since the ion thereafter travels on a straight line, the values y and z at the face of the tube are $y = L\tan\theta_y$ and $z = L\tan\theta_z$, where L is the length of the tube. Multiplying Eq. (36) by L, we find

$$y = \frac{E}{B^2 lL}\frac{m}{e}z^2 \tag{37}$$

This establishes that the locus of impact points on the screen is a parabola. Note that for a given magnetic field, the ions reach only one branch of this parabola ($z > 0$); but for a reversed magnetic field, the ions reach the other branch ($z < 0$). In practice, Thomson found it convenient to use both field directions, so he could see the axis of symmetry of the parabola.   ■

Since the shape of the parabola depends on the ratio e/m [see Eq. (37)], ions of different masses will show up on different parabolas. This permits the detection and the measurement of small mass differences between ions. For instance, while using a chemically pure sample of neon gas, Thomson found that this gas produced ions of two kinds, with a mass difference of about 10% between

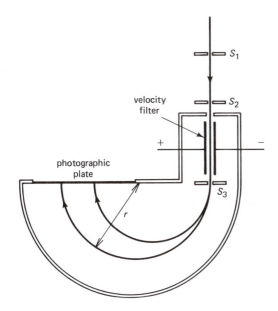

Fig. 1.9 Bainbridge's mass spectrograph. The magnetic field is perpendicular to the plane of the page. The strength of the magnetic field is B_1 in the velocity filter and B_2 in the semicircular chamber. The entire spectrograph is evacuated. [After K. T. Bainbridge, *Phys. Rev.* **40,** 130 (1932).]

them. In this way, Thomson demonstrated the existence of **isotopes,** or atoms that are chemically identical, but differ in mass.* In modern notation, the isotopes discovered by Thomson were ^{20}Ne and ^{22}Ne, where the superscript on the chemical symbol, called the **mass number,** is the rounded-off value of the mass of the isotope in atomic mass units (see below).

After World War I, F. W. Aston, A. J. Dempster, K. T. Bainbridge,[†] and others developed improved versions of Thomson's apparatus. All of these devices, called **mass spectrographs** or **mass spectrometers,** use electric and magnetic fields to discriminate between ions of different masses. For instance, Fig. 1.9 shows the mass spectrograph of Bainbridge. The beam of ions, collimated

---

* The existence of isotopes was first proposed by **Frederick Soddy,** 1877–1956, British chemist, who received the 1921 Nobel Prize in Chemistry for this.

† **Francis William Aston,** 1877–1945, English physicist, fellow at Cambridge, received the Nobel Prize in Chemistry in 1922. **Kenneth Tompkins Bainbridge,** 1904— , American physicist, professor at Harvard. **Arthur Jeffrey Dempster,** 1886–1950, American physicist, professor at Chicago.

by the slits S_1 and S_2, first enters a region with "crossed" electric and magnetic fields, i.e, electric and magnetic fields at right angles. This region acts as a velocity selector, or velocity "filter." We know from Eq. (24) that only the ions having a velocity $v = E/B_1$ will pass straight through; all others will be deflected right or left and will be stopped at S_3. The ions of the selected velocity then enter a region of uniform magnetic field where they travel on a circular arc. If the ions are singly ionized, then the radius of the circular arc is given by Eq. (16) with $q = e$:

$$r = \frac{mv}{eB_2} \tag{38}$$

Finally, the ions strike a photographic plate on which the radius of impact can be measured. This radius determines the mass,

$$m = \frac{eB_2}{v} r \tag{39}$$

Note that the mass difference between two ions is directly proportional to the distance between the impact points measured on the photographic plate,

$$\Delta m = \frac{eB_2}{v} \Delta r \tag{40}$$

The small distance Δr along the photographic plate can be measured much more precisely than the large distance r. Hence, in the operation of such a mass spectrograph, it is customary to employ one kind of ion as a standard of mass and to compare any other ion to it, by means of Eq. (40).

Modern spectrographs and spectrometers achieve very high precision: they can compare the masses of two ions to within seven significant figures. Even higher precision—up to nine significant figures—is attained by some recent devices that compare the cyclotron frequencies of ions orbiting in a uniform magnetic field. This cyclotron frequency is inversely proportional to the mass [see Eq. (17)], and hence a comparison of the frequencies of two ions amounts to a comparison of their masses. The frequencies of ions can be measured with high precision by a resonance method: a tangential radio-frequency electric field is applied at one location along the orbit, as in a cyclotron; the repeated pushes of this field throw the ion out of the original orbit, unless the radio frequency is exactly equal to one half the cyclotron frequency (or an odd multiple of this), in which case the acceleration suffered by the ion in one pass through the radio-frequency field will be compensated by a deceleration suffered in the next pass. The advantage of this ion cyclotron resonance method over the mass spectrometric method is that frequencies, and small frequency shifts, are much easier to measure accurately than are small distances.

Comparable precision—typically eight significant figures—is attained in mass comparisons that rely on energy measurements in nuclear reactions (energy differences are related to mass differences by Einstein's formula $\Delta E = c^2 \Delta m$; see Section 1.7). The extreme precision of all these methods of mass determination testifies to the remarkable accuracy obtainable by classical physics under suitable conditions. Only in celestial mechanics does classical physics attain comparable accuracy.

Conventionally, the isotope ^{12}C of carbon is employed as a standard to which other isotopes are compared. The mass of this isotope is assigned a value of exactly 12 **atomic mass units** (u),*

$$(\text{mass of } {}^{12}\text{C atom}) = 12.0000000 \text{ u} \qquad (41)$$

Thus, by definition, 1 u is $\frac{1}{12}$ of the mass of this carbon isotope.

The values of the masses listed in tables of isotopes (see, e.g., Fig. 9.1 and Appendix 4) were obtained by judicious combination of data from mass spectroscopy and from nuclear reactions. The lightest atom is that of hydrogen. The most abundant isotope of this atom is ^{1}H. It has a mass of

$$M_{^1\text{H}} = 1.00782503 \text{ u} \qquad (42)$$

The ion of this atom is a naked atomic nucleus. This hydrogen nucleus, called a **proton,** is the smallest of all atomic nuclei. Its mass is

$$m_{\text{p}} = 1.00727647 \text{ u} \qquad (43)$$

For the conversion of the atomic mass unit into kilograms, we need to know Avogadro's number N_A (the number of atoms in one mole). By definition, one mole of the isotope ^{12}C is exactly 12 grams.† The mass of a single atom of this isotope is then $12 \text{ g}/N_A$,

---

* This is the mass of the neutral atom. To obtain the mass of the ^{12}C$^+$ ion used in a mass spectrograph, we must subtract the mass of an electron ($m_e = 5.4854 \times 10^{-4}$ u) and add the mass equivalent of the binding energy, or ionization energy, of an electron in a carbon atom ($\Delta m = \Delta E/c^2 = 11.2 \text{ eV}/c^2 = 1.20 \times 10^{-8}$ u).

† One mole of any chemical element (or compound) is that amount of matter that contains exactly as many atoms (or molecules) as 12 g of carbon. The "atomic mass" of a chemical element (or the "molecular mass" of a compound) is the mass of one mole.

which must coincide with $12\,\mathrm{u}$. Consequently,

$$1\,\mathrm{u} = \frac{1\,\mathrm{g}}{N} = \frac{10^{-3}\,\mathrm{kg}}{N}$$

Recent determinations (see Section 8) of Avogadro's number give

$$N_A = 6.022098 \times 10^{23}\,\text{atoms/mole} \tag{44}$$

so that

$$1\,\mathrm{u} = \frac{10^{-3}\,\mathrm{kg}}{6.022098 \times 10^{23}} = 1.66055 \times 10^{-27}\,\mathrm{kg} \tag{45}$$

If we use this to convert the mass of the proton into kilograms, we find

$$m_{\mathrm{p}} = 1.67265 \times 10^{-27}\,\mathrm{kg} \tag{46}$$

Note that the number of significant figures in (46) is only six, whereas in (43) it is nine. The loss of precision in the conversion from atomic mass units to kilograms is due to the relatively poor precision of the determinations of N_A.

The most abundant isotope of helium is ^{4}He. This isotope has a mass of

$$M_{^4\mathrm{He}} = 4.0026032\,\mathrm{u}$$

$$= 6.64658 \times 10^{-27}\,\mathrm{kg} \tag{47}$$

The *doubly* ionized ion of this atom is a naked atomic nucleus, usually called an **alpha particle.** Its mass is

$$m_\alpha = 4.0015061\,\mathrm{u} \tag{48}$$

$$= 6.64476 \times 10^{-27}\,\mathrm{kg} \tag{49}$$

For later reference, we also list here the mass of the **neutron:**

$$m_{\mathrm{n}} = 1.00866501\,\mathrm{u} \tag{50}$$

$$= 1.67495 \times 10^{-27}\,\mathrm{kg} \tag{51}$$

Since the neutron is electrically neutral, its mass cannot be determined directly with a mass spectrograph. Instead, its mass has been determined from energy measurements in nuclear reactions that involve the release or the capture of a neutron by isotopes of known mass.

Note that the masses of the proton and the neutron are nearly equal (the mass of the neutron is slightly larger). Also note that the mass of the alpha particle is nearly four times the mass of a proton or a neutron. This is no accident—as we will see, the alpha particle consists of two protons and two neutrons tightly bound together.

1.4 Kinetic Theory

In principle, Newton's laws permit the calculation of the motion of any system of classical particles; but in practice, such a calculation becomes impossible if the number of particles is very large. An example will make this obvious: the number of molecules in one cubic centimeter of air is 2.7×10^{19}—this is such a large number that even if we knew the initial positions and velocities of all these molecules, we could not calculate the motion because there is no computer large enough and fast enough to handle the arithmetic. In order to deal with a system of a very large number of particles, we must forgo the calculation of the detailed and precise behavior of each particle. Instead, we must have recourse to statistical, probabilistic methods, which permit the calculation of the average or most probable behavior of the system. **Kinetic theory** uses such methods to calculate the mechanical and thermal properties of gases and other systems of a large number of well-separated, nearly independent particles. **Statistical mechanics** uses similar methods to deal with more general systems of a large number of interacting constituents.

A simple example of the use of probabilistic methods in the study of a system of particles is the determination of the spatial distribution of the molecules of an ideal gas over the volume of a container. We take as a basic assumption that, in the absence of external forces, all locations in the container are equally probable for a molecule. We can then immediately conclude that the density of molecules must be uniform, since this is the only distribution consistent with equal probability for all locations. Note, however, that this uniform density applies only on a macroscopic scale, when we compare volume elements dV that are sufficiently large to contain a large number of molecules. On a smaller scale, the density is subject to fluctuations, which can amount to a large fraction of the average density if the volume element contains few molecules (a careful probabilistic analysis shows that if the volume element dV contains an average of N molecules, then the expected fluctuations are $\pm \sqrt{N}$).

A less trivial, and more interesting, example of the use of probabilistic methods is the calculation of the pressure that the gas

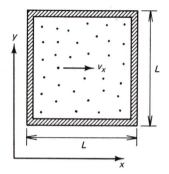

Fig. 1.10 A cube of side *L* filled with gas molecules.

exerts against the walls of the container. This pressure is due to the impact of the molecules on the walls. We can calculate this pressure by considering the average motion of the molecules. We assume that the container is a cube of side L, that the gas molecules only collide with the walls but not with each other, and that these collisions are elastic. These assumptions are not necessary, but they simplify the calculations.

Figure 1.10 shows the container filled with gas molecules. We can resolve the motion of each molecule into x-, y-, and z-components, and treat each component independently. Let us consider the x-motion of one molecule. The component of velocity in the x-direction is v_x, and the magnitude of this component remains constant since the collisions with the walls are elastic. The time between one collision with the face at $x = 0$ and the next collision with the same face is simply the time the molecule takes to move from $x = 0$ to $x = L$ and back to $x = 0$,

$$\Delta t = \frac{2L}{|v_x|}$$

When the molecule strikes the face at $x = 0$, its x-velocity is reversed from $-|v_x|$ to $|v_x|$. Hence, during each collision at $x = 0$, the x-momentum changes by

$$\Delta p_x = 2m|v_x| \tag{52}$$

where m is the mass of the molecule. The average rate at which the molecule transfers momentum to the face at $x = 0$ is therefore

$$\frac{\Delta p_x}{\Delta t} = \frac{2m|v_x|}{2L/|v_x|} = \frac{mv_x^2}{L} \tag{53}$$

This gives us the average force that the impacts of one molecule exert on the wall. To find the total average force exerted by the impacts of all the molecules, we must multiply the force (53) by the total number N of molecules; and to find the pressure we must divide by the area L^2 of the wall. This yields

$$P = \frac{N}{L^2} \frac{mv_x^2}{L} \tag{54}$$

or, in terms of the volume $V = L^3$,

$$P = \frac{Nmv_x^2}{V} \tag{55}$$

In our calculation we have made the implicit assumption that all the molecules have the same speed. This is of course not true—

the molecules of the gas have a distribution of speeds. To account for this distribution of speeds, we must replace the force (53) due to one given molecule by the mean force for all the molecules. Consequently, we must replace v_x^2 by a mean value for all the molecules in the container. We will designate this mean value of v_x^2 by $\overline{v_x^2}$. Equation (55) then becomes

$$P = \frac{Nm\overline{v_x^2}}{V} \tag{56}$$

To rewrite this equation, we note that on the average, molecules are just as likely to move in the x-, y-, or z-directions. Hence the mean values of v_x^2, v_y^2, and v_z^2 are equal,

$$\overline{v_x^2} = \overline{v_y^2} = \overline{v_z^2} \tag{57}$$

Since the sum of the squares of the components is the square of the magnitude of the velocity,

$$\overline{v_x^2} + \overline{v_y^2} + \overline{v_z^2} = \overline{v^2} \tag{58}$$

each of the terms on the left side of Eq. (58) must equal $\frac{1}{3}\overline{v^2}$. We can therefore rewrite Eq. (56) as

$$PV = \frac{Nm\overline{v^2}}{3} \tag{59}$$

This expresses the pressure in terms of the mean square of the velocity of the molecules.

It is instructive to compare this result with the ideal-gas law. For a gas of N molecules at (absolute) temperature T, this law states that

$$PV = NkT \tag{60}$$

where

$$k = 1.38066 \times 10^{-23} \, \text{J/K}$$

is *Boltzmann's constant*.* Obviously, the right sides of Eqs. (59) and (60) must be equal, i.e.,

$$\frac{m\overline{v^2}}{3} = kT \tag{61}$$

---

* **Ludwig Boltzmann,** 1844–1906, Austrian theoretical physicist, professor at Munich, Leipzig, and Vienna. He was one of the founders of statistical mechanics, and the leading defender of the atomic structure of matter in the debates in the early years of this century.

This shows that the mean square of the speed is directly proportional to the temperature. The square root of $\overline{v^2}$ is called the **root-mean-square speed** and it is usually designated by v_{rms}. According to Eq. (61),

$$v_{rms} = \sqrt{\overline{v^2}} = \sqrt{\frac{3kT}{m}} \qquad (62)$$

This root-mean-square speed may be regarded as the typical speed of the molecules of the gas. There are other ways of calculating a typical speed; for example, we may ask for the average of all the molecular speeds, or for the most probable of all the molecular speeds. These other typical speeds turn out to be roughly the same as v_{rms}, but their calculation requires some knowledge of the details of the distribution of molecular speeds. We will examine this distribution in the next section.

Example 4   What is the rms speed of oxygen molecules in air at 20°C?

Solution   The mass of an oxygen molecule is 32.0 u, or $32.0 \times 1.66 \times 10^{-27}$ kg and the absolute temperature is $T = 293$ K. Hence Eq. (62) gives

$$v_{rms} = \sqrt{\frac{3kT}{m}} = \sqrt{\frac{3 \times 1.38 \times 10^{-23} \, \text{J/K} \times 293 \, \text{K}}{32.0 \times 1.66 \times 10^{-27} \, \text{kg}}}$$

$$= 478 \, \text{m/s} \quad \blacksquare$$

1.5 The Maxwell Distribution of Molecular Speeds; the Boltzmann Factor

Measurements of the speeds of individual molecules of a gas at a given temperature reveal that the molecules have a wide variety of speeds. Mathematically, the distribution of speeds is specified by a distribution function $h(v)$ defined in such a way that $h(v) \, dv$ is the number of molecules with speeds in the range v to $v + dv$ [by analogy with the space density of molecules, we might call $h(v)$ the velocity density]. Figure 1.11 shows the measured distribution function of speeds for thallium atoms at a temperature of 870 K. The speeds were measured by aiming a beam of atoms emerging from a container (an oven) at a mechanical velocity "filter" consisting of a cylinder with helical grooves rotating about its axis (see

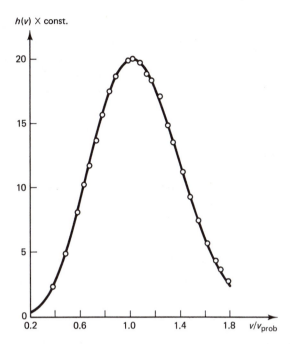

$h(v) \times$ const.

Fig. 1.11 Distribution of speeds in a sample of thallium atoms at a temperature of 870 K. The speed is expressed in units of the most probable speed v_{prob}; the measured value of this most probable speed at 870 K is 376 m/s. [After R. C. Miller and P. Kusch, *Phys. Rev.* **99**, 1314 (1955).]

(Fig. 1.12). The maximum in the distribution function corresponds to the most probable speed.

We can use probabilistic methods to derive an equation for the distribution function. For this purpose, we begin with the assumptions that the probability for a given speed is independent of the direction of the velocity, and that the x-velocities, y-velocities, and z-velocities are probabilistically independent. The latter assumption simply means that if we acquire information about, say, the x-velocity of a molecule, this conveys no information whatsoever about the y- or z-velocities. The distribution of x-velocities is described by a function $f(v_x)$ such that $f(v_x)\,dv_x$ is the number of molecules with x-velocities in the range v_x to $v_x + dv_x$. Thus, except for a normalization factor, $f(v_x)$ is the probability for a molecule to have an x-velocity in the range v_x to $v_x + dv_x$, and similarly for the y-velocities and the z-velocities. Since the probability is independent of the direction of the velocity, the distributions of the x-, y-, and z-velocities are described by the same functions $f(v_x)$,

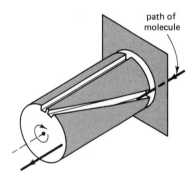

path of molecule

Fig. 1.12 Mechanical velocity filter used by Miller and Kusch. A molecule will pass through the helical groove on the rotating cylinder only if the translational speed of the molecule and the angular speed of the cylinder are precisely related, so that the transverse motion of the walls of the groove always keep them clear of the molecule in its longitudinal motion. (After R. C. Miller and P. Kusch, ibid.)

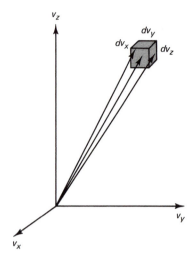

Fig. 1.13 The axes of this rectangular coordinate system represent the components of the velocity vectors. The tips of the velocity vectors shown all fall in a small volume element $dv_x\, dv_y\, dv_z$.

$f(v_y)$, and $f(v_z)$; and since the x-, y-, and z-velocities are probabilistically independent, the rules of probability theory tell us that the distribution function for the vector velocity is simply the product of the distribution functions for the components of the velocity. Hence the number of molecules whose velocity vectors fall within the three-dimensional volume element $dv_x\, dv_y\, dv_z$ (see Fig. 1.13) is

$$f(v_x)f(v_y)f(v_z)\, dv_x\, dv_y\, dv_z \tag{63}$$

Since the distribution for the velocity vectors is supposed to be independent of the direction, the product $f(v_x)f(v_y)f(v_z)$ must not depend on the separate components of the velocity, but only on the magnitude of the velocity, or the speed. Thus, the product $f(v_x)f(v_y)f(v_z)$ must equal a function of v or, equivalently, a function of v^2:

$$f(v_x)f(v_y)f(v_z) = F(v^2) = F(v_x^2 + v_y^2 + v_z^2) \tag{64}$$

To examine the implications of this equation, let us introduce the abbreviations $a = v_x^2$, $b = v_y^2$, and $c = v_z^2$, so that

$$f(a)f(b)f(c) = F(a + b + c) \tag{65}$$

We can determine the function F by setting $b = 0$ and $c = 0$; this gives

$$f(a)[f(0)]^2 = F(a) \tag{66}$$

from which we see that

$$F(a + b + c) = f(a + b + c)[f(0)]^2 \tag{67}$$

Substituting this expression for F into (65) we obtain

$$f(a)f(b)f(c) = f(a + b + c)[f(0)]^2 \tag{68}$$

Next, let us take the natural logarithm of both sides of this equation,

$$\ln f(a) + \ln f(b) + \ln f(c) = \ln f(a + b + c) + 2\ln f(0) \tag{69}$$

With abbreviation $g = \ln f$, this becomes

$$g(a) + g(b) + g(c) = g(a + b + c) + 2g(0) \tag{70}$$

We can now find the function g by differentiating (partially) with respect to a,

$$g'(a) = g'(a + b + c) \tag{71}$$

and then differentiating with respect to b,

$$0 = g''(a + b + c) \tag{72}$$

This says that the second derivative of the function g is identically zero. Consequently, this function must be a linear function (a polynomial of order one) of the form

$$g(a) = \alpha - \beta a \tag{73}$$

where α and β are constants.* The function $f(v_x)$ must then be

$$f(v_x) = e^{g(v_x^2)} = e^{\alpha - \beta v_x^2} \tag{74}$$

and similarly for $f(v_y)$ and $f(v_z)$. The distribution function for the vector velocity is then

$$f(v_x)f(v_y)f(v_z) = e^{3\alpha} e^{-\beta(v_x^2 + v_y^2 + v_z^2)} \tag{75}$$

We still have to determine the constants α and β. The constant α is related to the normalization of the distribution function. If we integrate the distribution function $f(v_x)f(v_y)f(v_z)$ over the full range of velocities from $-\infty$ to $+\infty$, the result must equal the total number of molecules:

$$N = \int_{-\infty}^{\infty} \int_{-\infty}^{\infty} \int_{-\infty}^{\infty} f(v_x)f(v_y)f(v_z) \, dv_x \, dv_y \, dv_z$$

$$= e^{3\alpha} \int_{-\infty}^{\infty} e^{-\beta v_x^2} \, dv_x \int_{-\infty}^{\infty} e^{-\beta v_y^2} \, dv_y \int_{-\infty}^{\infty} e^{-\beta v_z^2} \, dv_z \tag{76}$$

The integrals appearing in this equation are Gaussian integrals of the form

$$\int_{-\infty}^{\infty} e^{-\beta \xi^2} \, d\xi = \left(\frac{\pi}{\beta}\right)^{1/2} \tag{77}$$

Consequently, Eq. (76) becomes

$$N = e^{3\alpha} \left(\frac{\pi}{\beta}\right)^{3/2} \tag{78}$$

or

$$e^{3\alpha} = N \left(\frac{\beta}{\pi}\right)^{3/2} \tag{79}$$

Finally, we can determine the constant β by using the result (61) for the mean square of the speed:

$$\overline{v_x^2} = \frac{1}{3} \overline{v^2} = \frac{kT}{m} \tag{80}$$

---

* A negative sign has been inserted in front of β for the sake of later convenience.

In terms of the distribution function, the mean value of v_x^2 is given by the integral

$$\overline{v_x^2} = \frac{1}{N} \int_{-\infty}^{\infty} \int_{-\infty}^{\infty} \int_{-\infty}^{\infty} v_x^2 f(v_x) f(v_y) f(v_z) \, dv_x \, dv_y \, dv_z$$

$$= \left(\frac{\beta}{\pi}\right)^{3/2} \int_{-\infty}^{\infty} v_x^2 e^{-\beta v_x^2} \, dv_x \int_{\infty}^{\infty} e^{-\beta v_y^2} \, dv_y \int_{\infty}^{\infty} e^{-\beta v_z^2} \, dv_z \qquad (81)$$

The integral over v_x is related to the derivative of the Gaussian integral,

$$\int_{-\infty}^{\infty} \xi^2 e^{-\beta \xi^2} \, d\xi = -\frac{\partial}{\partial \beta} \int_{-\infty}^{\infty} e^{-\beta \xi^2} \, d\xi = -\frac{\partial}{\partial \beta} \left(\frac{\pi}{\beta}\right)^{1/2} = \frac{1}{2}\left(\frac{\pi}{\beta^3}\right)^{1/2}$$

$$(82)$$

Hence, Eq. (81) gives

$$\overline{v_x^2} = \left(\frac{\beta}{\pi}\right)^{3/2} \frac{1}{2}\left(\frac{\pi}{\beta^3}\right)^{1/2} \left(\frac{\pi}{\beta}\right)^{1/2} \left(\frac{\pi}{\beta}\right)^{1/2} = \frac{1}{2\beta} \qquad (83)$$

Comparing this with Eq. (80), we find $kT/m = 1/(2\beta)$, so that

$$\beta = \frac{m}{2kT} \qquad (84)$$

Thus, the distribution function for velocities is

$$f(v_x)f(v_y)f(v_z) = N\left(\frac{m}{2\pi kT}\right)^{3/2} e^{-\frac{1}{2}m(v_x^2 + v_y^2 + v_z^2)/kT} \qquad (85)$$

or

$$f(v_x)f(v_y)f(v_z) = N\left(\frac{m}{2\pi kT}\right)^{3/2} e^{-\frac{1}{2}mv^2/kT} \qquad (86)$$

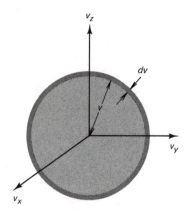

Fig. 1.14 Spherical shell of radius v and thickness dv.

This is called the **Maxwell distribution** of molecular velocities. Note that this distribution has a maximum at $v_x = v_y = v_z = 0$, i.e., the most probable velocity is zero. Furthermore, note that the average value of the velocity is also zero because, according to Eq. (85), positive and negative velocities are equally likely.

The distribution function for velocities implies the distribution function for speeds. If we want the number of molecules whose speed is in some interval dv (regardless of direction), we integrate the formula (86) over all the volume associated with this interval. Figure 1.14 shows this volume: it is a thin spherical shell of radius v and thickness dv. The magnitude of this volume is the product of area and thickness, $4\pi v^2\, dv$. The number of molecules with velocity vectors in this volume is the integral of the function $f(v_x)f(v_y)f(v_z)$ over the volume. Since the function is constant over the volume, the integral is simply the product of the function and the volume, i.e.,

$$f(v_x)f(v_y)f(v_z) \times 4\pi v^2\, dv = \left[N\left(\frac{m}{2\pi kT}\right)^{3/2} e^{-\frac{1}{2}mv^2/kT} \right] \times 4\pi v^2\, dv \tag{87}$$

According to the definition of the distribution function for speeds (see above), this must coincide with $h(v)\, dv$. Therefore,

$$h(v) = 4\pi N \left(\frac{m}{2\pi kT}\right)^{3/2} v^2\, e^{-\frac{1}{2}mv^2/kT} \tag{88}$$

The curve shown in Fig. 1.11 is a plot of this function. Obviously, the data points fit the theoretical curve extremely well.

Example 5   Find an expression for the most probable speed of a molecule.

Solution   The most probable speed corresponds to the maximum of $h(v)$. This maximum is determined by

$$0 = \frac{\partial h(v)}{\partial v} = 4\pi N \left(\frac{m}{2\pi kT}\right)^{3/2} \left[2v\, e^{-\frac{1}{2}mv^2/kT} - \frac{v^3 m}{kT} e^{-\frac{1}{2}mv^2/kT} \right] \tag{89}$$

Solving this for v, we find that the most probable speed is

$$v_{\text{prob}} = \sqrt{\frac{2kT}{m}} \quad \blacksquare \tag{90}$$

It is instructive to rewrite the Maxwell distribution function (86) in terms of the energy of a molecule. Since the energy of an ideal gas is purely kinetic,

$$E = K = \tfrac{1}{2}mv^2 \tag{91}$$

so that Eq. (86) becomes

$$f(v_x)f(v_y)f(v_z) = N \left(\frac{m}{2\pi kT}\right)^{3/2} e^{-E/kT} \tag{92}$$

The factor $e^{-E/kT}$ is called the **Boltzmann factor.** Apart from normalization, it gives the probability for a molecule to have a velocity in the interval $\mathbf{v}$ to $\mathbf{v} + d\mathbf{v}$:

$$\begin{pmatrix} \text{probability for velocity} \\ \text{between } \mathbf{v} \text{ and } \mathbf{v} + d\mathbf{v} \end{pmatrix} \propto e^{-E/kT} \qquad (93)$$

Although we have derived this proportionality in the context of the velocity distribution of an ideal gas, it turns out that it is of general validity. In any system in thermal equilibrium, the probability that a member of the system has a velocity or position in some specified interval is proportional to the Boltzman factor, regardless of whether E is translational kinetic energy, rotational kinetic energy, or potential energy. We will not attempt the general derivation of this theorem, which may be found in any treatise on statistical mechanics.*

By way of an illustration of the use of the Boltzmann factor, let us calculate the mean energy of a simple harmonic oscillator consisting of a particle moving in one dimension under the influence of an elastic force. If the equilibrium position is at $x = 0$, then the potential energy is $\frac{1}{2}\kappa x^2$, where κ is the spring constant. The total energy is

$$E = \tfrac{1}{2}mv_x^2 + \tfrac{1}{2}\kappa x^2 \qquad (94)$$

and the Boltzmann factor is

$$e^{-(\frac{1}{2}mv_x^2 + \frac{1}{2}\kappa x^2)/kT} \qquad (95)$$

The probability that the particle is in the interval x to $x + dx$ and simultaneously its velocity is in the interval v_x to $v_x + dv_x$ is proportional to this Boltzmann factor, and also proportional to the sizes of the intervals dx and dv_x,

$$\begin{pmatrix} \text{probability for position} \\ \text{between } x \text{ and } x + dx \\ \text{and velocity between } v_x \\ \text{and } v_x + dv_x \end{pmatrix} = A\, e^{-(\frac{1}{2}mv_x^2 + \frac{1}{2}kx^2)/kT}\, dx\, dv_x \qquad (96)$$

Here A is a constant of proportionality, which we can determine from the normalization condition for probability: the integral of the probability over the full range of values of x and v_x must be equal to 1. Thus,

$$1 = \int_{-\infty}^{\infty} \int_{-\infty}^{\infty} A\, e^{-(\frac{1}{2}mv_x^2 + \frac{1}{2}kx^2)/kT}\, dx\, dv_x$$

$$= A\left(\frac{2\pi kT}{m}\right)^{1/2} \left(\frac{2\pi kT}{\kappa}\right)^{1/2} \qquad (97)$$

---

* For example, L. D. Landau and E. M. Lifshitz, *Statistical Physics*, Chapter III.

and

$$A = \left(\frac{m}{2\pi kT}\right)^{1/2} \left(\frac{\kappa}{2\pi kT}\right)^{1/2} \tag{98}$$

The mean energy is then

$$\bar{E} = \int_{-\infty}^{\infty} \int_{-\infty}^{\infty} \left(\tfrac{1}{2}mv_x^2 + \tfrac{1}{2}\kappa x^2\right) A\, e^{-\left(\tfrac{1}{2}mv_x^2 + \tfrac{1}{2}\kappa x^2\right)/kT}\, dx\, dv_x \tag{99}$$

$$= A \int_{-\infty}^{\infty} e^{-\tfrac{1}{2}\kappa x^2/kT}\, dx \int_{-\infty}^{\infty} \tfrac{1}{2}mv_x^2\, e^{-\tfrac{1}{2}mv_x^2/kT}\, dv_x$$

$$+ A \int_{-\infty}^{\infty} \tfrac{1}{2}\kappa x^2\, e^{-\tfrac{1}{2}\kappa x^2/kT}\, dx \int_{-\infty}^{\infty} e^{-\tfrac{1}{2}mv_x^2/kT}\, dv_x \tag{100}$$

The first term on the right of this equation is the mean kinetic energy, $\tfrac{1}{2}mv_x^2$; the second term is the mean potential energy, $\tfrac{1}{2}\kappa x^2$. The integrals are of the same kind as those we encountered above, and they can be evaluated by the same methods. The result is that the mean kinetic energy and the mean potential energy are equal to $\tfrac{1}{2}kT$,

$$\tfrac{1}{2}m\overline{v_x^2} = \tfrac{1}{2}kT \tag{101}$$

and

$$\tfrac{1}{2}\kappa\overline{x^2} = \tfrac{1}{2}kT \tag{102}$$

The mean energy is then

$$\bar{E} = \tfrac{1}{2}kT + \tfrac{1}{2}kT = kT \tag{103}$$

Note that the mean kinetic energy and the mean potential energy in Eqs. (101) and (102) are independent of the mass and the spring constant; they depend only on the temperature. These results are special cases of a general theorem called the **equipartition theorem:** *each term in the energy proportional to the square of a velocity or a coordinate contributes $\tfrac{1}{2}kT$ to the mean energy at thermal equilibrium.* The proof of this theorem is a straightforward generalization of the above calculation for the simple harmonic oscillator.

In statistical mechanics, each term in the energy proportional to the square of a component of the velocity or a coordinate is called a **degree of freedom.** Hence the equipartition theorem can be concisely rephrased as follows: each degree of freedom contributes $\tfrac{1}{2}kT$ to the mean energy at thermal equilibrium.

Example 6   What is the mean energy of a three-dimensional harmonic oscillator at thermal equilibrium?

Solution   The energy of such an oscillator is

$$E = \tfrac{1}{2}m(v_x^2 + v_y^2 + v_z^2) + \tfrac{1}{2}\kappa(x^2 + y^2 + z^2)$$

The expression for the energy contains *six* terms proportional to the square of a coordinate or a velocity, i.e., six degrees of freedom. Hence the mean energy must be $6 \times \frac{1}{2}kT$, or $3kT$ at thermal equilibrium. Not surprisingly, this is three times as large as the mean energy of a one-dimensional oscillator. ∎

Example 7   In copper, the frequency of oscillation of individual atoms about their equilibrium positions in the crystal lattice is about 2×10^{13} radians/s. According to classical mechanics, what is the rms amplitude of oscillation of a copper atom at room temperature?

Solution   By Eq. (102), the rms amplitude of oscillation is

$$\sqrt{\overline{x^2}} = \sqrt{\frac{kT}{\kappa}} \tag{104}$$

For a simple harmonic oscillator, the frequency of oscillation is related to the spring constant by $\kappa = m\omega^2$, so that

$$\sqrt{\overline{x^2}} = \sqrt{\frac{kT}{m\omega^2}} \tag{105}$$

Since the atomic mass of copper is 63.5, the mass of a copper atom is 63.5 u, or $63.5 \times 1.66 \times 10^{-27}$ kg; room temperature is 20°C or 293 K. Hence

$$\sqrt{\overline{x^2}} = \sqrt{\frac{1.38 \times 10^{-23}\,\text{J/K} \times 293\,\text{K}}{63.5 \times 1.66 \times 10^{-27}\,\text{kg} \times (2 \times 10^{13}/\text{s})^2}} = 1 \times 10^{-11}\,\text{m}$$

This is about 3% of the separation between one copper atom and the next. ∎

1.6 Classical Waves

A wave is a propagating disturbance in a deformable, elastic medium. In the case of a sound wave or a water wave, the medium is the air or the water, respectively. In the case of a light wave or a radio wave propagating through vacuum, the medium consists of the electric and magnetic fields of this wave itself; these fields have the necessary properties of deformability and elasticity to support a propagating disturbance. Thus, an electromagnetic wave is its own medium.

All waves transport energy; some transport momentum, and some even angular momentum. For the sake of simplicity, in most of this section we will deal only with waves whose amplitude is a one-component quantity, such as the pressure of a sound wave or the height of a water wave. We will briefly deal with electro-magnetic waves—whose amplitude is a multicomponent vector quantity—at the end of this section.

At any given instant, the state of a wave is completely specified by its amplitude throughout space, and by the time derivative of this amplitude. The instantaneous amplitude of the wave is some function of the space coordinates, and the instantaneous time de-rivative of the amplitude is another, independent, function of the space coordinates.* These two quantities are analogous to the instantaneous position and the instantaneous velocity of a particle. In the case of a sound wave or a water wave, this analogy has an obvious physical basis: the amplitude of the wave is directly related to the displacement of the particles of the medium, and the time derivative of the amplitude is directly related to the velocity of the particles (if we ignore the random thermal motions of the particles).

The simplest kind of wave is a harmonic wave, which is a si-nusoidal or cosinusoidal function of the space coordinates and of time. For instance, a harmonic wave propagating in the positive x direction has the form

$$\Phi(x,t) = A \cos{(kx - \omega t + \delta)} \tag{106}$$

and a wave propagating in the negative x direction has the form

$$\Phi(x,t) = A \cos{(kx + \omega t + \delta)} \tag{107}$$

In these wavefunctions, the constant A is the **harmonic amplitude**[†] of the wave; k is the **wave number;** ω the **angular frequency;** and δ the **phase constant.** The wavelength of the wave is

$$\lambda = \frac{2\pi}{k} \tag{108}$$

and the frequency of the wave is

$$\nu = \frac{\omega}{2\pi} \tag{109}$$

Figure 1.15 shows a harmonic wave traveling toward the right, at successive instants of time.

---

* Mathematically, the amplitude at some fixed time t_0 is a function $\Phi(x, y, z, t = t_0)$; and the time derivative is the function $(\partial/\partial t)\Phi(x, y, z, t = t_0)$.

† We call it the harmonic amplitude to distinguish it from the instantaneous amplitude $\Phi(x,t)$. The word *amplitude* is used somewhat loosely in physics; the intended meaning must often be guessed from the context.

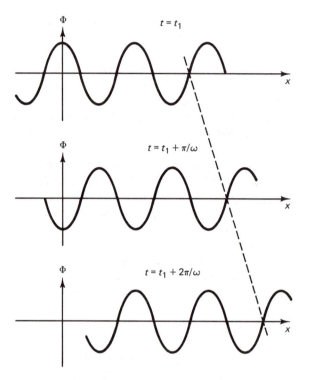

Fig. 1.15 A harmonic wave at successive instants of time. The wave is traveling toward the right.

The velocity of the wave crest or the wave trough of a harmonic wave is called the **phase velocity.** To find this velocity, consider the wave crest that corresponds to zero argument of the cosine in Eq. (106). The position of this crest as a function of time is given by

$$kx - \omega t + \delta = 0 \tag{110}$$

Taking the differential of this expression, we see that the increments of x and t at the position of the crest must be related by

$$k\,dx - \omega\,dt = 0 \tag{111}$$

from which we find a phase velocity

$$v_p = \frac{dx}{dt} = \frac{\omega}{k} \tag{112}$$

If the phase velocity depends on the wave number, or the wavelength, the medium in which the wave propagates is said to be **dis-**

persive. For instance, water waves in deep water have a frequency

$$\omega = \sqrt{gk} \tag{113}$$

where g is the acceleration of gravity, $g = 9.81 \, \text{m/s}^2$. The phase velocity of these waves is

$$v_p = \frac{\omega}{k} = \sqrt{\frac{g}{k}} = \sqrt{\frac{g\lambda}{2\pi}} \tag{114}$$

Thus water waves of long wavelength have a greater velocity than waves of short wavelength. Water—or more precisely, the water surface—is a dispersive medium.

Under the assumption that the amplitude of deformation of the medium does not exceed the elastic limits, waves obey a **principle of linear superposition:** when two or more waves arrive at any given point, the resultant instantaneous amplitude is the sum of the individual instantaneous amplitudes. Depending on the signs of these individual amplitudes, or the phases of the waves, the resultant wave can be larger (**constructive interference**) or smaller (**destructive interference**) than the individual waves.

Example 8   Two waves of equal harmonic amplitudes are simultaneously propagating along the x-axis in opposite directions. What is the net wave?

Solution   Suppose that the waves are

$$\Phi_1 = A \cos(kx - \omega t) \tag{115}$$

and

$$\Phi_2 = A \cos(kx + \omega t) \tag{116}$$

By the superposition principle, the net wave is simply the sum $\Phi_1 + \Phi_2$,

$$\Phi = \Phi_1 + \Phi_2 = A[\cos(kx - \omega t) + \cos(kx + \omega t)] \tag{117}$$

With the trigonometric identity

$$\cos \alpha + \cos \beta = 2 \cos \tfrac{1}{2}(\alpha + \beta) \cos \tfrac{1}{2}(\alpha - \beta) \tag{118}$$

this sum becomes

$$\Phi = 2A \cos kx \cos \omega t \tag{119}$$

This is a **standing wave.** The instantaneous amplitude has a maximum magnitude (corresponding to a wave crest or a wave trough) at the fixed positions given by

$$kx = n\pi \qquad n = 0, \pm 1, \pm 2, \ldots \tag{120}$$

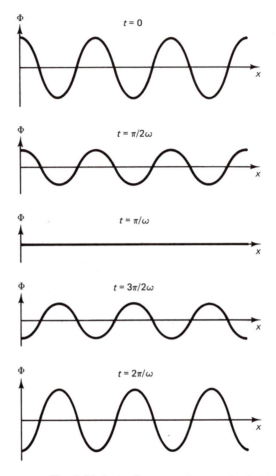

Fig. 1.16 A standing wave at successive instants of time.

At these positions the two waves interfere constructively. The instantaneous amplitude has minimum magnitude ($\Phi = 0$) at the fixed positions given by

$$kx = (n + \tfrac{1}{2})\pi \qquad n = 0, \pm 1, \pm 2, \ldots \qquad (121)$$

At these positions the two waves interfere destructively, and they cancel exactly at all times. Figure 1.16 shows a standing wave at successive instants of time. ■

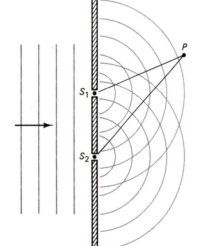

Fig. 1.17 Circular waves spreading out from two very narrow slits which act as point sources.

Example 9   A water wave with parallel wave fronts is incident on two very narrow slits in a wall (see Fig. 1.17). Small portions of the wave spill through the slits and, consequently, each of the slits acts as a source of a wave with concentric circular wave fronts. In what directions do these two sets of circular waves interfere constructively? Assume that the waves are observed at a large distance from the sources.

Solution   The waves propagating outward from the two sources and arriving at the point P are

$$\Phi_1 = A \cos (kr_1 - \omega t) \tag{122}$$

and

$$\Phi_2 = A \cos (kr_2 - \omega t) \tag{123}$$

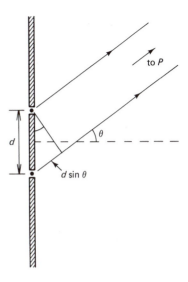

Fig. 1.18 The point of observation (*P*) is at large distance from the sources. The position angle θ is measured relative to the midline (dashed).

where r_1 and r_2 are the distances S_1P and S_2P measured from the sources to the point of observation (see Fig. 1.18).* The net wave is the sum of these two waves, $\Phi = \Phi_1 + \Phi_2$. The two waves interfere constructively wherever the arguments of the cosine functions in Eqs. (122) and (123) differ by 0, or 2π, or 4π, etc. Thus, the condition for constructive interference is

$$kr_2 - \omega t - (kr_1 - \omega t) = 2\pi n \qquad n = 0, \pm 1, \pm 2, \ldots \tag{124}$$

or

$$(r_2 - r_1)k = 2\pi n \tag{125}$$

If r_1 and r_2 are much larger than the separation d between the sources, the lines S_1P and S_2P are approximately parallel; the difference between these distances is then approximately $d \sin \theta$ (see Fig. 1.18). Hence Eq. (125) becomes

$$(d \sin \theta)k = 2\pi n \tag{126}$$

or, in terms of the wavelength,

$$d \sin \theta = n\lambda \qquad n = 0, \pm 1, \pm 2, \ldots \tag{127}$$

As expected from symmetry, we find constructive interference along the midline ($\theta = 0$); but we also find constructive interference along discrete directions on each side of the midline. The photograph in Fig. 1.19 shows the fanlike pattern of strong beams corresponding to the directions of constructive interference. Between these strong beams, we find nodal lines corresponding to the directions of destructive interference [the latter directions are given by $d \sin \theta = (n + \frac{1}{2})\lambda$]. ∎

In the preceding example we assumed that the waves emerging from the slits spread out more or less uniformly in all directions

---

* The amplitude A of a circular wave spreading out on a water surface decreases with distance, $A \propto 1/\sqrt{r}$. But we will ignore this; it does not affect our calculation.

Fig. 1.19 Interference pattern of circular water waves in a ripple tank. (From *PSSC Physics*, 2nd edition, 1965; D. C. Heath and Company with Education Development Center, Inc., Newton, MA.)

beyond the slits. This spreading of a wave upon passage through a narrow slit is called **diffraction.** Instead of propagating straight through the slit in the forward direction, the wave spreads sideways and spills into the shadow zone behind the walls. The amount of diffraction depends on the size of the slit. If the size of the slit is large compared to the wavelength of the wave, there is very little diffraction; most of the wave propagates in the forward direction, with very little lateral spread. But if the size of the slit is roughly equal to or smaller than the wavelength, then the diffraction is very pronounced (see Fig. 1.20) and gives rise to a fairly complicated fanlike pattern of beams of strong and of weak waves beyond the slit. The pattern displays a strong central beam in the forward direction and several weaker, secondary beams on both sides. From a mathematical analysis of diffraction* we learn that the half-width of the central maximum, or the position of the first lateral

---

* See any introductory textbook on classical physics such as, e.g., H. C. Ohanian, *Physics*, Chapter 39. The general formula for the minima of the diffraction pattern is $a\sin\theta = n\lambda, n = \pm 1, \pm 2, \pm 3, \ldots.$

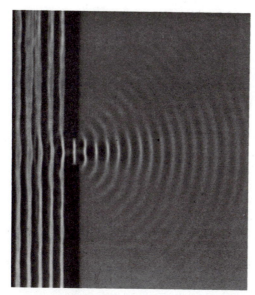

Fig. 1.20 Diffraction pattern of water waves in a ripple tank. Upon passage through the narrow slit the waves spread out laterally. (From *PSSC Physics*, 2nd edition, 1965; D. C. Heath and Company with Education Development Center, Inc., Newton, MA.)

minimum, is given by

$$a \sin \theta = \lambda \qquad (128)$$

where a is the width of the slit. If a is equal to or smaller than the wavelength, then the central maximum is very wide, covering the entire space beyond the slit. In general, the wave pattern produced by two or more slits involves a combination of diffraction effects at each slit and interference effects among all the slits. But if the slits are very narrow, as in Example 9, then the diffraction pattern is uniform and featureless, and only the interference pattern is evident.

The principle of linear superposition is an important tool in the mathematical study of wave pulses of finite extent, or wave packets. According to the **Fourier theorem** of advanced mathematics, any wave packet of arbitrary shape can be expressed as a sum of a large (infinite) number of harmonic waves of different wavelengths. Hence, if the superposition principle is valid, we can deduce all the properties of a wave packet from the properties of the harmonic waves that compose it. For instance, we can deduce the velocity of a wave packet consisting of a finite sequence of wave peaks and

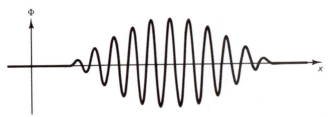

Fig. 1.21 A wave group.

wave troughs (see Fig. 1.21). Such a wave packet is called a **wave group.** It may be regarded as a harmonic wave modulated by an envelope (shown by the dashed line in Fig. 1.21). The velocity of the wave group is the velocity of the envelope; as we will see, this velocity differs from that of the wave crests of the harmonic wave, i.e., the individual wave crests move in relation to the envelope. This strange behavior comes about because the wave group is a superposition of many harmonic waves, whose interference pattern shifts as a function of time. The simplest way to construct a wave packet resembling Fig. 1.21 is by the superposition of two harmonic waves differing slightly in wave number and in frequency:

$$\Phi_1 = A \cos\left[(k + \tfrac{1}{2}\Delta k)x - (\omega + \tfrac{1}{2}\Delta\omega)t\right] \tag{129}$$

and

$$\Phi_2 = A \cos\left[(k - \tfrac{1}{2}\Delta k)x - (\omega - \tfrac{1}{2}\Delta\omega)t\right] \tag{130}$$

Again using the trigonometric identity (118), we find

$$\Phi_1 + \Phi_2 = 2A \cos\tfrac{1}{2}(\Delta k\, x - \Delta\omega\, t)\cos(kx - \omega t) \tag{131}$$

Figure 1.22 shows a plot of this wave at a given instant of time. Obviously, this wave consists of a succession of wave groups, each

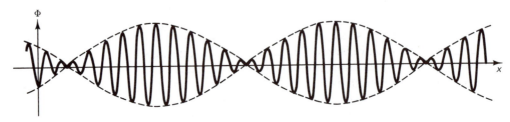

Fig. 1.22 A sequence of wave groups constructed by superposition of two harmonic waves of slightly different wave number and frequency.

resembling the wave group shown in Fig. 1.21. If we wanted to eliminate all of these extra wave groups except the central one, we would have to add many more waves to the superposition (131), in such a way that all the wave groups to the right and left of the central one cancel by destructive interference. But for our purposes, it is sufficient to simply ignore the extra wave groups, and to concentrate on the motion of the central group. If the phase velocity of both harmonic waves were the same (nondispersive medium), then the waves would always remain in step. The entire pattern in Fig. 1.22 would then move to the right rigidly. But if the phase velocities are different (dispersive medium), then the two waves move relative to each other. The peak of the envelope—where the two waves interfere constructively—then also moves relative to the waves. To obtain the velocity of the peak, or the **group velocity,** we note that the first factor in Eq. (131) represents the envelope shown by the dashed line in Fig. 1.22. The position of the peak is given by

$$\Delta k x - \Delta \omega t = 0 \tag{132}$$

and therefore the increments of x and of t at the position of the peak are related by

$$\Delta k \, dx - \Delta \omega \, dt = 0 \tag{133}$$

From this we find the group velocity:

$$v_g = \frac{dx}{dt} = \frac{\Delta \omega}{\Delta k} \tag{134}$$

We will assume that $\Delta \omega$ and Δk are very small, so that the wave group has a fairly well-defined frequency and wavelength. Then $\Delta \omega$ and Δk can be approximated by differentials and

$$v_g = \frac{d\omega}{dk}$$

For example, for water waves [see Eq. (113)], the group velocity is

$$v_g = \frac{d\omega}{dk} = \frac{d}{dk} \sqrt{gk} = \frac{1}{2} \sqrt{\frac{g}{k}} \tag{135}$$

Comparing this with Eq. (114), we see that the group velocity of a water wave is one-half the phase velocity.

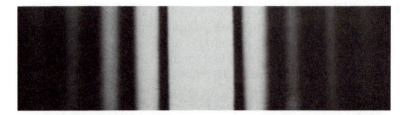

Fig. 1.23 Diffraction pattern of light registered on a photographic plate placed beyond a fine slit illuminated by a laser beam. (From H. C. Ohanian, *Physics*. Reprinted with permission of W. W. Norton & Co., Inc.)

The group velocity is the signal velocity. If we want to send a signal by means of a wave, we cannot use a harmonic wave. This kind of wave has no beginning and no end; it lasts forever, and therefore conveys no information, apart from the trivial information that the sender is in operation. (In the terminology of radio: the harmonic wave is the carrier wave, which merely gives a steady hum in the radio receiver.) To send a signal, we must switch the harmonic wave on and off, or we must vary its strength. (In the terminology of radio: we must modulate the carrier wave.) This means we must construct wave packets. Hence the signal velocity coincides with the group velocity of such a wave packet. Furthermore, it turns out that the velocity with which energy is transported by the wave also coincides with the group velocity.

We end this section with a few remarks on light waves and other electromagnetic waves. Newton was of the opinion that light is a stream of particles ejected from the light source. This opinion prevailed until the beginning of the nineteenth century when Thomas Young demonstrated that light displays interference phenomena, such as those described in Example 9, which are characteristic of waves. Young was also aware that light displays diffraction phenomena, i.e., that light passing through a small aperture deviates from rectilinear propagation and spills into the shadow zone (see Fig. 1.23), and he pointed out that such phenomena are another characteristic of waves. Shortly afterward, Augustin Fresnel developed a mathematical theory of interference and diffraction whose excellent agreement with the experimental observations convinced physicists that light is a wave. But this left unanswered the deeper question concerning the ultimate nature of light: just what are the light waves made of? Since it seemed incon-

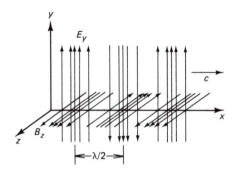

Fig. 1.24 Plane electromagnetic wave traveling toward the right, with a vertical direction of polarization.

ceivable to the physicists of Fresnel's days that light waves could propagate through a vacuum, they postulated the **ether,** a ghostly substance that was supposed to permeate all of space and serve as the medium for the propagation of light. The deeper question regarding the nature of light was finally answered in 1873 by James Clerk Maxwell, who formulated a set of equations describing electric and magnetic fields, and found that these equations had wave solutions with a speed $c = 1/\sqrt{\mu_0 \varepsilon_0}$, which is equal to the speed of light.* This led to the classical picture of light as a wave consisting of electric and magnetic fields. Since these fields can exist in a vacuum, light does not need any medium for its propagation.

Maxwell's theory also accounts for the **polarization** of light, discovered by Étienne Malus early in the nineteenth century. The electric and magnetic fields in a light wave are always perpendicular to each other, and perpendicular to the direction of propagation. Thus, the light wave is a **transverse** wave. The direction of polarization of the wave is the direction of the electric field. For instance, Fig. 1.24 shows the electric and magnetic fields of a plane wave propagating toward the right, with a vertical direction of

---

* With $\varepsilon_0 = 8.85 \times 10^{-12}$ F/m and $\mu_0 = 1.26 \times 10^{-6}$ H/m, Maxwell's theoretical formula for the speed of electromagnetic waves yields

$$c = \frac{1}{(8.85 \times 10^{-12}\,\text{F/m} \times 1.26 \times 10^{-6}\,\text{H/m})^{1/2}} = 2.99 \times 10^8\,\text{m/s}$$

in good agreement with the experimental value for the speed of light [see Eq. (136)]. The modern approach is to use the theoretical formula to calculate ε_0 from the given value of μ_0 and the measured value of the speed of light.

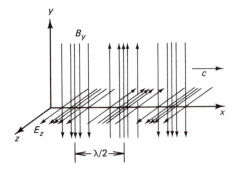

Fig. 1.25 Plane electromagnetic wave traveling toward the right, with a horizontal direction of polarization.

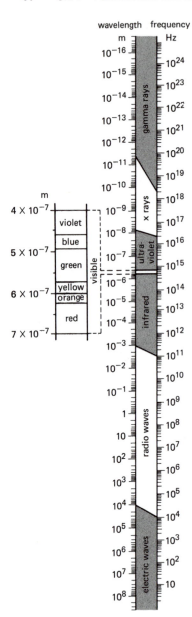

Fig. 1.26 The electromagnetic spectrum

polarization. Figure 1.25 shows the fields of a plane wave with the same direction of propagation, but with a horizontal direction of polarization. Any other direction of polarization, such as a polarization at 45° to the horizontal, can be regarded as a superposition of the two basic directions shown in these figures.

Radio waves, infrared radiation, ultraviolet radiation, and X rays are electromagnetic waves of the same kind as light, but of different wavelengths. Figure 1.26 is a summary of the electromagnetic spectrum. The speeds of all of these waves equal the speed of light. The best available modern value for the speed of light in vacuum is

$$c = 2.99792458 \times 10^8 \, \text{m/s} \tag{136}$$

According to the new definition of the meter officially adopted in 1984, the meter is the distance traveled by light in vacuum in $\frac{1}{299792458}$ of a second.* Thus, the above value of the speed of light is henceforth regarded as a fixed standard, and any future measurements can only bring about readjustments of the meter, not readjustments of the speed of light.

Maxwell's theory explains how electromagnetic waves are generated by accelerated charges. A moving charge is surrounded by both electric and magnetic fields. In essence, an electromagnetic wave is a propagating disturbance in these electric and magnetic fields of a charge. As long as the charge moves with uniform velocity, the electric and magnetic fields move with the charge, as though they were rigidly attached to it. But if the charge is sub-

---

* This definition of the meter can be conveniently implemented by means of the wavelength emitted by a laser; see Section 7.6.

jected to an acceleration, parts of the fields break away—they become independent of the charge and they travel outward as an
electric and magnetic disturbance, which carries energy away from
the charge.

Some of the most important properties of electromagnetic waves
are as follows:* The magnitudes of the electric and magnetic fields
are proportional; for a wave in vacuum,

$$|\mathbf{B}| = \frac{1}{c}|\mathbf{E}| \tag{137}$$

As in the case of light, the directions of these fields are perpendicular to each other and also perpendicular to the direction of propagation. These directions are related by the right-hand rule for cross
products: $\mathbf{E} \times \mathbf{B}$ is in the direction of propagation. The energy density of the electric and magnetic fields is

$$u = \frac{1}{\mu_0 c}|\mathbf{E}||\mathbf{B}| \tag{138}$$

The energy flux, or power per unit area, transported by the wave is

$$S = \frac{1}{\mu_0}|\mathbf{E}||\mathbf{B}| \tag{139}$$

or

$$S = cu \tag{140}$$

The magnitude and direction of the energy flux is given by the
Poynting vector,

$$\mathbf{S} = \frac{1}{\mu_0}\mathbf{E} \times \mathbf{B} \tag{141}$$

This direction is of course the same as the direction of propagation
of the wave. The momentum per unit area and unit time transported by the wave, or the pressure exerted by the wave on an
absorbing surface, is directly proportional to the energy flux,

$$P = \frac{S}{c} \tag{142}$$

From Eq. (142) we can deduce an interesting relation between
the energy and the momentum carried by the wave. In a time Δt,

---

* For the derivation of these properties of electromagnetic waves, see any
textbook on classical physics, e.g., H. C. Ohanian, *Physics*, Chapter 36.

the energy and the momentum flowing through a cross-sectional area A are $\Delta E = SA \, \Delta t$ and $\Delta p = PA \, \Delta t = (S/c)A \, \Delta t$. Hence the energy and the momentum carried by the wave obey the relation

$$\Delta E = c \, \Delta p \tag{143}$$

We will find this relation useful in the next section.

Besides the two linearly polarized waves illustrated in Figs. 1.24 and 1.25, there are waves with other kinds of polarization; all of these can be obtained by forming superpositions of the two linearly polarized waves. For instance, circularly polarized waves are obtained by superposition of the two linearly polarized waves with equal amplitudes but with a phase difference of 90°. At any fixed position x, the electric and magnetic fields of such a circularly polarized wave rotate around the x-axis with a frequency of rotation equal to the frequency of the wave (the direction of rotation depends on the sign of the phase difference). It can be shown from Maxwell's equations that such circularly polarized waves carry not only energy and momentum, but also angular momentum; this angular momentum is due to a circulating flow of momentum around the direction of propagation of the wave. The amount of angular momentum carried by the wave is equal to the amount of energy divided by the angular frequency,

$$\Delta L = \frac{\Delta E}{\omega} \tag{144}$$

1.7 The Equivalence of Energy and Mass

The equivalence of energy and mass is one of the most celebrated contributions that Albert Einstein made to physics. Although he first derived this equivalence from his theory of relativity, he later devised a much simpler derivation based on the relation between the energy and the momentum of a light wave. For this derivation Einstein relied on the following simple *Gedankenexperiment*.* Imagine that an emitter and an absorber of light are enclosed in a box of mass M. The emitter and the absorber are firmly attached to the ends of the box, separated by a distance L (see Fig. 1.27).

---

* German for "thought experiment," meaning an idealized experiment whose result can be predicted by thought alone.

The box is initially stationary, but is free to move. If the emitter sends a short light pulse of energy ΔE toward the right, the entire box will recoil toward the left. Since, according to Eq. (143), the momentum of the light pulse is $\Delta E/c$, conservation of momentum tells us that the recoil velocity must be

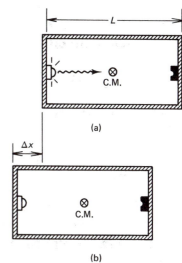

$$v_x = -\frac{p_x}{M} = -\frac{\Delta E}{Mc} \tag{145}$$

The light pulse takes approximately a time L/c to travel from the left end of the box to the right.* The light pulse is then absorbed; this stops the motion of the box. In the time L/c, the box suffers a displacement

$$\Delta x = v_x t = -\frac{\Delta E \, L}{Mc^2} \tag{146}$$

But in an isolated system, the center of mass must remain stationary. Thus, the transport of energy ΔE from one end of the box to the other must be accompanied by a transport of mass Δm. We can deduce the amount of mass from the requirement that the center of mass remain stationary,

Fig. 1.27 Emitter and absorber in a box: (a) before emission: (b) after emission and absorption of the light pulse.

$$M \, \Delta x + (\Delta m)L = 0 \tag{147}$$

which gives

$$\Delta m = -\frac{M \, \Delta x}{L} \tag{148}$$

or

$$\Delta m = \frac{\Delta E}{c^2} \tag{149}$$

This equation asserts that energy has mass, i.e., energy has inertia. It also asserts that mass has energy. At the absorber, radiant energy is converted into mass; and at the emitter, mass is converted into radiant energy.

Thus, mass is a form of energy, and energy is a form of mass. Any change of the energy of a body entails a corresponding change of mass, and vice versa. The most spectacular example of a change of mass that entails a change of energy is found in the process of particle–antiparticle **annihilation** (this example was not known to Einstein when he gave his derivation). Particles and antiparticles

---

*We neglect the speed of the box compared to the speed of light.

have exactly the same masses, but their electric charges are opposite.* For instance, the electron has a charge $-e$ and the **antielectron,** or **positron,** has a charge $+e$. Antielectrons are not found in ordinary matter, but they can be produced in high-energy collisions. Antielectrons usually do not last very long: when an antielectron encounters ordinary matter and comes into contact with an electron, the two particles immediately annihilate one another. This process destroys the masses of both particles and releases a corresponding amount of energy. Since the mass of an antiparticle is exactly the same as the mass of the particle, the energy released in electron–antielectron annihilation is

$$E = m_e c^2 + m_e c^2$$
$$= 2 \times 9.11 \times 10^{-31}\,\text{kg} \times (3.00 \times 10^8\,\text{m/s})^2$$
$$= 1.64 \times 10^{-13}\,\text{J} = 1.02\,\text{MeV} \tag{150}$$

The energy mc^2 that a particle has when at rest (with zero kinetic energy) is called the **rest-mass energy.**

Since the above derivation of the energy–mass equivalence does not explicitly use the theory of relativity, we might be tempted to suppose that the result has nothing to do with relativity. But this would be a mistake. The derivation relies on the relation between the energy and the momentum of a light wave obtained from Maxwell's equations, and these equations already contain relativity.

The equivalence of energy and mass implies that the laws of conservation of energy and of mass are not independent—each implies the other. When examining the energy–mass balance in a reaction, we can either express all the masses as rest-mass energies and use the law of conservation of energy, or else express all the absorbed and released energies as masses and use the law of conservation of mass. The alternative equations differ only by an overall factor of c^2.

Example 10 One method for the determination of the mass of the neutron relies on the measurement of the energy released in the spontaneous reaction of decay of the neutron into a proton and an electron,[†]

$$n \rightarrow p + e \tag{151}$$

---

* Their magnetic moments are also opposite; this can be regarded as a consequence of the opposite charges.

† This reaction is the **beta decay** of the neutron. It also involves the release of a neutrino, but the energy measurements concentrate on those reactions in which the neutrino has zero energy and can therefore be ignored.

The measured value of this energy is 0.78 MeV. Using this and the known masses of the proton and electron, deduce the mass of the neutron by appealing to the conservation of the total energy, including the rest-mass energies.

Solution   The total energy before the reaction is $m_n c^2$; the total energy after the reaction is $m_p c^2 + m_e c^2 + 0.78$ MeV. Hence

$$m_n = m_p + m_e + 0.78 \,\mathrm{MeV}/c^2$$

$$= 1.67265 \times 10^{-27}\,\mathrm{kg} + 9.1095 \times 10^{-31}\,\mathrm{kg}$$

$$+ \frac{0.78 \times 10^6 \times 1.60 \times 10^{-19}\,\mathrm{J}}{(3.00 \times 10^8\,\mathrm{m/s})^2}$$

$$= 1.67494 \times 10^{-27}\,\mathrm{kg} \quad \blacksquare \tag{152}$$

1.8 X Rays

X rays were discovered by W. C. Röntgen* in 1895. He found that the impact of cathode rays on the walls of the discharge tube generated invisible rays of great penetrating power, which could be registered on a photographic plate. These rays proved capable of passing through thick layers of opaque materials, including human tissues (see Fig. 1.28). For the efficient generation of X rays it is best if the electrons strike a target made of a heavy metal. Figure 1.29 shows a modern X-ray tube in which an energetic electron beam emerging from a hot filament on the cathode strikes a tungsten target and produces intense X rays.

The suspicion that X rays are electromagnetic waves of very short wavelength was confirmed by a "double" scattering experiment performed by C. G. Barkla[†] in 1906. In this experiment, a beam of X rays from an X-ray tube was allowed to strike a block of carbon, scattering the X rays in all directions. Those scattered

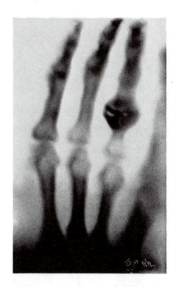

Fig. 1.28 X-ray photograph of a hand prepared in 1895 by Röntgen. (Courtesy Deutsches Museum.)

---

* **Wilhelm Konrad Röntgen,** 1845–1923, German experimental physicist, professor at Würzburg and at Munich. He discovered X rays while experimenting with a cathode-ray tube, and immediately recognized their remarkable penetrating power and their possible medical applications. For this discovery, he was awarded the first of the Nobel Prizes, in 1901.

† **Charles Glover Barkla,** 1877–1944, British physicist, professor at Edinburgh. He performed extensive experimental investigations into the properties of X rays, and was awarded the Nobel Prize in 1917 for the discovery of characteristic spectral lines in the X rays emitted by the elements.

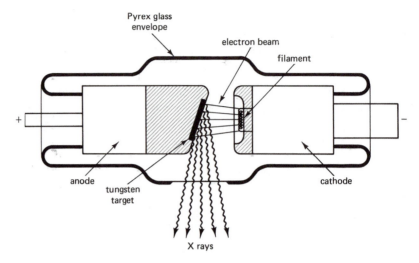

Fig. 1.29 Coolidge X-ray tube. (Courtesy General Electric Co.)

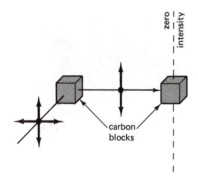

Fig. 1.30 Arrangement of scatterers in Barkla's experiment.

X rays that emerge horizontally at right angles to the original direction of propagation strike another block of carbon and scatter again. Barkla found that the second scattering gave zero intensity in the up-and-down direction (see Fig. 1.30). This experimental result can be easily understood as due to the polarization in an electromagnetic wave. The initial beam of X rays is unpolarized, i.e., it contains a random mixture of waves of all possible polarizations. However, since the direction of polarization must necessarily be perpendicular to the direction of propagation (x-direction in Fig. 1.30), all possible initial polarizations are in the y-z plane. The electric field of the wave striking the block of carbon accelerates the electrons in the block and causes them to emit radiation of the same frequency as that of the wave, i.e., the electrons emit scattered X rays. The possible accelerations of the electrons are in the y-z plane. From electromagnetic theory it is known that the direction of acceleration of a charge and the direction of polarization of the wave emitted by that charge are always in the same plane. Thus, the waves emitted at a right angle, toward the second carbon block, are polarized in the z-direction (see Fig. 1.30). When these waves strike the second block, they accelerate the electrons up and down in the z-direction. Again, from electromagnetic theory it is known that an accelerated charge does not emit in the direction of its acceleration. Hence the intensity of the scattered waves emerging from the second block is expected to be zero in the up-and-down direction, exactly as observed by Barkla.

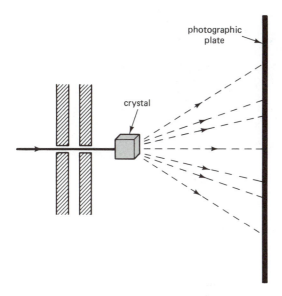

Fig. 1.31 Arrangement of crystal and photographic plate for an X-ray interference experiment.

The conclusive proof of the wave nature of X rays was supplied in 1912 by M. von Laue,* who proposed a method for measuring the wavelengths of X rays. It occurred to Laue that the separations between the atoms in a crystal might be of the same order of magnitude as the wavelengths of the X rays. If so, X rays passing through a crystal will exhibit distinctive interference effects, like the interference effects of light passing through a grating. The experiment was performed by W. Friedrich and P. Knipping. They aimed a fine beam of X rays at a crystal (see Fig. 1.31), and found that the X rays emerging on the other side had sharp, narrow maxima of intensity in selected directions. These maxima are so narrow that they take the form of beams emerging from the crystal; if intercepted by a photographic plate, the beams register as a pattern of bright spots, called **Laue spots** (see Fig. 1.32). From the angular positions of the beams and the separation of the atoms in the crystal, Friedrich and Knipping calculated that the wavelength of

---

* **Max von Laue,** 1879–1960, German physicist, professor at Munich and at Berlin, director of the Kaiser Wilhelm Institute. Laue worked out the theory of interference of waves by crystals and suggested to **Walter Friedrich,** 1883– , and **Paul Knipping,** 1883–1935, that they perform the experiment. For his discovery of interference of X rays, Laue was awarded the Nobel Prize in 1914.

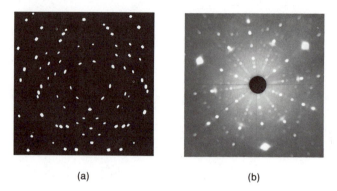

(a)   (b)

Fig. 1.32 (a) Laue spots produced with a silicon crystal. (Courtesy C. C. Jones, Union College.) (b) Laue spots produced with a ruby crystal. (Courtesy C. Bielan, Xerox Corporation.)

their X rays was about 0.5 Å.* Later investigations showed that X rays range in wavelength from 10 to 10^{-2} Å. The long-wavelength end of the X-ray spectrum merges into the ultraviolet, and the short-wavelength end merges into γ rays (see Fig. 1.26).

We can derive a simple formula relating the wavelength of the X rays to the angular positions of the beams and the separation of the atoms in the crystal, as follows: The atoms in the crystal may be thought of as located on sets of parallel planes, such that each plane in a given set has the same distribution of atoms. These planes are called **Bragg planes.** Every crystal has several sets of Bragg planes; for instance, Fig. 1.33a shows one set of Bragg planes in a cubic crystal, and Fig. 1.33b shows another set of Bragg planes in the same crystal. When an incident beam of X rays passes through a crystal, each atom scatters the rays in all directions. Strong emerging beams will be produced in those directions that correspond to constructive interference of the contributions of a large number of atoms, such as the atoms in a set of Bragg planes. To discover the condition for constructive interference we make use of Fig. 1.34, which shows a set of Bragg planes separated by a distance d. A beam of X rays, graphically represented by the parallel paths 1 and 2, is incident on these planes from the left, and a scattered beam emerges toward the right. If the contributions of two adjacent atoms in the *same* plane are to interfere constructively (see Fig. 1.34a), the path lengths AQ and BR must be equal so that

---

* The **angstrom** (Å) is a unit of length, 1 Å $= 10^{-10}$ m. This is a convenient unit for the wavelengths of X rays and for the wavelengths of light, as well as for atomic sizes.

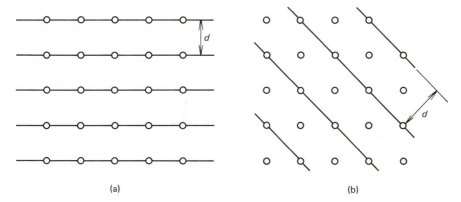

Fig. 1.33 Different sets of Bragg planes in a cubic crystal.

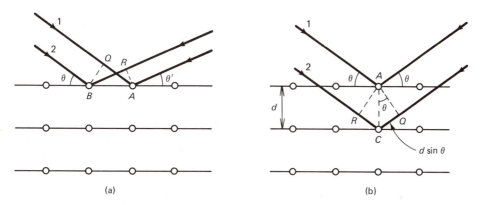

Fig. 1.34 (a) X rays incident on atoms in the same Bragg plane. The source of X rays and the point of observation are at large distances on the left and the right, respectively; thus, the paths of the incident rays and the paths of the scattered rays are approximately parallel. (b) X rays incident on atoms in adjacent Bragg planes.

the parallel rays 1 and 2 have the same length.* Consequently, the angles θ and θ' must be equal (this is the familiar law of reflection: the angle of the incident ray equals the angle of the emerging ray). If the contributions of two adjacent atoms in *different* planes are to

---

* It may be objected that constructive interference also occurs if AQ and BR differ by one, two, etc. wavelengths. This is true, but irrelevant. Constructive interference with unequal angles θ and θ' can always be regarded as constructive interference with *equal* angles produced by scattering from a different set of Bragg planes, inclined relative to the set of Bragg planes shown in Fig. 1.34 by some appropriate angle. The proof of this statement may be found in texts on crystallography.

interfere constructively (see Fig. 1.34b), the length RCQ must be an integral multiple of a wavelength, so that the path difference between the parallel rays 1 and 2 is an integral multiple of a wavelength. Since the length RCQ is $2d \sin \theta$, the condition for constructive interference is

$$2d \sin \theta = n\lambda \qquad n = 1, 2, 3, \ldots \tag{153}$$

This formula is called **Bragg's law,** after W. L. Bragg,* who first derived it.

Bragg's law can be applied in two ways in X-ray interference experiments.† First, if the separation between the atoms in the crystal is known, then Eq. (153) permits the calculation of the wavelength of the X rays. The crystal can therefore be used to analyze the spectrum of the X rays, just as an optical grating is used to analyze the spectrum of light. (We will postpone the discussion of X-ray spectra to Sections 3.4 and 4.7, because an understanding of the shape of the X-ray spectrum requires some knowledge of quantum theory.) Second, if the wavelength of the X rays is known, then Eq. (153) permits the calculation of the distance between the atoms. This has led to the most precise determinations of atomic distances, and also the most precise determination of Avogadro's number. To use this method it is necessary to make a preliminary measurement of the wavelength of the X rays, which can be done by means of an optical grating. In view of the high penetrating power and the small wavelength of the X rays, we might expect that an optical grating of glass would not produce an interference pattern—the X rays presumably pass straight through, unaffected by the glass; and even if the glass could affect them, the distance between the rulings of an optical grating is much too large to yield noticeable interference effects. But a clever trick, first exploited by A. H. Compton, avoids these limitations. The index of refraction of X rays in glass is slightly less than 1. Thus, if the X rays strike the glass at a very

---

* **William Henry Bragg,** 1862–1942, British physicist, professor at Adelaide and at Leeds. In collaboration with his son, **William Lawrence Bragg,** 1890–1971, he used X rays to investigate the structure of crystals. For these investigations, the Braggs were awarded the Nobel Prize in 1915.

† Such experiments are often called X-ray *diffraction* experiments. But, strictly, this is a misnomer since the formation of the strong scattered beams is an interference phenomenon, not a diffraction phenomenon (spreading of a wave upon passage through an aperture). Laue did not commit this mistake; the paper in which he proposed the measurement of X-ray wavelengths with crystals was entitled "Interference Phenomena for X-Rays."

shallow angle ("grazing incidence"; see Fig. 1.35), they will be strongly affected by the glass—they will be totally reflected. Furthermore, the projected distances between the rulings will be small as seen from the direction of incidence of the X rays. Under these conditions, an optical grating will produce an interference pattern with X rays. The wavelength of the X rays can then be calculated from the angular position of the interference maxima and the grating separation, obtained by yet another preliminary measurement of the interference pattern produced with light of known wavelength. If these same X rays are next used in an interference experiment with a crystal, we obtain a precise value of the separation between the atoms, via Eq. (153).

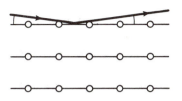

Fig. 1.35 Grazing incidence of X rays on a crystal.

To determine Avogadro's constant, we measure the (macroscopic) size and mass of the crystal. From the size and the separation between the atoms, we can then calculate the number of atoms in the crystal; and from the mass and the "atomic mass" we can calculate the number of moles. The ratio of these two numbers is Avogadro's number, N_A. Recent determinations of Avogadro's number have combined the separate X-ray and optical interference experiments (with grating and with crystal) into one single, simultaneous X-ray and optical interference experiment, employing a new kind of interferometer.* The value of Avogadro's number quoted in Section 1.3 was obtained by these means.

X-ray interference experiments are used extensively in crystallography to investigate the arrangement of atoms in crystals. The symmetry of the Laue spots (see Fig. 1.32) reveals the symmetry of the crystal structure. X-ray interference experiments are also used to investigate the arrangement of atoms in molecules. For instance, the structure of DNA was discovered by analysis of X-ray interference patterns.

Example 11   In one of their early experiments, Friedrich and Knipping used a crystal of rock salt (NaCl), which has a spacing of 2.81 Å between its principal Bragg planes. If the first-order beam ($n = 1$) emerged at an angle of 10° relative to the incident beam, what was the wavelength of the X rays?

Solution   The angle between the undeflected beam (shown dashed in Fig. 1.34) and the first-order beam is twice the Bragg angle; thus, $2\theta = 10°$ and $\theta = 5°$. With this, Bragg's law gives

$$\lambda = 2a\sin\theta = 2 \times 2.81\,\text{Å} \times \sin 5° = 0.49\,\text{Å} \quad \blacksquare$$

* R. D. Deslattes, *Atomic Masses and Fundamental Constants*, **5**, 552 (1976).

Chapter 1:   SUMMARY

Newton's Second Law:

$$m\mathbf{a} = \mathbf{F}$$

Kinetic energy:

$$K = \tfrac{1}{2}mv^2$$

Momentum:

$$\mathbf{p} = m\mathbf{v}$$

Angular momentum:

$$\mathbf{L} = \mathbf{r} \times \mathbf{p}$$

Force on charged particle:

$$\mathbf{F} = q\mathbf{E} + q\mathbf{v} \times \mathbf{B}$$

Electron volt:

$$1\,\text{eV} = e \times 1\,\text{V} = 1.6 \times 10^{-19}\,\text{J}$$

Radius of orbit in magnetic field:

$$r = \frac{p}{qB}$$

Cyclotron frequency:

$$v = \frac{qB}{2\pi m}$$

Ideal-gas law:

$$PV = \frac{Nm\overline{v^2}}{3} = NkT$$

Root-mean-square speed:

$$v_{\text{rms}} = \sqrt{\overline{v^2}} = \sqrt{\frac{3kT}{m}}$$

Maxwell distribution of velocities:

$$f(v_x)f(v_y)f(v_z) = N\left(\frac{m}{2\pi kT}\right)^{3/2} e^{-\frac{1}{2}mv^2/kT}$$

Boltzmann factor:

$$e^{-E/kT}$$

Equipartition theorem: Each term in the energy proportional to the square of a velocity or a coordinate contributes $\frac{1}{2}kT$ to the mean energy at thermal equilibrium.

Waves:

Angular frequency: $\omega = 2\pi\nu$ Phase velocity: $v_p = \dfrac{\omega}{k}$

Wave number: $k = \dfrac{2\pi}{\lambda}$

Group velocity: $v_g = \dfrac{d\omega}{dk}$

Interference maxima generated by two slits:
$$d\sin\theta = n\lambda \qquad n = 0, \pm 1, \pm 2, \ldots$$

Diffraction minima generated by single slit:
$$a\sin\theta = n\lambda \qquad n = \pm 1, \pm 2, \pm 3, \ldots$$

Electric and magnetic fields in electromagnetic wave:
$$|\mathbf{B}| = \frac{1}{c}|\mathbf{E}|$$

Poynting vector:
$$\mathbf{S} = \frac{1}{\mu_0}\mathbf{E} \times \mathbf{B}$$

Pressure of wave:
$$P = \frac{S}{c}$$

Einstein's mass–energy relation:
$$\Delta m = \frac{\Delta E}{c^2}$$

angstrom:
$$1\,\text{Å} = 10^{-10}\,\text{m}$$

Bragg's law:
$$2d\sin\theta = n\lambda \qquad n = 1, 2, 3, \ldots$$

Chapter 1:   PROBLEMS

1. What (uniform) electric field is required to give an electron a kinetic energy of 2×10^4 eV within a distance of 2.0 cm?

2. What was the cyclotron frequency for the cyclotron described in Example 1?

3. The strength of the Earth's magnetic field at Hawaii is 3.7×10^{-5} T. What is the radius of the orbit of an electron of energy 60 eV moving in a plane perpendicular to this magnetic field?

4. In the AGS accelerator at the Brookhaven National Laboratory, protons are held in a circular orbit of radius 128 m by a vertical magnetic field. The maximum field that the magnets can supply is 1.3 T. Calculate the maximum permissible momentum of the protons [note that if Eq. (16) is expressed in terms of momentum, it remains valid even for relativistic particles]. Taking into account that the speed of the protons is nearly the speed of light, calculate the orbital frequency of the protons.

5. Prove that the kinetic energy of a charged particle moving in an arbitrary magnetic field remains constant.

6. In the main ring at the Fermilab accelerator, protons are held in a circular orbit of diameter 2.0 km by a vertical magnetic field while they are accelerated by horizontal electric fields until they attain a final momentum of 5.3×10^{-16} kg·m/s. What magnetic field is required to hold protons of this momentum? [Note that if Eq. (16) is expressed in terms of momentum, it remains valid even for relativistic particles.]

7. A thin flexible wire hangs in a uniform magnetic field. A weight attached to the wire provides a tension T. The wire carries a current I, and the magnetic force on this current causes the wire to deflect sideways, adopting the shape of an arc of circle (see Fig. 1.36).

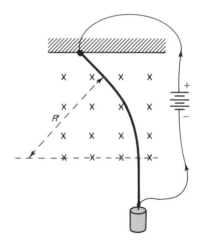

Fig. 1.36 Wire hanging in a magnetic field.

(a) Show that the radius of the arc of circle is T/BI.

(b) Show that if we remove the wire and launch a particle of charge q from point P with momentum $p = qT/I$ in the direction of the wire, it will move along the same arc of circle.

8. In his book, *The Electron*, Millikan reports an early series of measurements he performed on a single oil drop. The motion of the drop was timed with a stopwatch over a vertical distance of 10.21 mm. With the electric field switched off, the time of fall was $t_1 = 11.880$ s. With the electric field switched on, Millikan found that the time of rise of the drop over the same distance varied on successive trials because the drop sometimes had more electric charge, sometimes less. The following table gives the time of rise for several trials:

t_2	Relevant Data:
80.708 s	Plate distance $= 16$ mm
140.565	Volts $= 5085$ V
79.600	Oil density $= 0.9199$ g/cm^3
137.308	Air viscosity $= 1.824 \times 10^{-4}$ g/s·cm
34.638	
500.1	
19.69	
42.302	

To the right of this table are some relevant data, as reported by Millikan.

(a) Calculate the mass of the oil drop.

(b) For each value of t_2, calculate the electric charge on the drop.

(c) Check that all these values of the electric charge are (nearly) integral multiples of the fundamental electric charge $e = 1.66 \times 10^{-19}$ C.

9. A modified version of the Millikan experiment by Hopper and Laby determines the electric charge on an oil drop by allowing it to fall in a *horizontal* electric field. Show that under these conditons an oil drop of charge q falls at an angle θ with the vertical (see Fig. 1.37), such that $\sin \theta = qE/6\pi\eta Rv_2$, where all the

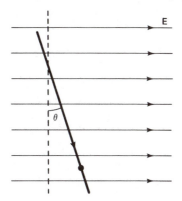

Fig. 1.37 Trajectory of a positively charged oil drop falling in a horizontal electric field.

symbols have the same meaning as in Section 1.2. Explain how the charge on the oil drop can be calculated from measurements of the speed v_1 in the absence of the electric field and the speed v_2 and the angle θ in the presence of the electric field.

10. A velocity "filter" with crossed electric and magnetic fields is to permit the passage of singly ionized ^{22}Ne atoms of energy 2.0×10^4 eV. The electric field is 3.0×10^5 V/m. What magnetic field is required? Appendix 4 gives a list of masses of isotopes.

11. We want to use a Dempster mass spectrograph to discriminate between the isotopes ^{56}Fe and ^{57}Fe of masses 55.935 u and 56.935 u, respectively. Assume that these isotopes are singly ionized (^{56}Fe$^+$ and ^{57}Fe$^+$) and that they enter the spectrograph with a speed of 3.25×10^5 m/s.
   (a) What magnetic field is required to give their orbits a radius of (approximately) 1.0 m?
   (b) What will be the separation Δr between their impact points on the photographic plate?

12. Suppose that a beam of singly ionized ^{22}Ne atoms strikes a certain point on the photographic plate of a mass spectrograph. What *doubly* ionized isotope will strike at approximately the same point? Use the table of masses given in Appendix 4.

13. The alpha particle consists of two protons and two neutrons closely bound together. Compare the mass of the alpha particle with that of two protons and two neutrons, and deduce the binding energy of the alpha particle. Express your answer in MeV.

14. The theoretical expression for the speed of sound in a gas is $v = \sqrt{1.4p/\rho}$, where ρ is the mass density. By what factor does this differ from the rms speed?

15. What is the rms speed of nitrogen molecules in air at 20°C? What is the most probable value of the speed? How do these speeds compare with the speed of sound?

16. Starting with the Maxwell distribution of molecular speeds [Eq. (88)], show that the mean speed of a molecule is

$$\bar{v} = \sqrt{\frac{8kT}{\pi m}}$$

17. Air consists of roughly 76% nitrogen molecules, 24% oxygen molecules, and 1% argon atoms (by mass). What is the overall mean speed of air molecules? (*Hint:* Use the formula given in Problem 16 for the mean speed of each kind of molecule.)

18. The gravitational potential energy of a molecule of air of mass m is $U = mgz$. According to the Boltzmann factor, what is the density of such molecules as a function of altitude (for constant temperature)? Show that this density satisfies the condition of hydrostatic equilibrium.

19. Each atom in a crystal can be regarded as a three-dimensional harmonic oscillator.

According to the equipartition theorem, what is the vibrational thermal energy of one mole of crystal?

20. Consider a pendulum of mass 0.4 kg and length 1.2 m hanging in its equilibrium position. If this pendulum is in thermal equilibrium with the surrounding air at 20°C, what is the thermal energy of oscillation of the pendulum? What is the rms speed of the pendulum? What is the rms angular displacement from the equilibrium position?

21. A torsional pendulum in a delicate Cavendish balance used in a gravity experiment has a period of oscillation 150 s and a moment of inertia 1.0×10^{-5} kg·m². What is the rms angular amplitude of oscillation of this pendulum when in thermal equilibrium at 300 K?

22. A parked automobile is in thermal equilibrium with the air and it therefore has a small random motion. Assume that the temperature is 20°C and the mass of the automobile is 1500 kg.
(a) Calculate the vertical rms speed of the automobile.
(b) Calculate the rms amplitude of the vertical oscillations of the automobile. Treat the automobile as a mass sitting on four suspension springs of spring constant 4.0×10^4 N/m each.

23. According to classical mechanics, calculate the rms speed and the rms amplitude of oscillation of a hydrogen atom in a hydrogen molecule. Each hydrogen atom may be regarded as attached to the center of mass by a spring of spring constant 1.13×10^3 N/m. Assume room temperature (300 K).

24. An *LC* circuit is mathematically equivalent to a mass on a spring—the energy $\frac{1}{2}Q^2/C$ in the capacitor plays the role of potential energy, and the energy $\frac{1}{2}L(dQ/dt)^2$ in the inductor plays the role of kinetic energy. According to the equipartition theorem, what are the rms values of the charge Q and the current dQ/dt in an *LC* circuit at thermal equilibrium at temperature *T*? Evaluate numerically for $T = 20°C$, $L = 2.5 \times 10^{-3}$ H and $C = 2.0 \times 10^{-8}$ F.

25. A rectangular cavity with conducting walls measures 2 cm × 5 cm × 7 cm. What is the frequency of the electromagnetic wave of longest wavelength that will form a standing wave in this cavity?

26. Given that the separation between the two wave sources in Fig. 1.19 is 10 cm, use a protractor to measure the angular position of the first lateral maximum and from this calculate the wavelength of the waves. Compare the result of this calculation with the wavelength obtained by direct measurement.

27. Two very thin parallel slits in an opaque foil are separated by a distance of 2.0×10^{-6} m. If light of wavelength 6328 Å from a laser is incident on these slits, what are the angular directions in which strong beams of constructive interference will be observed?

28. In Example 9 we assumed that the plane wave was normally incident on the wall. Show that if the wave is incident at an angle ϕ (see Fig. 1.38), then the condition for constructive interference is

$$d \sin \theta - d \sin \phi = n\lambda$$

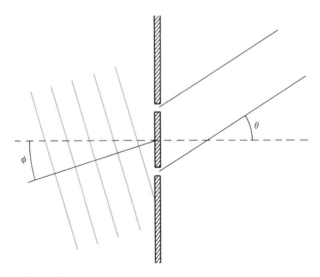

Fig. 1.38 Wave obliquely incident on a wall with two slits.

29. Diffraction of X rays has been observed when a beam strikes an *extremely* fine slit. What must be the width of the slit if X rays of wavelength 10 Å are to display a diffraction pattern with a central beam of angular width 0.1°?

30. Consider a wave packet formed by superposition of the two waves

$$\Phi_1 = \cos(1.00x - 2.00t)$$

$$\Phi_2 = \cos(1.01x - 2.03t)$$

where x and t are measured in meters and in seconds, respectively. What is the phase velocity of these waves? What is the group velocity?

31. What is the phase velocity and what is the group velocity of water waves of wavelength 30 m in the ocean? What is the frequency of such waves?

32. Water waves in shallow water have a phase velocity

$$v_p = \sqrt{gh}$$

where h is the depth of the water. What is the group velocity of these waves?

33. Show that the group velocity and the phase velocity obey the relation

$$v_g = v_p - \lambda \frac{dv_p}{d\lambda}$$

34. Small ripples on the surface of water are controlled by surface tension. The phase velocity of such ripples is

$$v_p = \sqrt{\frac{2\pi\gamma}{\lambda\rho}}$$

where $\gamma = 0.70$ N/m is the surface tension and $\rho = 10^3$ kg/m^3 the density of water.

(a) Find an equation for the group velocity of such ripples.

(b) What is the phase velocity and what is the group velocity for ripples of wavelength 1 mm?

35. The energy flux of sunlight incident on the Earth is 1.4×10^3 W/m^2. What is the pressure of sunlight? Express your answer in atmospheres.

36. Suppose that a spherical grain of dust is floating in interplanetary space at a distance from the Sun equal to the Sun–Earth distance. The energy flux of sunlight at this distance is 1.4×10^3 W/m^2. The grain has a diameter of 1.0×10^{-6} m and a density of 2.0×10^3 kg/m^3. Assuming that the grain completely absorbs all the sunlight striking it, what pressure force does sunlight exert on the grain? What gravitational attraction does the Sun exert on the grain? What are the magnitude and the direction of the acceleration of the grain?

37. The magnetic fields of the most powerful available magnets have a field strength of 30 T. What is the mass density in such a magnetic field? (The energy density in a magnetic field is $\frac{1}{2}B^2/\mu_0$.)

38. An alternative method for the determination of the mass of the neutron relies on the measurement of the energy released in the reaction

$$\text{n} + {}^1\text{H} \rightarrow {}^2\text{H}$$

The measured value of this energy is 2.23 MeV. From this, and from the known values of the masses of the hydrogen isotopes ^{1}H and ^{2}H (see Appendix 4), deduce the mass of the neutron.

39. The Sun is radiating heat and light at the rate of 3.9×10^{26} W. At what rate is the Sun losing mass due to this release of energy?

40. The hydrogen gas at the center of the Sun is at a temperature of 15×10^6 K. What is the thermal kinetic energy per kilogram of this gas? What is the mass associated with this thermal kinetic energy? What percentage contribution does this extra mass make to the net mass of the gas?

41. The crystal of potassium bromide (KBr) has a spacing of 3.30 Å between Bragg planes. What is the angle at which the first-order beam of 2.0-Å X rays emerges relative to the incident beam? The second-order beam?

42. Monochromatic X rays strike a set of Bragg planes with a spacing of 3.029 Å in a calcite crystal. Interference produces a sharp emerging beam if the angle of incidence is 3.455° and also at several larger angles of incidence, but not at any smaller angle. What is the wavelength of the X rays?

43. The index of refraction for X rays of wavelength 0.708 Å incident on calcite is $1 - 1.85 \times 10^{-6}$. What angle of incidence will give total reflection of such X rays?

The Theory of Special Relativity*

Newton's laws of motion are the same in all inertial reference frames. This is the Newtonian principle of relativity. It implies that no mechanical experiment can detect any intrinsic difference between two inertial reference frames. For instance, if we compare the behavior of racquet balls on a ship steaming away from the shore at constant velocity with the behavior of similar balls on the shore, we find no difference whatsoever. Experiments with racquet balls aboard the ship will not reveal the uniform motion of the ship. To detect this motion, the crew of the ship must take sightings of points on the shore or use some other navigational technique that fixes the position and velocity of the ship in relation to the shore. Hence, as concerns mechanical experiments, uniform translational motion of an inertial reference frame is always relative motion—it can only be detected as relative motion of the reference frame with respect to another reference frame. It is meaningless to speak of *absolute* motion through space.

This immediately raises the question of whether the relativity of mechanical experiments also applies to electric, magnetic, optical, and other experiments. Do any of these experiments permit us to detect an absolute motion of our reference frame through space?

---

* This chapter is optional. Its contents are not used in the later chapters, except in a few problems.

In 1905, Albert Einstein* answered this question in the negative. He laid down a principle of relativity for *all* laws of physics, which led to far-reaching consequences for our concepts of space and time.

Although relativity is usually thought of as a subject of interest only to physicists, it has some valuable practical applications of interest to engineers and physicians. The electrons in high-voltage TV tubes, in klystrons, and in specialized accelerators used for radiation therapy move at such high speeds that a knowledge of the relativistic equations of motion is needed for the design of these devices.

Relativity forms the basis of contemporary theoretical physics. It plays an essential role in the modern theories of elementary particles and their interactions. However, relativity played only a small role in the early development of quantum theory; hence we will have little occasion to use it in the following chapters.

2.1 **The Speed of Light and the Ether**

Since the laws of mechanics are the same in all inertial reference frames, it seems quite natural to assume that the laws of electricity and magnetism also are the same in all inertial reference frames. But this assumption immediately leads to a paradox concerning the velocity of light. From Maxwell's equations, we can deduce that the velocity of propagation of light always has the fixed value $1/\sqrt{\mu_0\varepsilon_0}$, or approximately 3.00×10^8 m/s. This deduction stands in contradiction to the Galilean addition law for velocities, according to which the velocity of light ought *not* to be the same in all reference frames. The Galilean addition law states that if the velocity of a particle or a signal is $\mathbf{v}$ in one reference frame and $\mathbf{v}'$ in another reference frame, then

$$\mathbf{v}' = \mathbf{v} - \mathbf{V} \tag{1}$$

---

* **Albert Einstein,** 1879–1955, German, Swiss, and later American theoretical physicist, professor at Zürich and at Berlin, director of the Kaiser Wilhelm Institute, and professor at the Institute for Advanced Study in Princeton. His theories of Special and General Relativity represent the greatest intellectual achievement of the twentieth century, but it took many years before these theories were fully understood and accepted by the scientific community. Einstein also made crucial contributions to modern quantum theory, introducing the concept of quanta of light (later called photons) and explaining the photoelectric effect. For this he was awarded the Nobel Prize in 1921. In his later years, Einstein devoted much effort to an unsuccessful quest for a unified theory of gravity and electromagnetism.

where **V** is the velocity of the second reference frame relative to the first. For example, imagine that we send a light signal toward a spaceship approaching the Earth with a velocity of, say, 1.00×10^8 m/s; if this light signal has a velocity of 3.00×10^8 m/s in the reference frame of the Earth, then Eq. (1) tells us that it ought to have a velocity of 4.00×10^8 m/s in the reference frame of the spaceship.

To escape this paradox, we must either sacrifice the notion that the laws of electricity and magnetism—and the values of the speed of light—are the same in all inertial reference frames, or else we must sacrifice the Galilean addition law for velocities. This means we must either abandon all hope for a principle of relativity embracing electricity and magnetism, or else we must overthrow the intuitively "obvious" notions of absolute time and absolute length on which the Galilean addition law is based.

The physicists of the nineteenth century adopted the first alternative. They assumed that there exists a preferred inertial reference frame in which the laws of electricity and magnetism take their simplest form. In this reference frame, the speed of light has its standard value $c = 3.00 \times 10^8$ m/s, while in any other reference frame it is larger or smaller according to the Galilean addition law. Thus, the propagation of light is analogous to the propagation of sound. There exists a preferred reference frame in which the wave equation describing the propagation of sound in, say, air takes its simplest form: the reference frame in which the air is at rest. In this reference frame the speed of propagation of sound is independent of direction and has a magnitude of 330 m/s, while in any other reference frame the speed depends on direction. For instance, on the deck of a ship moving north at 10 m/s relative to the air, the speed of sound is 320 m/s for propagation in the northward direction, and 340 m/s for propagation in the southward direction.

The analogy between the propagation of sound and of light suggested to the nineteenth-century physicists that light is some kind of undulation in an elastic medium filling all of space, even the interplanetary and interstellar space usually regarded as a vacuum. This pervasive, ghostly medium was supposed to have no effect whatsoever on the motion of planets or of particles—its only raison d'être was the propagation of light. Physicists called this ghostly medium the **luminiferous ether,** and they devoted much effort to contriving elastic and mechanical properties of this ether so that its undulations would behave like light waves. According to the ether theory, the preferred reference frame in which light has its standard speed is the reference frame in which the ether is at rest. The existence of such a preferred reference frame would imply that velocities are absolute—the ether frame sets an absolute standard of rest, and the velocity of any body can be referred to this frame.

The motion of the ether past the Earth constitutes an ether wind. If the Sun is at rest in the ether, then the ether wind would have a velocity opposite to that of the Earth around the Sun: its magnitude would be about 30 km/s. If the Sun is in motion, then the ether wind would vary with the seasons; its magnitude would be maximum when the velocity of the Earth is parallel to that of the Sun, and minimum six months later, when the velocities are antiparallel.

Experimenters sought to detect this ether wind by its effect on the propagation of light. In the reference frame of the Earth, the speed of a light wave would depend on direction: the light wave has maximum speed when propagating downwind, a reduced speed when propagating across the wind, and minimum speed when propagating upwind. According to the Galilean addition law, if the speed of the ether wind blowing through the laboratory is V, then the speed of light in this laboratory is $c + V$ in the downwind direction, $\sqrt{c^2 - V^2}$ in the direction perpendicularly across the wind, and $c - V$ in the upwind direction (see Fig. 2.1). If the magnitude of V is about 30 km/s, then the change of the speed of light from its standard value is about 1 part in 10^4 for the upwind or downwind direction, and about 1 part in 10^8 for the direction across the wind. Unfortunately, we cannot take full advantage of the substantial upwind or downwind change of speed because in an optical experiment comparing the speeds in different directions, the light signals must complete round trips, for which the upwind and downwind speed changes tend to average out, leaving only a small residual of about 1 part in 10^8.

The first experiment capable of detecting such a small change in the speed of light was performed by A. A. Michelson and E. W. Morley* in 1881, and often repeated thereafter. Their experiment used an interferometer (Fig. 2.2a) to split a light beam into two separate beams which are made to travel back and forth along two perpendicular paths, or arms, in the instrument. One beam travels

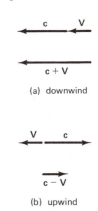

Fig. 2.1 The speed of light relative to the Earth is $|\mathbf{c} + \mathbf{V}|$, where **c** is the velocity of light relative to the ether and **V** is the velocity of the ether relative to the Earth.

(a) downwind

(b) upwind

(c) across the wind

---

* **Albert Abraham Michelson,** 1852–1931, American experimental physicist, professor at the Case School of Applied Science and at Chicago. He first performed the "Michelson–Morley" experiment alone during postgraduate studies at Berlin, and then repeated it several times in collaboration with **Edward Williams Morley,** 1838–1923, American physicist and chemist, professor at Western Reserve College. Michelson was well known for his development of sensitive optical equipment. He made a precise comparison between the length of the standard meter bar and the wavelength emitted by cadmium atoms, and measured the speed of light with unprecedented accuracy by sending light pulses back and forth over a distance of 35 km between two mountain peaks. He was awarded the Nobel Prize in 1907 for his investigations with optical precision instruments.

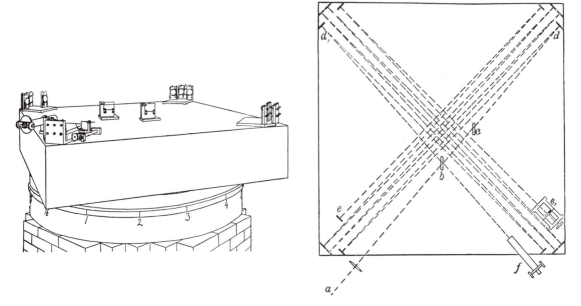

Fig. 2.2a The interferometer of Michelson and Morley. [From A. A. Michelson and E. W. Morley, *Phil. Mag.* **190**, 449 (1887).]

along an arm parallel to the direction of the (hypothetical) ether wind, and the other beam travels along an arm perpendicular to the direction of the ether wind. Sets of mirrors reflect the light beams at the ends of the arms, and make them travel repeatedly over the same paths; this multiplies the path length of the instrument, giving a greater sensitivity. Ultimately, the light beams are brought together, and they interfere constructively or destructively depending on the difference, if any, in travel time. This interference phenomenon serves as a sensitive indicator of any difference of travel time.

Example 1 The Michelson interferometer had two arms of equal length. Express the phase difference between the two beams of light emerging from the Michelson interferometer in terms of the speed of the ether wind.

Solution If the net path length is L (including the multiple back-and-forth reflections), the beam traveling parallel to the direction of the wind (horizontally in Fig. 2.2b) takes a time $L/(c - V)$ for its upwind motion and $L/(c + V)$ for its downwind motion. Hence the travel time for the round trip is

$$t_1 = \frac{L}{c - V} + \frac{L}{c + V} = \frac{L/c}{1 - V/c} + \frac{L/c}{1 + V/c} \tag{2}$$

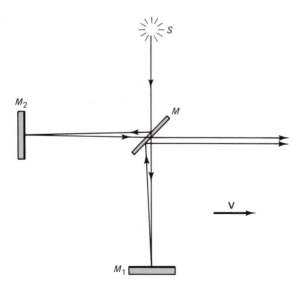

Fig. 2.2b Schematic diagram showing the path of light rays in the interferometer. This diagram shows only one pair (M_1, M_2) of the multiple mirrors.

The beam traveling perpendicularly across the wind (vertically in Fig. 2.2b) takes a time $L/\sqrt{c^2 - V^2}$ for each half of its trip. Hence the travel time for the round trip is

$$t_2 = \frac{2L}{\sqrt{c^2 - V^2}} = \frac{2L/c}{\sqrt{1 - V^2/c^2}} \tag{3}$$

If the wavelength of the light is λ, then the phase difference is $2\pi c/\lambda$ times the time difference between the beams. Thus, the phase difference is

$$\Delta\phi = \frac{2\pi c}{\lambda}(t_1 - t_2) \tag{4}$$

$$= \frac{2\pi}{\lambda} L\left(\frac{1}{1 - V/c} + \frac{1}{1 + V/c} - \frac{2}{\sqrt{1 - V^2/c^2}}\right) \tag{5}$$

Since $V^2/c^2 \ll 1$, we can make the approximations

$$\frac{1}{(1 \pm V/c)} \simeq 1 \mp \frac{V}{c} + \frac{V^2}{c^2} \quad \text{and} \quad \frac{1}{\sqrt{1 - V^2/c^2}} \simeq 1 + \frac{V^2}{2c^2}$$

so that the phase difference becomes

$$\Delta\phi \simeq \frac{2\pi}{\lambda} L\left(\frac{2V^2}{c^2} - \frac{2V^2}{2c^2}\right)$$

$$\simeq \frac{2\pi}{\lambda} L\frac{V^2}{c^2} \quad \blacksquare \tag{6}$$

The phase difference (6) caused by the ether wind can be most conveniently detected by rotating the interferometer by 90°, thereby reversing the roles of the parallel and the perpendicular arms. During the rotation, the phase difference gradually decreases from the value given in Eq. (6) to its negative; hence the emerging light beams will exhibit a sequence of alterations from constructive to destructive interference.

Michelson and Morley failed to detect any ether wind. The sensitivity of their original experiment was adequate to detect a wind of 5 km/s. Since the expected wind is about 30 km/s, their null result contradicted the ether theory of the propagation of light. This establishes that the propagation of light is *not* analogous to the propagation of sound. There is no preferred reference frame for the propagation of light. As the Earth orbits around the Sun, the Earth switches from one inertial reference frame to another as its velocity changes. According to the Michelson–Morley experiment, and according to later, even more precise versions of this experiment, all these inertial reference frames are completely equivalent in regard to the propagation of light.

Incidentally: Michelson interpreted his null result as evidence for an ether drag—the Earth was supposed to drag the ether along, eliminating or reducing the ether wind in the laboratory. But this interpretation is in conflict with **stellar aberration,** the apparent seasonal shift of the positions of stars in the sky. This shift, which can be as large as 40″ in extreme cases, arises from the component of velocity along the direction of motion of the Earth that the light from a star acquires when observed in the reference frame of the Earth; because of this extra velocity component, the direction of incidence of the starlight on the Earth is tilted forward, producing an apparent shift in the position of the star. Six months later, the direction of motion of the Earth and, consequently, the apparent shift in position of the star will be reversed. This stellar aberration would be absent if the ether were dragged along by the Earth.

2.2 Einstein's Principle of Relativity

As we noted in the preceding section, the invariance of the speed of light in all reference frames stands in contradiction to the Galilean addition law for velocities. In 1905, Einstein made the revolutionary proposal that this contradiction is to be resolved by overthrowing the Galilean addition law and the Newtonian concepts of space and time. Einstein was aware of the negative results of the Michelson–Morley experiment, but his real motivation was his firm belief that Maxwell's equations ought to be valid in all reference frames.

Einstein based his Theory of Special Relativity on a general hypothesis concerning all the laws of physics. This hypothesis is the

Principle of Relativity: *All the laws of physics are the same in all inertial reference frames.*

Since the laws for the propagation of light are included in the laws of physics, one immediate consequence of the Principle of Relativity is:

The velocity of light (in vacuum) is the same in all inertial reference frames; it always has the value $c = 3.00 \times 10^8$ m/s. *

The invariance of the speed of light requires that we sacrifice some of our intuitive, everyday notions of space and time. The fact that a light signal always has a velocity of 3.00×10^8 m/s, regardless of the motion of the source or of the observer, does violence to our intuition. This strange behavior of light is possible only because of a strange behavior of length and time. As we will see in a later section, neither length nor time is absolute—they both depend on the reference frame and suffer contraction and dilation when the reference frame changes.

Since the invariance of the speed of light is of such fundamental significance for the Theory of Relativity, many experimenters have sought to improve the Michelson–Morley experiment, in the hope of detecting some minute effect of the motion of the Earth on the propagation of light. The most precise of the interferometer experiments was that performed in 1930 by G. Joos, who used an interferometer with very long arms carefully isolated against vibrations of the ground. This experiment set a limit of 1.5 km/s on the maximum possible value of the ether wind.

Modern versions of the Michelson–Morley experiment have relied on the comparison of the frequencies of standing waves in two resonant cavities, oriented at right angles. The frequency of such a standing wave is directly proportional to the speed of the wave and inversely proportional to the length of the cavity [$v = vn/(2L)$, where n is the number of half-waves in the cavity]. If the

---

* For the sake of convenience, the value of the speed of light has been rounded off to three significant figures. The exact value, which has recently been adopted as the official standard of speed, is 2.99792458×10^8 m/s.

two cavities are of identical length, then any observed difference in the frequencies implies a difference in the speeds. An experiment by Jaseja et al.* employed lasers as resonant cavities and was able to set an upper limit of 30 m/s on the ether wind. A recent modified version of this experiment by Brillet and Hall[†] employed face-to-face parallel mirrors (Fabry–Perot etalons) as resonant cavities, which were driven by lasers; this experiment was able to set a tighter limit of 15 m/s.

Even more precise results have been obtained by measurements of the Doppler shift between a moving emitter and a receiver of light. According to Newtonian physics, the Doppler shift depends on the velocities of the emitter and the receiver relative to the ether, and therefore it can serve to detect the ether wind. The most precise of the Doppler-shift experiments used γ rays (essentially, short-wavelength X rays) emitted by a sample of radioactive ^{57}Fe. The apparatus of Champeney et al.[‡] consists of a turntable, rapidly spinning about its vertical axis (see Fig. 2.3). This turntable carries samples of ^{57}Fe on opposite points of its rim. In one of these samples, the iron nuclei are in an excited state of high energy and they emit γ rays by recoilless transitions to states of lower energy. The emission of γ rays without recoil, and therefore without reduction of the γ-ray energy, is called the **Mössbauer effect;**[§] in these emissions the iron crystal absorbs the recoil momentum. These γ rays can be captured by resonant absorption in the other sample of iron. According to Newtonian physics, the Doppler shift between emitter and absorber has one value when they are (instantaneously) on the north-south line, as shown in Fig. 2.3; and this Doppler shift has the opposite value one-half rotation later. The mismatch between the emitter and the absorber frequencies would inhibit the resonant absorption of the γ rays. In the experiment, no such inhibition of the absorption was found. This negative result set an upper limit of about 5 cm/s on the ether wind.[¶] The experimental evidence therefore establishes beyond all reasonable

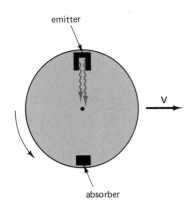

emitter

v

absorber

Fig. 2.3 Source and absorber of γ rays on a rotating turntable.

* T. S. Jaseja, A. Javan, J. Murray, and C. H. Townes, *Phys. Rev.* **133A,** 1221 (1964).

† A. Brillet and J. L. Hall, *Phys. Rev. Lett.* **42,** 549 (1979).

‡ D. C. Champeney, G. R. Isaak, and A. M. Khan, *Rev. Mod. Phys.* **36,** 469 (1964); *Nature* **198,** 1186 (1963). G. R. Isaak, *Phys, Bull.* **21,** 255 (1970).

§ **Rudolf Ludwig Mössbauer,** 1929– , German physicist, professor at the California Institute of Technology and at Munich. He discovered the Mössbauer effect while doing postgraduate research at Heidelberg. For this discovery, he received the Nobel Prize in 1961.

¶ If is of interest to note that this number is smaller than the experimental uncertainty in the best available measurements of the speed of light, ± 1 m/s.

doubt that the motion of the Earth has no effect whatsoever on the propagation of light.

2.3   The Lorentz Transformations

To measure the space and time coordinates of an event, we use a reference frame consisting of an array of measuring rods and clocks (see Fig. 2.4). The intersections of the measuring rods give the space coordinates and the time registered by the nearest clock gives the time coordinate. Of course, all the clocks in the reference frame must be synchronized with each other. The synchronization can be accomplished by sending out a flash of light from the origin of coordinates when the clock at this point indicates a given time $t = t_0$. The flash of light will take a time r/c to reach a clock at a distance r from the origin. The clock at this distance is therefore synchronized with the clock at the origin if it indicates a time $t_0 + r/c$ when the light reaches it (see Fig. 2.5). Alternatively, the synchronization can be accomplished by sending out a flash of light from a point exactly midway between the clock at the origin and the other clock. The two clocks are synchronized if both show exactly the same time when the light from the midpoint reaches them (see Fig. 2.6). Note that both of these synchronization procedures hinge on the invariance of the speed of light. If the speed of light were not the same in all reference frames, but were dependent on the reference frame and on direction (as in the ether theory), then we could not achieve synchronization by the simple procedures outlined above.

Throughout the following we will deal exclusively with inertial reference frames. Any two inertial reference frames are either at rest relative to one another, or else in uniform translational motion. Consider two inertial reference frames in uniform translational motion relative to one another. In Newtonian physics, the transformation equations that relate the space and time coordinates in one inertial reference frame to those in the other are the **Galilean transformations,**

$$x' = x - Vt \tag{7}$$

$$y' = y$$

$$z' = z$$

$$t' = t \tag{8}$$

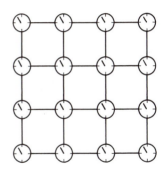

Fig. 2.4 A reference frame consists of a coordinate grid and synchronized clocks.

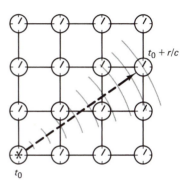

Fig. 2.5 Synchronization procedure. A flash of light is sent from one clock to the other.

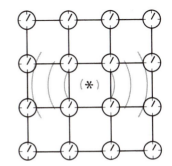

Fig. 2.6 Alternative synchronization procedure. A flash of light is sent from the midpoint toward each clock.

where we have assumed that the second reference frame moves with velocity V along the x-axis of the first frame, and that the origins of the two frames coincide at $t = t' = 0$. These transformation equations incorporate the absolute character of length and of time intervals.

In relativistic physics, we need a new set of transformation equations, which incorporate the invariance of the speed of light. These new transformation equations are the **Lorentz transformations.** As we will see, these transformation equations force us to give up the absolute character of length and of time intervals. They were originally discovered by H. A. Lorentz,* but their true meaning was established by Einstein. These transformation equations are an expression of the fundamental properties of space and time— they are an expression of the geometry of **spacetime.** In the following we will emphasize the geometric character of the Lorentz transformations by adopting a graphical representation of these transformations.

For the sake of simplicity, we will deal only with one-dimensional motion, so that the position x of a particle as a function of the time t gives a complete description of the motion. In relativistic physics, the plot of position vs. time is called the **worldline** of the particle. When dealing with a relativistic particle, with a speed comparable to the speed of light, the slope of the worldline in a plot of x vs. t would be inconveniently large. We will therefore find it advantageous to replace the time coordinate t by ct, and to plot x vs. ct. For example, Fig. 2.7 shows the x- and ct-axes of a reference frame attached to a laboratory on the Earth,[†] and it shows the worldline of a particle with one-dimensional motion along the x-direction. (In this figure, the x-axis has been arranged horizontally and the ct-axis vertically; this is the customary arrangement in relativistic physics.) Before time $t = 0$, the particle was at rest at $x = 0$ and its worldline coincided with the ct-axis; it then accelerated as indicated by the changing slope of the worldline; and it

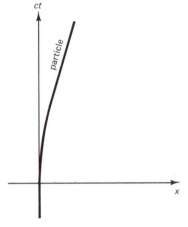

Fig. 2.7 Spacetime diagram showing the worldline of a particle.

* **Hendrik Antoon Lorentz,** 1853–1928, Dutch theoretical physicist, professor at Leiden. He discovered the Lorentz transformations in his mathematical studies of symmetries of Maxwell's equations, but failed to understand their significance as describing a change of inertial reference frame. Lorentz was highly regarded as an expert on electromagnetic theory. Several years before Thomson's discovery of the electron, Lorentz proposed that the generation of light by atoms arises from oscillating charges within them, and he perceived that the Zeemann effect (splitting of spectral lines by a magnetic field) provides proof of this hypothesis. For this insight he was awarded the Nobel Prize in 1902.

[†] We will pretend that this reference frame is inertial.

finally reached a constant velocity as indicated by the constant slope of the worldline.

The diagram shown in Fig. 2.7 is called a **spacetime diagram.** The points plotted in this diagram are called spacetime points, or **events.** Note that because the time axis in the spacetime diagram has been arranged vertically, the velocity of a particle is *inversely* proportional to the slope of its worldline.

$$(\text{slope}) = \frac{\Delta(ct)}{\Delta x} = \frac{c}{v}$$

The worldline of a light signal, with $v = \pm c$, has a slope of ± 1; thus, this worldline slopes at 45° to the right or to the left, depending on whether the light signal is propagating in the positive or the negative x-direction. For example, Fig. 2.8 shows the worldlines of several light signals that start at different points on the x-axis.

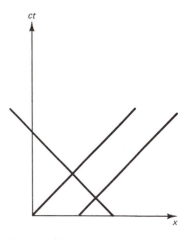

Fig. 2.8 Worldlines of several light signals propagating in the positive or negative x-directions. All these worldlines slope at 45°.

Instead of describing the motion of the particle in the reference frame of the laboratory, we can describe this motion in an inertial reference frame that has some (constant) velocity relative to the laboratory, say, the reference frame of a spaceship that moves with constant velocity V in the x-direction. In this new reference frame, the spacetime coordinates are x' and ct', and the motion of the particle is described by giving the position x' as a function ct'. We can then prepare a new spacetime diagram with axes x' and ct', and plot the worldline of the particle. This new plot represents the motion of the particle in the new reference frame. However, instead of tediously replotting all the points of the worldline, we will find it much more instructive to keep the worldline of Fig. 2.7, and to rearrange the coordinate axes so that this same worldline describes the motion of the particle in the new reference frame. The change of coordinate axes will then graphically represent the change of reference frame. Such a graphical representation of a change of reference frame is analogous to the familiar graphical representation of a rotation of the space coordinates by drawing a new set of space axes rotated by a suitable angle; the new coordinates can then be directly read off these new axes.

The details of the graphical construction of the new axes x' and ct' representing the new reference frame depend on whether we assume that the underlying physics is Newtonian or relativistic. To gain some familiarity with the graphical procedure, let us begin by assuming that Newtonian physics is valid, so that time intervals and lengths are absolute.

Figure 2.9 shows the worldline of the midpoint of the spaceship, plotted in a spacetime diagram with axes x and ct. The slope

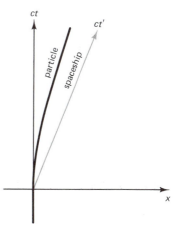

Fig. 2.9 Worldline of a particle and worldline of the midpoint of the spaceship (ct'-axis).

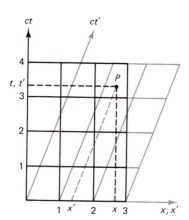

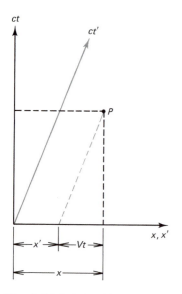

Fig. 2.10 Grids for the *x-ct* and *x'-ct'* coordinates. The coordinates of any spacetime point can be read off these grids.

Fig. 2.11 Relationship between *x* and *x'*.

of this worldline is c/V. The midpoint of the spaceship passes through the origin of the laboratory coordinates at time $t = 0$. We will regard this midpoint as the origin of the spaceship coordinates, which means that the midpoint is always at $x' = 0$. Hence the slanted worldline in Fig. 2.9 is the locus of points with $x' = 0$; this implies that this worldline coincides with the t'-axis, or the ct'-axis. To complete the graphical construction of the new axes, we need to discover the orientation of the x'-axis. This axis is the locus of points with $t' = 0$. Since in Newtonian physics time is absolute, the clocks in the reference frame of the laboratory and in the spaceship can be synchronized so that for all spacetime points the time coordinates t and t' agree, i.e., $t = t'$. Thus the locus of points with $t' = 0$ coincides with the locus of points with $t = 0$, i.e., the x'-axis coincides with the x-axis. It then only remains to fix the scales along the new x'- and ct'-axes. Since time is absolute and the clocks in the two reference frames have been synchronized, the line $ct = 1$ coincides with the line $ct' = 1$, and therefore the unit point on the ct'-axis must be horizontally aligned with the unit point on the ct-axis, as shown in Fig. 2.9. Since lengths are absolute, the unit points on the x'-axis must coincide with the unit point on the x-axis.

Figure 2.10 shows the grid of the new x'- and ct'-coordinates superimposed on the grid of the old x- and ct-coordinates; the new coordinate grid is slanted with respect to the old. This figure is a graphical representation of the transformation of coordinates from the laboratory frame to the spaceship frame. The old and the new coordinates x, ct and x', ct' of any spacetime point can be directly read off this diagram by drawing suitable lines intercepting the axes.

From Fig. 2.11 we readily see that the mathematical relationship between the values of the old and the new coordinates of an arbitrary spacetime point is

$$x' = x - Vt \tag{9}$$

$$t' = t \tag{10}$$

where V is the velocity of the spaceship relative to the laboratory. These equations are exactly the Galilean transformation equations. Thus, our graphical construction has permitted us to derive these equations.

We will now use a similar graphical construction to derive the Lorentz transformation equations. Whereas in the Newtonian case we relied on the absolute character of time intervals and lengths, we will now rely on the absolute value of the speed of light—the new axes x' and ct' must be constructed in such a way that the speed of light signals is the same in the old and the new reference frames. Figure 2.12 again shows the worldline of the midpoint of the spaceship. As before, this worldline coincides with the ct'-axis. To discover the orientation of the x'-axis, we will find it instructive to consider the worldline of a light signal that starts at the origin at time $t = 0$. Figure 2.12 shows this worldline. It lies exactly halfway between the x- and ct-axes; this makes the x- and ct-coordinates of any spacetime point on this worldline equal, which is equivalent to the condition that the slope equal 1. The invariance of the speed of light can be expressed graphically as the requirement that the worldline is also halfway between the x'- and the ct'-axes, so that its slope reckoned according to the x'- and ct'-coordinates is again equal to 1 (this assumes that the scales along the x'- and ct'-axes have been chosen to be equal, just as the scales along the x- and ct-axes have been chosen to be equal). Thus, we reach the conclusion that the ct'- and the x'-axes must be symmetrically placed with respect to the worldline of the light signal. Thus, both the x'- and the ct'-axes are slanted (see Fig. 2.13).

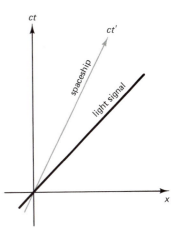

Fig. 2.12 Worldline of a light signal and worldline of the midpoint of the spaceship.

In view of the symmetry of the arrangement of the x'- and ct'-axis displayed in Fig. 2.13, we immediately recognize that the equations relating the old and the new coordinates must be of the form

$$x' \propto x - \frac{V}{c}ct \tag{11}$$

$$ct' \propto ct - \frac{V}{c}x \tag{12}$$

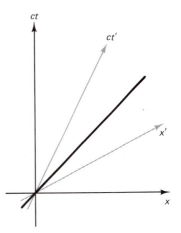

Fig. 2.13 The new x'-axis.

The first of these equations is essentially the same as Eq. (9); it expresses the slant of the x'-axis. The second equation expresses the slant of the ct'-axis; it follows from the first by symmetry, i.e., by exchange of the space and time coordinates. This symmetry ensures that the angles between the x, x' and ct, ct' axes are equal [this can also be verified by direct calculation of these angles from Eqs. (11) and (12)].

Note that Eqs. (11) and (12) have been stated as proportionalities rather than as equalities. The reason is that the requirement of symmetry with respect to the worldline of the light signal fixes the orientations of the axes, but not their scales. If we include the unknown scale factor in our equations, we can write them as

$$x' = \gamma\left(x - \frac{V}{c}ct\right) \tag{13}$$

$$ct' = \gamma\left(ct - \frac{V}{c}x\right) \tag{14}$$

Here, the same scale factor appears in each equation, in accord with our choice of equal scales for the x'- and ct'-axes.

The scale factor γ is some function of the velocity V. We can determine this function by a comparison of the transformation from x,t to x',t' coordinates with the inverse transformation from x',t' to x,t. This inverse transformation can be obtained by solving Eqs. (13) and (14) for x and ct:

$$x = \frac{1}{\gamma(1 - V^2/c^2)}\left(x' + \frac{V}{c}ct'\right) \tag{15}$$

$$ct = \frac{1}{\gamma(1 - V^2/c^2)}\left(ct' + \frac{V}{c}x'\right) \tag{16}$$

These two equations express the old coordinates x and ct in terms of the new coordinates x' and ct', whereas Eqs. (13) and (14) express the new coordinates in terms of the old. But there is an alternative way to express the old coordinates in terms of the new: instead of beginning with rectangular axes for the old coordinates and then constructing slanted axes for the new coordinates, we could equally well begin with rectangular axes for the new coordinates and construct slanted axes for the old coordinates. We would then obtain a pair of equations similar to Eqs. (13) and (14) with the new and the old coordinates exchanged, and with one further modification: the velocity V would be replaced by $-V$ (this change of sign of the velocity simply means that the velocity of the laboratory relative to the spaceship is opposite to the velocity of the spaceship relative to the laboratory). Thus, the pair of equations would be

$$x = \gamma\left(x' + \frac{V}{c}ct'\right) \tag{17}$$

$$ct = \gamma\left(ct' + \frac{V}{c}x'\right) \tag{18}$$

If we compare this pair of equations with Eqs. (15) and (16), we see

$$\gamma = \frac{1}{\gamma}\frac{1}{1 - V^2/c^2}$$

from which we find

$$\gamma = \frac{1}{\sqrt{1 - V^2/c^2}} \qquad (19)$$

With this value of γ, we obtain the transformation equations

$$x' = \frac{1}{\sqrt{1 - V^2/c^2}} \left(x - \frac{V}{c} ct \right) \qquad (20)$$

$$ct' = \frac{1}{\sqrt{1 - V^2/c^2}} \left(ct - \frac{V}{c} x \right) \qquad (21)$$

or

$$x' = \frac{1}{\sqrt{1 - V^2/c^2}} (x - Vt) \qquad (22)$$

$$t' = \frac{1}{\sqrt{1 - V^2/c^2}} (t - Vx/c^2) \qquad (23)$$

Similarly, we obtain the inverse transformation equations

$$x = \frac{1}{\sqrt{1 - V^2/c^2}} (x' + Vt') \qquad (24)$$

$$t = \frac{1}{\sqrt{1 - V^2/c^2}} (t' + Vx/c^2) \qquad (25)$$

These are the **Lorentz transformation** equations. Although in the derivation above we have labeled the two reference frames as "laboratory" and "spaceship," our equations are quite general and they are applicable to any two inertial reference frames. By construction, these equations ensure that the speed of light is the same in all inertial reference frames, as required by the Principle of Relativity.

Given any spacetime point with coordinates x and t, we can calculate the coordinates x', t' from Eqs. (22) and (23). Alternatively, we can read these coordinates directly off a carefully drawn spacetime diagram. For this purpose, we need to put the appropriate scale on the x'- and ct'-axes. We can do this by noting that, according to Eq. (22), the line $x' = 1$ intercepts the x-axis, or the line $t = 0$, at the point $x = \sqrt{1 - V^2/c^2}$. For instance, if $V/c = \frac{1}{2}$, then the intercept is $x = \sqrt{1 - \frac{1}{4}} = 0.866$ (see Fig. 2.14).

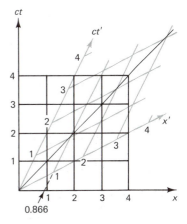

Fig. 2.14 Grids for the x-ct and x'-ct' coordinates.

By taking differences, we find that the Lorentz transformations for space and time intervals are

$$\Delta x' = \frac{1}{\sqrt{1 - V^2/c^2}} (\Delta x - V \Delta t) \tag{26}$$

$$\Delta t' = \frac{1}{\sqrt{1 - V^2/c^2}} (\Delta t - V \Delta x/c^2) \tag{27}$$

These equations show that what is a pure space interval or a pure time interval in one reference frame becomes a mixture of space and time intervals in another reference frame. This intimate relationship of space and time led H. Minkowski* to remark: "Henceforth space by itself, and time by itself, are doomed to fade into mere shadows, and only a kind of union of the two will preserve an independent reality." This union of space and time has come to be called spacetime.

One startling consequence of this mixing of space and time intervals is that simultaneity is relative—the simultaneity of two events depends on the reference frame. Suppose that two events occur at the same time in the reference frame of the laboratory ($\Delta t \neq 0$), but at different locations. Figure 2.15 is a spacetime diagram showing the two events and their separation Δx. As we can see from this diagram, the separation is purely spatial in the reference frame of the laboratory, but it has both space and time components in the reference frame of the spaceship. Thus, in the latter reference frame the two events are *not* simultaneous. We can calculate the time difference in this reference frame from Eq. (27):

$$\Delta t' = \frac{1}{\sqrt{1 - V^2/c}} (\Delta t - V \Delta x/c^2) = \frac{-V \Delta x/c^2}{\sqrt{1 - V^2/c^2}} \tag{28}$$

This shows that $\Delta t'$ is not zero, i.e., the events are not simultaneous in the reference frame of the spaceship.

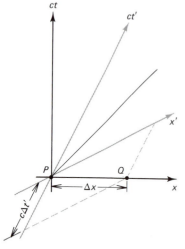

Fig. 2.15 The two spacetime points *P* and *Q* are simultaneous in the laboratory frame ($\Delta t = 0$), but they are not simultaneous in the spaceship frame ($\Delta t' \neq 0$).

Example 2   At 6:00 P.M. E.S.T., the street lights in New York and Boston are switched on simultaneously in the reference frame of the Earth. What is the time difference as reckoned in the reference frame of a spaceship traveling in the direction New York–Boston at a speed $V = 0.9c$? The distance between New York and Boston is 290 km.

---

* **Hermann Minkowski**, 1864–1909, German mathematician, professor at Göttingen.

Solution   With the x-axis along the direction New York–Boston, the displacement is $\Delta x = 290\,\text{km}$. According to Eq. (28):

$$\Delta t' = \frac{-V\,\Delta x/c^2}{\sqrt{1 - V^2/c^2}} = \frac{-0.9 \times 290 \times 10^3\,\text{m}/3.0 \times 10^8\,\text{m/s}}{\sqrt{1 - (0.9)^2}}$$

$$= -8.7 \times 10^{-4}\,\text{s}$$

The negative sign means that, in the reference frame of the spaceship, the lights in New York are switched on *later* than those in Boston.   ■

More generally, the time order of two events is relative whenever the spatial separation Δx between them has a larger magnitude than the temporal separation $c\,\Delta t$. Figure 2.16 shows two such events. In the reference frame of the laboratory, P occurs before Q; but in the reference frame of a sufficiently fast spaceship (with coordinate axes as in Fig. 2.16), Q occurs before P. Obviously, such a reversal of the time order is possible whenever the line connecting P and Q has a slope of less than 45°, so that we can find an x'-axis of slope *larger* than that of this line. The reversal of the time order is not possible if the line connecting P and Q has a slope of more than 45°; for such events, the time order is absolute.

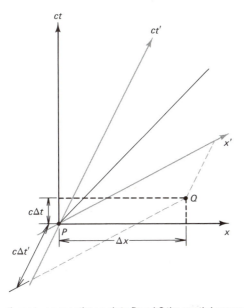

Fig. 2.16 For these two spacetime points P and Q the spatial separation is larger than the temporal separation ($\Delta x > c\,\Delta t$). In the laboratory frame P occurs before Q ($\Delta t > 0$), but in the spaceship frame Q occurs before P ($\Delta t' < 0$).

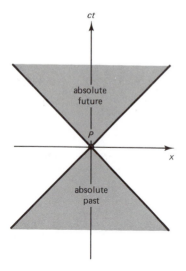

Fig. 2.17 The light cone.

For a given spacepoint P, we can therefore classify all other spacetime points Q according to whether the time order is absolute or relative. The set of points Q that occur later than P in all reference frames is called the **absolute future** of P (see Fig. 2.17). The set of points that occur earlier than P in all reference frames is called the **absolute past.** The set of points that lack an absolute time order in relation to P is called the **neutral region;** a point in this region will occur later than P, or earlier, or simultaneously depending on the choice of reference frame. The boundaries separating the absolute regions from the neutral regions are lines of slope $45°$ (see Fig. 2.17); these lines are the worldlines of light signals arriving at P or leaving P. This boundary is called the **light cone** (if we add the y-axis perpendicular to the x- and ct-axes in Fig. 2.17, then this boundary becomes a cone; if we add both the y- and the z- axes, then the boundary becomes a generalized cone in four dimensions).

As we will see in the next section, the Lorentz transformations have some other consequences that do violence to our intuition. We will see that motion affects the rate of a clock (time dilation) and the length of a body (length contraction). We must therefore give up the intuitively obvious notions of absolute time and absolute length on which Newtonian physics is based. The relativistic behavior of time and length compels us to adopt new definitions of momentum and energy, and new relativistic equations of motion. The differences between the Newtonian and the relativistic formulas are small if the speeds are small compared to the speed of light, but the differences become large when the speeds become comparable to the speed of light. This means that for the motion of the macroscopic bodies of our everyday experience, the formulas of Newtonian physics give an adequate approximation, but for the motion of high-speed particles, the Newtonian formulas must be replaced by the more accurate relativistic formulas.

If the velocity V in the Lorentz transformation Equations (22) and (23) is small compared to the speed of light, then $\sqrt{1 - V^2/c^2} \simeq 1$, and the equations become approximately

$$x' \simeq x - Vt \tag{29}$$

$$t' \simeq t - Vx/c^2 \tag{30}$$

The first of these coincides with the Galilean transformation equation for the x-coordinate. The second equation does not quite coincide with the Galilean equation for the t-coordinate—it differs by the extra term Vx/c^2. However, if V/c is small and if x is not very large, then this extra term usually can be neglected. For instance suppose that $V = 7.6 \, \text{km/s}$, the orbital speed of an Earth satellite in a low-altitude orbit, and that $x = 3.8 \times 10^8 \, \text{m}$, the distance from the Earth to the Moon; then $Vx/c^2 = 3.2 \times 10^{-5} \, \text{s}$, a quite small time interval, which can be neglected under ordinary circumstances. Thus, for sufficiently low velocities, the Lorentz transformations do indeed approximate the Galilean transformations.

In our derivation of the transformation equations we have focused on just one space coordinate, the x-coordinate, which is the coordinate measured parallel to the line of the relative motion of the reference frames. But we will also need to know the transformation equations for the y- and z-coordinates, which are measured perpendicular to the line of motion. The Lorentz transformation equations for these transverse coordinates are the same as the Galilean transformation equations:

$$y' = y \tag{31}$$

$$z' = z \tag{32}$$

These equations express the invariance of lengths perpendicular to the direction of motion. The proof of this invariance is by contradiction: Imagine that we have two identically manufactured pieces of pipe, one at rest on Earth and one at rest on the spaceship (see Fig. 2.18). If the motion of the spaceship relative to the Earth were to bring about a transverse contraction of the spaceship pipe, then, by the Principle of Relativity, the motion of the Earth relative to the spaceship would have to bring about a contraction of the Earth pipe. These contraction effects are contradictory since in one case the spaceship pipe would fit inside the Earth pipe and in the other case it would fit outside.

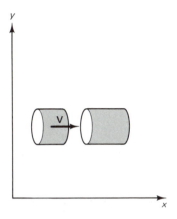

Fig. 2.18 Two equal pieces of pipe, one at rest in the reference frame of the laboratory, one at rest in the reference frame of the spaceship.

Example 3   **Tachyons** are hypothetical particles that travel faster than the speed of light. Show that if such particles exist, then we can use them to send signals into our own past.

Solution   In a spacetime diagram, the worldline of a particle traveling faster than the speed of light has a slope of less than 45°. Figure 2.19 shows the x- and ct-axes of the laboratory frame and it shows the worldline of a tachyon emitted in the positive x-direction from the origin of this reference frame. Suppose that at the spacetime point P, this tachyon reaches a detector on a high-speed spaceship traveling in the positive x-direction, away from the Earth. An emitter coupled to the detector on the spaceship immediately sends out another tachyon in the negative x-direction, toward the Earth. Figure 2.19 shows the x'- and ct'-axes of the spaceship frame and the worldline of the second tachyon. Note that in the reference frame of the spaceship this tachyon travels into the future ($\Delta t' > 0$), but in the

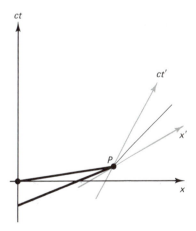

Fig. 2.19 Worldlines (dashed) of hypothetical tachyon signals.

reference frame of the laboratory it travels into the past ($\Delta t < 0$), and reaches the origin $x = 0$ at a time $t < 0$. Thus, the second tachyon arrives at the origin of the laboratory *before* the first tachyon is emitted, and we have succeeded in sending a tachyon signal into our own past. Such signals into our own past lead to logical contradictions (for instance, imagine what would happen if the laboratory detector were coupled to the emitter by logic circuits programmed to abort the emission of the first tachyon if and only if the second tachyon is detected). Because of these contradictions, we can confidently conclude that tachyons, or any other signals exceeding the speed of light, do not exist. ■

2.4 **The Time Dilation**

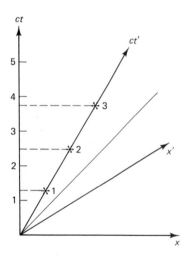

Fig. 2.20 Spacetime diagram showing ticks of a clock at rest in the spaceship frame.

Consider one of the clocks belonging to the spaceship frame, which is moving with velocity V relative to the laboratory. We want to compare the rate of this clock with the rate of the clocks belonging to the laboratory frame. Each tick of the clock can be regarded as an event. Figure 2.20 is a spacetime diagram showing these events. For the purpose of this diagram, the clock has been assumed to tick at intervals of $c\,\Delta t' = 1$ m, i.e., it ticks at $ct' = 1$ m, at $ct' = 2$ m, at $ct' = 3$ m, etc. In the spacetime diagram, we can compare this clock with the laboratory clocks by drawing horizontal lines from the tick events to the ct-axis. These horizontal lines intercept the ct-axis above the calibration marks—as measured in the laboratory, the interval $c\,\Delta t$ between the ticks is longer than 1 m. Hence, as compared to the laboratory clocks, the spaceship clock goes slow. From our discussion of the scale difference between the two sets of axes in the spacetime diagram, we know that the horizontal line $ct = 1$ m intercepts the ct'-axis at $\gamma \times 1$ m. Thus, we conclude that the rate of the spaceship clock is slow by a factor of γ. This is the **time-dilation** effect of special relativity.

We can, of course, obtain the same result from our Lorentz-transformation formulas. According to Eq. (25),

$$\Delta t = \frac{1}{\sqrt{1 - V^2/c^2}}\left(\Delta t' + V\Delta x'/c^2\right) \tag{33}$$

Since the clock is at rest in the spaceship frame, $\Delta x' = 0$ and

$$\Delta t = \frac{1}{\sqrt{1 - V^2/c^2}}\,\Delta t' \qquad \text{(clock at rest in spaceship)} \tag{34}$$

which, again, shows that the rate of the spaceship clock is slow by a factor of γ. The time registered by a clock in its own reference frame is called the **proper time,** usually designated by τ. In terms of proper time, Eq. (34) becomes

$$\Delta t = \frac{1}{\sqrt{1 - V^2/c^2}} \Delta\tau \tag{35}$$

The time-dilation effect is symmetric: as measured by the clocks on the spaceship, a clock in the laboratory runs slow by the same factor,

$$\Delta t' = \frac{1}{\sqrt{1 - V^2/c^2}} \Delta t \qquad \text{(clock at rest in laboratory)} \tag{36}$$

The derivation of Eq. (36) can be based on Eq. (27) with $\Delta x = 0$.

Figure 2.21 is a plot of the time-dilation factor $1/\sqrt{1 - V^2/c^2}$ as a function of V. The slowing down of the rate of lapse of time applies to all physical processes: atomic, nuclear, biological, etc. Reckoned in proper time, the rate of all such processes is independent of the motion of the spaceship relative to the laboratory; but reckoned in laboratory time, the rate of the processes is slowed down. At low speeds the effect is insignificant, but at high speeds it can become quite appreciable. For instance, muons are unstable particles which usually decay in about 2.2×10^{-6} s; but if they are moving at high speed through our laboratory, then the internal processes that produce the decay are slowed down and the muon lives a longer time as reckoned by the clocks of our laboratory. In a careful experiment performed at CERN by Bailey et al.,* muons of velocity $V = 0.9994c$ were found to have a lifetime 29.3 times as long as muons at rest, in excellent agreement with the prediction based on Eq. (34).

The time dilation of muons has a crucial effect on the secondary cosmic rays that penetrate through the atmosphere and reach sea level. Most of these secondary cosmic rays are muons, which are produced in the decay of pions created by the impact of primary cosmic rays on the top of the atmosphere. Without time dilation, a muon traveling at the maximum speed of 3×10^8 m/s would cover a distance of only 3×10^8 m/s $\times 2.2 \times 10^{-6}$ s $\simeq 660$ m during its lifetime. Thus, next to no muons would survive long enough to complete the trip through the atmosphere. The abundant flux of cosmic-ray muons found at sea level is a direct consequence of time dilation.

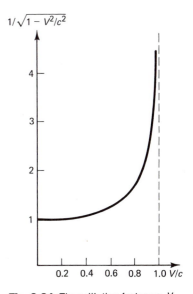

Fig. 2.21 Time-dilation factor vs. V.

* J. Bailey et al., *Nature* **268,** 301 (1977).

At everyday velocities, the time-dilation effect is extremely small. For example, consider a clock aboard an airplane traveling at 300 m/s over the ground. This corresponds to $V/c = 1.0 \times 10^{-6}$. To evaluate the time-dilation factor, we use the approximation

$$\frac{1}{\sqrt{1 - V^2/c^2}} \simeq 1 + \frac{1}{2} V^2/c^2$$

which gives

$$\frac{1}{\sqrt{1 - V^2/c^2}} \simeq 1 + 5 \times 10^{-13}$$

that is, the clock in the airplane will only slow by 5 parts in 10^{13}. However, such a small change is not beyond the reach of modern atomic clocks. Some years ago, Häfele and Keating[*] placed portable atomic clocks on board commercial airliners and kept them flying for several days, making a complete trip around the world. Before and after the trip the clocks were compared with an identical clock that was kept on the ground. The flying clocks were found to have lost time—in one instance the total time lost because of the motion of the clock was about 10^{-7} s.

The most precise test of the time dilation has emerged from a recent experiment by Kaivola et al.[†] which compared the frequency of the resonant absorption of light by a beam of neon atoms of speed $V = 4 \times 10^{-3} c$ with the frequency of the light absorbed by neon atoms at rest. In this case, the frequency of the moving atoms is shifted by the time-dilation effect and also by the ordinary Doppler shift associated with a moving source. The time-dilation effect is much smaller than the ordinary Doppler shift, and hence it is very difficult to notice with light absorbed from any one direction. However, if an atom absorbs equal amounts of light incident from opposite directions along the beam,[‡] then the ordinary Doppler shift averages to zero, and the only remaining frequency shift is that produced by the relativistic time dilation. For the sake of high accuracy, the frequencies of the moving and the stationary atoms were not compared directly, but instead two lasers were tuned to these frequencies, and these laser frequencies were compared by counting beats between their light waves. This experiment confirmed the expected time dilation to within 4 parts in 10^5.

[*] J. C. Häfele and R. E. Keating, *Science* **177**, 274 (1960).

[†] M. Kaivola et al. *Phys. Rev. Lett.* **54**, 255 (1985).

[‡] The atom absorbs two photons, one from each direction.

The time-dilation effect leads to the famous **twin paradox** of Special Relativity. Imagine that a pair of identical twins—Stella and Terra—celebrate their twentieth birthday together on Earth, and then Stella embarks on a fast spaceship, which carries her at a speed of, say, $V = \frac{1}{2}c$ on a round-trip to a nearby star. As reckoned in the reference frame of the Earth, the clocks and all the physiological processes on the spaceship suffer from time dilation. Hence Stella ages at a slower rate than Terra and at the completion of the trip, when the twins are reunited on the Earth, Stella will be younger than Terra. The paradox emerges when we attempt to reckon the aging of the twins from the point of view of the spaceship frame. Relative to this reference frame, the Earth is in motion, and hence the clocks and physiological processes on the *Earth* suffer from time dilation. This leads us to a contradictory conclusion: Terra ages at a slower rate than Stella, and when the twins are reunited, Terra will be younger than Stella.

Note that the return of Stella is a crucial ingredient in this paradox—if Stella were to keep on traveling away from Earth, then each twin could assert that she was older, without fear of contradiction. To resolve the paradox, we must take into account that the return of Stella requires a reversal of the motion of the spaceship, i.e., it requires an acceleration of Stella's reference frame. Our formula for time dilation is valid only if the moving clock is observed from the point of view of an *inertial*, i.e., unaccelerated, reference frame. Since Stella's reference frame is not inertial, it is not legitimate to use the simple formula (36) when reckoning the time dilation of the Earth clocks relative to this reference frame. (On the other hand, it turns out that it is legitimate to use the simple formula when reckoning the time dilation of the spaceship clocks relative to the Earth frame—for a moving accelerated clock observed from an inertial reference frame, Eq. (34) gives the *instantaneous* time-dilation factor, even if v is not constant.) A detailed analysis of the behavior of the Earth clocks relative to the spaceship frame* shows that the Earth clocks suffer from time dilations during the outbound and inbound portions of the trip, when the spaceship is moving with uniform velocity, but the Earth clocks suffer a strong time contraction when the spaceship accelerates while turning around at the target star. This time contraction more than compensates the time dilations, and the net result is that the Earth clocks will be ahead of the spaceship clocks at the end of the trip. Thus, the paradox disappears, and no matter what

---

* See, e.g., E. F. Taylor and J. A. Wheeler, *Spacetime Physics*, pp. 94, 95.

reference frame we adopt for reckoning the relative aging of the twins, the conclusion is the same: when the twins are reunited, Stella will be younger than Terra.

2.5 The Length Contraction

Suppose that a rigid body,* such as a meter stick, is at rest in the spaceship frame. The length of this body is $\Delta x' = 1$ m in the spaceship frame. To find the length in the laboratory frame, we must measure the position of the forward end and of the rearward end of this meter stick at the same instant of *laboratory* time. Figure 2.22 shows the worldlines of the forward and the rearward ends of the meter stick in a spacetime diagram. From this diagram we see that at one instant of time, say, $t = 0$, the meter stick extends over an x-interval somewhat less than 1 m. Hence, as measured in the laboratory, the meter stick in the spaceship is contracted. Since the slant line $x' = 1$ m intercepts the x-axis at $\gamma \times 1$ m, we conclude that the length of the meter stick is contracted by a factor of γ. This is the **length-contraction** effect.

Alternatively, we can obtain this result from Eq. (26),

$$\Delta x' = \frac{1}{\sqrt{1 - V^2/c^2}}(\Delta x - V\Delta t) \tag{37}$$

For our measurement of length, $\Delta t = 0$ and

$$\Delta x' = \frac{1}{\sqrt{1 - V^2/c^2}}\Delta x \tag{38}$$

or

$$\Delta x = \sqrt{1 - V^2/c^2}\,\Delta x' \qquad \begin{array}{l}\text{(body at rest} \\ \text{in spaceship)}\end{array} \tag{39}$$

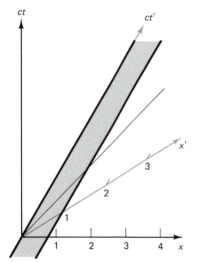

Fig. 2.22 Spacetime diagram showing the worldlines of the ends of a meter stick at rest in the spaceship frame. The shaded band is the spacetime region occupied by the body of the meter stick.

which indicates a length contraction by a factor of γ. This effect is of course also symmetric: a body at rest in the laboratory will suffer from length contraction when measured by instruments on board the spaceship.

The length contraction has not been tested directly by experiment. There is no practical method for the high-precision measurement of the length of a fast-moving body. The obvious method would seem to be high-speed photography, but this is not accurate

---

* In this context, *rigid* merely means that the length remains constant in the spaceship frame.

enough since the contraction is extremely small even at the highest speeds that we can impress on a macroscopic body.

If we could take a sharp photograph of a macroscopic body passing by at a speed of, say, $V = 0.8c$, the photographic image would not only show the contraction, but also a strong distortion of the apparent shape of the body. A photograph never shows the surface of the body as it is, but rather as it was at the instant the light from it began to travel to the camera. If light from near and far parts of the body is to arrive at the camera at the same instant, the light from the far parts must start out earlier than that from the near parts. Thus the photograph shows different parts of the body at different times—far parts at an early time and near parts at a late time. The photograph therefore is a distorted picture of the body. Figure 2.23 shows a computer simulation of the photograph that

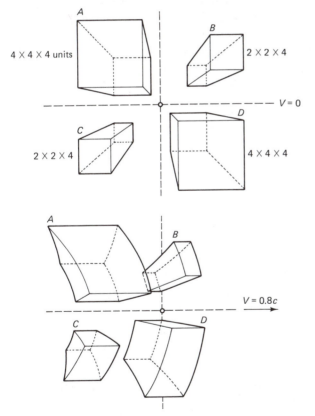

Fig. 2.23 Computer simulation of a photograph of a set of four rectangular boxes moving at very high speed. [From G. D. Scott and M. R. Viner, *Am. J. Phys.* **33**, 534 (1965). © American Association of Physics Teachers.]

a camera would record when aimed at some rectangular boxes traveling past the camera at $V = 0.8c$. The length contraction is clearly noticeable along the midline, but everywhere else the image is severely distorted.

2.6 The Combination of Velocities

The Galilean addition law for velocities is a consequence of the absolute properties of length and time in Newtonian physics. Likewise the relativistic combination law for velocities is a consequence of the relativistic properties of length and time contained in the Lorentz transformation equations. Since these transformation equations were specifically designed to incorporate the invariance of the speed of light, we expect that the relativistic combination law for velocities will return the speed of light whenever the speed of light is combined with any other speed.

To derive the relativistic combination law for the x-component of the velocity, we simply take differentials of the Lorentz transformation equations:

$$dx' = \gamma(dx - V\,dt) \tag{40}$$

$$dt' = \gamma(dt - V\,dx/c^2) \tag{41}$$

which yield

$$\frac{dx'}{dt'} = \frac{dx - V\,dt}{dt - V\,dx/c^2} = \frac{dx/dt - V}{1 - (V/c^2)\,dx/dt} \tag{42}$$

But dx/dt is the velocity v_x of the particle, light signal, or whatever with respect to the reference frame of the laboratory, and dx'/dt' is the velocity v'_x with respect to the reference frame of the spaceship. Hence we can write Eq. (42) as

$$v'_x = \frac{v_x - V}{1 - v_x V/c^2} \tag{43}$$

This relativistic combination law is to be compared with the Galilean addition law

$$v'_x = v_x - V \tag{44}$$

The difference lies in the denominator of Eq. (43), which has a drastic effect if the speeds are large. For instance, suppose that v is the velocity of a light signal along the x-axis in the reference frame of the laboratory. Then $v_x = c$ and Eq. (43) gives

$$v'_x = \frac{c - V}{1 - V/c} = c \tag{45}$$

Thus, as required, the velocity of the light signal has exactly the same magnitude in the reference frame of the spaceship.

The inverse combination law that expresses the velocity v_x in terms of v'_x is

$$v_x = \frac{v'_x + V}{1 + v'_x V/c^2} \tag{46}$$

The combination law for light velocities has been explicitly tested with high precision by Alväger et al.* in an experiment involving a beam of very fast π° mesons. These particles decay spontaneously by a reaction that emits two γ rays. Hence such a beam of pions can be regarded as a high-speed light source. In the experiment, the velocity of the pions relative to the laboratory was $V = 0.99975c$. The Galilean addition law would then predict a laboratory velocity of $v_x = 1.99975c$ for γ rays emitted in the forward direction and of $v_x = 0.00025c$ for γ rays emitted in the backward direction. But the experiment confirmed the relativistic addition law—the laboratory velocity of the rays was the same in both directions.

Of course, Eq. (43) gives only the combination law for the component of velocity parallel to the motion of the reference frame (x-component). The components of the velocity perpendicular to the motion of the reference frame (y- and z-components) also have combination laws different from the Galilean transformation laws. Omitting the derivations, we state the results

$$v'_y = \frac{v_y\sqrt{1 - V^2/c^2}}{1 - Vv_x/c^2}, \qquad v'_z = \frac{v_z\sqrt{1 - V^2/c^2}}{1 - Vv_x/c^2} \tag{47}$$

2.7 Relativistic Energy and Momentum

In Newtonian physics, the momentum of a particle is the product of mass and velocity,

$$\mathbf{p} = m\mathbf{v} \tag{48}$$

---

* T. Alväger et al. *Phys. Lett.* **12**, 260 (1964).

The momentum of a system of particles obeys an important conservation law: in the absence of external forces, the net momentum
of the system of particles is constant. This conservation law is a
direct consequence of Newton's Second and Third Laws. In the
simple case of a system with only two interacting particles (A
and B), the proof is as follows:

$$\frac{d}{dt}(\mathbf{p}_A + \mathbf{p}_B) = \mathbf{F}_A + \mathbf{F}_B = \mathbf{F}_A - (\mathbf{F}_A) = 0 \qquad (49)$$

where the first equality depends on the Second Law and the next
equality depends on the Third Law, $\mathbf{F}_A$ and $\mathbf{F}_B$ being an action–
reaction pair.

In relativistic physics we cannot derive the conservation of momentum in this way. The trouble is that Newton's Third Law
fails—the relativity of simultaneity makes the instantaneous balance of action and reaction meaningless. For example, consider two
charged particles exerting Coulomb forces on one another while
in motion. Action and reaction are then defined at two different
points of space (the positions of the particles), and they both
change with time. The permanent balance of action and reaction
would require that these changes are simultaneous, in all reference
frames; but this is impossible because the relativity of simultaneity
tells us that events at different points of space cannot be simultaneous in all reference frames. It turns out that the imbalance of
action and reaction, and the consequent imbalance of conservation
of momentum of the system of particles, is compensated by a transfer of momentum to the fields of force surrounding the particles—
these fields act as storehouses of momentum in such a way that the
net momentum of particles plus fields is conserved. However, to
prove this, we need a complete theory of momentum and energy
stored in fields. This would take us beyond the scope of this book.

Here we will be satisfied with a discussion of momentum conservation for the special, simple case of contact forces. In this
case action and reaction are defined at the same point of space
(the point of collision), and therefore the relativity of simultaneity
does not interfere with Newton's Third Law. In any collision involving such contact forces, we then expect that momentum will
be conserved. However, as we will see below, we can only satisfy
this expectation if we define momentum in relativistic physics by
a formula somewhat more complicated than the simple formula
[Eq. (48)] of Newtonian physics.

Consider a collision between two smooth, identical elastic balls.
The balls have momenta of equal magnitude (and opposite sign)
in the y-direction, but ball A differs from B in that it has a large
momentum in the x-direction. Figure 2.24 shows the balls in the
reference frame of the laboratory at a short time Δt before the

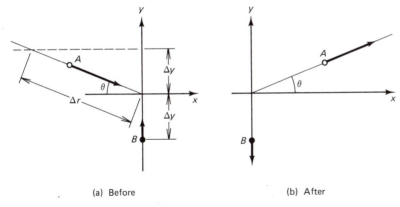

(a) Before   (b) After

Fig. 2.24 Elastic collision of two balls seen in the reference frame of the laboratory. (a) At a time Δt before the collision, ball B has a y-displacement Δy from the point of collision. At this time, ball A has a y-displacement smaller than this Δy. The distance Δr [used in Eq. (51)] is measured from the point of collision to that point on the trajectory of A which has a y-displacement Δy. (b) At a time Δt after the collision, ball B again has a vertical displacement Δy, and the y-components of the velocities of both balls are reversed.

collision, and at a short time Δt after the collision. Note that at these times the y-displacement of ball A is less than that of B; as we will see, the reason for this is that if the y-momenta are to be equal, then ball A must have a lower y-velocity than B. Figure 2.25 shows the same collision in the reference frame of a spaceship moving to the right with a velocity equal to the x-velocity of A; in the spaceship frame B has a large x-momentum. It is assumed

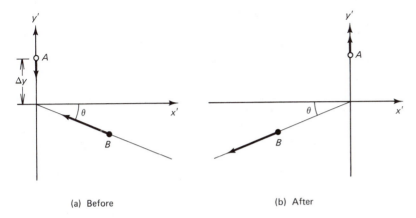

(a) Before   (b) After

Fig. 2.25 Collision seen in the reference frame of the spaceship: (a) before the collision; (b) after the collision. The times before and after the collision have been chosen so that the y-displacement of ball A shown here equals the y-displacement of ball B shown in Fig. 2.24.

that the y-velocities are all quite small (nonrelativistic); more precisely, in all the following arguments we consider the limiting case of a y-velocity that tends to zero.

As seen in the reference frame of the laboratory, the momentum change of ball B is

$$2m \frac{\Delta y}{\Delta t} \tag{50}$$

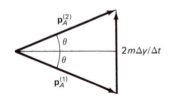

Fig. 2.26 Momentum conservation as seen in the reference frame of the laboratory. The momentum of ball A is $\mathbf{p}_A^{(1)}$ before the collision and $\mathbf{p}_A^{(2)}$ after the collision.

where Δy is the displacement that ball B suffers in the short (laboratory) time Δt. In this equation it is legitimate to use the Newtonian expression for momentum since $\Delta y/\Delta t \ll 1$ by hypothesis. By momentum conservation, the y-momentum change of ball A must be equal in magnitude to (50). From the momentum triangle in Fig. 2.26, we find that the magnitude of the momentum of A must be

$$p_A = \frac{m}{\sin \theta} \frac{\Delta y}{\Delta t} \tag{51}$$

From Fig. 2.24a we then see that this can be rewritten as

$$p_A = m \frac{\Delta r}{\Delta t}$$

where Δr is the distance from the point of collision to that point on the trajectory of A which is at a height Δy above the point of collision. Note that since at time Δt before the collision, ball A is *not* at height Δy but, as shown in Figure 2.24, at a somewhat lower height, the quantity $\Delta r/\Delta t$ cannot be interpreted as the speed of A.

In order to express $\Delta r/\Delta t$ in terms of quantities that refer to A at one given point in space and time along its trajectory, we observe that in the limiting case of small y-velocity, the laboratory time Δt is the same thing as the proper time $\Delta \tau_B$ of a clock carried by B. Furthermore, the proper time $\Delta \tau_B$ that B needs to cover a vertical distance Δy in the *laboratory* is the same as the proper time $\Delta \tau_A$, measured by a clock carried by A, that A needs to cover an equal distance Δy in the *spaceship*. This follows from the symmetry between Figs. 2.24 and 2.25: whatever B does in the laboratory frame, A does in the spaceship frame. Hence

$$\Delta t = \Delta \tau_B = \Delta \tau_A \tag{52}$$

and

$$p_A = m \frac{\Delta r}{\Delta \tau_A} \tag{53}$$

This last equation says that the momentum of A is equal to the mass multiplied by the displacement of A in a given amount of *proper* time. Dropping the A subscripts, and taking into account the vectorial nature of both momentum and displacement, we then have

$$\mathbf{p} = m\frac{d\mathbf{x}}{d\tau} \tag{54}$$

This is the relativistic formula for momentum. The mass that appears in Eq. (54) is the rest mass, i.e., the mass that the particle has when measured in a reference frame in which it is at rest. In this book we *never* use any other kind of "mass."

To compare Eq. (54) with the Newtonian expression, we rewrite it as

$$\mathbf{p} = \frac{m\mathbf{v}}{\sqrt{1 - v^2/c^2}} \tag{55}$$

If the velocity of the particle is small compared to the velocity of light, then $\sqrt{1 - v^2/c^2} \simeq 1$ and Eq. (55) becomes approximately

$$\mathbf{p} \simeq m\mathbf{v} \tag{56}$$

This shows that for low velocities the relativistic and nonrelativistic formulas coincide. At high velocities, the formulas differ drastically—the relativistic momentum becomes infinite as the velocity of the particle approaches the velocity of light. Figure 2.27 is a plot of the magnitude of p as a function of v.

The relativistic equation of motion has the same form as the Newtonian equation of motion when expressed in terms of momentum:

$$\frac{d\mathbf{p}}{dt} = \mathbf{F} \tag{57}$$

However, the momentum **p** appearing in this equation is the relativistic momentum of Eq. (55). Thus,

$$\frac{d}{dt}\frac{m\mathbf{v}}{\sqrt{1 - v^2/c^2}} = \mathbf{F} \tag{58}$$

For instance, the equation of motion for a high-speed particle of charge e moving under the influence of electric and magnetic fields is

$$\frac{d}{dt}\frac{m\mathbf{v}}{\sqrt{1 - v^2/c^2}} = e\mathbf{E} + e\mathbf{v} \times \mathbf{B} \tag{59}$$

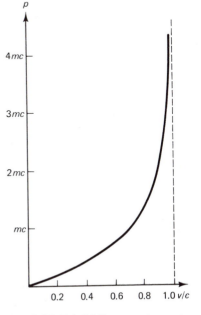

Fig. 2.27 Relativistic momentum vs v.

Example 4   What is the radius of the circular orbit of a charged particle of speed v in a constant magnetic field? What is the frequency of the orbital motion?

Solution   The equation of motion for the particle is

$$\frac{d}{dt} \frac{m\mathbf{v}}{\sqrt{1 - v^2/c^2}} = e\mathbf{v} \times \mathbf{B} \tag{60}$$

Since the particle moves with constant speed, the factor $1/\sqrt{1 - v^2/c^2}$ is a constant. Thus, Eq. (60) reduces to

$$\frac{m}{\sqrt{1 - v^2/c^2}} \frac{d\mathbf{v}}{dt} = e\mathbf{v} \times \mathbf{B} \tag{61}$$

Except for the extra, constant factor $1/\sqrt{1 - v^2/c^2}$, this is the same equation as for a nonrelativistic particle [see Eq. (1.14)]. Hence, except for the extra factor $1/\sqrt{1 - v^2/c^2}$, the formula for the radius must be the same as Eq. (1.16), i.e.,

$$r = \frac{1}{\sqrt{1 - v^2/c^2}} \frac{mv}{eB} \tag{62}$$

This is equivalent to

$$r = \frac{p}{eB} \tag{63}$$

Thus, when expressed in terms of momentum, the equation for the radius has the same form as in the Newtonian case, but the momentum is now the relativistic momentum.

The period of the orbital motion is

$$T = \frac{2\pi r}{v} = \frac{2\pi m}{\sqrt{1 - v^2/c^2}\, eB}$$

Hence the frequency is

$$\nu = \frac{1}{T} = \frac{eB}{2\pi m} \sqrt{1 - v^2/c^2} \quad \blacksquare \tag{64}$$

Example 5   What is the acceleration of a charged particle in a uniform electric field? Assume that the particle moves along a straight line parallel to the electric field

Solution   The equation of motion is

$$\frac{d}{dt} \frac{m\mathbf{v}}{\sqrt{1 - v^2/c^2}} = e\mathbf{E}$$

If the electric field and the motion are in the x-direction, this reduces to

$$\frac{d}{dt} \frac{mv_x}{\sqrt{1 - v_x^2/c^2}} = eE \tag{65}$$

Differentiating, we obtain

$$\frac{m}{(1 - v_x^2/c^2)^{1/2}} \frac{dv_x}{dt} + \frac{mv_x^2/c^2}{(1 - v_x^2/c^2)^{3/2}} \frac{dv_x}{dt} = eE \tag{66}$$

and combining the two terms by means of a common denominator, we obtain

$$\frac{m}{(1 - v_x^2/c^2)^{3/2}} \frac{dv_x}{dt} = eE \tag{67}$$

Hence the acceleration is

$$\frac{dv_x}{dt} = \frac{eE}{m} (1 - v_x^2/c^2)^{3/2} \tag{68}$$

This shows that the acceleration depends on the (instantaneous) speed. It is a simple mathematical exercise to integrate Eq. (68) and to show that, for a particle starting from rest at $x = 0$ at $t = 0$, the speed and the position are given by the following formulas:

$$v_x = \frac{eE}{m} \frac{t}{\sqrt{1 + [eEt/(mc)]^2}} \tag{69}$$

$$x = \frac{mc^2}{eE} \left[\sqrt{1 + \left(\frac{eEt}{mc}\right)^2} - 1 \right] \quad \blacksquare \tag{70}$$

We also need a new, relativistic formula for kinetic energy. To derive such a formula, we start with the usual expression for the change of kinetic energy produced by a force **F** acting over a displacement $d\mathbf{x}$,

$$dK = \mathbf{F} \cdot d\mathbf{x} = \frac{d\mathbf{p}}{dt} \cdot d\mathbf{x} = d\mathbf{p} \cdot \mathbf{v} \tag{71}$$

With the relativistic expression for momentum, this becomes

$$dK = d\mathbf{p} \cdot \mathbf{v} = \left[\frac{m \, d\mathbf{v}}{(1 - v^2/c^2)^{1/2}} + \frac{\frac{1}{2}m\mathbf{v}d(v^2)/c^2}{(1 - v^2/c^2)^{3/2}} \right] \cdot \mathbf{v} \tag{72}$$

$$= \left[\frac{\frac{1}{2}md(v^2)}{(1 - v^2/c^2)^{1/2}} + \frac{\frac{1}{2}mv^2 d(v^2)/c^2}{(1 - v^2/c^2)^{3/2}} \right] \tag{73}$$

$$= \frac{\frac{1}{2}md(v^2)}{(1 - v^2/c^2)^{3/2}} \tag{74}$$

If we integrate this from a lower limit $K = 0$ and $v^2 = 0$ to an upper limit $K \neq 0$ and $v^2 \neq 0$, we obtain, with the change of variable $\xi = v^2/c^2$,

$$K = \int_0^{v^2/c^2} \frac{\frac{1}{2}mc^2 \, d\xi}{(1 - \xi)^{3/2}} = \left[\frac{mc^2}{(1 - \xi)^{1/2}} \right]_0^{v^2/c^2} \tag{75}$$

or

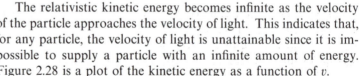

$$K = mc^2 \left(\frac{1}{\sqrt{1 - v^2/c^2}} - 1 \right) \tag{76}$$

This is the relativistic formula for kinetic energy. To compare this with the Newtonian formula, we again use the approximation

$$1/\sqrt{1 - v^2/c^2} \simeq 1 + \tfrac{1}{2}v^2/c^2$$

valid if $v^2/c^2 \ll 1$. Then Eq. (76) reduces to

$$K \simeq mc^2 \left(1 + \frac{v^2}{2c^2} \right) - mc^2 = \frac{1}{2} mv^2 \tag{77}$$

Again, for low velocities the relativistic and nonrelativistic formulas coincide.

The relativistic kinetic energy becomes infinite as the velocity of the particle approaches the velocity of light. This indicates that, for any particle, the velocity of light is unattainable since it is impossible to supply a particle with an infinite amount of energy. Figure 2.28 is a plot of the kinetic energy as a function of v.

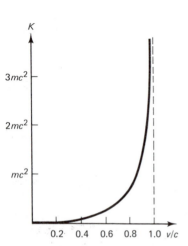

Fig. 2.28 Relativistic kinetic energy vs. v.

Example 6   The maximum speed attained by artificially accelerated particles is $0.99999999967c$, for electrons at the Stanford Linear Accelerator. Find the kinetic energy of an electron moving with this speed.

Solution   For v/c near 1, we can make the approximation

$$\sqrt{1 - v^2/c^2} = \sqrt{1 + v/c}\sqrt{1 - v/c} \simeq \sqrt{2}\sqrt{1 - v/c} \tag{78}$$

In our case, the value of $1 - v/c$ is 3.3×10^{-10}. Hence

$$K = mc^2 \left(\frac{1}{\sqrt{1 - v^2/c^2}} - 1 \right)$$

$$= 9.1 \times 10^{-31} \, \text{kg} \times (3.0 \times 10^8 \, \text{m/s})^2 \left(\frac{1}{\sqrt{2}\sqrt{3.3 \times 10^{-10}}} - 1 \right)$$

$$= 3.2 \times 10^{-9} \, \text{J} \quad \blacksquare$$

According to Einstein's energy–mass equivalence, a particle of mass m at rest has an energy mc^2. We have already given a simple derivation of this equivalence in Section 1.7. The total energy of a free particle in motion is the sum of its rest-mass energy mc^2 and its kinetic energy K,

$$E = mc^2 + K = mc^2 + \frac{mc^2}{\sqrt{1 - v^2/c^2}} - mc^2 \qquad (79)$$

This leads to a simple formula for the total relativistic energy of a free particle:

$$E = \frac{mc^2}{\sqrt{1 - v^2/c^2}} \qquad (80)$$

It is easy to verify that this relativistic energy can be expressed as follows in terms of the relativistic momentum:

$$E = \sqrt{c^2 p^2 + m^2 c^4} \qquad (81)$$

For a particle moving at a speed close to the speed of light, the first term within the square root sign in Eq. (81) is much larger than the second term. Hence, for such an ultrarelativistic particle we can neglect $m^2 c^4$ and obtain

$$E \simeq cp \qquad (82)$$

Thus, the momentum and the energy of an ultrarelativistic particle are directly proportional. For particles of zero mass, such as photons and neutrinos, Eq. (82) is an *exact* relation between the energy and the momentum. Note that Eq. (82) coincides with Eq. (1.143), which expresses the relation between the energy and the momentum of a light wave, as implied by Maxwell's equations.

Chapter 2: SUMMARY

Einstein's Principle of Relativity: All the laws of physics are the same in all inertial reference frames.

Lorentz transformations:

$$t' = \frac{t - Vx/c^2}{\sqrt{1 - V^2/c^2}}$$

$$x' = \frac{x - Vt}{\sqrt{1 - V^2/c^2}}$$

$$y' = y$$

$$z' = z$$

Time dilation:

$$\Delta t = \frac{\Delta t'}{\sqrt{1 - V^2/c^2}} \qquad \text{(clock at rest in spaceship)}$$

Length contraction:

$$\Delta x = \sqrt{1 - V^2/c^2}\,\Delta x' \qquad \text{(body at rest in spaceship)}$$

Combination of velocities:

$$v'_x = \frac{v_x - V}{1 - v_x V/c^2}$$

$$v'_y = \frac{v_y \sqrt{1 - V^2/c^2}}{1 - v_x V/c^2}$$

$$v'_z = \frac{v_z \sqrt{1 - V^2/c^2}}{1 - v_x V/c^2}$$

Momentum:

$$\mathbf{p} = \frac{m\mathbf{v}}{\sqrt{1 - v^2/c^2}}$$

Kinetic energy:

$$K = mc^2 \left(\frac{1}{\sqrt{1 - v^2/c^2}} - 1 \right)$$

Total energy:

$$E = \frac{mc^2}{\sqrt{1 - v^2/c^2}} = \sqrt{c^2 p^2 + m^2 c^4}$$

Chapter 2:   PROBLEMS

1. The interferometer of Michelson and Morley had arms of an effective length of 11 m (each arm was folded—the light traveled back and forth several times between pairs of mirrors) and it could detect a phase shift as small as 18°, or $\pi/10$ radians. The wavelength of the light was 5900 Å. Given these data, what limit does the null result of this experiment set on the ether wind?

2. G. F. Fitzgerald and H. A. Lorentz attempted to explain the null result of the Michelson–Morley experiment by postulating that the arm of the interferometer parallel to the ether wind suffers a contraction.
   (a) By what factor must the length L of this arm contract if the phase difference between the two perpendicular light beams is to be exactly zero?
   (b) The Lorentz–Fitzgerald contraction hypothesis was proved false by a modified version of the Michelson–Morley experiment with an interferometer with

arms of unequal length. Explain why the contraction factor proposed for the equal-arm experiment is of no help for the unequal-arm experiment.

3. Suppose that the velocity of the spaceship frame is $V = -0.6c$ (the minus sign means that the spaceship moves in the negative x-direction relative to the laboratory frame).
   (a) Draw a diagram analogous to Fig. 2.14. Draw the coordinate grids for the x-ct and x'-ct' coordinates and make sure that you have the correct angles between axes and the correct scales along the axes.
   (b) Plot the spacetime point $x = -2$, $ct = 3$ in your diagram. Read the values of the x'- and ct'-coordinates of this event directly off your diagram. Check that your graphical determination of the values of x' and ct' agrees with the computation of these values from Eqs. (22) and (23).

4. Show that the value of the expression $(c\,\Delta t)^2 - (\Delta x)^2$ is unchanged by Lorentz transformations, i.e., show that

$$(c\,\Delta t)^2 - (\Delta x)^2 = (c\,\Delta t')^2 - (\Delta x')^2$$

The expression $(c\,\Delta t)^2 - (\Delta x)^2$ is called the **spacetime interval;** it plays the role of a "distance" in spacetime.

5. Show that the Lorentz transformations (22) and (23) can be put in the form

$$x' = x\cosh\theta - ct\sinh\theta$$

$$ct' = -x\sinh\theta + ct\cosh\theta$$

where θ is a parameter defined by $V/c = \tanh\theta$.

6. A meter stick is at rest in the laboratory frame. It lies in the x-y plane, making an angle of 30° with the x-axis. What angle does this meter stick make with the x'-axis of a reference frame moving at $V = 0.8c$ in the x-direction?

7. At $9^h\,0^m\,0^s$ an aircraft touches down at the La Guardia airport in New York City. At $9^h\,0^m\,0.01^s$ another aircraft touches down at the San Francisco airport. The (straight) distance between these airports is 3.8×10^3 km.
   (a) Show that any signal that the pilot of the first aircraft sends after the instant of touchdown will reach the pilot of the second aircraft after his own touchdown.
   (b) Show that in the reference frame of a suitably chosen spaceship the San Francisco touchdown is earlier than the New York touchdown. What must be the minimum speed of such a spaceship relative to the Earth? For the purposes of this problem, pretend that the Earth is an inertial reference frame.

8. Prove that the existence of a rigid body, i.e., a body that suffers no deformation when pushed by a force, is inconsistent with special relativity.

9. Consider a clock belonging to the reference frame of the laboratory. Draw a spacetime diagram analogous to Fig. 2.20 to illustrate the time dilation suffered by this clock when observed from the spaceship frame.

10. The record for the fastest speed at which anybody has traveled relative to the Earth is held by the crew of the *Apollo X* module, who attained 24,791 mi/h on their return trip from the Moon. At this speed, what was the percent difference between the rates of their clocks and clocks of the Earth?

11. The decay of unstable particles, such as muons, obeys a probabilistic law: the probability that the particle does *not* decay within a given time t is $e^{-t/\tau}$, where τ is the *mean lifetime*. For muons, this mean lifetime is 2.20×10^{-6} s. Suppose that a cosmic ray creates a muon at an altitude of 40 km in the atmosphere. The muon has a downward velocity of magnitude $V = 0.9940c$; for the purposes of this problem pretend that the velocity remains constant
    (a) Without time dilation, what is the probability that this muon will reach sea level?
    (b) With time dilation, what is the probability?

12. The gestation period of elephants is 21 months. Suppose we place a freshly impregnated elephant cow in a spaceship and launch it into interstellar space at a constant speed of $V = 0.9c$.
    (a) At what distance from the Earth will the elephant give birth?
    (b) If we monitor the condition of the elephant by radio, at what time after launch will we hear the squeals of the newborn calf?

13. We want to send a spaceship from the Earth to the Large Magellanic Cloud, at a distance of 1.6×10^5 light-years from the Earth. By taking advantage of the time dilation, the cosmonauts on the spaceship plan to reach their destination in 5 years of their own time. At what speed, relative to the Earth, must the spaceship travel? When the spaceship reaches the Large Magellanic Cloud, how many years have elapsed on the Earth?

14. A spaceship is to be sent to the star α-Centauri, which is at a distance of 4.4 light-years from us. With its rocket motors, the spaceship is to accelerate at a constant rate of $1.0 \, \text{m/s}^2$ until it reaches the midpoint of the trip; the spaceship is then to decelerate at $1.0 \, \text{m/s}^2$ until it reaches its destination. The return trip is to be performed in the same manner.
    (a) Calculate the elapsed time for the roundtrip according to the clocks on the Earth. Ignore the time spent at the destination.
    (b) Calculate this elapsed time according to the clocks on the spaceship.

15. An accurate clock is kept in a laboratory located on the equator of the Earth, and two identical clocks are placed aboard two aircraft flying at 250 m/s (ground-speed) in the eastward and westward directions around the equator. Taking into account the rotation of the Earth, calculate the difference that accumulates between the laboratory clock and each of the two flying clocks upon one trip around the Earth. (*Hint:* Consider the three clocks from the point of view of an inertial reference frame centered on the Earth, but not rotating with it.)

16. Consider a meter stick at rest in the laboratory frame. Draw a spacetime diagram analogous to Fig. 2.22 to illustrate the length contraction of this meter stick when observed in the spaceship frame.

17. What is the percent length contraction of an aircraft traveling at Mach 2 (twice the speed of sound)?

18. The diameter of a proton is 2.0×10^{-15} m in its own reference frame. What is the longitudinal diameter of a proton when its speed is within 100 m/s of the speed of light?

19. Derive Eq. (47) for v'_y and v'_z.

20. Solve Eqs. (46) and (47) for v_y and v_z in terms of v'_x, v'_y, and v'_z, and thereby find the inverses of the combination laws for the y- and z-components of the velocity.

21. In the reference frame of the laboratory, a particle moves in the x-y plane with a velocity of $0.7c$ at an angle of $40°$ with the x-axis. Find the velocity (magnitude and direction) of this particle in the reference frame of a spaceship moving at $V = 0.9c$ in the positive x-direction.

22. Consider a distant star located directly above the orbit of the Earth.
    (a) Use the Lorentz transformation equations for the velocity components to deduce the direction of the velocity of the starlight incident on the (moving) Earth. What is the shift in the angular position of the star in six months?
    (b) Repeat the calculations using the Galilean transformation equations instead of the Lorentz transformation equations, and comment on the differences in your results.

23. Consider a light signal propagating in some arbitrary direction, with

$$v_x \neq 0, \quad v_y \neq 0, \quad v_z \neq 0, \quad \text{and} \quad v_x^2 + v_y^2 + v_z^2 = c^2$$

Use the Lorentz transformation equations for the components of the velocity to show that

$$v'^2_x + v'^2_y + v'^2_z = c^2$$

24. From the Lorentz transformation equations for the components of the velocity, derive the transformation equations for the components of the acceleration.

25. Calculate the speed of an electron for each of the following kinetic energies: 0.1 MeV, 0.5 MeV, 1.0 MeV, and 2.0 MeV.

26. The electron in the hydrogen atom has a speed of 2.2×10^6 m/s. What is the percent deviation between the Newtonian and the relativistic values of the kinetic energy of this electron?

27. The Sun moves around the center of the Galaxy at a speed of about 200 km/s. Calculate the percent difference between the Newtonian and the relativistic values for the momentum of the Sun. Calculate the percent difference between the Newtonian and the relativistic values for the kinetic energy of the Sun.

28. At the Fermilab Tevatron accelerator, protons are given a kinetic energy of 1000 GeV. By how much (in m/s) does the speed of such a proton differ from the speed of light? What is the momentum of such a proton?

29. At the Brookhaven AGS accelerator, protons of kinetic energy 33 GeV are made to orbit around a circle of radius 128 m. What magnetic field is required to hold the protons in this orbit?

30. A K_S^0 particle at rest decays into a muon and an antimuon. What will be the speed of each of the latter? The mass of the K_S^0 is 498 MeV/c^2 and the mass of the

muon is $106 \,\text{MeV}/c^2$. (*Hint:* use Einstein's equation $\Delta E = \Delta mc^2$ to find the energy released in this decay.)

31. A K_S^0 particle with a speed $0.80c$ relative to the laboratory decays into two pions, π^+ and π^-.

(a) Calculate the speed of each pion in the rest frame of the K_S^0. The mass of the K_S^0 is $498 \,\text{MeV}/c^2$ and the masses of the pions are $140 \,\text{MeV}/c^2$. (*Hint:* Use Einstein's equation $\Delta E = \Delta mc^2$ to find the energy released in the decay.)

(b) Assume that the pions move along a line parallel to the original direction of motion of the K_S^0. Calculate the speeds of the pions in the laboratory.

32. The pion decays spontaneously into a muon and a neutrino,

$$\pi^- \rightarrow \mu + \nu$$

(a) Assuming that the pion is originally at rest, calculate the energies and the momenta of the muon and of the neutrino. The mass of the pion is $140 \,\text{MeV}/c^2$ and the mass of the muon is $106 \,\text{MeV}/c^2$. (*Hint:* Use Einstein's equation $\Delta E = \Delta mc^2$ to find the energy released in the decay. The neutrino has zero mass, and is always ultrarelativistic; use Eq. (82).)

33. A proton of kinetic energy $3 \,\text{GeV}$ is about to collide with a proton at rest.

(a) What is the speed of the moving proton in the laboratory reference frame?

(b) What is the speed of a reference frame in which both of the colliding protons have the same speed (and are moving in opposite directions)?

(c) What is the energy of each proton in the latter reference frame?

34. In a given reference frame, a particle of mass m has a momentum with components p_x, p_y, p_z and an energy E. Show that the Lorentz transformation equations for the momentum and the energy are

$$p_x' = \frac{p_x - VE/c^2}{\sqrt{1 - V^2/c^2}}$$

$$p_y' = p_y$$

$$p_z' = p_z$$

$$E' = \frac{E - Vp_x}{\sqrt{1 - V^2/c^2}}$$

35. Integrate Eq. (68) and derive Eqs. (69) and (70).

36. Show that for $eEt/(mc) \ll 1$, the relativistic equations (69) and (70) agree approximately with the corresponding Newtonian equations for a particle moving with constant acceleration.

37. The Stanford Linear Accelerator accelerates electrons in a beam tube of length $3 \,\text{km}$. Pretend that the electric field along this tube is uniform, of magnitude $7 \times 10^6 \,\text{V/m}$.

(a) Calculate the time an accelerated electron takes to travel the full length of the tube.

(b) Calculate the distance at which an electron attains 90% of the speed of light.

38. Provided the acceleration and the force are parallel, we can define a relativistic effective mass as the constant of proportionality between force and acceleration.
(a) What is the effective mass according to Eq. (61)? This is called the **transverse mass,** since the instantaneous velocity is transverse to the acceleration.
(b) What is the effective mass according to Eq. (67)? This is called the **longitudinal mass,** since the instantaneous velocity is along the acceleration.
(c) From the relativistic equation of motion

$$\frac{d(m\mathbf{v}/\sqrt{1-v^2/c^2})}{dt} = \mathbf{F}$$

show that if the velocity is not perpendicular and not parallel to the acceleration, then the acceleration is *not* parallel to the force, and therefore the concept of effective mass lacks general validity.

Quanta of Energy

According to the view of classical physics, light is a wave consisting of electric and magnetic fields with a smooth distribution of energy. The interference and diffraction phenomena described in Chapter 1 give direct experimental evidence for the wave properties of light. However, in the early years of the twentieth century, theoretical and experimental investigations established that light sometimes has particle properties. According to this new view, light acts like a stream of particlelike energy packets. These energy packets are called **quanta** of light, or **photons.** Further investigations soon established that energy quantization is a pervasive feature of the microscopic world—the energy of atoms and the energy of electrons and other subatomic particles is packaged in energy quanta.

The discovery of energy quantization, described in this chapter, gave the first indication of the inadequacy of classical physics. At first, physicists sought to limit the damage—instead of discarding the classical laws, they attempted to supplement these laws by an extra set of quantum conditions restricting the energy to certain preferred, discrete values. But it soon became evident that classical physics was damaged beyond repair, and that a completely new set of laws of quantum physics was required. In this chapter and in the next, we will deal with the early attempts at a quantum modification of classical physics. In later chapters we will raze the ruins of classical physics and erect the new structure of quantum physics on fresh foundations.

3.1 Blackbody Radiation

The first indication of the inadequacy of classical physics emerged around 1900 from the study of thermal radiation, or "radiant heat" emitted by hot bodies. This is electromagnetic radiation containing a wide variety of wavelengths. The intensity and the predominant wavelength of the radiation varies with the temperature of the body. For instance, at a temperature of 5800 K (the surface temperature of the Sun), much of the thermal radiation takes the form of visible light—the body therefore looks bright white. At a temperature of 1200 K, the intensity is lower and most of the thermal radiation lies in the near infrared, with only a small amount of visible light—the body exhibits a deep red glow. At room temperature, the intensity is very low and most of the radiation lies in the far infrared—the body gives off no light visible to the human eye.

The spectrum of thermal radiation is continuous, i.e., the energy of the radiation is smoothly distributed over all wavelengths or, equivalently, over all frequencies. For instance, Fig. 3.1 shows the spectrum of thermal radiation emitted by the Sun as a function of frequency. The quantity S_ν plotted in this figure is called the **spectral emittance.** This is the energy flux (or power per unit area) emitted by the glowing surface per unit frequency interval; thus $S_\nu \, d\nu$ is the energy flux emitted in a small frequency interval from ν to $\nu + d\nu$. Note that the spectral emittance is small at very high and at very low frequencies, and that it has a broad peak with a maximum near the visible region, at $\nu = 3.4 \times 10^{14}$ Hz.

The thermal radiation emerging from the surface of a glowing body is generated within the volume of the body by the random thermal motions of atoms and electrons. Before the radiation reaches the surface and escapes, it is absorbed and reemitted many times, and it attains thermal equilibrium with the atoms and electrons. This equilibration process shapes the continuous spectrum of the radiation, completely washing out all of the original spectral features of the radiation. Hence the spectrum of the radiation within the volume of the body depends only on the temperature, not on the kind of atoms in the body.

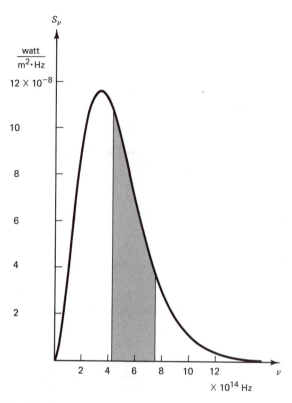

Fig. 3.1 Spectral emittance of the Sun (5800 K). In this plot the discrete dark spectral lines (Fraunhofer lines), which result from the blocking out of some of the thermal radiation by gas in the solar atmosphere, have been ignored. The shaded band indicates the visible region.

The flux of thermal radiation emerging from the surface of a glowing body depends to some extent on the characteristics of the surface. The surface usually permits the escape of only a fraction of the flux reaching it from inside the body. Likewise, if the body is immersed in a bath of thermal radiation of the same temperature as the body, then the surface permits the ingress of only an equal fraction of the flux reaching it from outside, reflecting the rest. This equality of the emissive and absorptive characteristics of the surface can readily be established by appealing to the Second Law of Thermodynamics, according to which heat cannot flow spontaneously from a colder system to a hotter system. If the body were to emit more radiation than it absorbs from the bath of radiation, then its temperature would decrease, and that of the bath of radiation would increase. Heat would therefore be flowing from a colder system to a hotter system, in contradiction to the Second Law.

The equality of the emissive and absorptive characteristics of the surface of a body implies the following general rule: a good absorber is a good emitter, and a poor absorber is a poor emitter. A body with a perfectly absorbing (and emitting) surface is called a **blackbody;** such a body would look black under outside illumination. When a blackbody is hot, its surface emits more thermal radiation than that of any other hot body at the same temperature. In practice the characteristics of an ideal blackbody are most easily achieved by a trick: take a body with a cavity, such as a hollow cube, and drill a small hole in one side of the cube (see Fig. 3.2). The hole then acts like a blackbody—any radiation incident on the hole from outside will be completely absorbed. Because of this equivalence between a blackbody and a hole in a cavity, blackbody radiation is often called cavity radiation.

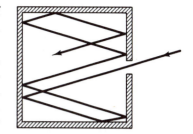

Fig. 3.2 A cavity with a small hole. Any radiation entering the hole will remain trapped; it will suffer repeated reflections, and it will ultimately be absorbed by the walls.

The blackbody plays a preferential role in the study of thermal radiation because its spectral emittance does not depend on the material of which it is made or on any other characteristics of the body—the spectral emittance depends *exclusively on the temperature* of the body. We can prove this by appealing, again, to the Second Law of Thermodynamics. Consider two cavities at equal temperatures with holes of equal size (see Fig. 3.3). The cavity on the left radiates into the cavity on the right, and vice versa. If the flux emitted by the cavity on the left were larger than that of the cavity on the right, the radiative heat transfer would produce an increase of temperature on the right and a decrease on the left, in contradiction to the Second Law. This argument leads us to the conclusion that the fluxes emitted by both cavities are the same. Furthermore, by a refinement of this argument we can demonstrate that the fluxes in any given small frequency interval $d\nu$ are also the same. We need only make a slight alteration in the arrangement shown in Fig. 3.3 by inserting a filter for light between the two cavities, a filter that only permits the passage of radiation of frequencies in an interval from ν to $\nu + d\nu$. Our thermodynamic argument then leads to the conclusion that the fluxes emitted by the two cavities in this chosen wavelength interval must be the same. Thus for a blackbody, the spectral emittance S_ν must be a universal function of the wavelength λ and of the temperature T, and of nothing else.

Although thermodynamics does not tell us the detailed form of the function S_ν, it does impose a few extra restrictions on this function: it tells us something about the total energy flux emitted by the surface of the blackbody, and it tells us something about the location of the maximum in the spectral emittance. The total energy flux (or power per unit area) is the integral of the spectral

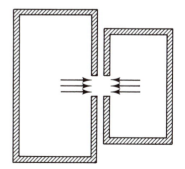

Fig. 3.3 Two cavities with holes of equal size exchanging radiation.

emittance over all frequencies,

$$S = \int_0^\infty S_v \, dv \tag{1}$$

Graphically, this is the area under the curve in Fig. 3.1. By an elegant thermodynamic argument, which we will not attempt to reproduce here,* it can be established that this total energy flux is proportional to the fourth power of the temperature,

$$S = \sigma T^4 \tag{2}$$

This statement is called the **Stefan–Boltzmann law.**[†] The constant of proportionality σ is called the Stefan–Boltzmann constant; its value is

$$\sigma = 5.6703 \times 10^{-8} \, \text{watt}/(\text{m}^2 \cdot \text{K}^4) \tag{3}$$

By a similar thermodynamic argument it can be established that the frequency v_{max} at which the spectral emittance attains its maximum is directly proportional to the temperature,

$$v_{\text{max}} = \text{const.} \times T \tag{4}$$

This is called **Wien's displacement law,**[‡] so named because it shows that as the temperature increases, the maximum in the emittance shifts to higher frequencies. The numerical value of the constant in Eq. (4) is

$$\text{const.} = 5.880 \times 10^{10} \, \text{Hz/K} \tag{5}$$

Example 1   At the Earth's surface, the energy flux in sunlight is $1.0 \times 10^3 \, \text{watt/m}^2$. If a black sheet of paper faces the Sun, what is the equilibrium temperature of the paper? Assume that the bottom of the

---

* See, e.g., M. Born, *Atomic Physics*, pp. 251, 252.

[†] **Josef Stefan,** 1835–1893, Austrian experimental physicist, professor at Vienna. He discovered the Stefan–Boltzmann law empirically. Boltzmann later provided its theoretical explanation.

[‡] **Wilhelm Wien,** 1864–1928, German physicist, professor at Würzburg and at Munich. Both Wien's displacement law and the Stefan–Boltzmann law are corollaries of a general law

$$S_v = v^3 f\left(\frac{v}{T}\right)$$

where f is some function of the ratio v/T, about which nothing further can be said on the basis of thermodynamics (see Born, ibid.). For his discovery of this general law, Wien was awarded the Nobel Prize in 1911.

paper is insulated so that the only heat loss is by blackbody radiation from the top surface.

Solution   According to the Stefan–Boltzmann law, the power radiated by the paper per unit area is σT^4. At equilibrium, this must match the power of sunlight incident per unit area, i.e.,

$$\sigma T^4 = 1.0 \times 10^3 \, \text{watt/m}^2$$

from which

$$T = \left[\frac{1.0 \times 10^3 \, (\text{watt/m}^2)}{\sigma} \right]^{1/4}$$

$$= 364 \, \text{K} = 91^\circ \text{C}$$

(In practice the temperature will be somewhat lower because the paper reflects some of the sunlight and suffers heat losses by conduction and convection.)   ■

All these general facts about blackbody radiation had been discovered during the nineteenth century. In the last years of that century, physicists engaged in an intensive experimental and theoretical effort to uncover the precise form of the spectral emittance function S_ν.

One theoretical attempt was made by Lord Rayleigh. Instead of dealing with the spectral emittance S_ν, he chose to deal with the spectral energy density u_ν; these two quantities are directly proportional,

$$S_\nu = \frac{c}{4} u_\nu \tag{6}$$

We can readily understand this proportionality if we recall that for a plane electromagnetic wave the energy flux (or Poynting vector) **S** and the energy density u are related by $S = cu$; for thermal radiation, consisting of many plane waves traveling in random directions, it can be shown that the energy flux in any one direction is reduced by an extra factor of $\frac{1}{4}$.* Rayleigh began by noting that the radiation in a cavity is made up of a large number of standing waves; Fig. 3.4 shows some of these standing waves. Each of these standing waves is a mode of vibration of the cavity, and can be regarded as mathematically equivalent to a simple harmonic oscillator. Rayleigh then appealed to the equipartition theorem according to which, at thermal equilibrium, each simple harmonic oscillator has an average thermal energy of kT (see Section 1.4).

---

* See Problem 7.

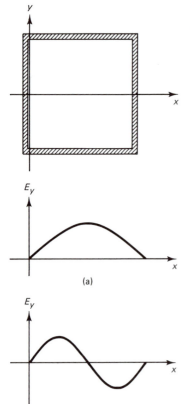

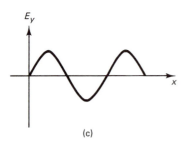

Fig. 3.4 Some of the possible standing electromagnetic waves in a closed cavity. For the sake of simplicity this figure only shows waves with a horizontal direction of propagation. The plots (a), (b), and (c) give possible electric fields as a function of *x* at one instant of time.

Thus, each of the standing waves of Fig. 3.4 ought to have an energy kT, and from this he immediately calculated the spectral emissivity. Although his calculation gave reasonable results at the long-wavelength end of the blackbody spectrum, it gave disastrous results at the short-wavelength end: the number of possible standing-wave modes of very short wavelength is infinitely large and, if each of these modes had an energy kT, the total energy in the cavity would be infinite! This disastrous failure of classical theory came to be called the **ultraviolet catastrophe.**

3.2 Planck's Quantization of Energy

The correct mathematical law for the spectral emittance of a blackbody was finally obtained by Max Planck* in 1900. By some inspired guesswork, based on thermodynamics, Planck hit upon the following formula:

$$S_v = \frac{2\pi h}{c^2} \frac{v^3}{e^{hv/kT} - 1} \tag{7}$$

where h is a new universal constant, called **Planck's constant.** The best modern value of this constant is

$$h = 6.62618 \times 10^{-34}\,\text{J}\cdot\text{s} \tag{8}$$

In his calculations, Planck treated this constant as an adjustable parameter, chosen to fit the data on blackbody radiation; but, as we will see, this constant plays a vital role throughout the microscopic world, and its modern value is obtained from experiments in atomic physics not directly related to blackbody radiation. Experimental investigations soon established that the spectral emittance calculated from Eq. (7) agreed very precisely with the measurements. Figure 3.5 shows the spectral emittance according to Eq. (7) for several values of the temperature.

---

* **Max Karl Ernst Ludwig Planck**, 1858–1947, German theoretical physicist, professor at Berlin, and president of the Kaiser Wilhelm Institute (now the Max Planck Institute). Planck made significant contributions to thermodynamics before he became involved with the study of blackbody radiation. After proposing the quantization of energy as an explanation of his blackbody radiation formula, he tried unsuccessfully to find an alternative explanation without quantization. He was awarded the Nobel Prize in 1918 for his discovery of energy quanta.

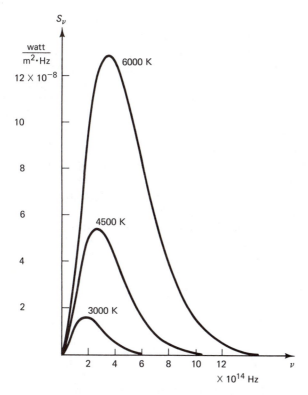

Fig. 3.5 Spectral emittance as a function of frequency according to Planck's law, at three different temperatures.

At first, Planck had proposed his formula merely as an empirical law that happens to give a good fit to the experimental data. But then he searched for a theoretical justification of this law. The search led Planck to the quantization of energy. This revolutionary discovery marks the birth of modern quantum physics. For Planck, the postulate of the quantization of energy was "an act of desperation" into which he was forced because "a theoretical explanation had to be found at any cost, whatever the price".*

Planck began his derivation of the cavity-radiation law by making a theoretical model of the walls of the cavity. He regarded the atoms in the walls as small harmonic oscillators with electric charges. Although this is a rather crude model of an atom, it was adequate for his purposes since, by the thermodynamic argument

---

* Unpublished letter by M. Planck; quoted in M. Jammer, *The Conceptual Development of Quantum Mechanics*, p. 22.

of the preceding section, the radiation in a cavity is completely independent of the characteristics of the wall. The random thermal motions of the oscillators cause the emission of electromagnetic radiation. This radiation fills the cavity and acts back on the oscillators. When thermal equilibrium is attained, the average rate of emission of radiation energy by the oscillators matches the rate of absorption of radiation energy. Because the oscillators share their energy with the radiation in the cavity, Planck was able to show that—under equilibrium conditions—the average radiation energy in a mode of vibration of the cavity at some frequency v is equal to the average energy $\bar{E}$ of an oscillator of frequency v.* The radiation energy within some small frequency interval dv from v to $v + dv$ is therefore $\bar{E}\,dn$, where dn is the number of modes of vibration within the frequency interval dv. In terms of the spectral energy density u_v, this radiation energy can be written $Vu_v\,dv$, where V is the volume of the cavity. Thus,

$$Vu_v\,dv = \bar{E}\,dn \tag{9}$$

or, since $u_v = (4/c)S_v$,

$$\frac{4}{c}VS_v\,dv = \bar{E}\,dn \tag{10}$$

A straightforward counting argument, which we omit for the sake of brevity,[†] shows that the number of modes of vibration in the small frequency interval dv is $dn = V(8\pi v^2/c^3)\,dv$. Hence

$$\frac{4}{c}S_v\,dv = \bar{E}\,\frac{8\pi v^2}{c^3}\,dv \tag{11}$$

from which

$$S_v = \frac{2\pi v^2}{c^2}\,\bar{E} \tag{12}$$

These steps of Planck's calculation involved nothing but classical statistical mechanics. But in the next step of the calculation, Planck departed radically from classical physics. He postulated that the energies of the oscillators are quantized according to the following rule: *In an oscillator of frequency v, the only permitted*

---

* See, e.g., M. Jammer, *The Conceptual Development of Quantum Mechanics*, Appendix A.

† The argument will be spelled out in Section 8.2 and in Problem 8.11.

values of the energy are

$$E = 0, hv, 2hv, 3hv, \ldots \qquad (13)$$

All other values of the energy are forbidden. The constant h in Eq. (13) is Planck's constant, the same as appears in Eqs. (7) and (8). The energy hv is called an **energy quantum;** according to the quantization rule, the energy of an oscillator is always some integer multiple of the fundamental energy quantum hv:

$$E = nhv \qquad n = 0, 1, 2, 3, \ldots \qquad (14)$$

The integer n is called the **quantum number** of the oscillator.

With this quantization condition, Planck calculated the average energy of the oscillator as follows: The probability that the oscillator is in a state of energy nhv is proportional to the Boltzmann factor $e^{-E/kT}$, or $e^{-nhv/kT}$. Therefore, the average energy of the oscillator is

$$\bar{E} = \sum_n nhv \times (\text{probability for } nhv)$$

$$= \sum_n nhv \times \frac{e^{-nhv/kT}}{\sum_{n'} e^{-n'hv/kT}} \qquad (15)$$

where the summation in the denominator is the constant of proportionality that converts the Boltzmann factor into a probability, normalized to 1. For convenience, we write $x = e^{-hv/kT}$, so that

$$\bar{E} = hv \frac{\sum_n nx^n}{\sum_{n'} x^{n'}} \qquad (16)$$

To evaluate this, we note that the infinite sum $\sum x^n$ is simply a geometric series. As is well known, such a geometric series has the value

$$\sum x^n = \frac{1}{1-x} \qquad \text{for } |x| < 1 \qquad (17)$$

If we differentiate this equation with respect to x, we obtain

$$\sum nx^{n-1} = \frac{1}{(1-x)^2}$$

and if we multiply by x, we obtain

$$\sum nx^n = \frac{x}{(1-x)^2} \qquad (18)$$

Substituting (17) and (18), respectively, into the denominator and the numerator of Eq. (16), we find

$$\bar{E} = hv \frac{x/(1-x)^2}{1/(1-x)} = hv \frac{x}{1-x} = hv \frac{1}{1/x - 1} \tag{19}$$

or

$$\bar{E} = \frac{hv}{e^{hv/kT} - 1} \tag{20}$$

With this value for the average energy, our expression (12) for the spectral emissivity becomes

$$S_v = \frac{2\pi v^2}{c^2} \frac{hv}{e^{hv/kT} - 1} \tag{21}$$

which is exactly Planck's law.

It is now a simple mathematical exercise to check that Stefan–Boltzmann's law and Wien's law are corollaries of Planck's law. By integrating Eq. (21), we find

$$\int_0^\infty S_v \, dv = \frac{2\pi h}{c^2} \int_0^\infty \frac{v^3}{e^{hv/kT} - 1} \, dv \tag{22}$$

This integral can be evaluated analytically, albeit not by elementary means.* The result is

$$\int_0^\infty S_v \, dv = \frac{2}{15} \frac{\pi^5 k^4}{h^3 c^2} T^4 \tag{23}$$

By differentiating Eq. (21) we find that the frequency at the maximum of S_v is given by

$$3 - \frac{hv}{kT} \frac{e^{hv/kT}}{e^{hv/kT} - 1} = 0 \tag{24}$$

Solving this numerically, we obtain

$$v_{max} = 2.822 \frac{kT}{h} \tag{25}$$

These equations show how the constants of proportionality in Stefan's and Wien's laws are related to fundamental physical constants.

Qualitatively, we can see that Planck's calculation avoids the ultraviolet catastrophe because the quantization of energy prevents

---

* See, e.g., L. D. Landau and E. M. Lifshitz, *Statistical Physics*, p. 164.

equipartition. Consider an oscillator of very high frequency; the energy quantum hv is then very large ($hv \gg kT$). If this oscillator is initially quiescent ($n = 0$), it cannot begin to move unless it acquires one energy quantum; but since this energy quantum hv is very large, the random thermal disturbances will be insufficient to provide it—the oscillator will remain quiescent. Thus, the quantization of energy inhibits the thermal excitation of the high-frequency oscillators and prevents equipartition. If the high-frequency oscillators remain quiescent, then they will not supply energy to the corresponding high-frequency standing waves in the cavity, and there will be no ultraviolet catastrophe. Mathematically, we can recognize this absence of equipartition from Eq. (20): the average energy of an oscillator is not kT, but $hv/(e^{hv/kT} - 1)$; if $hv \gg kT$, then this average energy is much smaller than kT.

Example 2 Suppose that $v = 5.0 \times 10^{14}$ Hz and T = 5000 K. According to Planck, what is the average energy of the oscillator?

Solution   With these values of the frequency and the temperature, we have

$$hv = 6.63 \times 10^{-34} \, \text{J·s} \times 5.0 \times 10^{14}/\text{s}$$

$$= 3.32 \times 10^{-19} \, \text{J}$$

and

$$kT = 1.38 \times 10^{-23} \, \text{J/K} \times 5000 \, \text{K}$$

$$= 6.90 \times 10^{-20} \, \text{J}$$

Hence

$$\bar{E} = \frac{hv}{e^{hv/kT} - 1} = \frac{3.32 \times 10^{-19} \, \text{J}}{\exp(3.32 \times 10^{-19} \, \text{J}/6.90 \times 10^{-20} \, \text{J}) - 1}$$

$$= 2.7 \times 10^{-21} \, \text{J}$$

Obviously, this average energy is much smaller than kT. ∎

Note that for an oscillator with a frequency of $v = 5.0 \times 10^{14}$ Hz, which is typical for atomic vibrations, the energy quantum is $hv = 3.3 \times 10^{-19}$ J. Since this is a very small amount of energy, quantization does not make itself felt on a macroscopic scale. But it does play an essential role on the atomic scale.

Unfortunately, Planck could not offer any basic justification for his postulate of quantization of energy. His postulate took care of the radiation law, but raised serious questions about classical physics. Obviously, quantization of energy does not make sense in classical physics—there is nothing in the laws of Newton that would prevent an oscillator from acquiring energy in any amount

whatsoever. Planck could only justify his postulate by its consequences. A deeper explanation of the quantization of energy only
emerged much later, with the development of quantum mechanics
(see Chapter 6).

3.3 Photons and the Photoelectric Effect

In 1905, the same year he formulated the theory of Special Relativity, Einstein showed that the high-frequency portion of the spectrum of blackbody radiation could be understood very simply in
terms of a direct quantization of the energy of the radiation. Planck
had postulated that the oscillators in the walls of the cavity have
discrete, quantized energies, but he had treated the electromagnetic
radiation in the cavity as a smooth, continuous distribution of
energy, exactly as demanded by classical electromagnetic theory.
In contrast, Einstein proposed that all electromagnetic radiation
consists of discrete particle-like packets of energy, each with an
energy $h\nu$. An electromagnetic wave then has an energy $h\nu$ if it
contains only one such packet, $2h\nu$ if it contains two, $3h\nu$ if it contains three, etc. Einstein called these discrete energy packets **quanta
of light;** they later came to be called **photons.** The thermal radiation in a cavity, with waves traveling in all directions at random,
can thus be regarded as a gas of photons. At first, Einstein was
able to establish this new picture of thermal radiation only for the
high-frequency portion of the blackbody spectrum, i.e., for high-
frequency photons. Because of this and other reasons, Planck and
most of his contemporaries remained skeptical of Einstein's concept
of photons. But some twenty years later, S. Bose* introduced a
modification of statistical mechanics (the so-called Bose–Einstein
statistics, which takes into account that photons are intrinsically
indistinguishable and identical), by means of which he was able to
calculate the complete energy distribution of the photon gas. He
demonstrated that the energy spectrum of blackbody radiation
deduced from the photon-gas picture is in agreement with Planck's
law over the full range of frequencies.

The number of photons in sunlight and other common light
sources is very large; hence we do not directly perceive the grainy

---

* **Satyendranath Bose,** 1894– , Indian physicist, professor at Dacca and
Calcutta.

character of the energy distribution in light on the macroscopic scale.

Example 3 The energy flux of sunlight incident on the surface of the Earth is 1.0×10^3 watt/m². How many photons reach the surface of the Earth per cm² per second? For the purposes of this calculation make the rough approximation that all the photons in sunlight have an average wavelength of 5500 Å.

Solution The energy of a photon of wavelength 5500 Å is

$$E = h\nu = \frac{hc}{\lambda} = \frac{6.63 \times 10^{-34}\,\text{J}\cdot\text{s} \times 3.00 \times 10^8\,\text{m/s}}{5.5 \times 10^{-7}\,\text{m}}$$

$$= 3.6 \times 10^{-19}\,\text{J}$$

The energy incident per cm² per second is 1.0×10^{-1} J. To obtain the number of photons, we must divide this by the energy per photon, which gives us $1.0 \times 10^{-1}\,\text{J}/3.6 \times 10^{-19}\,\text{J} = 2.8 \times 10^{17}$ photons per cm² per second. ■

Note that as a consequence of Planck's hypothesis, the energy of an oscillator can change only by $h\nu$ or some multiple of this; hence the energy of the radiation emitted by the oscillator must necessarily be $h\nu$ or some multiple of this *immediately after emission*. However, if classical electrodynamics were valid, the wave pulse carrying this energy could split when it scatters off some obstacle, and therefore the energy packets of the radiation would have no permanence. The essence of Einstein's proposal is that the energy packets of the radiation are permanent; they retain their identity from the moment they are emitted to the moment they are absorbed, without ever splitting. In the modern view, the quantization postulates of *both* Planck and Einstein are true: the oscillators (or atoms) in the wall have quantized energies, *and* the radiation in the cavity consists of quantized photons.

Einstein immediately recognized that his concept of the photon also held the clue to the explanation of the **photoelectric effect,** which had been posing somewhat of a puzzle for classical physicists. During the early experiments on the production of radio waves by spark discharges, H. R. Hertz had noticed that light shining on an electrode tended to facilitate the formation of sparks. Subsequent careful experiments by P. Lenard* established that the impact of light on the electrode can eject electrons. These electrons emerge

---

* **Philipp Eduard Anton von Lenard,** 1862–1947, German experimental physicist, professor at Heidelberg. He received the Nobel Prize in 1905 for work on cathode rays.

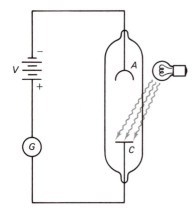

Fig. 3.6 Schematic diagram of apparatus used for the investigation of the photoelectric effect. The light from the lamp ejects electrons from the emitting electrode C (cathode), and they travel to the collecting electrode A (anode).

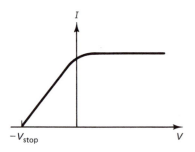

Fig. 3.7 Photoelectric current as function of the potential difference between emitter and collector. If the collector is negative and the potential difference exceeds V_{stop}, the current is zero.

with a kinetic energy which increases linearly with the frequency of the light.

Figure 3.6 is a schematic diagram of the apparatus later developed by Millikan for the investigation of the photoelectric effect. Monochromatic light from a lamp illuminates an electrode of metal (C) enclosed in an evacuated glass bulb. Electrons ejected from this electrode travel to the collecting electrode (A) and then flow around the external circuit, forming an electric current. A galvanometer (G) detects this electric current. The kinetic energy of the ejected photoelectrons can be determined by applying a potential difference between the emitting and the collecting electrodes by means of an adjustable source of emf (V). If the polarity is as shown in the diagram, the collector has a negative potential relative to the emitter; thus, the collector exerts a repulsive force on the photoelectrons. If the potential energy at the collector matches or exceeds the initial kinetic energy of the photoelectrons, then the flow of these electrons will stop; the corresponding potential is called the **stopping potential** (see Fig. 3.7). The measured magnitude of this stopping potential gives us the kinetic energy of the ejected electrons:

$$K = e|V_{stop}| \tag{26}$$

Experimentally, it is found that the kinetic energy determined in this way increases linearly with the frequency of the incident light but does not depend at all on the intensity. For example, Fig. 3.8 shows a plot of the kinetic energy vs. the frequency of the light for photoelectrons ejected from sodium. Note that for a frequency of 4.4×10^{14} Hz, the kinetic energy is zero; at lower frequencies, the light is incapable of ejecting electrons. This critical frequency is called the **threshold frequency** of sodium.

The classical wave theory of light fails to account for these features of the photoelectric effect. According to the wave theory, the crucial parameter that determines the ejection of a photoelectron should be the intensity of light. If an intense electromagnetic wave strikes an electron, it should be able to jolt it loose from the metal, regardless of the frequency of the wave. Furthermore, the kinetic energy of the ejected electron should be a function of the intensity of the wave. The observational evidence contradicts these predictions of the wave theory: a wave with a frequency below the threshold frequency never ejects an electron, regardless of its intensity. And furthermore, the kinetic energy depends on the frequency, and not on the intensity—light of high intensity ejects a larger number of photoelectrons, but does not give the individual electrons more kinetic energy.

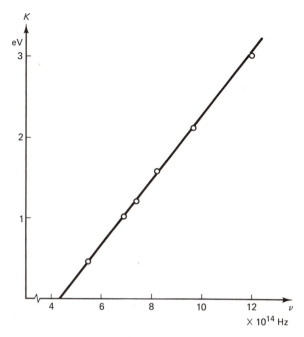

Fig. 3.8 Kinetic energy (in eV) of photoelectrons ejected from sodium vs. frequency of the incident light. [After R. A. Millikan, *Phys. Rev.* **7,** 355 (1916).]

How does Einstein's quantum theory of light account for the features of the photoelectric effect? The electrons in the illuminated electrode absorb photons, one at a time. When an electron absorbs a photon, it acquires an energy hv. But before this electron can emerge from the electrode, it must overcome the restraining forces that bind it within the metal of the electrode. The energy required for this is called the **work function** of the metal, designated by ϕ. The kinetic energy of the emerging electron is then the difference between hv and ϕ,

$$K = hv - \phi \tag{27}$$

Some electrons suffer extra energy losses in collisions within the metal before they emerge. Thus, the expression (27) actually gives the *maximum* kinetic energy with which electrons can emerge; this maximum kinetic energy is an experimentally relevant quantity— the measurement of the stopping potential gives us this quantity. The expression (27) is **Einstein's photoelectric equation.** It shows that the kinetic energy is indeed a linearly increasing function of the frequency, in agreement with the data in Fig. 3.8. According

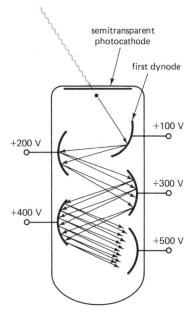

semitransparent
photocathode

first dynode

+100 V

+200 V

+300 V

+400 V

+500 V

Fig. 3.9 Schematic diagram of a photomultiplier tube. The secondary electrodes are called dynodes. For the purposes of this diagram it has been assumed that each electron impact on a dynode releases two electrons. The arrows show an avalanche of electrons.

to Eq. (27), the slope of the straight line in Fig. 3.8 should equal Planck's constant. Note that for $hv = \phi$, the kinetic energy is zero and the light becomes incapable of ejecting electrons, regardless of its intensity.

Einstein's photoelectric equation was verified in detail by a long series of meticulous experiments by Millikan, completed in 1916 (the data in Fig. 3.8 are due to him). In order to obtain reliable results, Millikan found it necessary to take extreme precautions to avoid contamination of the surface of the photosensitive electrode. Since the surfaces of metals exposed to air quickly accumulate a layer of oxide, he developed an ingenious technique for shaving the surface of his metals in a vacuum by means of a magnetically operated knife. The results of Millikan's experiments provided strong experimental evidence for Einstein's quantum theory of light.

The photoelectric effect has many practical applications in sensitive electronic devices for the detection of light, such as photomultipliers, television cameras, charge-coupling devices, etc. For instance, in a photomultiplier tube, an incident photon ejects an electron from an electrode; this electron is accelerated toward a second electrode, or dynode (see Fig. 3.9), where its impact ejects several secondary electrons; these are accelerated towards a third electrode, where their impact ejects tertiary electrons, etc. The result is that each electron from the first electrode generates an avalanche of electrons. In a high-gain photomultiplier tube, a pulse of 10^9 electrons emerges from the last electrode, delivering a measurable pulse of current to an external circuit. Thus, the photomultiplier tube can detect the arrival of individual photons. Some sensitive television cameras, such as the image orthicon, rely on the same multiplier principle to convert the arrival of a single photon at a photosensitive faceplate into a measurable pulse of current.

Example 4   The work function for zinc is 6.8×10^{-19} J. What is the threshold frequency for the ejection of photoelectrons from zinc?

Solution   When an electron absorbs a photon at the threshold frequency, it will just barely have enough energy to overcome the binding forces holding it in the metal, and it will emerge with zero kinetic energy. According to Eq. (27), this implies

$$0 = hv - \phi \tag{28}$$

or

$$v = \frac{\phi}{h} = \frac{6.8 \times 10^{-19}\,\text{J}}{6.63 \times 10^{-34}\,\text{J}\cdot\text{s}} = 1.0 \times 10^{15}\,\text{Hz} \quad \blacksquare$$

3.4 The Compton Effect

Conclusive experimental evidence for the particlelike behavior of photons was discovered by A. H. Compton* in 1922, during his investigations of the scattering of X rays by targets of graphite. When Compton bombarded graphite with a beam of monochromatic X rays, he found that the scattered X rays emerged with a somewhat longer wavelength than that of the incident X rays. For instance, X rays scattered at 90° gained about 0.02 Å in wavelength. This **Compton effect** is in complete disagreement with a classical wavelike behavior of the X rays. If X rays were simply classical waves, they would shake the electrons in the atoms of the target with a frequency equal to that of the incident wave. The wavelength of the scattered waves radiated by the electrons would then match the incident wavelength.

Compton quickly recognized that the observed wavelength change could be explained as due to collisions between particlelike photons and electrons. In such a collision, the electron of a carbon atom can be regarded as free, because the energy available from the photon is much larger than the binding energy of the electron within the atom. When the photon strikes the electron, the electron recoils and thereby acquires some of the photon's energy—the scattered photon is left with reduced energy. Since the energy of the photon is inversely proportional to the wavelength ($E = h\nu = hc/\lambda$), a reduction of energy implies an increase of wavelength. Qualitatively, we expect that those photons scattered through the largest angles should lose the most energy and therefore emerge with the longest wavelength. Such an increase of wavelength is indeed observed in the Compton effect.

For a quantitative discussion of the photon–electron collision, we need an expression for the momentum of a photon. We know that an electromagnetic wave of energy E carries a momentum $p = E/c$ [see Eq. (1.143)]. If the energy is that of a single photon, $E = h\nu$, then the momentum is

$$p = \frac{h\nu}{c} \tag{29}$$

* **Arthur Holly Compton,** 1892–1962, American physicist, professor at Chicago and chancellor at Washington University. For his discovery of the Compton effect he received the Nobel Prize in 1927. During the war, he was one of the leaders of the Manhattan Project.

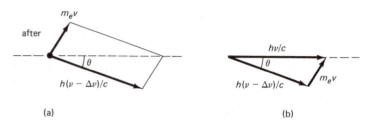

Fig. 3.10 (a) Momentum vector of photon before the collision, and momentum vectors of photon and electron after the collision. Before the collision the photon has momentum $h\nu/c$, and the electron is at rest. After the collision, the photon has momentum $h(\nu - \Delta\nu)/c$, and the electron has momentum $m_e v$. (b) The triangle of the momentum vectors of photon and electron.

Before the collision, the electron is at rest and the photon has a frequency ν and momentum $h\nu/c$. During the collision, the electron recoils with a momentum $m_e v$, and the photon is deflected through some angle with a reduced frequency $\nu - \Delta\nu$ and momentum $h(\nu - \Delta\nu)/c$. Figure 3.10a shows the initial momentum vector of the photon, and the final momentum vectors of the electron and the photon. Conservation of momentum demands that the sum of the two final momentum vectors equals the initial momentum vector; the three momentum vectors therefore form a triangle (see Figure 3.10b). If we apply the law of cosines to this triangle, we find

$$(m_e v)^2 = \left(\frac{h\nu}{c}\right)^2 + \left[\frac{h(\nu - \Delta\nu)}{c}\right]^2 - \frac{2h\nu}{c}\frac{h(\nu - \Delta\nu)}{c}\cos\theta \qquad (30)$$

or

$$(m_e v)^2 = \left(\frac{h\nu}{c}\right)^2 \left[1 + \left(1 - \frac{\Delta\nu}{\nu}\right)^2 - 2\left(1 - \frac{\Delta\nu}{\nu}\right)\cos\theta\right] \qquad (31)$$

$$= \left(\frac{h\nu}{c}\right)^2 \left[2 - 2\frac{\Delta\nu}{\nu} + \left(\frac{\Delta\nu}{\nu}\right)^2 - 2\left(1 - \frac{\Delta\nu}{\nu}\right)\cos\theta\right] \qquad (32)$$

Neglecting the term $(\Delta\nu/\nu)^2$, which is small compared to $\Delta\nu/\nu$, we can rewrite this as

$$\frac{1}{2}m_e v^2 \simeq \frac{1}{m_e}\left(\frac{h\nu}{c}\right)^2\left(1 - \frac{\Delta\nu}{\nu}\right)(1 - \cos\theta) \qquad (33)$$

Conservation of energy demands that the kinetic energy of the electron equals the energy lost by the photon:

$$\tfrac{1}{2}m_e v^2 = h\,\Delta v \tag{34}$$

Combining Eqs. (33) and (34), we find

$$\Delta v = \frac{hv^2}{m_e c^2}\left(1 - \frac{\Delta v}{v}\right)(1 - \cos\theta) \tag{35}$$

or

$$\frac{c\,\Delta v}{v(v - \Delta v)} = \frac{h}{m_e c}(1 - \cos\theta) \tag{36}$$

By some rearrangement, we recognize that the left side of this equation is simply the change of wavelength,

$$\frac{c\,\Delta v}{v(v - \Delta v)} = \frac{c}{v - \Delta v} - \frac{c}{v} = (\lambda + \Delta\lambda) - \lambda = \Delta\lambda \tag{37}$$

Here we have taken into account that a decreased frequency $(v - \Delta v)$ implies an increased wavelength $(\lambda + \Delta\lambda)$. We therefore obtain the following result for the change of wavelength of the photon as a function of the angle of deflection:

$$\Delta\lambda = \frac{h}{m_e c}(1 - \cos\theta) \tag{38}$$

In the calculation above we have implicitly assumed that the motion of the electron is nonrelativistic, with $v_e \ll c$; however, it turns out that the result (38) remains valid even if the motion is relativistic [in fact, the relativistic calculation shows that Eq. (38) is *exact*].

Figure 3.11 gives a plot of $\Delta\lambda$ as a function of θ. For $\theta = 180°$ (a head-on collision), the wavelength shift reaches its maximum value $\Delta\lambda = 2h/m_e c = 0.0485\,\text{Å}$; but even this maximum is quite small. Thus, for X rays of medium wavelength, the maximum wavelength shift amounts to only a few percent of the wavelength. For instance, in his experiments, Compton used X rays of a wavelength $\lambda = 0.71\,\text{Å}$; thus the maximum wavelength shift amounted to about 7% of the wavelength.

The excellent agreement between the theoretical prediction (38) and the experimental results confirms that the collision between a photon and an electron is like a collision between two particles. Thus the Compton effect provides clear proof of the presence of particle-like quanta in light.

Quantum effects also play a crucial role in the production of X rays. These rays are simply a form of light of extremely short wavelength, in the range 0.01 to $100\,\text{Å}$. As we know from Section 1.7,

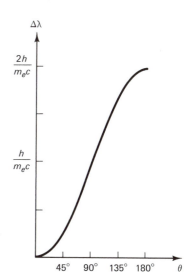

Fig. 3.11 Wavelength shift of photon as a function of deflection angle.

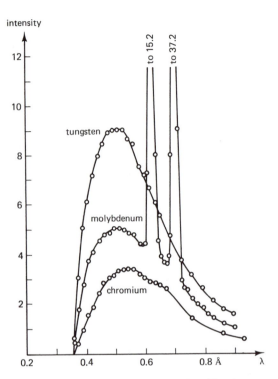

intensity

to 15.2   to 37.2

tungsten

molybdenum

chromium

0.2   0.4   0.6   0.8 Å   λ

Fig. 3.12 Spectra of X rays produced by electrons of 35 keV incident on targets of tungsten, molybdenum, and chromium. [From C. Ulrey, *Phys. Rev.* **11**, 401 (1911).]

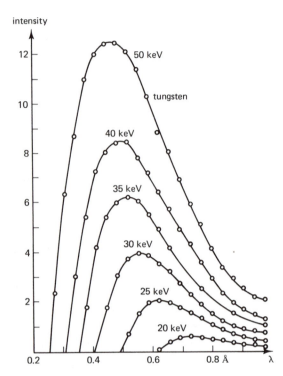

intensity

50 keV

tungsten

40 keV

35 keV

30 keV

25 keV

20 keV

0.2   0.4   0.6   0.8 Å   λ

Fig. 3.13 Spectra of X rays emitted by electrons of diverse energies incident on a target of tungsten. (From C. Ulrey, ibid.)

they are readily produced by bombarding a target, usually a small piece of metal, with high-speed electrons. Upon impact on the target, the electrons suffer sudden decelerations in collisions with the atoms of the target. As expected from classical electrodynamics, the electrons emit high-frequency electromagnetic radiation during such decelerations. The X rays produced by this mechanism are called **Bremsstrahlung.*** Figure 3.12 shows the spectra of X rays produced by electrons of 35 keV incident on targets of tungsten, molybdenum, and chromium. Each of the smooth continuous spectral curves in this figure represents *Bremsstrahlung*. Besides the continuous spectrum, each target material displays a discrete set of sharp spikes of intensity at a set of wavelengths characteristic of the material. Figure 3.12 shows two of these sharp spikes for the case of molybdenum (the sharp spikes for tungsten and chromium

---

* German for "braking radiation."

lie outside the range of wavelengths covered by Fig. 3.12). These sharp spikes constitute the line spectrum, or the **characteristic spectrum** of the target material. The discrete spectral lines are due to radiation emitted by the atoms of the target material when stimulated by electron impacts. For now we will ignore the line spectrum and concentrate on the continuous *Bremsstrahlung* spectrum.

A remarkable feature of the *Bremsstrahlung* spectrum is that for each given electron energy, there is a cut-off wavelength, i.e., a minimum wavelength below which no radiation is emitted (see Fig. 3.13). This minimum wavelength is called the **Duane-Hunt limit;** it is inversely proportional to the electron energy. For instance, Fig. 13 shows that for $K = 20\,\text{keV}$, $\lambda_{\text{cut}} = 0.62\,\text{Å}$, and for $K = 40\,\text{keV}$, $\lambda_{\text{cut}} = 0.31\,\text{Å}$. This feature of the spectra is easily explained in terms of quanta of light. The emission of X rays by a decelerating electron is essentially an inverse photoelectric effect—instead of absorbing a photon and gaining energy, the electron emits a photon (or photons) and loses energy. The cutoff in the *Bremsstrahlung* spectrum corresponds to the conversion of all the kinetic energy of an electron into a single photon. The frequency of this photon will then be given by

$$h v_{\text{cut}} = K \tag{39}$$

or

$$v_{\text{cut}} = \frac{K}{h} \tag{40}$$

and

$$\lambda_{\text{cut}} = \frac{hc}{K} \tag{41}$$

For instance, if $K = 20\,\text{keV} = 3.2 \times 10^{-15}\,\text{J}$, then

$$\lambda_{\text{cut}} = \frac{6.63 \times 10^{-34}\,\text{J} \cdot \text{s} \times 3.0 \times 10^{8}\,\text{m/s}}{3.2 \times 10^{-15}\,\text{J}} = 0.62\,\text{Å}$$

in agreement with the experimental result quoted above.

Note that the cutoff frequency is independent of the target material—it depends only on the kinetic energy of the electrons. Strictly, this independence is an approximation which arises from neglecting the work function ϕ in Eq. (39). If we wanted to be more precise, we would have to include ϕ in Eq. (39), exactly as we did in Eq. (27). However, in the calculation of the frequency the value of a few eV for ϕ is completely negligible compared to the value of several thousand eV for K.

Note that if classical electrodynamics were valid for the generation of X rays, there would not be any sharp cutoff in the spectrum. Instead, the violent decelerations of electrons in collisions with atoms would generate radiation with a spectrum that extends to very high frequencies and only fades gradually and smoothly as the frequency tends to infinity, just as the collision of, say, two automobiles generates a bang of sound with a spectrum that extends to very high sound frequencies.

3.5 **The Attenuation of X rays**

Röntgen and other early investigators were impressed with the penetrating power of X rays. Beams of X rays can penetrate through materials opaque to light, such as paper, thin sheets of metal, or human tissues. However, all materials attenuate X rays to some extent; as a general rule, the attenuation increases with the density of the material. The X-ray photograph of a hand shown in Fig. 1.28—one of the first photographs made by Röntgen—illustrates this dependence of the attenuation on the density: the dense bones attenuate the X rays strongly and throw a dark shadow on the photograph, whereas the surrounding soft tissues attenuate the X rays only weakly.

The photoelectric effect and the Compton effect play a crucial role in the attenuation of a beam of X rays by a layer of material. When the X-ray photons encounter the electrons of the material, they produce photoelectrons and they engage in Compton collisions. Any photon that produces a photoelectron gives up all of its energy, i.e., it disappears. Any photon that engages in a Compton collision gives up part of its energy and, furthermore, it is deflected sideways. We will assume that the beam is fairly narrow, so that any deflected photon is effectively lost from the beam.* Thus, both the photoelectric effect and the Compton effect lead to a reduction of the intensity of the beam. Moreover, if the energy of the photons is sufficiently large, another attenuation mechanism comes into

---

* The deflected photons contribute to the attenuation of the beam, but they do not contribute equally to the absorption of energy by the material, since they do not (immediately) deposit their energy in the material. Hence it is necessary to make a distinction between **attenuation** and **absorption.** In this section we will deal only with attenuation since this is what is relevant in simple experiments with narrow beams whose intensity is measured with detectors placed directly in the beam.

play: a high-energy photon can create an electron–antielectron pair in a collision with an atomic nucleus. In this reaction the photon disappears, giving up all of its energy to the electron and antielectron. The minimum photon energy required to permit such a pair creation is simply the sum of the rest-mass energies of the electron and the antielectron, i.e., $m_e c^2 + m_e c^2$, or 1.02 MeV. Pair creation is the dominant attenuation mechanism for photons of high energies, whereas the photoelectric effect dominates at low energies, and the Compton effect at intermediate energies.

Regardless of the attenuation mechanism, the removal of the photons from a beam of X rays is a probabilistic process: within a given layer of material of thickness dx, each photon has some chance of being absorbed or deflected out of the beam by an encounter with an electron. Hence the decrease dN of the number of photons in the beam is proportional to the number N of photons multiplied by the number of electrons. Taking into account that the number of electrons in the layer of material is proportional to the thickness dx, we see that

$$dN = -\mu N \, dx \tag{42}$$

where μ is a constant of proportionality, called the **linear attenuation coefficient.** The attenuation coefficient depends on the photon energy. Consequently, to proceed with our calculation, we will have to assume that the beam is monochromatic, consisting of photons of equal energies, so that the attenuation coefficient has a unique value. (If the beam is not monochromatic, then each of the monochromatic components in the beam must be treated separately.) Since the intensity of a monochromatic beam is proportional to the number of photons, Eq. (42) immediately translates into an equation for the decrease of intensity:

$$dI = -\mu I \, dx \tag{43}$$

We can rewrite this equation as

$$\frac{dI}{I} = -\mu \, dx \tag{44}$$

and integrate it, to obtain

$$\ln I - \ln I_0 = -\mu x \tag{45}$$

where $\ln I_0$ is a constant of integration, chosen so that I_0 is the intensity at $x = 0$. Upon taking the exponential function of both sides of Eq. (45), we find

$$I = I_0 e^{-\mu x} \tag{46}$$

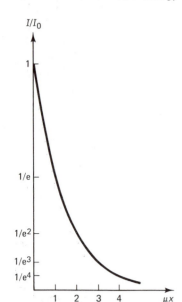

Fig. 3.14 Intensity as a function of penetration.

Thus, the intensity decreases exponentially as the beam penetrates into the material (see Fig. 3.14). Note that for $x = 1/\mu$, the remaining intensity is $1/e$ of the original intensity.

Figure 3.15 shows the energy dependence of the linear attenuation coefficient for a beam of X rays penetrating through lead. For photon energies above ~ 4 MeV, the attenuation is mainly due to pair creation; below ~ 1 MeV, it is due mainly to the photoelectric effect; and in the intermediate range, it is due mainly to the Compton effect. The sharp peaks in the attenuation coefficient

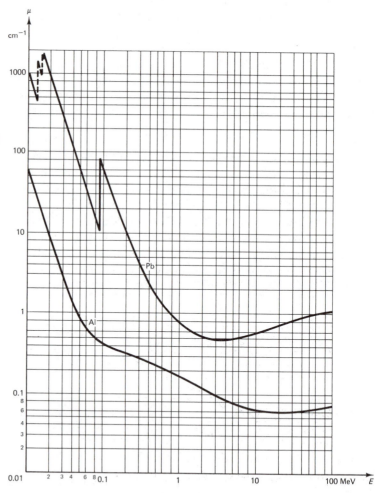

Fig. 3.15 Linear attenuation coefficients for lead and aluminum as a function of photon energy. (After R. D. Evans, *The Atomic Nucleus*. © 1955 by McGraw-Hill Book Co. Reproduced with permission.)

at 0.09 MeV and at 0.015 MeV are called the K and the L edges. These peaks are due to resonant absorption of photons by the electrons orbiting around the nuclei of the lead atoms.

Figure 3.15 also shows the energy dependence of the linear attenuation coefficient for aluminum. The attenuation in aluminum is smaller than that in lead. Broadly speaking, we can attribute the smaller attenuation in aluminum to the smaller density—aluminum has fewer electrons per unit volume than lead, and hence a photon is less likely to encounter an electron in a given thickness of material. However, the differences in detail between the shapes of the curves in Fig. 3.15 must be attributed to the differences between the electronic arrangements in aluminum and lead atoms.

Example 5   What thickness of lead will attenuate a beam of 0.4-MeV X rays by a factor of 2?

Solution   From Eq. (46), we obtain

$$\ln\left(\frac{I}{I_0}\right) = -\mu x$$

For $I/I_0 = \frac{1}{2}$, this gives

$$x = -\frac{1}{\mu}\ln\left(\frac{1}{2}\right) \tag{47}$$

According to Fig. 3.15, the linear attenuation coefficient at 0.4 MeV is 2.3/cm. Hence.

$$x = -\frac{1}{2.3/\text{cm}}\ln\left(\frac{1}{2}\right) = 0.30\,\text{cm} \quad \blacksquare$$

3.6 The Specific Heat of Solids

The quantization postulate also proved capable of providing a theoretical explanation of an old puzzle concerning the specific heat of crystalline solids. At high temperature, all crystalline solids have a specific heat of about 6 cal/K per mole, i.e., they require 6 calories per mole to raise their temperature by 1 K. This observed universality of the specific heats is called the **Law of Dulong and Petit,** and it is readily explained by the equipartition theorem. We can think of the atoms in the solid as particles held in their places by harmonic-oscillator forces or "springs" (see Fig. 3.16). These springs permit an atom to oscillate about its equilibrium position, but they prevent it from straying too far. Thus, each atom

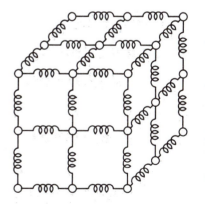

Fig. 3.16 Schematic model of a crystalline solid. The atoms are held in their positions by springs.

acts like a three-dimensional simple harmonic oscillator. Heat energy stored in the solid is merely the energy of the random thermal motion of the atoms. According to the equipartition theorem (see Example 1.6), the average thermal energy of each atom is $3kT$. Thus, the thermal energy per mole is $3kN_AT$. The specific heat is then

$$c = \frac{dE}{dT} = 3kN_A = 3 \times 1.38 \times 10^{-23} \text{ J/K} \times 6.02 \times 10^{23} \quad (48)$$

$$= 24.9 \text{ J/K}$$

$$= 5.94 \text{ cal/K} \quad (49)$$

Unfortunately, this neat agreement between observation and classical theory breaks down if the temperature is not high. Observations show that at room temperature and below, the specific heat of crystalline solids is not a universal constant. For instance, Fig. 3.17 is a plot of the specific heats of a few materials as a function of temperature. In all of these materials, the specific heat asymptotically approaches about $6 \text{ cal}/(\text{K} \cdot \text{mole})$ at high tempera-

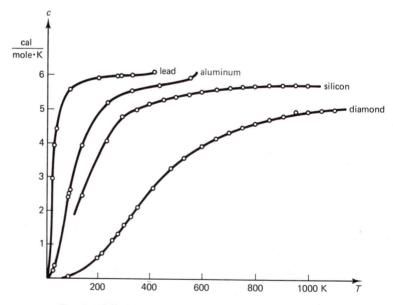

Fig. 3.17 Specific heats of lead, aluminum, silicon, and diamond as a function of temperature. (From F. K. Richtmeyer et al., *Introduction to Modern Physics.* ©1955 by McGraw-Hill Book Co. *Reproduced with permission.*)

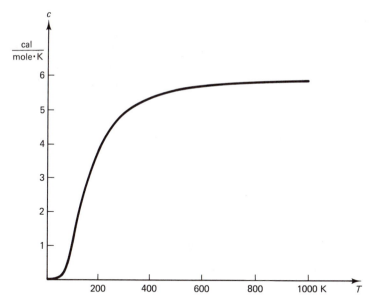

Fig. 3.18 Specific heat of a crystalline solid according to Einstein's formula, with an assumed frequency of 1×10^{13} Hz. At high temperature, the specific heat asymptotically approaches $3kN_A$.

ture. But at low temperature, the specific heat decreases toward zero—in complete contradiction with the result (49) deduced from classical equipartition.

This puzzle was solved in 1907 by Einstein. He treated the atoms as three-dimensional harmonic oscillators, with quantized energies. As we have seen in Section 2, the average thermal energy of a quantized one-dimensional harmonic oscillator is $hv/(e^{hv/kT} - 1)$. A three-dimensional harmonic oscillator may be regarded as consisting of three independent one-dimensional oscillators. Thus, its average thermal energy is $3hv/(e^{hv/kT} - 1)$. The thermal energy per mole is then $3hvN_A/(e^{hv/kT} - 1)$, and the specific heat is

$$c = \frac{dE}{dT} = 3kN_A \left(\frac{hv}{kT}\right)^2 \frac{e^{hv/kT}}{(e^{hv/kT} - 1)^2} \tag{50}$$

Figure 3.18 is a plot of the specific heat calculated from this formula. As a function of temperature, this plot exhibits the general features of the experimental curves: at high temperature c approaches $3kN_A$, and at low temperature c approaches zero. Hence, Einstein's formula for the specific heat is in qualitative agreement with the observations. This established that quantization is not only a feature of blackbody radiation, but a general feature of the microscopic world. A few years later, P. Debye* introduced some refinements into Einstein's formula by taking into account that the atoms in the crystal lattice form a system of interacting particles, rather than a system of independent particles. With these refinements, Debye was able to bring theory and observation into quantitative agreement.

Chapter 3: SUMMARY

Spectral emittance of blackbody:

$$S_v = \frac{2\pi h}{c^2} \frac{v^3}{e^{hv/kT} - 1}$$

$$h = 6.63 \times 10^{-34} \,\text{J} \cdot \text{s}$$

Stefan–Boltzmann law:

$$S = \sigma T^4 \qquad \sigma = 5.67 \times 10^{-8} \,\text{watt}/(\text{m}^2 \cdot \text{K}^4)$$

Wien's law:

$$v_{max} \propto T$$

Energy quantization of oscillator:

$$E = nhv \qquad n = 0, 1, 2, \ldots$$

Einstein's photoelectric equation:

$$K = hv - \phi$$

Energy and momentum of photon:

$$E = hv \qquad p = hv/c$$

Wavelength shift of photon (Compton effect):

$$\Delta\lambda = \frac{h}{m_e c}(1 - \cos\theta)$$

Duane–Hunt limit:

$$\lambda_{cut} = \frac{hc}{K}$$

---

* **Peter Joseph Wilhelm Debye,** 1884–1966, Dutch, and later American, physicist, professor at Zürich, Utrecht, Gottingen, Leipzig, Berlin, and Cornell University. He received the 1936 Nobel Prize in Chemistry for research on molecular structure.

Attenuation of X rays:

$$I = I_0 e^{-\mu x}$$

Law of Dulong–Petit:

$$c \simeq 6\,\text{cal/K per mole}$$

Chapter 3:   PROBLEMS

1. The temperature of the tungsten filament of an incandescent light bulb is typically 3200 K. At what frequency is the spectral emittance S_ν maximum? Assume that the filament acts like a blackbody.

2. The tungsten filament of an incandescent light bulb is a wire of diameter 0.080 mm and length 5.0 cm at a temperature of 3200 K. What is the power radiated by the filament? Assume that the filament acts like a blackbody.

3. (a) Show that for $h\nu/kT \ll 1$, Planck's law yields the approximation

$$S_\nu \propto \nu^2 T$$

   (b) Show that for $h\nu/kT \gg 1$, Planck's law yields the approximation

$$S_\nu \propto \nu^3 e^{-h\nu/kT}$$

   This expression is called Wien's spectral law.

4. Interplanetary and interstellar space is filled with thermal "fireball" radiation of a temperature of 2.9 K, a relic of the Big Bang at the beginning of the universe.
   (a) At what frequency is the spectral energy density u_ν maximum?
   (b) What is the power incident on the surface of the Earth due to this radiation?

5. Determine the radius of the star Procyon B from the following data: the flux of starlight reaching us is 1.7×10^{-12} watt/m², the distance of the star is 11 light-years, and its surface temperature is 6600 K. Assume the star radiates like a blackbody.

6. At the top of the Earth's atmosphere, the flux of sunlight per unit area facing the Sun is 1.37×10^3 watt/m². What equilibrium temperature of the Earth can you deduce from this? Assume that the Earth radiates like a blackbody at uniform temperature.

7. Prove that for electromagnetic radiation consisting of waves of random directions of propagation, the power flowing across a unit area from one side to the other is related to the energy density by

$$S_\nu = \frac{c}{4} u_\nu$$

(*Hint:* Suppose that the unit area is horizontal. At any given instant, one half of the waves will be crossing this area in the upward direction, one half in the downward direction. The power carried by the half crossing in the, say, upward direction, is equal to the sum of the vertical components of the Poynting vectors of these waves, which is proportional to the average of $\cos\theta$ over the upper hemisphere of solid angles.)

8. Show that according to Rayleigh's calculation, the spectral emittance should be

$$S_\nu = \frac{2\pi\nu^2}{c^2} kT$$

and show that $\int_0^\infty S_\nu \, d\nu = \infty$.

9. An atom in a crystal lattice can be regarded as a mass of 2.0×10^{-26} kg attached to a spring. The frequency of this oscillator is 1.0×10^{13} Hz. What is the amplitude of oscillation if the energy of oscillation is one energy quantum? Two energy quanta?

10. Show that if $kT \gg h\nu$, Planck's formula for the average energy of an oscillator [Eq. (20)] gives approximately $\bar{E} \simeq kT$.

11. Derive the Stefan–Boltzmann Law

$$S \propto T^4$$

from Planck's law. [*Hint:* Consider the integral $\int S_\nu \, d\nu$; change the variable of integration to $x = h\nu/kT$ and show that the result has the form (const.) $\times T^4$; you do not have to evaluate the constant.]

12. After changing the variable of integration to $x = h\nu/kT$, evaluate the integral given in Eq. (22) numerically on a computer or on a programable calculator, and show that the result agrees with Eq. (23).

13. Solve Eq. (24) numerically, by trial and error, or by some other method.

14. Show that the flux radiated by a blackbody in a wavelength interval $d\lambda$ is

$$S_\lambda \, d\lambda = \frac{2\pi c^2 h}{\lambda^5} \frac{1}{e^{hc/kT\lambda} - 1} d\lambda$$

Does the maximum of S_ν coincide with the maximum of S_λ?

15. A powerful radar transmitter radiates 10^4 kW at a wavelength of 23 cm. How many photons does the transmitter radiate per second?

16. The energy flux in the starlight reaching us from a sixth-magnitude star (the faintest that can be seen by the naked eye) is 1.4×10^{-10} W/m². If you are looking at this star, how many photons per second enter your eye? The diameter of your pupil is 0.70 cm.

17. The average energy density of radio waves in intergalactic space is about 10^{-20} J/m³. What is the corresponding density of photons? Assume the average wavelength of the photons is 1 m.

18. An incandescent light bulb radiates 40 watt of thermal radiation from a filament of temperature 3200 K.
(a) Roughly how many photons are radiated per second? Assume that the photons have an average frequency equal to the ν_{max} given by Wien's law.

(b) If you are looking at this light bulb from a distance of 2.0 m, how many photons per second enter your eye? Assume that the diameter of your pupil is 0.50 cm.

19. The binding energy (ionization energy) of an electron in a lithium atom is 5.4 eV. Suppose that a photon of wavelength 1500 Å strikes the atom and gives up all of its energy to the electron. With what kinetic energy will the electron be ejected from the atom?

20. Figure 3.7 shows a plot of the measured photoelectric current as a function of the potential difference between the emitting and the collecting electrodes for light of a given intensity and frequency incident on the surface of a sample of sodium.
    (a) Explain why the current levels off at high values of V.
    (b) Roughly plot the current for light of twice the intensity.
    (c) Roughly plot the current for light of twice the frequency.

21. According to the data supplied with Fig. 3.8, what is the work function of sodium? Express your answer in eV.

22. The work function of zinc is 4.24 eV. What is the threshold frequency for the photoelectric effect in zinc?

23. By inspection of Fig. 3.8, find the slope of the straight line in eV/Hz. Convert these units into J·s and verify that the slope equals Planck's constant, as claimed in Section 3.

24. Show that the flux of *photons* emitted by a blackbody in the frequency interval dv is

$$\frac{2\pi}{c^2}\frac{v^2}{e^{hv/kT}-1}dv$$

25. Calculate the energy and the momentum of the following kinds of photons: radio wave of wavelength 3.0 m, visible light of wavelength 5500 Å, X rays of wavelength 0.80 Å.

26. Some of the characteristic X rays emitted by molybdenum have a wavelength of 0.72 Å. If these X rays are scattered by a block of graphite, what is the wavelength of the X rays emerging at 45°? At 90°? At 135°?

27. A photon of initial energy 2.4×10^3 eV is deflected by 120° in a collision with a free electron, initially stationary. What energy does the electron acquire in this collision?

28. A photon of initial energy 2.4×10^3 eV collides with a free electron, initially stationary. What is the maximum energy that the electron can acquire in this collision?

29. Suppose that a photon of initial wavelength 0.60 Å suffers two collisions with two electrons. In the first collision, the photon is deflected by 30°, and in the second collision by 60°. What is the final wavelength of the photon?

30. A photon collides with a free *proton*, initially at rest. The photon is deflected by 90°. What is its change of wavelength?

31. A photon of energy 3.0×10^3 eV collides with a free electron, initially at rest. The photon emerges at an angle of $60°$. At what angle and with what energy does the electron emerge?

32. Show that the kinetic energy of the recoiling electron in the Compton effect is given by the general formula

$$\frac{1}{2} m_e v^2 = \frac{2(h/m_e c)\sin^2(\theta/2)}{c/v + 2(h/m_e c)\sin^2(\theta/2)} \, h\nu$$

†33. Using the exact relativistic formulas for the energy and momentum of an electron (see Chapter 2), derive the wavelength shift of the photon in the Compton effect. [Your final result should agree with Eq. (38).]

34. What thickness of aluminum will attenuate a beam of 0.20 MeV photons by the same factor as a layer of lead 1.0 mm thick?

35. What thickness of water gives the same attenuation of an X-ray beam as 1 m of water vapor at 100°C?

36. You want to use a layer of lead to provide shielding against X rays of wavelength 0.30 Å. What thickness of lead do you need to attenuate the X rays by a factor of 100?

37. The soft tissues in your body are mostly water. Estimate the factor by which these tissues in your hand attenuate a beam of X rays of wavelength 1 Å. The linear attenuation coefficient for X rays of 1 Å in water is about 3/cm.

38. By what factor will a beam of X rays of 0.06 MeV be attenuated when penetrating through a sandwich consisting of a 0.5-mm layer of lead and a 8.0-mm layer of aluminum?

39. A beam of X rays initially consists of a 50–50 mixture of photons of energies 1.0 MeV and 0.5 MeV. What will be the photon composition of this beam after it has penetrated through 0.10 cm of lead? By what factor is the intensity (energy flux!) of the beam reduced?

40. By inspection of Fig. 3.17, decide which material has the highest frequency of oscillation of the atoms about their equilibrium positions, and which has the lowest.

41. The specific heat of silver at $T = 47$ K is 2.58 cal/mole. According to Einstein's formula for the specific heat, what (approximate) value of the frequency of oscillation of the atoms can you deduce from this?

Atomic Structure

and Spectral Lines

Solids, liquids, and gases are made of atoms, small granules of matter with a diameter of about 10^{-10} m. The earliest evidence for the existence of atoms came from the study of chemical reactions and the behavior of gases. But this evidence was circumstantial, and until well into the twentieth century, some critics—such as the physicist and philosopher Ernst Mach—held to the opinion that atoms were merely an unproved hypothesis. We will not bother to review the cumulative circumstantial evidence for the existence of atoms. Such a review would be a superfluous exercise in historiography, since in recent years physicists have obtained direct visual proof of the existence of atoms by means of new kinds of powerful microscopes (see Fig. 4.1). However, not even these microscopes are sufficiently powerful to reveal the inside of the atom. Hence, for the exploration of the interior structure of the atom, we still have to rely on the same techniques first used by Ernest Rutherford and his associates around 1910.

By that time, most physicists had come to believe that atoms are made of some combination of positive and negative electric charges, and that the attractions and repulsions between these electric charges are the basis for all the chemical and physical phenomena observed in solids, liquids, and gases. Since electrons were known to be present in all of these forms of matter, it seemed reasonable to suppose that each atom consists of a combination of

Fig. 4.1 Tip of a platinum needle viewed with an ion microscope invented by E. W. Müller. The magnification is approximately $5 \times 10^5 \times$. The tip of the needle is a hemisphere of radius 1000 Å. The dots show the positions of individual atoms but do not faithfully represent their sizes. To prepare such a picture, the needle is placed in a vessel containing helium at low pressure, and a high voltage is applied to the needle. This produces very strong local electric fields near the protruding atoms on the surface of the needle. These local electric fields ionize atoms of helium, which are then accelerated by the fields toward a fluorescent screen. The photograph shows the pattern of impacts of helium ions on the screen. This pattern is a distorted representation of the pattern of protruding atoms on the surface of the tip of the needle. (Courtesy T. T. Tsong, Pennsylvania State University.)

electrons and positive charge. The vibrational motions of the electrons within such an atom would then result in the radiation of electromagnetic waves; this was supposed to account for the emission of light by the atom. However, both the arrangement of the electric charges within the atom and the factors determining the characteristic colors of the emitted light remained mysteries until the discovery of the nucleus by Rutherford and the discovery of the quantization of atomic states by Niels Bohr.

In this chapter we will look at these two great discoveries. Historically and logically, Bohr's "old" quantum theory of the atom stands at the borderline between classical physics and quantum physics. Although this theory has been superseded by a "new" quantum theory based on wave mechanics, it provides an excellent introduction to some of the basic concepts of quantum physics: quantization of angular momentum and of energy, stationary states, and transitions between these states.

4.1 Spectral Lines

Any putative theory of atomic structure must explain the characteristic colors of the light emitted by atoms. Light with characteristic colors is emitted whenever a sample of atoms is stimulated by heat or by an electric current. For instance, if we seal some neon gas at low pressure into a discharge tube and apply a high voltage to the terminals (see Fig. 4.2), the gas will glow and the emitted light will display the distinctive red-orange color of neon signs. If instead of neon gas we use mercury vapor, the emitted light will display a distinctive greenish color.

In order to obtain distinctive colors, we must use a sparse sample of atoms. When we use a bulky sample—such as a rod of iron—and heat it to incandescence, it will emit white light, a continuous mixture containing all colors. This is thermal radiation, whose properties we have examined in Chapter 3. In such thermal radiation, the distinctive color of the light emitted by the individual iron atoms has been destroyed by multiple scatterings that the light suffered on its way to the surface of the rod, and the energy of the light has been spread out smoothly over all colors, or wavelengths. If we want the light to retain its original features, we must make sure that the light can escape directly from the emitting atom.

For the precise analysis of the wavelengths of the light, we need a prism or a grating. In the usual experimental arrangement (see Fig. 4.3) the source of light is placed behind a narrow slit; the narrow beam of light emerging from this slit strikes a prism or a grating and breaks up into several beams, according to color. The apparatus often also contains several lenses to focus the beams of light; it is then called a **spectroscope** (see Fig. 4.4). On a screen or on a photographic film, the several emerging beams show up as a set of discrete, narrow colored lines. These are the **spectral lines.** Each kind of atom has its own discrete spectral lines. Figure 4.5 shows the spectral lines of hydrogen, as recorded on photographic film; the numbers next to the spectral lines give the wavelength in Å. Hydrogen has a red line, a blue-green line, several violet lines,

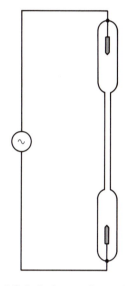

Fig. 4.2 A discharge tube used as a light source. The tube contains gas at very low pressure. When the terminals are connected to a high-voltage generator, an electric current flows through the gas and makes it glow.

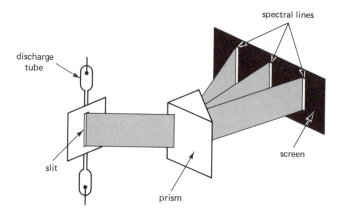

Fig. 4.3 Analysis of light by means of a prism. Each separate color of light emerging from the slit gives rise to a separate spectral line on the screen.

and many ultraviolet and infrared lines not visible to the human eye. Figures 4.6–4.8 show the spectral lines of sodium, magnesium, and mercury.

Discrete wavelengths show up both in the emission of light and in the absorption of light. If we place a sample of gas in the beam

Fig. 4.4 Prism spectroscope built by Fraunhofer and used by him for the measurement of the Fraunhofer lines in the spectrum of the Sun. The light source is an illuminated slit placed far to the left. (Courtesy Deutsches Museum.)

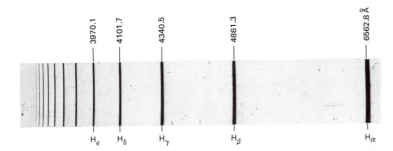

Fig. 4.5 Spectrum of hydrogen in the visible and near-ultraviolet region. The numbers give the wavelengths in air, in angstroms. (From G. Herzberg, *Atomic Spectra and Atomic Structure*. Reproduced by permission of Dover Publications.)

Fig. 4.6 Spectrum of sodium. (From G. Herzberg, ibid.)

Fig. 4.7 Spectrum of magnesium. (From G. Herzberg, ibid.)

Fig. 4.8 Spectrum of mercury. (From G. Herzberg, ibid.)

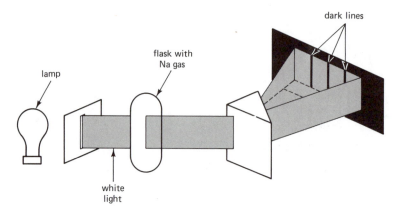

Fig. 4.9 Experimental arrangement for the observation of the absorption lines of a gas.

of white light from an incandescent lamp, the gas will selectively absorb discrete wavelengths. When we analyze the remaining light with a prism (see Fig. 4.9), the absorbed wavelengths will show up as dark lines. These are the **absorption lines.** Figure 4.10 shows the absorption lines of sodium. These lines coincide with some of the emission lines of sodium. It is a general rule that any spectral line observed in emission can also show up in absorption. However, if the absorbing gas is cool, only some absorption lines will show up; if the gas is hot, additional lines will show up. (We will see in Section 4 that this has to do with the initial states of the atoms: if the gas is cool, all atoms are in the state of lowest internal energy; if the gas is hot, some atoms are in states of higher energy.)

The distinctive spectral lines produced by a chemical element can be used as an analytical test for the presence of the element in an unknown sample. We need only heat a small portion of the sample and examine its light with a spectroscope—the spectral lines are the "fingerprints" of an atom. Tables of all the spectral lines of all the elements are a valuable tool in this identification

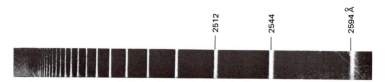

Fig. 4.10 Absorption spectrum of sodium. [From H. Kuhn, *Zeitschr. f. Physik* **76**, 782 (1932).]

TABLE 4.1 A Page from the M.I.T. Wavelength Tables

5901.9—5873.1 A.

Wave-length	Element	Arc	Spk.,[Dis.]	R
5901.911	Ta	80	–	–
5901.68	Fe	3	–	–
5901.577	Tm	5	–	–
5901.472	Mo	30	–	–
5901.421	Sm	12	–	–
5901.321	Ce	5 d	–	–
5901.227	W	8	–	–
5901.21	Cu II	–	5	Sh
5901.20	Sb II	–	[5]	Lg
5901.086	Zr I	4	–	–
5900.93	Rn I	–	[20]	Rs
5900.89	Kr II	–	[8 whl]	Me
5900.75	La I	3	–	Me
5900.674	Ce	3 w	–	–
5900.616	Cb	200	200	–
5900.430	Nd	5	–	–
5900.00	Ho	8	–	Ed
5899.757	Sm	2	–	–
5899.74	Te	–	[25]	Bl
5899.678	Mo	12	–	–
5899.576	Ir I	2	–	–
5899.491	Nd	3	–	–
5899.465	Tm	10	–	–
5899.414	Th	6	–	–
5899.40	Tb	15	–	Ed
5899.322	Ti I	150	150	–
5899.171	Nd	2	–	–
5899.02	I I	–	[25]	Db
5898.971	Ru	6	–	–
5898.962	Sm	25	–	–
5898.94	Rh	3	–	Me
5898.867	Nd	3	–	–
5898.84	Tb	25	–	Ed
5898.825	Mo	8	–	–
5898.80	Yb	3	50 h	Me
5898.785	U	8	–	–
5898.785	Mo	8	[8]	Me
5898.56	Xe I	–	[20]	Ps
5898.406	Ne I	–	25	Sh
5897.986	Cu II	–	–	–
5897.929	Ta	2 l	–	–
5897.865	Mo	5	–	–
5897.59	Gd	7	–	Ks
5897.544	V	–	30	Me
5897.47	Kr	–	[2 whl]	Me
5897.379	Sm	100	–	–
5897.22	Yb	7	100 h	Me
5896.872	Sm	3	–	–
5896.7	bh La	80	–	Me
5896.65	Te	–	[25]	Bl
5896.61	Hf	2	–	Me
5896.61	Yb	5	–	Me
5896.278	Sm	5	–	–
5896.02	In	–	5	Sq
5895.923	Na I	5000 R	500 R	Hz
5895.70	Pb	20 hl	2	Wt
5895.626	Tm	80	20	–
5895.62	Xe I	–	[2 h]	Me
5895.578	Nd	2	–	–
5895.497	Fe	4	–	–
5895.288	Eu	25	–	–
5895.196	Ta	2 h	–	–
5895.154	Sm	6	–	–
5895.09	Sb II	–	[150 wh]	Lg
5894.988	Xe I	–	[100]	IMe
5894.847	La I	25	–	–
5894.718	Sm	15 d	–	–
5894.63	Sc I	5	–	Me
5894.6	Bi II	–	6	Ml
5894.56	Kr II	–	[8 whl]	Me
5894.47	Dy	2	–	Me
5894.4	Rn	–	[30]	Ny
5894.351	Zn II	3	[30]	IHz
5894.291	Pr	15 w	–	–
5894.065	Ir I	20	–	–
5894.05	I I	–	[60]	Mu
5893.9	bh Yt	10	–	Me
5893.738	Mo	5	–	–
5893.6	bh La	60	–	Me
5893.498	Th	8	–	–
5893.46	Ge II	–	100	Lg
5893.442	Ce	3	–	–
5893.43	I II	–	[8]	Bl
5893.376	Mo	70 h	15	–
5893.29	Xe II	–	[150]	Hu

Wave-length	Element	Arc	Spk.,[Dis.]	R
5893.190	Ce	2	–	–
5892.878	Ni I	6	–	IKs
5892.670	Nd	2	–	–
5892.66	La II	1	4	Me
5892.633	U	4	–	–
5892.56	Ho	50	–	Ed
5892.448	Ta	2 l	–	–
5892.401	Sm	5 d	–	–
5892.294	Mo	20	–	–
5892.231	Pr	10	1	–
5891.906	Sm	2	–	–
5891.65	C II	–	30	Fl
5891.614	W	12	–	–
5891.562	Mo	25 h	–	–
5891.528	Nd	20	–	–
5891.43	Te	–	[15]	Bl
5891.416	Sm	15	–	–
5891.303	Ru	4	–	–
5891.29	Se	–	[8]	Bt
5891.269	Eu	200	–	–
5891.10	Tb	10	–	Ed
5890.98	S	–	[8]	Bl
5890.626	Sm	4	–	–
5890.503	Nd	5	–	–
5890.484	Co I	7	–	–
5890.45	Hf	8	–	Me
5890.333	W	7	–	–
5890.26	In	–	10	Sq
5890.16	Hg	–	[40]	Wd
5889.989	Cr	12	–	–
5889.978	Mo	50 h	–	–
5889.97	C II	–	60	Fl
5889.953	Na I	9000 R	1000 R	Hz
5889.75	S I	–	[5]	Ms
5889.74	Se II	–	[15]	Bt
5889.695	Sm	20	–	–
5889.12	Xe I	–	[20]	Me
5889.06	Tb	10	–	Ed
5888.94	Hg II	–	[20]	Ps
5888.89	Te	–	[8]	Bl
5888.78	Cr	3 l	–	–
5888.675	Ti	15	–	–
5888.6	Rn	–	[80]	Ny
5888.592	A I	–	[300]	IMe
5888.493	Ta	5	–	–
5888.326	Mo	150	100	–
5888.283	Sm	8	–	–
5888.248	Sm	2	–	–
5888.15	Tb	10	–	Ed
5888.008	Cr	20	–	–
5887.907	Nd	25	–	–
5887.758	Sc	4 wh	–	–
5887.68	Kr I	–	[3]	Me
5887.58	Ho	12	–	Ed
5887.4	bh Sc	20	–	Me
5887.364	Ir	10	–	–
5887.268	Sm	3	–	–
5887.23	Lu	1	6 h	Me
5886.952	U	3	–	–
5886.471	Mo	5	–	–
5886.458	Er	30	–	–
5886.44	Gd	6	–	Ks
5886.34	Se II	–	[20]	Bl
5886.30	Er	12	–	Ed
5886.235	Nd	3	–	–
5886.068	Sm	3	–	–
5885.714	Th	10	5	–
5885.619	Zr I	25	–	–
5885.58	Hf II	–	2	Me
5885.164	Eu	6	–	–
5884.815	Eu	6	–	–
5884.701	Pr	4	–	–
5884.625	Sm	2 h	–	–
5884.452	Cr I	18	–	–
5884.332	Mo	12	–	–
5883.848	Fe I	15	10	–
5883.663	Sm	3	–	–
5883.66	Hf	3	1	Me
5883.410	Co I	3	–	–
5883.292	Nd	15	–	–
5882.99	Ho	200	–	Ed
5882.916	Os	7	–	–
5882.81	Yb	2	10	Me
5882.784	Nd	10	–	–
5882.724	Mo	15	–	–

Wave-length	Element	Arc	Spk.,[Dis.]	R
5882.67	Kr	–	[2 whl]	Me
5882.625	A I	–	[100]	Ms
5882.493	Sm	5	–	–
5882.33	I	–	[8]	Ev
5882.30	Ir I	10	–	–
5882.295	Ta	80	–	–
5882.16	Gd	10	–	Ks
5881.8950	Ne I	–	[1000]	S
5881.526	Mo	20	–	–
5881.33	Nd	3	–	Ks
5881.18	Kr I	–	[2]	Me
5881.14	Er	30	–	Ed
5881.077	Co I	4	–	–
5880.784	Eu	3	–	–
5880.647	La II	30	50	–
5880.54	Tb	15	–	Ed
5880.306	Ti I	60	125	–
5880.25	Nd	3	–	Ks
5880.223	W	15	–	–
5880.19	Cd II	4	3	Vs
5879.994	Fe I	6	–	–
5879.94	Yt I	3	–	Me
5879.900	Kr I	–	[50]	IJa
5879.85	Tb	10	–	Ed
5879.797	Zr I	60	–	–
5879.782	Fe	8	–	–
5879.585	Sm I	2	–	–
5879.264	Sm	2	1	–
5879.253	Pr	10	–	–
5878.92	Xe I	–	[6]	Me
5878.896	Ce	3	–	–
5878.70	Se I	–	[15]	Rd
5878.378	Sm	15	–	–
5878.265	Th	8	–	–
5878.111	Pr	10 w	–	–
5878.078	Ce	2	–	–
5878.070	Sm II	20	–	–
5878.008	La I	6	–	–
5878.002	Fe	5 h	–	–
5877.827	Nd	3	–	–
5877.8	bh Sc	10	–	Me
5877.786	Cb	5	5	–
5877.77	Ti I	15	–	Rl
5877.634	La I	8	–	–
5877.56	Rn I	–	[10]	Rs
5877.425	Co I	4 h	–	–
5877.355	Ta	100	–	–
5877.24	Gd	7	–	Ks
5876.9	bh Sc	3	–	Me
5876.894	Sm II	10	–	–
5876.72	Gd	3 h	–	Ks
5876.7	Pb II	–	[40]	Ea
5876.587	Mo	25	–	–
5876.564	Cr	3	–	–
5876.344	Nd	2	–	–
5876.315	Cb	10	1	–
5876.103	Co I	4 h	–	–
5876.1	bh Yt	10	–	Me
5875.930	Sm	8	–	–
5875.989	He I	–	[10]	Ps
5875.813	Nd	2	–	–
5875.663	W	8	–	–
5875.618	He I	–	[1000]	IMr
5875.372	Fe	15 h	–	–
5875.258	Cb	5	3 wh	–
5875.13	I	–	[15]	Bl
5875.090	Sm	5	–	–
5875.018	Xe I	–	[100]	IMe
5874.736	La I	20	–	–
5874.732	Pr	4	–	–
5874.700	Cb	30	5	–
5874.70	Yb	1	30 h	Me
5874.6	Te I	–	[3 s]	Rd
5874.392	Nd	2	–	–
5874.229	W	8	–	–
5874.225	Mo	5	–	–
5874.194	Sm	40	–	–
5874.003	La II	2	6	–
5874.0	bh Yt	5	–	Me
5873.882	Ce	3	–	–
5873.52	Er	8	–	Ed
5873.500	Ir I	8	–	–
5873.342	Nd	2	–	–
5873.219	Fe I	8	2	–
5873.15	Tb	10	–	Ed

[a] Wavelengths are measured in air, with an index of refraction 1.0002765. Neutral atoms are labeled with a Roman numeral I; singly ionized atoms with II. Intensities are given on a rough scale from 1 to 9000. (From G. R. Harrison, *MIT Wavelength Tables*, © 1969 by The Massachusetts Institute of Technology. Reproduced by permission of MIT Press.)

procedure; for example, Table 4.1 is a list of all the known spectral lines in a small range of wavelengths.

Spectroscopic analysis of unknown samples has largely superseded the traditional chemical "wet" analysis. One important advantage of spectroscopic analysis is that even extremely small traces of an element can be detected. Furthermore, spectroscopic analysis plays a crucial role in astronomy—astronomers cannot pluck samples of material from the surface of a distant star, but they can examine its light and detect chemical elements by their telltale spectral lines. Usually, astronomers look for absorption lines; these are produced by layers of somewhat cooler gas lying just above the incandescent surface of a star. In the spectrum of our Sun we find many such absorption lines, called the **Fraunhofer lines;** they indicate the presence of hydrogen, sodium, iron, calcium, etc. (see Fig. 4.11). In the spectra of other stars we sometimes find both absorption and emission lines, indicating the presence of a variety of elements (see Fig. 4.12).

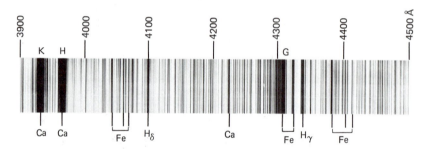

Fig. 4.11 Absorption lines in the spectrum of the Sun. (Mount Wilson and Las Campanas Observatory.)

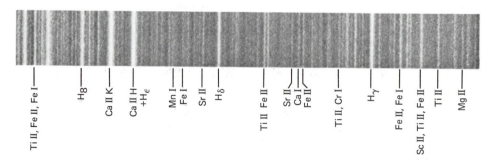

Fig. 4.12 Absorption lines in the spectrum of the star ϕ Cassiopeiae. (From W. W. Morgan, H. A. Abt, and J. W. Tapscott, *Revised MK Spectral Atlas*.)

Inspection of the spectral lines in photographs of spectra reveals that some sets of these lines exhibit systematic regularities in spacing and in intensity. Such regularities are strikingly obvious in the hydrogen spectrum and in the sodium absorption spectrum. In Figs. 4.5 and 4.10 we see that spacings between the lines and the intensities of the lines decrease systematically with wavelength. In each of these spectra the wavelengths of the lines form an infinite sequence, converging to a lower limit in wavelength. Such sets of spectral lines exhibiting systematic regularities are called **spectral series:** the Balmer series of hydrogen, and the principal series of sodium. Of course, we also find the principal series in the emission spectrum of sodium (see Fig. 4.6), but there it overlaps with other spectral series, and is therefore somewhat harder to recognize. In the spectra of other elements, we find many other spectral series, usually overlapping one another.

4.2 **The Spectral Series of Hydrogen**

The Balmer series of hydrogen is named after J. Balmer* who, in 1885, speculated that the systematic regularities in the spacings of the spectral lines should be represented by some simple mathematical formula. Examining the numbers in a table of measured wavelengths, he discovered that these wavelengths accurately fit the formula

$$\lambda = 911.76 \, \text{Å} \times \frac{4n^2}{n^2 - 4} \tag{1}$$

with $n = 3, 4, 5$, etc. Note that for $n \to \infty$, the wavelength approaches the asymptotic value $\lambda = 3647 \, \text{Å}$; this is the series limit of the Balmer series. In terms of frequencies, Eq. (1) becomes

$$v = \frac{c}{\lambda} = \frac{c}{911.76 \, \text{Å}} \left(\frac{1}{4} - \frac{1}{n^2} \right) \tag{2}$$

or

$$v = cR_{\text{H}} \left(\frac{1}{2^2} - \frac{1}{n^2} \right) \tag{3}$$

---

* **Johann Jakob Balmer,** 1825–1898, Swiss mathematics teacher and lecturer at the University of Basle. Besides finding the formula for the Balmer series, he predicted the other spectral series of hydrogen.

TABLE 4.2 The Balmer Series in the Hydrogen Spectrum

Wavelength, λ^a
6564.7 Å
4862.7
4341.7
4102.9
3971.2
3890.2
3836.5
3799.0

[a] Wavelengths are measured in vacuum.

where R_H is called the **Rydberg constant,**[*] $R_H = 1/(911.76 \text{ Å}) = 109{,}678 \text{ cm}^{-1}$.

Balmer's formula does not provide any explanation of the atomic mechanism responsible for the production of the spectral lines; the formula is merely descriptive, or phenomenological. However, Balmer recognized that his formula lends itself to a generalization, and he suggested that there might be other spectral series in the hydrogen spectrum, with 2^2 in Eq. (3) replaced by 1^2, or 3^2, or 4^2, or 5^2, etc. This yields several new spectral series with frequencies

$$v = cR_H\left(\frac{1}{1^2} - \frac{1}{n^2}\right) \qquad n = 2, 3, 4, \ldots \qquad (4)$$

$$v = cR_H\left(\frac{1}{3^2} - \frac{1}{n^2}\right) \qquad n = 4, 5, 6, \ldots \qquad (5)$$

$$v = cR_H\left(\frac{1}{4^2} - \frac{1}{n^2}\right) \qquad n = 5, 6, 7, \ldots \qquad (6)$$

$$v = cR_H\left(\frac{1}{5^2} - \frac{1}{n^2}\right) \qquad n = 6, 7, 8, \ldots \qquad (7)$$

These four new spectral series were discovered some years after Balmer proposed them. They are called, respectively, the Lyman, Paschen, Brackett, and Pfund series. The Lyman series lies in the far ultraviolet, and the other series lie in the infrared (see Fig. 4.13).

Example 1 What are the series limits of the Lyman, Paschen, Brackett, and Pfund series?

---

[*] **Johannes Robert Rydberg,** 1854–1919; Swedish physicist, professor at Lund. Independently of Balmer, he discovered the general formula for the spectral series of hydrogen.

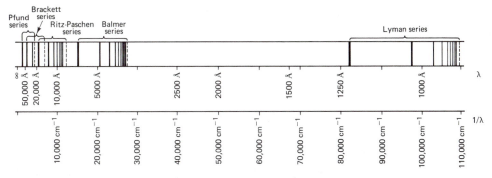

Fig. 4.13 Schematic diagram of the spectrum of hydrogen. In this diagram, the intensity of the spectral lines is roughly indicated by thickness. (From G. Herzberg, ibid.)

Solution   To obtain the series limits, we must let $n \to \infty$ in Eqs. (4)–(7). This gives

$$v = cR_{\mathrm{H}}, \quad v = \frac{cR_{\mathrm{H}}}{9}, \quad v = \frac{cR_{\mathrm{H}}}{16}, \quad \text{and} \quad v = \frac{cR_{\mathrm{H}}}{25}$$

respectively. The numerical values of the corresponding wavelengths are 911.76 Å, 9 × 911.76 Å, 16 × 911.76 Å, and 25 × 911.76 Å. ■

We can combine all the frequencies of all the spectral series of hydrogen in one concise general formula:

$$v = cR_{\mathrm{H}} \left(\frac{1}{n_2^2} - \frac{1}{n_1^2} \right) \tag{8}$$

where n_1 and n_2 are positive integers, and $n_1 > n_2$. Note that this formula expresses the frequencies as differences between two terms, cR_{H}/n_2^2 and cR_{H}/n_1^2. This is found to be a general feature of the frequencies of the spectral lines of all the atoms: in all cases the frequencies can be expressed as differences between pairs of terms, although the terms for these other atoms do not have the simple mathematical form of the terms for the hydrogen atom. The tabulation of the larger number of different frequencies emitted by an atom can therefore be replaced by a tabulation of a smaller number of terms. Not only is every frequency represented by the difference between two terms, but conversely, every difference between two terms corresponds to an observed frequency. The latter assertion is called the **Rydberg–Ritz Combination Principle.** As we will see in Section 4, the terms are proportional to the permitted (quantized) energies of the atom, and the term differences are proportional to energy differences.

4.3 **The Nuclear Atom**

In the early theoretical investigations of the structure of the atom, the electromagnetic character of light provided an important clue. According to Maxwell's equations, light is an electromagnetic wave, and, like all electromagnetic waves, it must be generated by electric charges executing some kind of periodic motion. This led H. A. Lorentz to the conclusion that there must be electric charges within the atom. As a test of this hypothesis, he proposed placing atoms in a strong magnetic field, which would disturb the motions of the charges within the atom and cause alterations of the frequency of the emitted light. The first experiment in search of this effect was performed by P. Zeeman* in 1896, and it verified Lorentz' prediction. The shift of frequency or wavelength of spectral lines emitted by atoms in a magnetic field came to be called the **Zeeman effect;** we will deal with it in detail in Chapter 7. On the basis of classical theory, Lorentz had concluded that the frequency shift should be proportional to the strength of the magnetic field and to the charge-to-mass ratio of the radiating charges within the atom;[†] comparing this theoretical result with the experimental measurements of Zeeman, he deduced a charge-to-mass ratio of about 10^{11} C/kg.

A year later, Thomson discovered the electron in his experiments with cathode rays, and he found that it had this same charge-to-mass ratio of about 10^{11} C/kg. From this coincidence, Thomson immediately drew the correct conclusion that the charged particle in cathode rays and the charged particle in the atom were one and the same. Using the electron as an ingredient, Thomson then attempted to design a model of the atom and to explain spectral lines on the basis of the internal structure of the

---

* **Pieter Zeeman,** 1865–1943, Dutch experimental physicist, professor at Amsterdam. For the discovery of the Zeeman effect, he and H. A. Lorentz shared the Nobel Prize in 1902.

[†] In retrospect, we now know that Lorentz' calculation was not quite correct—he took into account the orbital motion of the charges, but not their intrinsic rotation, or "spin." The spin makes an important contribution to the Zeeman effect.

atom. He adopted a model first proposed by Lord Kelvin: an atom consists of several electrons, say Z electrons, embedded in a cloud of positive charge. The cloud is heavy, carrying almost all of the mass of the atom. The charge of the cloud is $+Ze$, so that it exactly neutralizes the charge $-Ze$ of the electrons. In an undisturbed atom the electrons will sit at their equilibrium positions, where the attraction of the cloud on the electrons balances their mutual repulsion (see Fig. 4.14). But if the electrons are disturbed by, say, a collision, then they will vibrate around their equilibrium positions and emit light. Thomson's model of the atom, sometimes called the "plum-pudding model," does yield frequencies of vibration of the same order of magnitude as the frequency of light, but it does not yield the observed spectral series; for instance, on the basis of this model, hydrogen should have only one single spectral line, in the far ultraviolet (see Problem 9).

Direct experimental evidence against the Thomson model of the atom emerged in 1910 from the work of Ernest Rutherford* and his collaborators. Rutherford had been studying the emission of alpha particles by radioactive substances. As we saw in Section 1.3, alpha particles carry a positive charge $2e$ and they have a mass of 6.6×10^{-27} kg. Some radioactive substances, such as radioactive polonium and radioactive bismuth (^{218}Po and ^{214}Bi or, in the terminology of Rutherford's days, RaA and RaC), emit alpha particles with energies of several MeV. Such energetic alpha particles readily pass through thin foils of metal, or thin sheets of glass, or other materials. Rutherford was much impressed by the penetrating power of these alpha particles and it occurred to him that a beam of these particles can serve as a probe to "feel" the interior of the atom. When energetic alpha particles strike a foil of metal, they penetrate the atoms and are deflected by collisions with the subatomic structures; the observed magnitude and angular distribution of these deflections yield clues about the subatomic structures. For example, if the interior of the atom had the "plum-pudding" structure proposed by J. J. Thomson, then the alpha particles would suffer only very small deflections since neither the

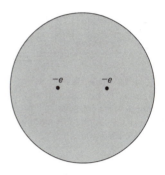

Fig. 4.14 Thomson's model of the helium atom.

---

* **Ernest Rutherford,** first baron Rutherford, 1871–1937, British experimental physicist, professor at McGill University, Manchester, Cambridge, and director of the Cavendish Laboratory, where he succeeded J. J. Thomson. He received the Nobel Prize in Chemistry in 1908 for his discovery of the transmutation of elements by radioactive decay and for the identification of alpha particles as ions of helium. After establishing the existence of the nucleus, Rutherford laid the foundations of nuclear physics with his discovery of artificial nuclear reactions initiated by bombardment with alpha particles.

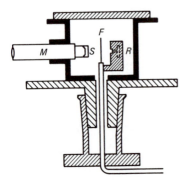

Fig. 4.15 Apparatus of Geiger and Marsden. *R*, radium source; *F*, foil; *S*, zinc sulfide screen; *M*, microscope. [After H. Geiger and E. Marsden, *Phil. Mag.* **25**, 604 (1913).]

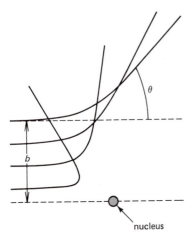

Fig. 4.16 Hyperbolic trajectories of several alpha particles deflected by the nucleus. The distance *b* is the impact parameter of one of these alpha particles.

electrons, with their small masses, nor the diffuse cloud of positive charge would be able to disturb the motion of a massive and energetic alpha particle.

The crucial experiments were performed by H. Geiger and E. Marsden* working under Rutherford's direction. They used thin foils of gold and of silver as targets and bombarded these with a beam of alpha particles from a RaC source. The alpha particles that passed through the foil were detected with a zinc sulfide screen on which the impact of each particle becomes visible as a faint scintillation (see Fig. 4.15). To Rutherford's amazement, some of the alpha particles were deflected by angles of nearly 180°, emerging in the backward direction. In Rutherford's own words: "It was quite the most incredible event that has ever happened to me in my life. It was almost as incredible as if you fired a 15-inch shell at a piece of tissue paper and it came back and hit you." Rutherford soon recognized that the large deflection must be produced by a close encounter between the alpha particle and a very small but very massive kernel inside the atom. He therefore adopted the following nuclear model: an atom consists of a very small nucleus of charge $+Ze$ containing almost all of the mass of the atom; this nucleus is surrounded by a swarm of Z electrons. Such nuclear models, sometimes called planetary models because of their resemblance to the solar system, had been proposed earlier by several physicists, but Rutherford deserves the credit for placing the model on a firm experimental foundation.

Taking his nuclear model of the atom as a starting point and relying on classical mechanics, Rutherford calculated what fraction of the alpha particles in the incident beam should be deflected through what angle, and he established that these calculations were in close agreement with the experimental observations. If an alpha particle passes near the nucleus it will experience a large electric repulsion and it will be deflected by a large angle; if it passes far from the nucleus it will only be deflected by a small angle. Figure 4.16 shows the trajectories of several alpha particles approaching a nucleus; these trajectories are hyperbolas. The perpendicular distance between the nucleus and the original (undeflected) line of motion is called the **impact parameter**. According to classical mechanics, the angle of deflection can be expressed as a function of the

* **Hans Wilhelm Geiger,** 1882–1945, German physicist, head of the Radioactivity Laboratories at the Physikalische-Technische Reichsanstalt, professor at Kiel, Tübingen, and Berlin-Charlottenburg; one of Geiger's first contributions to science was a radiation detector, now usually called the Geiger–Müller counter. Sir **Ernest Marsden,** 1889–1970, physicist from New Zealand, professor at Victoria University College, Secretary of the New Zealand Department of Research.

impact parameter. A fairly simple calculation* establishes that for an alpha particle of energy E and impact parameter b incident on a stationary nucleus of charge Ze, the angle of deflection is

$$\theta = 2 \cot^{-1} \left(\frac{4\pi\varepsilon_0 Eb}{Ze^2} \right) \tag{9}$$

In order to suffer a large deflection, the alpha particle must strike an atom with a very small impact parameter, 10^{-13} m or less; since the alpha particles in the beam impact on the foil of metal at random, only very few of them will score such a close hit.

Example 2   What impact parameter will give a deflection of 1° for an alpha particle of 7.7 MeV incident on a gold nucleus? What impact parameter will give a deflection of 30°?

Solution   With $E = 7.7\,\text{MeV} = 1.2 \times 10^{-12}\,\text{J}$ and $Z = 79$, Eq. (9) yields

$$b = \frac{Ze^2}{4\pi\varepsilon_0 E} \cot\left(\frac{\theta}{2}\right)$$

$$= \frac{79 \times (1.6 \times 10^{-19})^2}{4\pi \times 8.85 \times 10^{-12}\,\text{F/m} \times 1.2 \times 10^{-12}\,\text{J}} \cot\left(\frac{\theta}{2}\right) \tag{10}$$

$$= 1.48 \times 10^{-14} \cot\left(\frac{\theta}{2}\right) \tag{11}$$

where the impact parameter is measured in meters. For $\theta = 1°$ and for $\theta = 30°$, this becomes, respectively, $b = 1.7 \times 10^{-12}\,\text{m}$ and $b = 5.5 \times 10^{-14}\,\text{m}$. ∎

Example 3   Compare the probabilities for deflection by angles in excess of 1° with the probability for deflection by angles in excess of 30° for an alpha particle of 7.7 MeV incident on a gold atom.

Solution   According to Example 2, if the alpha particle is to be deflected by more than 1°, it must have an impact parameter of less than 1.7×10^{-12} m, i.e., its original line of motion must strike within a circular area $\pi \times (1.7 \times 10^{-12}\,\text{m})^2$ centered on the nucleus (see Fig. 4.17). To be deflected by more than 30°, its original line of motion must strike within a circular area $\pi \times (5.5 \times 10^{-14}\,\text{m})^2$. Since the particles strike at random over the entire frontal area of the atom, the probabilities for these deflections are in the same proportion as these target areas,

$$\frac{(\text{probability for } \theta \geq 1°)}{(\text{probability for } \theta \geq 30°)} = \frac{\pi \times (1.7 \times 10^{-12}\,\text{m})^2}{\pi \times (5.5 \times 10^{-14}\,\text{m})^2}$$

$$= 940 \tag{12}$$

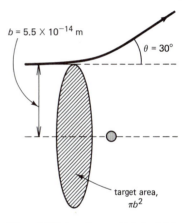

$b = 5.5 \times 10^{-14}$ m

$\theta = 30°$

target area, πb^2

Fig. 4.17 Imaginary target area within which the alpha particle must be aimed to suffer a deflection in excess of the specified value.

---

* The calculation exploits conservation of energy and angular momentum along the orbit. For details see, e.g., M. Born, *Atomic Physics*.

Thus, whenever 940 incident particles are deflected by angles larger than 1°, only one of these will be deflected by an angle larger than 30° (on the average). This shows very clearly that large deflections are very rare. ■

The target area that corresponds to a given deflection is called the **cross section** for that deflection, usually designated by σ. For instance, according to the calculation of Example 3, the cross section for deflections greater than 30° is

$$\sigma = \pi \times (5.5 \times 10^{-14}\,\mathrm{m})^2 = 9.6 \times 10^{-27}\,\mathrm{m}^2$$

Rutherford also estimated the size of the nucleus. He found that large-angle deflections of alpha particles by the nucleus of aluminum differed from the deflections calculated on the basis of purely electric repulsion—in these deflections the particles must have approached the nucleus so closely that they came in contact with the nuclear surface and suffered some extra deviation. The on-set of these extra deviations indicated that the nucleus had a diameter of about $10^{-14}\,\mathrm{m}$.

The deflection of particles of a beam by impact on a target is called **scattering.** Rutherford's method of using scattering experiments to explore subatomic structures has found widespread application in modern physics. In 1910, physicists did not have available projectiles of sufficient energy to penetrate the nucleus—Rutherford's alpha particles barely managed to reach the nuclear surface. But starting in 1930, physicists began to build accelerators capable of producing intense beams of high-energy particles. These projectiles served to probe the nucleus and even to smash the nucleus apart. The largest of contemporary accelerators produce projectiles of sufficient energy to probe the interior of protons and neutrons. We will discuss these explorations of the interiors of nuclei and of protons and neutrons in later chapters.

4.4 Bohr's Theory

According to classical physics, the nuclear atom of Rutherford suffers from a serious lack of stability. Since the negative electrons experience the electric attraction of the positive nucleus, they cannot remain at rest in space at some distance from the nucleus; instead they must orbit around the nucleus, like planets around the Sun. But according to classical electrodynamics, an accelerated electron will radiate electromagnetic waves and thereby lose energy. A simple calculation shows that as a consequence of this energy loss, an electron orbiting around the nucleus of, say, a hydrogen atom will spiral inward and impact on the nucleus in less than 10^{-10} s. This means that the atom is unstable—it collapses almost instantaneously. Furthermore, during this process of collapse, the electron emits a burst of electromagnetic waves of continuously ascending frequency (a glissando "scream") as it orbits faster and faster around the nucleus. This is in stark conflict with the facts: hydrogen atoms are very stable, they have no tendency to collapse; and when they do emit radiation, the spectrum has a discrete set of frequencies, not a continuous range of frequencies.

In 1913, Niels Bohr* took a bold step toward resolving these difficulties. He proposed that the orbits and energies of the hydrogen atom are quantized, i.e., only certain discrete orbits and energies are permitted. This quantization of the hydrogen atom is analogous to Planck's quantization of the harmonic oscillator. The quantization implies that the laws of classical mechanics and of classical electromagnetism are inapplicable at the atomic level, and that they must be replaced, or supplemented, by new laws. Bohr summarized these new laws of atomic mechanics as follows:[†]

1. *When an electron is in one of the quantized orbits, it does not emit any electromagnetic radiation; thus, the electron is said to be in a **stationary state**. The electron can make a discontinuous transition, or quantum jump, from one stationary state to another. During this transition it does emit radiation.*

---

* **Niels Bohr,** 1885–1962, Danish theoretical physicist, director of the Institute of Theoretical Physics (now the Niels Bohr Institute) in Copenhagen. He studied under J. J. Thomson and Rutherford in England. After formulating his "old" theory of the structure of the atom—for which he received the Nobel Prize in 1922—he played a leading role in the development and the interpretation of the "new" quantum theory based on wave mechanics. Later, Bohr's interests turned to nuclear physics, to which he contributed the liquid-drop model of the nucleus, the compound model for nuclear reactions, and, with John A. Wheeler, the first theoretical analysis of the process of nuclear fission. He was influential in the foundation of the Centre Européen de la Recherche Nucléaire (CERN). At Bohr's death, his son Aage Niels Bohr, 1922–   , succeeded him as director of the Institute, and won the 1975 Nobel Prize (jointly with Ben Mottelson, 1926–   , and James Rainwater, 1917–   ), for theoretical work on the structure of the nucleus.

[†] These postulates are taken, with slight editorial modifications, from the concluding summary of Bohr's three papers in *Phil. Mag.* **26,** 1, 476, 857 (1913). In the first of these papers, Bohr begins by deriving the quantization of energy from a rather muddled argument about the frequency radiated when the electron is initially unbound and then settles into a bound orbit; the quantization of angular momentum makes its appearance only as an afterthought.

2. *The laws of classical mechanics apply to the orbital motion of the electrons in a stationary state, but these laws do not apply during the transition from one state to another.*
3. *When an electron makes a transition from one stationary state to another, the energy difference ΔE is released as a single photon of frequency $v = \Delta E/h$.*
4. *The permitted orbits are characterized by quantized values of the orbital angular momentum. This angular momentum is always an integer multiple of $h/2\pi$, i.e.,*

$$L = n\frac{h}{2\pi} = n\hbar \qquad n = 1, 2, 3, \ldots \tag{13}$$

where

$$\hbar = \frac{h}{2\pi} = 1.054589 \times 10^{-34}\,\text{J}\cdot\text{s}$$

For the simple case of circular orbits, it is quite easy to use the above postulates to find the sizes of the quantized orbits and their quantized energies. The attractive Coulomb force between the electron (charge $-e$) and the nucleus (charge $+e$) is $e^2/(4\pi\varepsilon_0 r^2)$. According to postulate 2, the equation of motion for an electron in a circular orbit is then

$$\frac{m_e v^2}{r} = \frac{1}{4\pi\varepsilon_0}\frac{e^2}{r^2} \tag{14}$$

According to postulate 4, the orbital angular momentum must be $\hbar$ multiplied by an integer,

$$L = m_e v r = n\hbar \qquad n = 1, 2, 3, \ldots \tag{15}$$

The number n is called the **quantum number,** or, more precisely, the **angular-momentum quantum number.** From Eq. (15) we find

$$v^2 = \frac{n^2\hbar^2}{m_e^2 r^2} \tag{16}$$

which, when substituted into Eq. (14), gives us the radius of the orbit:

$$r = \frac{4\pi\varepsilon_0}{e^2}\frac{n^2\hbar^2}{m_e}$$

We will label this radius with the subscript n,

$$r_n = \frac{4\pi\varepsilon_0}{e^2}\frac{n^2\hbar^2}{m_e} \tag{17}$$

The radius of the smallest permitted orbit ($n = 1$) is

$$r_1 = \frac{4\pi\varepsilon_0}{e^2} \frac{\hbar^2}{m_e} \qquad (18)$$

This is called the **Bohr radius,** usually designated by a_0,

$$a_0 = \frac{4\pi\varepsilon_0}{e^2} \frac{\hbar^2}{m_e} = 0.5291771 \times 10^{-10}\, m = 0.5291771\ \text{Å} \qquad (19)$$

Figure 4.18 shows the permitted circular orbits, drawn to scale.

The energy of the electron in one of these orbits is the sum of the kinetic energy $\frac{1}{2}m_e v^2$ and the potential energy $-e^2/4\pi\varepsilon_0 r$,

$$E_n = \frac{1}{2}m_e v^2 - \frac{e^2}{4\pi\varepsilon_0 r}$$

$$= \frac{1}{2}m_e\left(\frac{n^2\hbar^2}{m_e^2}\right)\left(\frac{e^2 m_e}{4\pi\varepsilon_0 n^2 \hbar^2}\right)^2 - \frac{e^2}{4\pi\varepsilon_0}\left(\frac{e^2 m_e}{4\pi\varepsilon_0 n^2 \hbar^2}\right) \qquad (20)$$

or

$$E_n = -\frac{e^4}{2(4\pi\varepsilon_0)^2} \frac{m_e}{\hbar^2} \frac{1}{n^2} \qquad n = 1, 2, 3, \ldots \qquad (21)$$

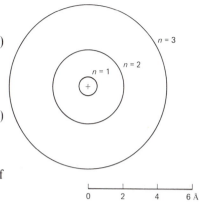

Fig. 4.18 Circular orbits in the hydrogen atom according to Bohr's theory.

The energy of the stationary state of least energy, or **ground state** ($n = 1$), is

$$E_1 = -\frac{e^4}{2(4\pi\varepsilon_0)^2} \frac{m_e}{\hbar^2} = -2.18 \times 10^{-18}\, J = -13.6\,\text{eV} \qquad (22)$$

The energies of the other stationary states, or **excited states,** are

$$E_n = -\frac{13.6\,\text{eV}}{n^2} \qquad (23)$$

Figure 4.19 displays the quantized energies of the hydrogen atom in an **energy-level diagram.** If the electron is in the ground state, its binding energy, or ionization energy, is 13.6 eV; this is the energy required to remove the electron from the atom. Whenever the atom is left undisturbed, the electron settles into the ground state. To raise the electron to an excited state, we must subject the atom to some external disturbance in the form of heat or an electric current. Collisions between the atom and other atoms, or between the atom and the current carriers, then perturb the motion of the electron and occasionally push it into an excited state. From there, the electron makes a transition into a lower excited state or

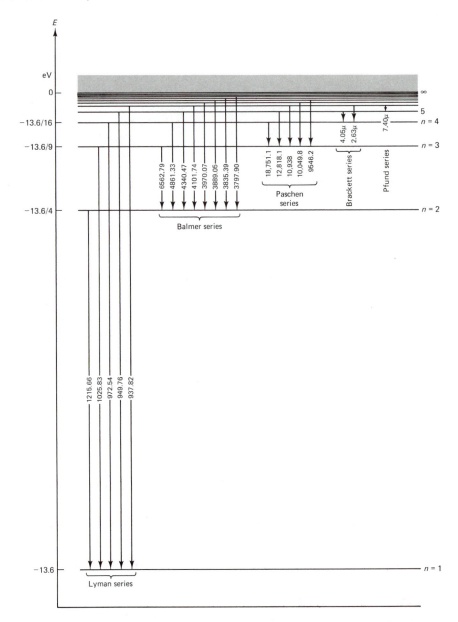

Fig. 4.19 Energy-level diagram for the hydrogen atom. The arrows indicate the transitions that give rise to different spectral series; the numbers give the wavelengths in air.

into the ground state. These downward transitions usually occur *spontaneously*, without the need of any further external disturbance. According to Bohr's postulate 1, such downward transitions involve the emission of radiation. (But Bohr's theory does not tell us how soon the electron makes its spontaneous downward jump, nor does it tell us into which of the available lower states the electron will jump; hence the theory does not permit us to calculate the intensities of the spectral lines.)

Figure 4.19 shows that the transitions can be grouped into several series according to the lower state into which the electron jumps. These series of transitions give rise to the Lyman, Balmer, Paschen, Brackett, Pfund, etc., series of spectral lines. In a transition from an initial state i to a final state f, the electron releases an energy

$$\Delta E = E_i - E_f = \frac{e^4}{2(4\pi\varepsilon_0)^2} \frac{m_e}{\hbar^2} \left(\frac{1}{n_f^2} - \frac{1}{n_i^2} \right) \tag{24}$$

According to postulate 3, this energy is radiated as a single photon of frequency

$$\nu = \frac{\Delta E}{h} = \frac{E_i - E_f}{h} = \frac{e^4}{4\pi(4\pi\varepsilon_0)^2} \frac{m_e}{\hbar^3} \left(\frac{1}{n_f^2} - \frac{1}{n_i^2} \right) \tag{25}$$

This equation coincides with Balmer's formula,

$$\nu = cR_{\mathrm{H}} \left(\frac{1}{n_f^2} - \frac{1}{n_i^2} \right) \tag{26}$$

Comparison of Eqs. (25) and (26) yields the following theoretical formula for the Rydberg constant:

$$R_{\mathrm{H}} = \frac{e^4 m_e}{4\pi(4\pi\varepsilon_0)^2 \hbar^3 c} \tag{27}$$

With the best values of the physical constants given in Appendix 1, we obtain

$$R_{\mathrm{H}} = \frac{(1.602189 \times 10^{-19}\,\mathrm{C})^4 \times 9.10953 \times 10^{-31}\,\mathrm{kg}}{4\pi(4\pi \times 8.854188 \times 10^{-12}\,\mathrm{F/m})^2 \times (1.054589 \times 10^{-34}\,\mathrm{J\cdot s})^3 \times 2.997925 \times 10^8\,\mathrm{m/s}}$$

$$= 109{,}737\,\mathrm{cm}^{-1}$$

This theoretical value of R_{H} agrees quite well with the experimental value $R_{\mathrm{H}} = 109{,}678\,\mathrm{cm}$ quoted in Section 2. And the agreement can be improved by recognizing that our calculation is in need of a small correction. In our calculation we have assumed that the electron orbits around a nucleus which remains fixed, but actually both the electron and the nucleus orbit about their common center of mass. The following example shows how to take into account this motion of the nucleus about the center of mass.

Example 4   Repeat the calculation above, including the motion of the nucleus.

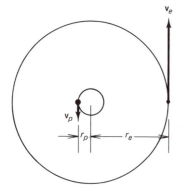

Solution   Figure 4.20 shows the orbits of the electron and the nucleus around their center of mass. The radii of these orbits are in inverse proportion to the masses,

$$r_e = r \frac{m_p}{m_p + m_e}, \qquad r_p = r \frac{m_e}{m_p + m_e} \tag{28}$$

where r is the distance from nucleus to electron and m_p is the mass of the nucleus (mass of the proton). Likewise, the speeds are in the same proportion,

$$v_e = v \frac{m_p}{m_p + m_e}, \qquad v_p = v \frac{m_e}{m_p + m_e} \tag{29}$$

where v is the speed of the electron *relative* to the nucleus ($v = v_e + v_p$).
The energy of the system is then

$$E = \frac{1}{2} m_e v_e^2 + \frac{1}{2} m_p v_p^2 - \frac{e^2}{4\pi\varepsilon_0 r}$$

$$= \frac{1}{2} m_e \left(\frac{v m_p}{m_p + m_e} \right)^2 + \frac{1}{2} m_p \left(\frac{v m_e}{m_p + m_e} \right)^2 - \frac{e^2}{4\pi\varepsilon_0 r}$$

$$= \frac{1}{2} \left(\frac{m_e m_p}{m_p + m_e} \right) v^2 - \frac{e^2}{4\pi\varepsilon_0 r} \tag{30}$$

The angular momentum of the system is

$$L = m_e v_e r_e + m_p v_p r_p$$

$$= m_e v r \left(\frac{m_p}{m_p + m_e} \right)^2 + m_p v r \left(\frac{m_e}{m_p + m_e} \right)^2$$

$$= \left(\frac{m_e m_p}{m_p + m_e} \right) v r \tag{31}$$

Finally, the equation of motion of the electron is

$$m_e \frac{v_e^2}{r_e} = -\frac{e^2}{4\pi\varepsilon_0 r}$$

which, with the substitution (28) and (29) for v_e and r_e, becomes

$$\left(\frac{m_e m_p}{m_p + m_e} \right) \frac{v^2}{r} = -\frac{e^2}{4\pi\varepsilon_0 r} \tag{32}$$

Fig. 4.20 Circular orbits of the electron and the nucleus around their common center of mass. The size of the orbit of the nucleus has been exaggerated; it is actually 1800 times smaller than that of the electron.

Equations (30), (31), and (32) have a mathematical form similar to Eqs. (20), (15), and (14); the only difference between these two sets of equations is that the former equations have the **reduced mass**

$$\mu = \frac{m_e m_p}{m_e + m_p} \tag{33}$$

in place of the electron mass m_e. Hence, the final result of our calculation will involve the same replacement,

$$E_n = -\frac{e^4}{2(4\pi\varepsilon_0)^2} \frac{\mu}{\hbar^2} \frac{1}{n^2} \tag{34}$$

The theoretical expression for the Rydberg constant then becomes

$$R_{\mathrm{H}} = \frac{e^4 \mu}{4\pi(4\pi\varepsilon_0)^2\hbar^3 c} = 109{,}678 \text{ cm}^{-1} \tag{35}$$

This revised theoretical value agrees with the experimental value to within six significant figures!*

The result (34) is of particular interest in connection with the spectrum of heavy hydrogen, or deuterium. This atom has a nucleus with about twice the mass of a proton. Hence the reduced mass for this atom is somewhat larger than that for ordinary hydrogen—and all the spectral lines have slightly higher frequencies. This is called the **isotope shift** of spectral lines. The existence of deuterium was first demonstrated by observation of the isotope shift in spectral lines emitted by samples of natural hydrogen. ∎

Thus, we see that Bohr's theory gives a successful explanation of the spectral lines emitted by the hydrogen atom on the basis of a model of the internal structure of this atom. Besides, the theory gives a value for the atomic radius which agrees quite well with typical values of atomic radii deduced from measurements on crystals and gases.

As a further refinement of Bohr's theory, A. Sommerfeld[†] laid down rules for the quantization of the periodic components of the motion in any general system of particles. He dealt with the

---

* This means that these values agree to within the uncertainties in our knowledge of the constants m_e and $\hbar$. Actually, this comparison is not quite fair because R_{H} is an ingredient in the determination of the accepted values of m_e and $\hbar$.

[†] **Arnold Sommerfeld**, 1868–1951, German theoretical physicist, professor at Munich, and director of the Institute of Theoretical Physics. Sommerfeld was a gifted teacher and he attracted many brilliant students, among them Debye, Pauli, Heisenberg, and Bethe. He was the author of *Atombau und Spektrallinien* (Atomic Structure and Spectral Lines), the earliest and most influential treatise on quantum mechanics.

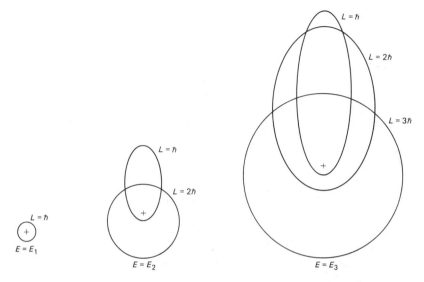

Fig. 4.21 Elliptical orbits in the hydrogen atom according to Sommerfeld. Ellipses with the same major axis have the same energy, except for small relativistic corrections.

elliptical orbits of the hydrogen atom, and also took into account relativistic corrections to the energy of the electron. A well-known theorem of Newtonian mechanics tells us that, for a particle moving under the influence of an inverse square force, circular and elliptical orbits with the same major axis have the same energy (see Fig. 4.21). However, the relativistic corrections to the kinetic energy gives rise to a small difference between the energies of circular and elliptic orbits. When Sommerfeld calculated these relativistic corrections, he found that he could account for the **fine structure** of the hydrogen spectrum: when examined with a spectroscope of very high resolving power, all the hydrogen spectral lines are seen to be **multiplets,** i.e., groups of lines of nearly identical wavelength.

But we must not let these spectacular successes of Bohr's theory blind us to the defects. The theory is an incongruous medley of classical and nonclassical ideas. On the one hand, we are asked to accept that the forces within the atom are given by the classical laws of electromagnetism; on the other hand, we are asked to discard the formula for the emission of radiation implied by these very same laws. Bohr's theory did not explain *why* angular momentum is quantized, nor *why* atoms make transitions. Furthermore, this theory failed miserably in all attempts at calculating the spectrum of helium and other atoms with more than one electron.

All of these defects are remedied by the new quantum theory, based on wave mechanics, that we will develop in the next chapters. Bohr's great and lasting contribution was his recognition that atoms have stationary states with quantized energies and that radiation is due to transitions between these states. The new quantum theory retains this aspect of Bohr's old theory; but the new theory gives us a completely different description of the stationary states, a description having nothing to do with classical orbits.

4.5 The Correspondence Principle

As is obvious from the energy-level diagram for the hydrogen atom (see Fig. 4.19), the energy difference between one energy level and the next is very small for large values of the quantum number n. Likewise, the difference between the radius of one Bohr orbit and the next is also very small. Thus, if an electron in the stationary state with, say, $n = 4000$ jumps down, step by step, to the next lowest state ($n = 3999$), and then to the next ($n = 3998$), etc., the steps in energy and radius are very small, and the changes in energy and radius will appear to proceed almost continuously. The difference in angular momentum between one state and the next has the fixed value $\hbar$, which does not become smaller for large values of n. Nevertheless, the changes in angular momentum will also appear to proceed almost continuously because each step in angular momentum is very small compared with the remaining angular momentum, i.e., the fractional decrement from, say, $L = 4000\hbar$ to $L = 3999\hbar$ is very small. Obviously, under these conditions the electron will behave pretty much as a classical particle—the effects of quantization will be hardly noticeable.

The foregoing arguments establish a concordance between classical and quantum mechanics in the limiting case of large quantum numbers. Bohr formulated a useful generalization of this concordance by requiring that it apply not only to the mechanical behavior of the electron, but also to its electromagnetic behavior in the emission of radiation. He expressed this requirement in his **Correspondence Principle:** *in the limiting case of large quantum numbers, the frequencies and the intensities of radiation calculated from classical theory must agree with those of quantum theory.*

It is easy to verify explicitly that the frequencies of the light emitted by an electron in a hydrogen atom are consistent with this Correspondence Principle, i.e., in the limiting case of large n, the frequency calculated from quantum theory approaches that calculated from classical theory. According to Eq. (25), the frequency for a transition from n to $n - 1$ is

$$\nu = \frac{e^4}{4\pi(4\pi\varepsilon_0)^2} \frac{m_e}{\hbar^3} \left[\frac{1}{(n-1)^2} - \frac{1}{n^2}\right] \tag{36}$$

$$= \frac{e^4}{4\pi(4\pi\varepsilon_0)^2} \frac{m_e}{\hbar^3} \frac{2n-1}{n^2(n-1)^2} \tag{37}$$

If n is very large, $(2n - 1)/n^2(n - 1)^2 = 2n/n^4 = 2/n^3$, so that

$$\nu \simeq \frac{e^4}{2\pi(4\pi\varepsilon_0)^2} \frac{m_e}{\hbar^3} \frac{1}{n^3} \tag{38}$$

This is the frequency according to quantum theory. To find the frequency according to classical theory, we note that for an accelerated charge, classical electrodynamics predicts that the frequency of the emitted light coincides with the frequency of the motion. For our electron in a circular orbit, the frequency of the motion is $v/2\pi r$. From Eqs. (16) and (17) we then find that the frequency of the emitted light is

$$\nu_{class} = \frac{v}{2\pi r} = \frac{n\hbar/m_e r}{2\pi r} = \frac{n\hbar}{2\pi m_e} \frac{1}{r^2} = \frac{n\hbar}{2\pi m_e} \left(\frac{e^2 m_e}{4\pi\varepsilon_0 n^2 \hbar^2}\right)^2 \tag{39}$$

If we simplify the right side of this equation, we recognize that it coincides with the right side of Eq. (38), i.e., the result of the classical calculation agrees with the result of the quantum-mechanical calculation. Note, however, that this agreement only holds for large values of n; if n is *not* large, then the classical frequency [Eq. (39)] is smaller than the quantum-mechanical frequency [Eq. (37)].

More generally, it can be shown that for any periodic motion, the agreement between the frequencies of radiation calculated from quantum theory and from classical theory in the limiting case of large quantum numbers is always a direct consequence of the quantization postulate. Thus, what the Correspondence Principle asserts about frequencies is always fulfilled automatically. The practical value of the Correspondence Principle lies in what it asserts about the intensities of the emitted light. In the early days of quantum theory, Bohr and his collaborators exploited this aspect of the Correspondence Principle to calculate the intensities of spectral lines, and to formulate selection rules for permitted lines (of nonzero intensity) and for forbidden lines (of zero or nearly zero intensity). We will not attempt to duplicate these somewhat complicated calculations, but we will summarize and use some of the selection rules in Chapter 7.

4.6 Absorption of Energy; the Franck–Hertz Experiment

Absorption of light by an atom is the reverse of emission. When an electron absorbs a quantum of light (supplied by some external source of light), it jumps from a state of low energy to a state of higher energy. The energy of the quantum must, of course, match the energy difference between the states. The frequencies of light that the atom can absorb are therefore exactly the same as the frequencies that the atom can emit. For instance, if a hydrogen atom is initially in its ground state, it can absorb any of the frequencies of the Lyman series; if initially in its first excited state, it can absorb any of the frequencies of the Balmer series, etc. The typical initial state of the atom depends on the temperature. Under ordinary laboratory conditions, hydrogen is found in its ground state and it will then only display the Lyman absorption lines (see Fig. 4.13); however, in the hot atmospheres of stars, hydrogen is sometimes found in its first excited state and it will then display the Balmer absorption lines.

The absorption of light can be regarded as a totally inelastic collision between a photon and an atom. In this collision the entire energy of the photon is transferred to the atom, and the photon ceases to exist. For a hydrogen atom initially in its ground state, such a totally inelastic collision requires a photon with an energy of at least 10.2 eV, since this is the energy difference between the ground state and the first excited state,

$$\Delta E = 13.6\,\text{eV}\left(\frac{1}{1^2} - \frac{1}{2^2}\right) = 13.6\,\text{eV} \times \frac{3}{4} = 10.2\,\text{eV} \qquad (40)$$

If the energy of the photon is below this threshold, absorption is impossible and the collision between the photon and the atom will be completely elastic—the photon will merely bounce off the atom (scatter). This inability of an atom to accept energy in an amount below a certain threshold is a general feature of energy absorption during collisions.* This behavior is a direct consequence of energy quantization and it makes no difference whether the projectile striking the atom is a photon, another atom, or an electron.

---

* In this context we leave out of account the translational kinetic energy that the atom acquires by recoil during the collision; we will not regard such an energy transfer as absorption. For atoms bombarded with photons or electrons, the recoil energy is usually quite insignificant.

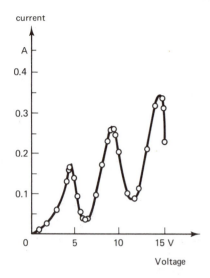

current

Fig. 4.22 Results obtained by Franck and Hertz for current vs. accelerating voltage for electrons passing through a cylinder of mercury vapor. For low voltages, the electron current increases with voltage. The sharp decrease of the current at ∼4.9 V marks the onset of inelastic collisions. Another sharp decrease at ∼9.8 V marks the onset of twice-repeated inelastic collisions—the electron loses half of its energy in a first collision, and the other half in a second collision. [From J. Franck and G. Hertz, *Verhand. Deut. Phys. Ges.* **16**, 457 (1914).]

Soon after Bohr proposed the quantization of atomic energies, his theory received support from an experiment by J. Franck and G. Hertz* on the absorption of energy by mercury atoms in collisions with electrons. They passed a current of electrons through a cylinder filled with mercury vapor and monitored the current as a function of the accelerating voltage applied to the electrons. If the accelerating voltage and the acquired kinetic energy were low, the electrons suffered only completely elastic collisions; but if the kinetic energy reached 4.88 eV, then the electrons suffered inelastic collisions with a loss of speed and the mercury atoms began to emit ultraviolet light at a wavelength of 2537 Å. The onset of inelastic collisions shows up clearly in a plot of electron current vs. accelerating voltage (see Fig. 4.22)—the loss of speed of the electrons in inelastic collisions produces a sudden reduction of the current. The threshold energy of 4.88 eV corresponds to the energy difference between the ground state and first excited state of mercury; and the wavelength of 2537 Å corresponds to photons emitted by spontaneous transitions of excited atoms returning to the ground state (the energy of a photon of 2537 Å is

$$E = h\nu = h\frac{c}{\lambda}$$

$$= 6.63 \times 10^{-34}\,\text{J}\cdot\text{s} \times \frac{3.00 \times 10^8\,\text{m/s}}{2537\,\text{Å}} = 4.88\,\text{eV})$$

If the kinetic energy of the electrons is larger, then several competing absorption processes come into play. An electron may suffer a partially inelastic collision retaining some of its energy, but losing just enough to push the atom into its first excited state; or the electron may lose enough energy to push the atom into one or another of its higher excited states. Spontaneous transitions from these excited atomic states back into the ground state will then give rise to the emission of light with several spectral lines. This is the mechanism by which a gas is made to emit light by electric excitation in a gas discharge tube (see Fig. 4.2). The electric current

* **James Franck**, 1882–1964, German and later American physicist, professor at Göttingen and Chicago. For his work on energy changes in electron collisions he received the Nobel Prize in 1925. During the war he worked on the Manhattan Project, but he advocated that the bomb be used merely for a demonstration. **Gustav Hertz**, 1887–1975, German physicist, collaborated with Franck and shared the Nobel Prize with him. Hertz became director of the research laboratories at Siemens Company, and after the war he worked on atomic physics in the USSR and received the Stalin Prize.

fed through the tube consists of a stream of very energetic electrons that push the atoms into a diversity of excited states, leading to the emission of an abundance of different spectral lines.

A gas can also be made to emit light by thermal excitation. In this case the atoms must be heated to such a high temperature that inelastic atom–atom collisions push one or both of the colliding atoms into an excited state. For a crude estimate of the temperature required to excite a transition, we recall that the average thermal kinetic energy of an atom of a gas is $\frac{3}{2}kT$ (where k is Boltzmann's constant). If an inelastic collision between, say, two hydrogen atoms of kinetic energies $\frac{3}{2}kT$ is to supply the energy of 10.2 eV required to excite one atom from its ground state, we must have $2 \times \frac{3}{2}kT = 10.2$ eV, which yields $T = 3.9 \times 10^4$ K. However, this simple-minded calculation overestimates the temperature because many of the atoms of the gas have kinetic energies larger than the average kinetic energy $\frac{3}{2}kT$. A temperature of 1×10^4 K is already sufficient to produce appreciable excitation of hydrogen atoms— stars with such temperatures display the spectral lines of hydrogen.

In recent years, scientists at the Oak Ridge Laboratory have developed an extremely sensitive technique for the detection and identification of a single atom in a sample of gas. For instance, they have been able to detect the presence of just one atom of cesium in a chamber containing about 10^{19} other atoms. Their technique, called **resonance ionization spectroscopy,** * relies on the high selectivity that the atom displays when absorbing photons—the atom will readily absorb a photon of the right frequency for a transition, but it will refuse to absorb a photon of the wrong frequency. Photons of the right frequency are said to be in resonance with the atomic transition. Figure 4.23 shows the experimental arrangement. The cylindrical chamber of a Geiger counter contains argon gas and a few cesium atoms. A flash of laser light, consisting of photons of exactly the right frequency for a transition of the cesium atom, is sent through the chamber. The cesium atoms will absorb photons and make transitions to an excited state, but the argon atoms will remain in their ground state. Furthermore, since the laser beam contains very many photons, an excited cesium atom almost immediately will absorb a second photon; this second absorption of energy is large enough to ionize the atom. The electron released in this ionization is then counted by the Geiger counter. Thus, each cesium atom signals its presence by means of a released electron.

---

* "Counting the Atoms," G. S. Hurst et al., *Physics Today*, September 1980.

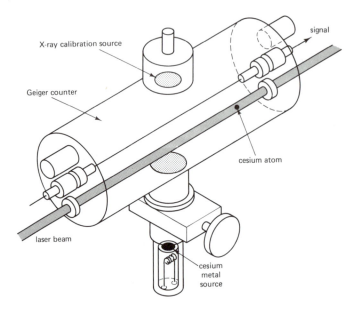

Fig. 4.23 Single-atom detector. (After G. S. Hurst et al., *Physics Today*, September 1980.)

To achieve high sensitivity with this technique, one needs an intense laser beam, so that every single cesium atom will be ionized; and one needs an efficient electron detector, so that every released electron will be counted. Resonance ionization spectroscopy can be used to identify and count the atoms of all the known elements, with the exception of helium and neon for which no suitable laser beams are available.

4.7 The Characteristic X-Ray Spectrum; Moseley's Law

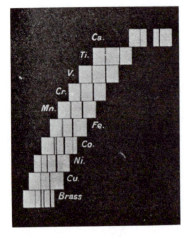

Fig. 4.24 X-ray spectra of several elements. These spectral lines belong to the K series. Note that the wavelength decreases sytematically as the atomic number increases. (From H. G. J. Moseley, *Phil. Mag.* **26**, 1024 (1913).]

As we mentioned in Section 3.4, the X rays generated by the impact of high-energy electrons on a target display a characteristic spectrum of discrete spectral lines whose wavelength depends on the target material. When a crystal is used to analyze the X rays (see Section 1.8), these spectral lines can be recorded on photographic film in much the same way as optical spectral lines can be recorded when light is analyzed with a grating. Figure 4.24 shows some spectral lines of a few elements. For each element, the spectral lines in these photographs belong to a series, called the K series. The individual spectral lines in the K series are designated K_α, K_β, K_γ,

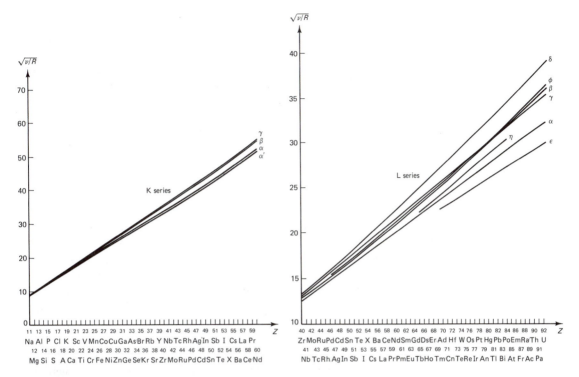

Fig. 4.25 Moseley plots for the K series and the L series. (After A. Sommerfeld, *Atombau und Spektrallinien*.)

etc. Besides this series, there are several other series starting at longer wavelengths, called the L series, the M series, etc.

In 1913, H. G. J. Moseley* carefully measured the wavelengths of the spectral lines in the K series and the L series of some 40 different elements and found that these wavelengths vary systematically with the atomic number of the element. Figure 4.25 is a plot of the square root of the frequency of the emitted X rays vs. the atomic number of the emitting material, called a Moseley plot. According to this plot, the square root of the frequency is linearly related to the atomic number; this linear relationship is called **Moseley's law.**

Bohr's theory explains Moseley's law as follows: Consider an atom of fairly large atomic number, say $Z > 20$. The charge on the

---

* **Henry Gwyn Jeffreys Moseley,** 1887–1915, English physicist, lecturer at Manchester, where he worked under Rutherford. Moseley was a skilful experimenter, and his brilliant investigations of the characteristic spectral lines of the elements led to firm assignments of the atomic numbers. He was killed in action in the Gallipoli campaign, at age 28.

nucleus of the atom is then Ze. For an electron in an inner orbit, close to this nucleus, the dominant force will be the attractive Coulomb force of the nucleus, whereas the repulsive Coulomb force of the other electrons will be insignificant. The electron therefore moves in a hydrogenlike orbit. The energy of this orbit is given by Eq. (21), with one modification: the product $e \times e$ of the electron and the proton charges must be replaced by the product $e \times Ze$ of the electron and the nuclear charges, and thus e^4 must be replaced by $e^4 Z^2$, leading to

$$E_n = -\frac{Z^2 e^4}{(4\pi\varepsilon_0)^2} \frac{m_e}{2\hbar^2} \frac{1}{n^2} \tag{41}$$

Figure 4.26 shows the energy-level diagram for such an electron. The K series is due to transitions from higher states into the $n = 1$ state; the L series is due to transitions from higher states into the $n = 2$ state. The electrons in the $n = 1$ state are said to form the K shell of the atom; electrons in the $n = 2$ state form the L shell; electrons in the $n = 3$ state form the M shell, etc. Thus, the K series is due to transitions into the K shell, the L series is due to transitions into the L shell, etc. (see Fig. 4.26).

In an undisturbed atom, all of the inner shells are full of electrons. The K shell holds two electrons, the L shell eight, etc. (the numbers of electrons in each shell are restricted by the Exclusion Principle; we will discuss this in Chapter 7). If the atom is subjected to the impact of a high-energy projectile, one of the inner electrons may be ejected from its orbit. This permits an electron from a higher orbit to make a transition into the available, vacant slot, with the emission of a characteristic X ray.

The frequency of the X ray emitted in such a transition can be calculated in the usual way. For instance, the frequencies of the K series are

$$v = \frac{Z^2 e^4}{4\pi(4\pi\varepsilon_0)^2} \frac{m_e}{\hbar^3} \left(1 - \frac{1}{n_i^2}\right) \tag{42}$$

This establishes that the square root of the frequency is linearly proportional to the atomic number. To achieve better agreement with the experimental data, Moseley replaced the factor Z in Eq. (42) by $Z - 1$:

$$v = \frac{(Z-1)^2 e^4}{4\pi(4\pi\varepsilon_0)^2} \frac{m_e}{\hbar^3} \left(1 - \frac{1}{n_i^2}\right) \tag{43}$$

The justification for this sleight of hand is that a vacancy in the K shell still leaves one electron in this shell, and this electron shields one nuclear charge from the view of any other electron.

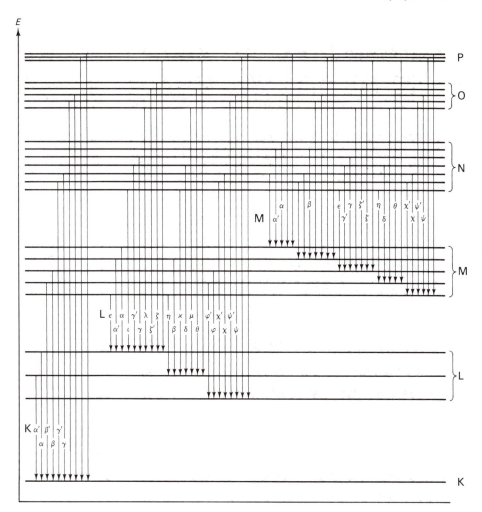

Fig. 4.26 Energy-level diagram for an electron in an inner orbit in an atom. The arrows indicate the transitions that give rise to the different spectral series of X rays.

Bohr's theory also explains why a projectile of fairly large energy is required to trigger the emission of characteristic X rays. The incident projectile must remove an electron from the K shell or the L shell of an atom, i.e., it must supply the (large) ionization energy for this electron.

Moseley's work not only lent additional support to Bohr's theory but also gave a precise method for the determination of the atomic numbers of the elements. This led to some rearrangements in the periodic table of the elements. For instance, before Moseley's days, nickel was listed before cobalt in the periodic table, because the atomic mass of nickel is smaller than that of cobalt. But the comparison of the characteristic X-ray spectra established conclusively that the atomic number of cobalt is smaller than that of nickel.

Example 5   According to Fig. 3.12, the X-ray spectrum of molybdenum displays characteristic spectral lines at 0.72 Å and at 0.61 Å. Compare these wavelengths with those of the K_α and K_β lines calculated from Eq. (43).

Solution   For molybdenum, $Z = 42$. Hence

$$\lambda = \frac{c}{v} = \frac{4\pi(4\pi\varepsilon_0)^2}{(41)^2 e^4}\frac{\hbar^3 c}{m_e}\frac{1}{1 - 1/n_i^2}$$

$$= \frac{1}{(41)^2 R_H}\frac{1}{1 - 1/n_i^2}$$

For $n_i = 2$ and for $n_i = 3$ this gives, respectively, $\lambda = 0.723$ Å and $\lambda = 0.610$ Å, in good agreement with the observed values.   ∎

Chapter 4:   SUMMARY

Spectral series of hydrogen:

$$v = cR_H\left(\frac{1}{n_2^2} - \frac{1}{n_1^2}\right)\qquad R_H = 109{,}678\,\mathrm{cm}^{-1}$$

Deflection of alpha particle by nucleus:

$$\theta = 2\cot^{-1}\left(\frac{4\pi\varepsilon_0 Eb}{Ze^2}\right)$$

Quantization of angular momentum:

$$L = n\hbar$$

Bohr radius:

$$a_0 = \frac{4\pi\varepsilon_0\hbar^2}{e^2 m_e} = 0.529\,\text{Å}$$

Energies of stationary states of hydrogen:

$$E_n = -\frac{e^4 m_e}{2(4\pi\varepsilon_0)^2\hbar^2}\frac{1}{n^2}\qquad n = 1, 2, 3, \ldots$$

$$= -\frac{13.6\,\text{eV}}{n^2}$$

Frequency of photon emitted in transition:

$$v = \frac{E_i - E_f}{h}$$

Reduced mass, hydrogen:

$$\mu = \frac{m_e m_p}{m_p + m_e}$$

Correspondence Principle: In the limiting case of large quantum numbers, the results obtained from quantum theory must agree with those from classical theory.

Chapter 4:   PROBLEMS

1. Table 4.1 gives the wavelengths of the spectral lines *in air*. The spectral lines at 5889.95 Å and 5895.92 Å are the famous *yellow doublet* of sodium. What are the wavelengths of these lines in vacuum? What are their frequencies?

2. When excited in an electric discharge tube, an unknown sample of gas emits a strong spectral line at 5875.6 Å as well as some weaker spectral lines. What is the color of the emitted light? Use Table 4.1 to identify the gas.

3. What are the wavelengths of the first four lines of the Lyman series?

4. Which of the spectral lines of the Brackett series is closest in wavelength to the first spectral line of the Paschen series? By how much do the wavelengths differ?

5. What is the range of wavelengths in the Lyman series, i.e., what are the shortest and the longest wavelengths in this series? The Balmer series? The Paschen series? The Brackett series? The Pfund series?

6. The spectral lines in the light of all distant galaxies exhibit a redshift, i.e., the wavelengths of these spectral lines are longer than the wavelengths of the corresponding spectral lines in light from sources on the Earth by a common multiplicative factor. For instance, the light of a galaxy beyond the constellation Virgo contains spectral lines of wavelength 4117 Å and 4357 Å due to hydrogen. (a) Identify these two spectral lines. What is the factor by which these wavelengths are longer than the normal wavelengths of the two spectral lines? (b) What is the speed of recession of the galaxy?

7. One of the spectral series of the lithium atom is the **principal series** with the following wavelengths (in vacuum): 6709.7 Å, 3233.5 Å, 2742.1 Å, 2563.3 Å, 2476.1 Å. Show that these wavelengths approximately fit the formula

$$\frac{1}{\lambda} = R\left[\frac{1}{(1+s)^2} - \frac{1}{(n+p)^2}\right] \qquad n = 2, 3, 4, \ldots$$

Where $R = 109{,}729 \text{ cm}^{-1}$ is the Rydberg constant for lithium, and s and p are constants characteristic of the series. What values of s and p must you use to make the wavelengths fit the formula?

8. Another of the spectral series of the lithium atom is the **diffuse series** with the following wavelengths (in vacuum): 6105.2 Å, 4604.3 Å, 4133.5 Å, 3916.1 Å, 3795.8 Å. These wavelengths approximately fit the formula

$$\frac{1}{\lambda} = R\left[\frac{1}{(2+p)^2} - \frac{1}{(n+d)^2}\right] \qquad n = 3, 4, 5, \ldots$$

where, as in the preceding problem, $R = 109,729 \text{ cm}^{-1}$ and p and d are constants.
(a) What values of p and d must you use to make the wavelengths fit the formula?
(b) The principal series and the diffuse series of lithium are analogous to two spectral series of hydrogen. Which two series?
(c) Lithium has other spectral series analogous to the other spectral series of hydrogen. Can you guess a formula describing one of these other spectral series?

9. According to the Thomson model, the atom consists of a cloud of positive charge within which electrons sit at equilibrium positions. Assume that in the case of hydrogen, the positive cloud has a charge e uniformly distributed over the volume of a sphere of radius 0.50 Å. The equilibrium position of the electron is then at the center of this sphere. What is the frequency of small radial oscillations of the electron about this equilibrium position? What would be the wavelength of the light radiated by the electron? Use classical mechanics and classical electromagnetism for this calculation.

10. According to the Thomson model, a helium atom consists of a cloud of positive charge, within which two electrons sit at equilibrium positions. Assume that the positive cloud has a charge $2e$ uniformly distributed over the volume of a sphere of radius 0.50 Å.
(a) Find the equilibrium positions of the two electrons. Assume that the electrons are symmetrically placed with respect to the center.
(b) What is the frequency of small radial oscillations of the electrons about their equilibrium positions? Assume the electrons move symmetrically, with identical amplitudes.

11. What is the distance of closest approach for a 5.5-MeV alpha particle in a head-on collision with a gold nucleus?

12. The nucleus of lead has a radius 7.10×10^{-15} m and an electric charge of $82e$. What must be the minimum energy of an alpha particle in a head-on collision if it is to just barely reach the nuclear surface? Assume the alpha particle is pointlike.

13. An alpha particle of energy 5.5 MeV is incident on a silver nucleus with an impact parameter 8.0×10^{-15} m. Use conservation of energy and of angular momentum to find the distance of closest approach of the particle, i.e., the distance from the apex of the hyperbolic orbit to the center of the nucleus.

14. Suppose that 1000 alpha particles of energy 7.7 MeV suffer deflections by more than 30° when a beam of such particles strikes a very thin foil of gold.
(a) How many of these will suffer deflections by more than 40°? How many will suffer deflections between 30° and 40°?
(b) How many will suffer deflections between 40° and 50°? Between 50° and 60°?

15. A foil of gold, 2.1×10^{-5} cm thick, is being bombarded by alpha particles of energy 7.7 MeV. The particles impact at random over an area of 1 cm^2 of the foil of gold.
    (a) How many atoms are there within the volume 1 cm$^2 \times 2.1 \times 10^{-5}$ cm under bombardment? The density of gold is 19.3 g/cm^3 and the mass of one atom is 3.27×10^{-25} kg.
    (b) To suffer a deflection of more than 30°, an alpha particle must strike within 5.5×10^{-14} m of the center of a gold nucleus (see Example 2). What is the probability for this to happen?
    (c) If 10^{10} alpha particles impact on the foil, how many will suffer deflections of more than 30°?

16. Show that for an alpha particle incident on a stationary nucleus the cross section for a deflection angle in the infinitesimal interval θ to $\theta + d\theta$ is

$$d\sigma = \left(\frac{e^2}{4\pi\varepsilon_0}\right)^2 \frac{1}{2E^2} \frac{\pi \sin\theta}{\sin^4\theta/2} d\theta$$

This is called the **differential cross section.**

17. Alpha particles of energy 8.0 MeV are incident on a gold atom. What is the cross section for deflections larger than 50°?

18. Suppose that alpha particles of energy 6.1 MeV are incident on a silver atom. What is the cross section for deflections of more than 20°? For deflections of more than 30°? For deflections of more than 20° but less than 30°?

19. When an alpha particle moving at *high speed* passes by a stationary nucleus of charge Ze, the main effect of the electric force is to give the particle a transverse impulse. Figure 4.27 shows the alpha particle moving at (almost) constant velocity v along the x-axis in an (almost) straight line, and shows the charge Ze sitting at a distance b below the origin. The transverse impulse that the electric field of the charge Ze gives to the alpha particle is

$$\Delta p_y = \int_{-\infty}^{\infty} F_y \, dt = \frac{2e}{v} \int E_y \, dx$$

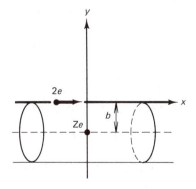

Fig. 4.27 High-speed alpha particle passing by a nucleus of charge *Ze*.

(a) Evaluate the integral $\int E_y \, dx$ by means of Gauss' law and prove that

$$\int E_y \, dx = \frac{Ze}{2\pi\varepsilon_0 b}$$

(*Hint:* Consider $2\pi b \int E_y \, dx$; show that this is the electric flux that the charge Ze produces through the infinite cylindrical surface shown in Fig. 4.27.)
(b) The deflection angle of the alpha particle is approximately $\theta = \Delta p_y / p = \Delta p_y / m_\alpha v$. Show that this leads to

$$\theta = \frac{Ze^2}{\pi\varepsilon_0 m_\alpha v^2 b}$$

This result is valid for small deflections. Does Eq. (9) give the same result for small deflections?

20. What is the speed of an electron in the smallest ($n = 1$) Bohr orbit? In the next ($n = 2$) orbit? Express your answers as a fraction of the speed of light.

21. What is the frequency of the orbital motion for an electron in the smallest ($n = 1$) Bohr orbit? In the next ($n = 2$) Bohr orbit? Compare these frequencies with the frequency of the light emitted during the transition $n = 2$ to $n = 1$.

22. What is the *longest* wavelength a hydrogen atom will absorb if initially in the ground state?

23. In the singly ionized helium atom (He II) one electron orbits around a nucleus of mass 6.65×10^{-27} kg and charge $2e$.
(a) According to Bohr's theory, what are the energies of the stationary states? What is the value of the ionization energy (in eV)?
(b) Show that for every spectral line of the hydrogen atom, the ionized helium atom has a spectral line of almost identical wavelength. Taking into account the reduced-mass correction for hydrogen and for helium, calculate the difference between the wavelength for the $n = 2$ to $n = 1$ transition of hydrogen and the wavelength for the $n = 4$ to $n = 2$ transition of helium.

24. The nucleus of deuterium has a mass of 3.34×10^{-27} kg, as compared to 1.67×10^{-27} kg for ordinary hydrogen. Calculate the wavelength difference between the first Balmer line emitted by hydrogen and the first Balmer line emitted by deuterium. Repeat the calculation for the second Balmer lines. For which of these two spectral lines is this isotope shift of wavelength larger?

25. For deuterium and for tritium, find the isotope shift of the ground-state energy of the electron relative to the ground-state energy of the electron in hydrogen. Express your answer in eV.

26. In the doubly ionized lithium (Li III), one electron orbits around a nucleus of charge $Z = 3e$. Find the radius of the smallest Bohr orbit in doubly ionized lithium. Find the energy of the electron in this orbit.

27. The atom of **positronium** consists of an electron and a positron (or antielectron) orbiting about each other. Find the Bohr radius of this system. Find the wavelength of the photon released in the transition $n = 2$ to $n = 1$.

28. The muon (or μ-meson) is a particle similar to an electron; it has a charge $-e$ and a mass 206.8 times as large as the mass of the electron. When such a muon and a proton go into orbit around each other, they form a muonic hydrogen atom, similar to an ordinary hydrogen atom, but with the muon playing the role of the electron. Calculate the Bohr radius of this "muonium" atom and calculate the energies of the stationary states (take into account the reduced mass of the system). What is the wavelength of the light emitted when the muon makes a transition from $n = 2$ to $n = 1$?

29. Assume that, as proposed by Thomson, the hydrogen atom consists of a cloud of positive charge e, uniformly distributed over a sphere of radius R. However, instead of placing the electron in static equilibrium at the center of the sphere, assume that the electron orbits around the center with uniform circular motion under the influence of the electric centripetal force $(e^2/4\pi\varepsilon_0)(r/R^3)$. Assume that the angular momentum of this orbiting electron is quantized according to Bohr's theory. What are the radii and the energies of the quantized orbits? What are the wavelengths of the Lyman series? Give numerical answers, taking $R = 1.0\,\text{Å}$.

30. In principle, Bohr's theory can also be used to describe the motion of the Earth around the Sun. The Earth plays the role of the electron, the Sun that of the nucleus, and the gravitational force that of the electric force.
(a) Find a formula analogous to Eq. (17) for the radii of the permitted circular orbits of the Earth around the Sun.
(b) The actual radius of the Earth's orbit is $1.50 \times 10^{11}\,\text{m}$. What value of the quantum number n does this correspond to?
(c) What is the difference between the radius of the Earth's actual orbit and the radius of the next larger orbit?

31. According to classical electrodynamics an electron in an elliptical orbit emits not only radiation at the orbital frequency, but also radiation at harmonic multiples of the orbital frequency: if the orbital frequency is v, the emitted radiation contains the fundamental frequency v and also the harmonic frequencies $2v$, $3v$, $4v$, etc. Show that this agrees with the Correspondence Principle. (*Hint:* According to Bohr's theory, consider the frequencies the electron emits in the transitions n to $n - 2$, n to $n - 3$, n to $n - 4$, etc.)

32. According to classical statistical mechanics the spectral emittance of a blackbody ought to be

$$S_v = \frac{8\pi v^2 kT}{c^3}$$

This classical formula, called the **Rayleigh–Jeans law,** rests on the assumption that each of the standing-wave modes in the cavity has an energy kT. Since the quantum numbers of the modes of low frequency are large (each of these modes contains a large number of quanta), the Correspondence Principle requires that for low frequencies ($hv \ll kT$) the Rayleigh–Jeans law should coincide with the Planck law. Show that this is indeed the case.

33. Suppose that in a Franck–Hertz experiment you use electrons of energy 13.0 eV to excite hydrogen atoms. What spectral lines will the hydrogen atoms emit under these conditions?

34. What is the difference in wavelength between the K_α lines of cobalt and of nickel? Which element has the longer wavelength?

35. Calculate the wavelengths of the K_α, K_β, and K_γ lines in the X-ray spectrum of platinum.

36. When a sample of an unknown element is used as target for the electron beam in an X-ray tube, it emits a series of characteristic spectral lines in which the two longest wavelengths are 0.228 Å and 0.192 Å. What is the element?

37. What is the longest wavelength in the K series of X rays emitted by cobalt?

38. The frequencies of the X rays of the L_α lines are given by the formula

$$\nu = \frac{e^4}{(4\pi\varepsilon_0)^2} \frac{(Z-b)^2 m_e}{4\pi h^3} \left(\frac{1}{4} - \frac{1}{n_i^2}\right)$$

where b is a constant. What value of b can you deduce from the data given in Fig. 4.25?

Wave Mechanics I—

Free Particles

Bohr's quantum theory of the atom, as elaborated by Sommerfeld, marks the highest state of perfection of classical mechanics and also its final decline. Sommerfeld was able to calculate the spectrum of the hydrogen atom—including the very small relativistic corrections—with astonishing accuracy, but he was totally unable to extract any sensible results for the helium atom or the hydrogen molecule. Soon after, classical mechanics was superseded by wave mechanics, a completely new approach based on the discovery that electrons exhibit wave properties. Thus electrons, and all other "particles" are endowed with both wave aspects and particle aspects, just like photons. On the atomic scale, the wave aspect dominates the picture and there exist no well-defined orbits. Bohr had retained classical kinematics and merely supplemented the laws of classical dynamics with extra quantum conditions. Upon the discovery of the wave properties of electrons, it became necessary to discard the classical concept of a particle, and both classical kinematics and dynamics. What took their place was a new concept of a quantum-mechanical "particle" whose state is described by a wave, and a new set of laws of wave mechanics. It turns out that classical mechanics is a special, limiting case of wave mechanics, just as geometrical optics is a limiting case of wave optics. The orbits of classical mechanics, just as the rays of geometrical optics, provide a good approximation for the motion of the underlying waves whenever the wavelength is extremely small compared to other relevant distances, so that the wavelike aspects of matter remain well hidden.

In this chapter we will deal with the experimental and conceptual foundations of wave mechanics. For a start, we will concentrate on free particles, i.e., particles without external forces. According to classical mechanics, such a particle moves along a straight line, with a well-defined position as a function of time. We will see how this classical description breaks down, and why it must be replaced by a wave description.

5.1 The de Broglie Wavelength

As we saw in Chapter 3, the energy of a photon is related to the frequency of the light wave ($E = h\nu$), and the momentum of the photon is related to the wavelength ($p = h\nu/c = h/\lambda$). We can write these relationships as

$$\nu = \frac{E}{h} \tag{1}$$

$$\lambda = \frac{h}{p} \tag{2}$$

In 1924, L. de Broglie* suggested that since light waves have particle properties, particles should have wave properties. He proposed that the relationships (1) and (2) are valid not only for photons, but for all particles. Thus, the wavelength associated with a (nonrelativistic) particle of mass m and velocity v is

$$\lambda = \frac{h}{p} = \frac{h}{mv} \tag{3}$$

This is called the **de Broglie wavelength.**

Example 1 What is the de Broglie wavelength associated with an electron of energy 1 eV?

---

* **Louis Victor, prince de Broglie** (broy), 1892– , French theoretical physicist, professor at the University of Paris. His thesis containing the proposal of wave properties of matter was at first viewed with some scepticism. He was awarded the Nobel Prize in 1929, after these wave properties were confirmed experimentally.

Solution   With an energy $1\,eV = 1.6 \times 10^{-19}\,J$, the momentum of the electron is

$$p = \sqrt{2m_e E} = (2 \times 9.1 \times 10^{-31}\,kg \times 1.6 \times 10^{-19}\,J)^{1/2}$$

$$= 5.4 \times 10^{-25}\,kg \cdot m/s \tag{4}$$

and hence the de Broglie wavelength is

$$\lambda = \frac{h}{p} = \frac{6.66 \times 10^{-34}\,J \cdot s}{5.4 \times 10^{-25}\,kg \cdot m/s} = 1.2 \times 10^{-9}\,m \tag{5}$$

This is much smaller than the wavelength of visible light.   ■

In principle, the de Broglie wavelength associated with an electron could be large if the velocity is very low. However, in practice it is impossible to produce electron beams of energies much below $1\,eV$. As Example 1 shows, for an electron of that energy the de Broglie wavelength is quite small. For a macroscopic "particle," such as a grain of dust or a billiard ball, the wavelength given by Eq. (3) is even smaller. Thus, on a macroscopic scale, particles do not display any noticeable wave properties.

However, on a microscopic scale the wave properties play a crucial role. For instance, de Broglie was able to show that Bohr's quantization condition for the angular momentum of an electron in a circular orbit can be interpreted as a condition for periodically repeating waves around this orbit. We recognize this immediately by rearranging Eq. (4.15),

$$m_e v r = n\frac{h}{2\pi} \tag{6}$$

or

$$2\pi r = n\frac{h}{m_e v} \tag{7}$$

In terms of the de Broglie wavelength this is

$$2\pi r = n\lambda \tag{8}$$

which states that the length of the circumference of the orbit is an integral number of wavelengths. Under these conditions a wave traveling around the orbit will repeat itself after a complete circuit— it will have a rigid periodic pattern that rotates steadily around the orbit.* Figure 5.1 shows the peaks and troughs of such a rotating

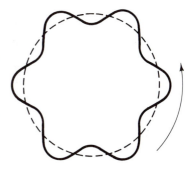

Fig. 5.1 De Broglie wave traveling around a circular orbit. The wave is shown at one instant of time.

---

* Note that this is not a standing wave—it does not have fixed nodes and antinodes.

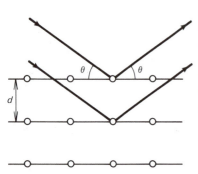

Fig. 5.2 Parallel Bragg planes. The electron wave is incident from the left. Constructive interference of scattered waves gives a strong emerging beam toward the right.

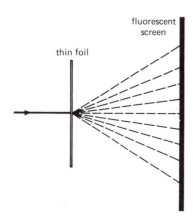

Fig. 5.3 Thomson's experimental arrangement for electron interference.

wave with $n = 6$ at one instant of time. Thus, de Broglie's proposal of wave properties for all particles supplied a rudimentary explanation of Bohr's quantization condition, an explanation that was later to be refined and expanded into wave mechanics.

In view of the small wavelengths of electrons and other particles, the direct experimental observation of the wave properties requires delicate interference or diffraction experiments using gratings with extremely small spacings between the slits. Since electrons of an energy of a few eV have a wavelength of a few Å, the same as that of ordinary X rays, we expect that such electrons will exhibit interference effects like those of X rays. Thus, we expect that electrons incident on a crystal will generate Laue spots like those of X rays.

The first experimental observation of such interference effects[*] with electrons was made by C. J. Davisson and L. A. Germer in 1927.[†] By serendipity, in the course of an investigation of the elastic scattering of electrons by the surface of a nickel crystal, Davisson noticed that electrons emerged from the crystal at unexpected angles. The angles of the emerging electron beams correspond to constructive interference according to the Bragg condition

$$2d \sin \theta = n\lambda \qquad (9)$$

where d is the spacing between parallel Bragg planes in the crystal, and θ the angle of reflection (measured from the reflecting plane toward the incident direction; see Fig. 5.2). From the known value of the spacing d, and from the measured value of the angle θ, Davisson determined the wavelength of the electrons and found it to agree with the de Broglie wavelength calculated from Eq. (3).

At nearly the same time, G. P. Thomson performed similar experiments on electron interference. Instead of a single, thick crystal, such as Davisson's, Thomson used a thin metallic foil containing many microcrystals. Such a thin foil is more or less transparent to electrons, so that the geometry of the experiment can be arranged to imitate that for X-ray interference, with the electrons emerging in the forward direction (see Fig. 5.3). We would then expect to see a set of Laue spots on the fluorescent screen. However, since the many microcrystals in the metallic foil are oriented at random, for every microcrystal that produces a Laue spot at

---

[*] The experiment of Davisson and Germer is often called a diffraction experiment; it really is an interference experiment.

[†] **Clinton Joseph Davisson,** 1881–1958, and **Lester Germer,** 1896–1971, American physicists at the Western Electric Company (later Bell Telephone). Davisson and **George Paget Thomson,** 1892–1975, British physicist, professor at Imperial College, shared the 1937 Nobel Prize for their independent experimental verifications of electron interference.

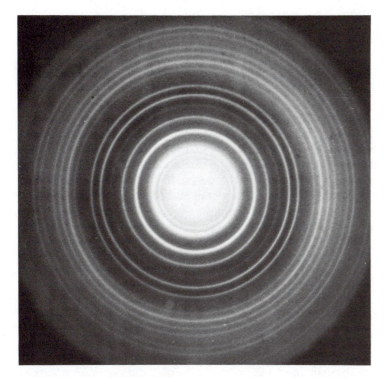

Fig. 5.4 Interference pattern produced by electrons incident on a foil of magnesium oxide. (Courtesy H. F. Meiners, Rensselaer Polytechnic Institute.)

some angle above the beam, there will be another microcrystal that produces a corresponding spot at an equal angle below the beam, and another that produces a spot on the right, and another that produces a spot on the left, and others that produce spots at intermediate locations lying at equal angles with respect to the incident beam. Thus, these replicas of a given Laue spot trace out a bright circle on the screen. The set of all the possible Laue spots of the crystal gives rise to a set of concentric bright circles (see Fig. 5.4).

The wavelike behavior of particles has also been demonstrated with beams of hydrogen atoms and beams of helium molecules scattered by crystals. Very clear Laue patterns with isolated spots have been obtained with neutron beams scattered by crystals of NaCl (see Fig. 5.5). Neutrons have great penetrating power, and it is therefore possible to use quite thick crystals that give strong and sharp spots. These experiments verify the validity of de Broglie's relation for all kinds of particles.

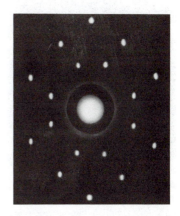

Fig. 5.5 Laue pattern produced by neutrons incident on a crystal of NaCl. (Courtesy C. G. Shull, Massachusetts Institute of Technology.)

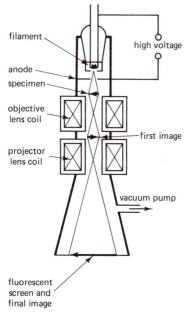

filament

high voltage

anode

specimen

objective
lens coil

first image

projector
lens coil

vacuum pump

fluorescent
screen and
final image

Fig. 5.6 Schematic diagram of an electron microscope.

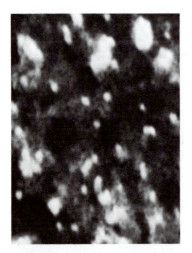

Fig. 5.7 Micrograph of uranium atoms fixed on a thin film of carbon, taken with a high-power electron microscope at the University of Chicago. The magnification is about $10^7 \times$. (Courtesy A. V. Crewe and M. Utlaut, University of Chicago.)

The wave properties of electrons find an important practical application in electron microscopes which use electron waves to illuminate an object, in much the same way as ordinary optical microscopes use light waves. The lenses in the electron microscope consist of magnets that deflect the electrons so as to produce the same geometry of rays as in an ordinary microscope (see Fig. 5.6). As is well known from optics, the ultimate limit of resolution of an ordinary microscope is set by diffraction effects at the aperture— roughly, the size of the smallest detail that the microscope can resolve equals one wavelength. Since electron microscopes operate with wavelengths much shorter than that of ordinary light, they can resolve much finer detail and achieve much higher magnification. Figure 5.7 is a photograph of uranium atoms taken with an electron microscope of very high power—the magnification is about 10^7!

5.2 Particle vs. Wave; Duality

Classical physics teaches us that electrons are particles. However, from the experiments described in the preceding section we learned that electrons have wave properties. We are forced to concede that electrons are neither classical particles nor classical waves. They are some new kind of quantum-mechanical particle, unknown to classical physics, with a subtle combination of both particle and wave properties. B. Hoffmann has coined the word **wavicle** for this new kind of quantum-mechanical particle. It is difficult to achieve a clear understanding of the concept of wavicle because these objects are very remote from our everyday experience. We all have an intuitive grasp of the concepts of a classical particle and a classical wave from our experience with, say, billiard balls and water waves. But we have no such experience with wavicles.

We can gain some insight into the interplay between particle behavior and wave behavior of an electron by a detailed examination of a simple idealized experiment. Although this experiment has never been actually performed, the results of the experiment are known by extrapolation from actual experiments; in the following we will feel free to make use of such knowledge. Let us take a source of electrons and shoot a beam toward a fluorescent screen, which indicates the impact of an electron by emitting a small flash of light. We assume the electrons all have the same energy. Between the source and the fluorescent screen we place a

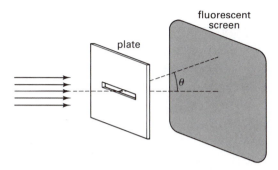

Fig. 5.8 Arrangement for an electron diffraction experiment. The angle θ measures the deflection of an electron from the straight path.

plate capable of stopping electrons; this plate has a slit cut into it. The whole arrangement is placed in a vacuum (Fig. 5.8). If the electrons travel along straight orbits like particles in the absence of forces, they would strike the fluorescent screen in the geometrical projection of the slit; they would not strike anywhere else. Hence, on the fluorescent screen we would simply see a bright band indicating the zone of impacts of electrons. But when we perform the experiment we do not find this. What we find depends on the width of the slit. If the slit is reasonably wide, the electrons indeed strike in the geometrical projection of the slit and we find a bright band on the screen (see Fig. 5.9). However, if we make the slit very narrow, instead of a single narrow bright band we find a series of wide parallel bands on the fluorescent screen (see Fig. 5.10). This widely distributed pattern of electron impacts shows that the electrons do not travel along straight paths; in fact, as we will see later, they do not travel along *any* well-defined paths. The width of the bands in Fig. 5.10 grows *inversely* as the width

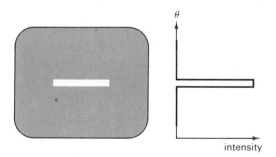

Fig. 5.9 Pattern seen on screen if slit is wide.

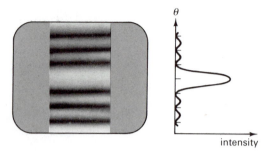

Fig. 5.10 Pattern seen on screen if slit is very narrow.

of the slit—if we use an even narrower slit, the band pattern becomes wider. In the case of an extremely narrow slit, the band pattern can become so wide that the central maximum covers the entire fluorescent screen and the illumination over all of the screen becomes fairly uniform (see Fig. 5.11). Such behavior is characteristic not of particles, but rather of waves: when the electrons are made to go through narrow slits, they spill around the edge of the slit into the shadow zone. The electrons exhibit diffraction, like waves. The angular half-width of the central diffraction band is

$$\Delta\theta \simeq \frac{\lambda}{a} \tag{10}$$

where λ is the de Broglie wavelength and a the width of the slit. This formula is the same as for the diffraction of water waves or of light waves [see Eq. (1.128), with $\theta \ll 1$].

To obtain noticeable diffraction bands, we need a *very* narrow slit. For electrons of an energy of approximately 10^2 eV the slit must be less than 10^{-8} cm in width. We cannot construct a plate with such a narrow slit and therefore the experiment described above has never been carried out. But, as we have seen in the pre-

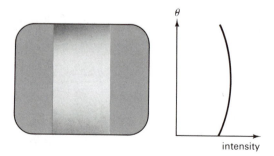

Fig. 5.11 Pattern seen on screen if slit is extremely narrow.

ceding section, it is possible to perform a similar experiment using a piece of a crystal instead of the plate.

Are electrons waves? Although our experiment with the slit seems to show they are waves, there is a difficulty. Suppose we reduce the flow of electrons so that only one electron per minute reaches the fluorescent screen. If electrons were classical waves, the fluorescent screen would be initially dark. Then, when an electron arrives, the band pattern would appear on the screen for a a moment and then fade away. The screen would then remain dark until the next electron arrives. But this is not what we actually find. If only one electron arrives per minute, we see one small flash of light somewhere on the fluorescent screen each minute. The electron seems to arrive at the screen as a particle—it is not spread out but appears in one spot only. So the electron is not a classical wave. The idea of electrons as classical waves is absurd in any case. For suppose we let such a wave pass through a shutter, and when half of it has passed we suddenly close the shutter. A classical wave is split in two pieces by such a chopping operation: one half of the wave is reflected or absorbed by the shutter; the other half moves on. But electrons always make their appearance as whole electrons—they always have the same mass and the same charge (and also the same spin, same magnetic moment). One-half electrons do not exist.

Are electrons classical particles or waves? They are neither. They are wavicles, or quantum-mechanical particles—they have some properties of a classical particle and some properties of a classical wave. We say that electrons exhibit **duality:** they sometimes behave like classical particles, and sometimes like classical waves. The electrons behave in one way or another depending on the experimental arrangement. In our idealized experiment with the slit, the electrons behave like waves when passing through the slit and like particles when striking the screen and giving off the small, pointlike flash of light. This dual particle-wave behavior also holds for protons, neutrons, and other quantum-mechanical particles. It also holds for photons. In the usual diffraction experiments photons behave like classical waves; in the Compton effect they behave more like classical particles (but their behavior has some additional peculiarities because photons are never at rest, they are always relativistic, and they have zero mass).

Let us return to our idealized experiment with the low-intensity source. Since the impact of one electron does not give the diffraction pattern, how does the pattern come into being? We can see how the pattern forms by aiming a camera on the fluorescent screen, leaving the shutter open for a while, and thereby recording the points of impact of a number of electrons. Figure 5.12a shows a typical pattern of impacts for 30 electrons; the pattern seems quite random. Figure 5.12b shows the pattern of accumulated impacts for 300 electrons; and Fig. 5.12c shows it for 3000 electrons. In these figures we can recognize a tendency of the electrons to cluster in bandlike zones. These zones correspond to the maxima of the diffraction pattern predicted by the wave theory. Finally, Fig. 5.12d shows the pattern of accumulated impacts for a very large number of electrons; this is simply the familiar intensity pattern of light diffracted by a slit (see Fig. 1.23). We see that although one electron does not give the pattern, the *cumulative* impacts of many do. Hence we are led to the conclusion that the behavior of the electrons is governed by a probabilistic law. The point of impact of electrons; this is simply the familiar intensity pattern of light average distribution of impacts of a large number of electrons is predictable.

Since electrons have wave properties, we will have to introduce a wavefunction Ψ for the mathematical description of the motion of an electron. In general, this wavefunction depends on x, y, z, and t, and it represents a wave moving in three dimensions, $\Psi = \Psi(x, y, z, t)$. However, for the purposes of our idealized experiment with the slit, we are interested only in the value of the wavefunction at the screen. There, the wavefunction depends only on the vertical coordinate y measured upward along the screen, $\Psi = \Psi(y, t)$. This wavefunction is supposed to determine the distribution of electrons found at the screen. Since the pattern of impacts of the electrons on the screen looks like the intensity pattern of a dif-

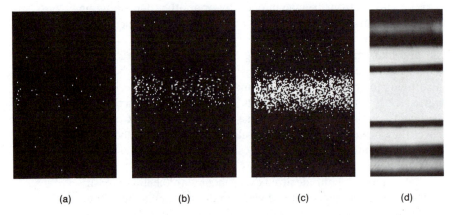

(a)   (b)   (c)   (d)

Fig. 5.12 Pattern of impact points of a number of electrons on the screen: (a) 30 electrons; (b) 300 electrons; (c) 3000 electrons; (d) a very large number of electrons. [(a)–(c) are simulations generated by a computer; (d) is a diffraction pattern obtained with light.]

fracted wave, we are led to the hypothesis that the probability for an electron to be found at any given point is proportional to the intensity of the electron wave at this point, i.e.,

$$\begin{pmatrix} \text{probability for electron} \\ \text{to be found between} \\ y \text{ and } y + dy \end{pmatrix} \propto (\text{intensity of wave}) \times dy$$

$$\propto |\Psi(y,t)|^2 \, dy \qquad (11)$$

Note that here the intensity of the wave has been expressed as the square of the absolute value ($|\Psi|^2$), rather than simply the square (Ψ^2); this is necessary because it turns out that Ψ is complex (see the next section). Provided we include a suitable normalization factor in Ψ, the proportionality can be written as an equality,

$$\begin{pmatrix} \text{probability for electron} \\ \text{to be found between} \\ y \text{ and } y + dy \end{pmatrix} = |\Psi(y,t)|^2 \, dy \qquad (12)$$

If a large number of electrons arrive at the screen, the probability for one electron to be found in the interval dy permits us to calculate the average number of electrons to be found in the interval dy:

$$\begin{pmatrix} \text{number of electrons} \\ \text{found between} \\ y \text{ and } y + dy \end{pmatrix} = \begin{pmatrix} \text{total number} \\ \text{of electrons} \end{pmatrix} \times |\Psi(y,t)|^2 \, dy \qquad (13)$$

This probability interpretation of the wavefunction is due to Max Born.* Note that the probability interpretation provides a link between the wave and the particle aspects of the electrons: the *wave* tells us the probability for finding the electron as a *particle* at some point when we perform a suitable experiment with a particle detector such as, e.g., a fluorescent screen.

The description of the motion of an electron by a wavefunction implies a complete break with classical mechanics, both in kinematics and in dynamics. Instead of describing the motion by position as a function of time ($x(t)$, $y(t)$, and $z(t)$), we are now describing it by a wavefunction ($\Psi(x, y, z, t)$) in which x, y, z, and t appear merely as independent variables. And instead of Newton's laws of dynamics we will have to construct a new set of laws for the wavefunction. Note that the wavefunction does not correspond to a

---

* **Max Born,** 1882–1970, German and later British theoretical physicist, professor at Göttingen and at Edinburgh. He was tardily awarded the Nobel Prize in 1954 for the probabilistic interpretation of the wavefunction, which he had proposed in 1926.

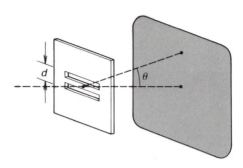

Fig. 5.13 Slits and screen. The separation between the slits is assumed to be much smaller than the distance to the screen.

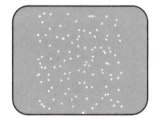

Fig. 5.14 Pattern of impact points if only one (extremely narrow) slit is open.

Fig. 5.15 Pattern of impact points if both slits are open.

vibration in some kind of underlying medium. Like an electromagnetic wave, the electron wave is a thing in itself—it does not require a medium for its propagation.

We now turn to another idealized experiment for which it is very easy to calculate the positions of the maxima of the interference pattern of the wavefunction. Instead of a plate with a single slit, we take a plate with two parallel, very narrow slits (see Fig. 5.13). Our first experiment showed that as electrons go through narrow slits they suffer diffraction; they do not propagate along straight paths. We can check with our new experiment that the diffraction pattern is not the result of electrons behaving like classical particles deflected by some complicated collisions with the atoms at the edge of the slit.

First we suppose that one of our two slits is covered. Assuming that the slits are extremely narrow, we find that the electron impacts for only one open slit are fairly uniformly distributed over the fluorescent screen (see Fig. 5.14). If the electrons were classical particles somehow deflected by the edges of the slit, what would we expect to see when both slits are open? Obviously the same kind of uniform distribution that we get with only one open slit, but the intensity at each point of the screen should now be twice as large. However, this is not what we actually find. With both slits open, we find a pattern of evenly spaced bright and dark bands (Fig. 5.15). This makes no sense at all from the point of view of a particle picture: if the electron goes through one slit, it cannot be affected by the presence or absence of the second slit; and the intensity at a point on the fluorescent screen cannot *decrease* upon opening the second slit. But the pattern is easy to explain if the behavior of an electron is controlled by a wave. What we are seeing in Fig. 5.15 is an interference pattern resulting from the super-position of waves originating at the two slits. To calculate the positions of the maxima of the band pattern, suppose that the wavefunction contributed by one of the slits is Ψ_1, and that con-tributed by the other is Ψ_2. The amplitudes of these (complex*) wavefunctions are equal, $|\Psi_1| = |\Psi_2|$, but their phases are different. We assume that the wavefunctions obey the Principle of Linear Superposition: if both slits are open, the net wavefunction at the screen is the sum of Ψ_1 and Ψ_2,

$$\Psi = \Psi_1 + \Psi_2$$

The amplitude of this wave has a maximum wherever the two waves Ψ_1 and Ψ_2 interfere constructively. If the screen is at a large distance from the slits, then the condition for constructive interference is [see Eq. (1.127)]

$$d \sin \theta = n\lambda \qquad n = 0, \pm 1, \pm 2, \ldots \qquad (14)$$

where λ is the wavelength of the electrons, d the separation between the slits, and θ the angular position of the point on the screen (see Fig. 5.13). Likewise, the amplitude of the net wave has a minimum wherever the two waves interfere destructively; the condition for this is

$$d \sin \theta = (n + \tfrac{1}{2})\lambda \qquad (15)$$

Thus, the maxima and the minima form bright and dark parallel bands on the screen, exactly as shown in Fig. 5.15. Note that the Superposition Principle plays a crucial role in this calculation, and the experimental observation of the interference bands may there-fore be regarded as firm evidence for this Superposition Principle.

5.3 Heisenberg's Uncertainty Relations; Complementarity

The wave properties of wavicles impose some ultimate limitations on the precision of measurements of position and momentum. We can discover these limitations by appealing to a simple *Gedanken-experiment*. Suppose we have an electron wave propagating in the

---

* For the following discussion it is not necessary to assume that Ψ is complex. Why Ψ has to be complex will be seen in the next section.

x-direction and we want to measure the position of the electron. Of course, the position has *x*-, *y*-, and *z*-components; we will concentrate on the *y*-component, perpendicular to the direction of propagation. Figure 5.16 shows the wave propagating in the *x*-direction, which is horizontal; the *y*-direction is vertical. To determine the vertical position of the electron, we use a plate with a narrow slit placed in the path of the wave (see Fig. 5.16). If the electron succeeds in passing through this slit, then we will have achieved a determination of the vertical position to within an uncertainty

$$\Delta y = a \tag{16}$$

where *a* is the width of the slit. If the electron fails to pass through this slit, then our measurement is inconclusive and will have to be repeated.

By making the slit very narrow, we can make the uncertainty of our determination of the *y*-coordinate very small. But this has a surprising consequence for the *y*-component of the momentum of the electron: if we make the uncertainty in the *y*-coordinate small, we will make the uncertainty in the *y*-component of the momentum large. To see how this comes about, let us recall that according to our discussion of the single-slit experiment in the preceding section, the electron suffers diffraction by the slit and emerges at some angle θ (see Fig. 5.16). This angle θ is unpredictable; all we can say about the electron is that after it emerges from the slit it will be heading toward some point within the diffraction pattern. Thus, the direction of motion of the electron is uncertain. As a rough measure of the magnitude of this uncertainty in direction, we can take the angular width of the central diffraction maximum (most of the in-

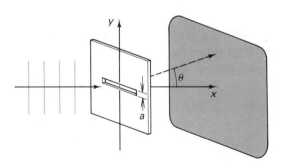

Fig. 5.16 Electron wave incident on a slit.

tensity of the electron wave is gathered within the region of this central maximum, and hence the electron is most likely to be found in this region). This estimate of the uncertainty of the angle gives us

$$\Delta\theta \gtrsim \frac{\lambda}{a} \qquad (17)$$

The y-component of the momentum is $p_y = p\sin\theta$; since we are concerned with a small angle, we can approximate this by $p_y = p\theta$. The uncertainty in p_y is then

$$\Delta p_y = p\,\Delta\theta \gtrsim p\,\frac{\lambda}{a} \qquad (18)$$

But $p = h/\lambda$ and consequently

$$\Delta p_y \gtrsim \frac{h}{a} \qquad (19)$$

Thus, if the slit is very narrow, the uncertainty in the y-component of the momentum will be very large! Comparing Eqs. (19) and (16), we find that

$$\Delta y\,\Delta p_y \gtrsim h \qquad (20)$$

This equation states that Δy and Δp_y cannot both be small; if one is small then the other must be large, so that their product is greater than or equal to Planck's constant.

Equation (20) is one of **Heisenberg's uncertainty relations.*** There are corresponding relations for the other components of position and momentum,

$$\Delta x\,\Delta p_x \gtrsim h \qquad (21)$$

$$\Delta z\,\Delta p_z \gtrsim h \qquad (22)$$

Although we have obtained the uncertainty relations (20)–(22) by examining the special case of a position measurement by means of a slit, it turns out that these relations are actually of general validity

---

* **Werner Heisenberg,** 1901–1976, German theoretical physicist, professor at Leipzig and director of the Max Planck Institute. Heisenberg gave the first complete formulation of the fundamental equations of quantum mechanics. His equations relied on matrices, and they were soon shown to be mathematically equivalent to the Schrödinger wave equation. For his formulation of quantum mechanics, he received the Nobel Prize in 1932.

for any kind of position measurement.* The Heisenberg uncertainty relations tell us that there exist ultimate, insuperable limitations in the precision of our measurements. On the macroscopic scale, the quantum uncertainties in our measurements can be neglected because h is small. But on the atomic scale, these quantum uncertainties are often so large that it is completely meaningless to speak of position or momentum of a wavicle.

Example 2   If we assume that in the ground state of the hydrogen atom the position of the electron along the Bohr orbit is not known and not knowable, then the uncertainty in the position is about $\Delta x = a_0 = 0.53$ Å. What is the corresponding quantum uncertainty in the momentum?

Solution   According to Eq. (21),

$$p \gtrsim \frac{h}{\Delta x} = \frac{h}{a_0}$$

$$= 6.6 \times 10^{-34}\,\text{J}\cdot\text{s}/0.53\,\text{Å}$$

$$= 1.2 \times 10^{-23}\,\text{kg}\cdot\text{m/s}$$

This uncertainty in the momentum is *larger* than the magnitude of the momentum of the electron ($p = L/a_0 = \hbar/a_0$). Hence it is meaningless to speak of the position and the momentum of the electron in such a Bohr orbit—and it is also meaningless to speak of an orbit!   ■

Since the momentum is directly related to the wavelength ($\lambda = h/p$), an uncertainty in p implies an uncertainty in λ. Hence the Heisenberg relations imply that the particle and the wave aspects of a quantum-mechanical wavicle are to some extent mutually exclusive: if the position is sharply defined (small Δx), then the wavelength is necessarily very poorly defined (large $\Delta\lambda$) and conversely. The particle and wave aspects of a wavicle exhibit **complementarity.** They never appear together; which one appears depends on what experiment is performed. An experiment that measures particle properties will not detect wave properties, and conversely. This principle of complementarity was formulated by Bohr.

The intimate connection between the Heisenberg relations and complementarity can be illustrated by some further discussion of the interference experiment with two slits. Suppose we use some apparatus to detect whether the electron passes through the upper slit or the lower slit. Obviously, if it goes through, say, the upper

---

* The exact form of the Heisenberg relations is $\Delta x\,\Delta p_x \geqslant \hbar/2$, etc.

slit, then the presence of the lower slit can make no difference. We therefore expect then even though both slits are open, there will be no interference effects and the intensity on the fluorescent screen will simply be twice the intensity due to one slit. Somehow the apparatus that is used to detect the electron at the slit disturbs the wave enough so that the interference is destroyed.

To understand how such a disturbance comes about, we must examine the interaction of the apparatus with the electron. Figure 5.17 shows a suitable apparatus: the two parallel slits are cut in two separate plates with independent suspensions. If the electron passes through a slit and is deflected, the change in the momentum of the electron will be equal and opposite to the change in the momentum of the plate. Hence a measurement of the momentum changes of the plates will tell us through which of the slits the electron went. There are some limitations to be taken into account. The recoil momentum is $p \sin \theta$ and therefore the measurement of the y-component of the momentum of the plate must be precise enough to detect this recoil:

$$\Delta p'_y \ll p \sin \theta \qquad (23)$$

(Here we have inserted a prime to distinguish the momentum of the plate from that of the electron.) Furthermore, any uncertainty in the position of the plate will tend to spoil the interference pattern since it implies an uncertainty in the distance between slits, and hence an uncertainty in the phase difference between the waves from the two slits. If we want to preserve the interference pattern we must keep the error in phase much smaller than 2π. According to Eq. (14), the maximum permissible uncertainty in d is

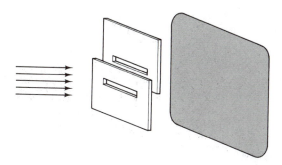

Fig. 5.17 Two parallel slits in two separate plates with independent suspensions.

then given by

$$\Delta y' \sin \theta \ll \lambda \qquad (24)$$

Combining Eqs. (23) and (24), we see that if we want to discover through which slit the electron went and also avoid the destruction of the interference pattern, the position and momentum of each plate must have uncertainties limited by

$$\Delta y' \, \Delta p_y' \ll \frac{\lambda}{p} \qquad (25)$$

i.e.,

$$\Delta y' \, \Delta p_y' \ll h \qquad (26)$$

However, (25) is forbidden by the Heisenberg uncertainty condition for the coordinate and momentum of the *plate*. Thus, although h is small and the uncertainties in a large object—such as our slotted plate—can often be neglected, for the purposes of our very delicate electron-detection experiment these uncertainties are crucial.

The conclusion is that if we use the apparatus of Fig. 5.17 to detect through which slit the electron went, then the interference pattern will be destroyed. The quantum-mechanical fluctuations in our measuring instrument (plate) are so large that an uncontrollable and random phase difference is introduced between the two waves from the two slits. Upon performing the experiment with a large number of electrons, one after the other, we find an accumulated intensity distribution that does not contain any interference term. The probability distribution over the screen is simply the sum of the two single distributions. From this *Gedankenexperiment* we see that electrons can behave like waves or like particles depending on what experiment we perform. As long as we do not attempt to localize the electrons at the slit, they will behave like waves and display interference. If we do observe their position, they will behave like particles and not display interference.

It is an essential feature of the quantum-mechanical measurement process that measurement produces a disturbance of the measured system. This is in contrast to the classical case, where we can, in principle, make the disturbance as small as we please. Quantum mechanically, the magnitude of the smallest disturbance is determined by the magnitude of h. The unavoidable quantum-mechanical uncertainties in the measuring device produce unpredictable and uncontrollable changes in the wavefunction of the measured system. The wavefunction after measurement is usually very different from that before.

For the consistency of quantum mechanics it is necessary that *all* systems in our universe satisfy the uncertainty relation. The existence of a classical system with no intrinsic uncertainty would permit the determination of the position and momentum of a quantum system with an uncertainty smaller than that given by the uncertainty relations. Suppose that the classical system suffers an elastic collision with the quantum system. From the exact position and velocity of the classical system before and after the collision we could then determine the *exact* point at which the collision took place. Hence the position of the quantum-mechanical system at the instant of collision would be determined with a precision which exceeds that permitted by the uncertainty relation.* To avoid this inconsistency we must assume that no classical system can ever interact with a quantum-mechanical system. This is equivalent to assuming that classical systems do not exist. Incidentally, the same argument can be used to show that all systems must satisfy the uncertainty relation with exactly the same value of *h*; only then is the uncertainty relation self-consistent.

There is another important principle necessary for the consistency of quantum mechanics: only *observed* quantities have meaning. Physical quantities must be defined operationally by means of experimental procedures; these procedures, if not possible in practice (because of technological limitations), must at least be possible in principle. To say that an electron passes through this slit, or that one, is meaningless if no experiment can be set up to check the statement.

We can summarize our conclusions thus:

1. Electrons and other bodies in the real world are not classical particles or classical waves. In some experiments they behave nearly like particles, in others nearly like waves (duality). Experiments that measure particle properties give no information about wave properties and conversely (complementarity).
2. An electron can be described by a wavefunction whose intensity gives the probability for finding the electron.
3. The wavefunctions obey the **Superposition Principle:** If the amplitude due to one source is Ψ_1 and the amplitude due to a second source is Ψ_2, then the total amplitude for both sources acting together is $\Psi_1 + \Psi_2$. Alternatively, if Ψ_1 is a solution of the quantum-mechanical equations and Ψ_2 is another solution, then $\Psi_1 + \Psi_2$ is also a solution.

---

* This argument assumes that the interaction is of extremely short range. This is not essential. In general, the instantaneous acceleration of the classical system determines the position of the quantum-mechanical system (if the force law is known).

5.4 Schrödinger's Wave Equation for a Free Particle

We now want to establish some of the mathematical properties of the wave describing a quantum-mechanical particle, such as an electron. In this chapter we will assume that the quantum-mechanical particle is free, i.e., external forces are absent. Furthermore, for the sake of simplicity, we will assume that the motion is one-dimensional, say, along the x-axis.

The wave describing a particle of momentum p and energy E has a wavelength $\lambda = h/p$ and a frequency $v = E/h$. For convenience, we define the wave number*

$$k = \frac{2\pi}{\lambda} = \frac{p}{\hbar} \tag{27}$$

and the angular frequency

$$\omega = 2\pi v = \frac{E}{\hbar} \tag{28}$$

Since for our free particle $E = p^2/(2m)$, we can also write Eq. (27) as

$$k = \frac{\sqrt{2mE}}{\hbar} \tag{29}$$

If the particle travels in the $+x$ direction, the wave will also have to travel in that direction. The possible harmonic waves satisfying this condition are

$$\sin(\omega t - kx), \qquad \cos(\omega t - kx) \tag{30}$$

$$e^{-i(\omega t - kx)}, \qquad e^{i(\omega t - kx)} \tag{31}$$

The waves (30) are *real* and the waves (31) are *complex*. According to the mathematical rules for evaluating complex exponentials we have the identities

$$e^{i\alpha} = \cos\alpha + i\sin\alpha \tag{32}$$

and

$$\cos\alpha = \frac{e^{i\alpha} + e^{-i\alpha}}{2}, \qquad \sin\alpha = \frac{e^{i\alpha} - e^{-i\alpha}}{2i} \tag{33}$$

Thus

$$e^{-i(\omega t - kx)} = \cos(\omega t - kx) - i\sin(\omega t - kx) \tag{34}$$

---

* In three dimensions we would define the wave vector as $\mathbf{k} = \mathbf{p}/\hbar$.

and

$$e^{+i(\omega t - kx)} = \cos(\omega t - kx) + i\sin(\omega t - kx) \qquad (35)$$

This shows that the complex waves are linear combinations of the real waves, with a complex coefficient. To settle which of the waves (30) or (31) we want, let us appeal to the Superposition Principle of quantum mechanics: whenever Ψ_1 and Ψ_2 are possible quantum-mechanical waves, then so is $\Psi_1 + \Psi_2$. Suppose we try to describe a particle traveling in the positive x-direction by the wave $\sin(\omega t - kx)$. Then a particle traveling in the negative x-direction will be described by $\sin(\omega t + kx)$. According to the Superposition Principle, the sum

$$\sin(\omega t - kx) + \sin(\omega t + kx) \qquad (36)$$

should then be a possible quantum-mechanical wavefunction. But (36) equals

$$2\sin\omega t \cos kx \qquad (37)$$

Obviously this is not a satisfactory wavefunction because at $t = 0$ it vanishes identically everywhere, i.e., the probability for finding the particle is zero everywhere! Since the particle cannot disappear, we must regard the wavefunction (36) as unacceptable, and we conclude that $\sin(\omega t - kx)$ is not the wavefunction we want. By a similar argument, we can rule out $\cos(\omega t - kx)$.

Let us now consider $e^{-i(\omega t - kx)}$. The superposition of this with $e^{-i(\omega t + kx)}$ gives

$$e^{-i(\omega t - kx)} + e^{-i(\omega t + kx)} = 2e^{-i\omega t}\cos kx \qquad (38)$$

This never vanishes identically everywhere, and is an acceptable wavefunction. Thus, the wavefunction $e^{-i(\omega t - ikx)}$ is consistent with the Superposition Principle. A wavefunction of the type $e^{i(\omega t - kx)}$ is also consistent with the Superposition Principle; but according to a convention laid down by the founders of quantum mechanics, this wavefunction is to be discarded in favor of $e^{-i(\omega t - kx)}$.* Thus we are led to the conclusion that the correct wavefunction is

$$\boxed{\Psi(x, t) = e^{-i(\omega t - kx)}} \qquad \begin{array}{l}\text{for motion in the positive} \\ x\text{-direction}\end{array} \qquad (39)$$

---

* We have to choose one or the other of these. Any attempt at using both leads to trouble, since the superposition $e^{-i(\omega t - kx)} + e^{i(\omega t - kx)}$ equals $2\cos(\omega t - kx)$, which is an unacceptable wavefunction.

and

$$\Psi(x, t) = e^{-i(\omega t + kx)}$$ for motion in the negative
x-direction   (40)

Next, we want to discover the **wave equation** obeyed by these wavefunctions. In general, the wavefunction describing any physical system obeys a wave equation, which permits us to calculate the time evolution of the system. Mathematically, the wave equation is a differential equation involving Ψ and the derivatives of Ψ with respect to x and t. To discover this differential equation for Ψ, we begin by expressing ω and k in terms of p:

$$\Psi = e^{-i(\omega t - kx)} = e^{-i(Et/\hbar - \sqrt{2mE}\,x/\hbar)} \tag{41}$$

If we take the first derivative of Ψ with respect to time and the second derivative of Ψ with respect to x, we find

$$\frac{\partial \Psi}{\partial t} = -i\frac{E}{\hbar}\Psi \tag{42}$$

and

$$\frac{\partial^2 \Psi}{\partial x^2} = -\frac{2mE}{\hbar^2}\Psi \tag{43}$$

Comparing these expressions, we see that these derivatives are proportional. By adjusting the constant of proportionality, we obtain an equality:

$$-\frac{\hbar^2}{2m}\frac{\partial^2}{\partial x^2}\Psi(x, t) = i\hbar\frac{\partial}{\partial t}\Psi(x, t) \tag{44}$$

This is the wave equation for a free particle; it is called **Schrödinger's equation.*** In the next chapter we will discover how to include forces—or, more precisely, potentials—in Schrödinger's equation. This equation plays the same role in quantum mechanics as Newton's Second Law does in classical mechanics. From Schrödinger's equation we can calculate the motion of a quantum-mechanical wave just as from Newton's Law we can calculate the motion of a classical particle.

---

* **Erwin Schrödinger,** 1887–1961, Austrian theoretical physicist, professor at Zürich, Berlin, Dublin, and Vienna. For his formulation of the Schrödinger equation, with which he laid the foundation for wave mechanics, he was awarded the Nobel Prize in 1933.

Example 3   Show that the (unacceptable) wavefunctions $\sin(\omega t - kx)$, $\cos(\omega t - kx)$, and $e^{i\omega(t-kx)}$ are *not* solutions of Schrödinger's equation.

Solution   For the sake of brevity, we will here only spell out the calculation for $\sin(\omega t - kx) = \sin(Et/\hbar - \sqrt{2mE}\, x/\hbar)$. The derivatives of this wavefunction are

$$\frac{\partial}{\partial t} \sin\left(\frac{Et}{\hbar} - \sqrt{2mE}\,\frac{x}{\hbar}\right) = \frac{E}{\hbar} \cos\left(\frac{Et}{\hbar} - \sqrt{2mE}\,\frac{x}{\hbar}\right) \qquad (45)$$

and

$$\frac{\partial^2}{\partial x^2} \sin\left(\frac{Et}{\hbar} - \sqrt{2mE}\,\frac{x}{\hbar}\right) = -\frac{2mE}{\hbar^2} \sin\left(\frac{Et}{\hbar} - \sqrt{2mE}\,\frac{x}{\hbar}\right) \qquad (46)$$

Substitution of these into Eq. (44) does *not* give an equality, i.e., this wavefunction is *not* a solution of Schrödinger's equation.   ∎

The arguments presented in this section illustrate the power of the Superposition Principle. We have seen that this principle demands that the wavefunction be *complex*, and determines what differential equation must be obeyed. The Superposition Principle is one of the most important principles of quantum mechanics.

The intensity of the wave gives us the probability for finding the particle. Since we are now contemplating a particle that moves along the x-axis, the rule for calculating the probability is

$$\left(\begin{array}{l} \text{probability for} \\ \text{finding particle} \\ \text{between } x \text{ and } x + dx \end{array}\right) = |\Psi|^2\, dx \qquad (47)$$

Here the intensity $|\Psi|^2$ is the absolute value squared of the complex number Ψ. In terms of the real and the imaginary parts of Ψ this is

$$|\Psi|^2 = (\text{Re}\,\Psi)^2 + (\text{Im}\,\Psi)^2 \qquad (48)$$

Because the probability for finding the particle *somewhere* must equal 1, we require that Ψ be normalized so that

$$\int_{-\infty}^{\infty} |\Psi|^2\, dx = 1 \qquad (49)$$

There is one difficulty with this normalization requirement. The wavefunction $e^{-i(\omega t - kx)}$ of a free particle of momentum p cannot be normalized since

$$\int_{-\infty}^{\infty} \left|e^{ikx - i\omega t}\right|^2 dx = \int_{-\infty}^{\infty} 1\, dx = \infty \qquad (50)$$

The trouble with this wavefunction is that it assigns equal probability to all points in space. This is clearly not realistic since we always know that the particle is somewhere in a room, or somewhere on the Earth, or at least somewhere in the galaxy. A particle therefore never has the wavefunction $e^{ikx-i\omega t}$, and therefore never has an absolutely precise momentum. It is possible to construct a normalizable wavefunction (a wave packet; see the next section) only if the momentum is left somewhat uncertain. The amount by which the momentum must be uncertain can be estimated from Heisenberg's relation. For example, if it is known that the particle is somewhere in a room ($\Delta x \sim 10\,\text{m}$), we find

$$\Delta p \sim \frac{h}{\Delta x} \sim 7 \times 10^{-35}\,\text{kg·m/s}$$

This is an extremely small momentum uncertainty (for an electron, it corresponds to a velocity $v \sim 10^{-4}\,\text{m/s}$) and therefore imposes no practical limitation on the precision with which the momentum of, say, a beam of electrons can be determined. Although $e^{-i(\omega t - kx)}$ is strictly speaking not a normalizable wavefunction, we can do many calculations with it as though it were. This involves using some tricks which we will consider in later sections.

5.5 **Wave Packets; Group Velocity**

Although electrons and other quantum-mechanical particles are always described by waves, it is nevertheless sometimes possible for them to move along a fairly well-defined classical path. This can happen if the wave describing the system has the form of a wave packet, i.e., a wave that has an appreciable amplitude only in a small region and is zero, or nearly zero, everywhere else (see Fig. 5.18). Such a wave packet can move along a path and, as long as it remains small, it will behave very much like a classical particle. However, the wave packet must have some finite size and this means that there necessarily exists an uncertainty in the position of the system. There must also exist an uncertainty in the momentum of the system, since a wave packet of finite size does not have a single, well-defined wavelength. Figure 5.19 shows examples of several wave packets with different values of Δx and Δp_x. Note that the packet with a large value of Δx has a small value of Δp_x, and conversely. This reciprocal relationship between Δx and Δp_x

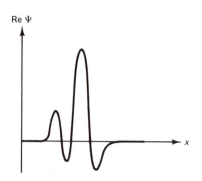

Re Ψ

x

Fig. 5.18 A wave packet.

Re Ψ

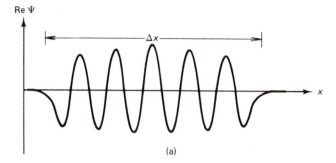

(a)

Re Ψ

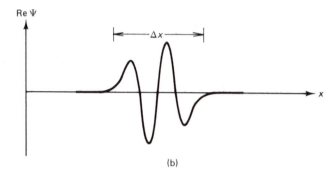

(b)

Re Ψ

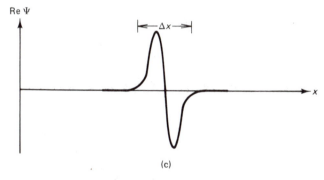

(c)

Fig. 5.19 Several wave packets of different widths. The packet (a) has a clear periodicity and therefore a sharply defined wavelength (and a sharply defined momentum). The packet (c) has no periodicity and therefore a poorly defined wavelength (and a poorly defined momentum).

is in accord with the Heisenberg relation

$$\Delta x \, \Delta p_x \gtrsim h \tag{51}$$

In fact, it can be proved rigorously that the inequality (51) is always satisfied for any wave packet. The proof hinges on the Fourier theorem: any wave packet can be expressed as a superposition of an infinite number of harmonic waves. The decomposition of a wave packet into its constituent harmonic waves is called **Fourier analysis.** The harmonic waves required to construct a narrow wave packet (small Δx) span a wide range of wavelengths (large $\Delta\lambda$, hence large Δp_x), exactly as expected from the Heisenberg relation.

Although the proof of Fourier's theorem is beyond the mathematical limitations imposed on this book, we can gain some qualitative insights into the construction of wave packets by looking at the following simple example of the superposition of two harmonic waves. Consider two harmonic waves of slightly different frequencies and wave numbers,

$$e^{-i[(\omega + \frac{1}{2}\Delta\omega)t - (k + \frac{1}{2}\Delta k)x]} \tag{52}$$

and

$$e^{-i[(\omega - \frac{1}{2}\Delta\omega)t - (k - \frac{1}{2}\Delta k)x]} \tag{53}$$

The superposition of these waves gives

$$\Psi = e^{-i[(\omega + \frac{1}{2}\Delta\omega)t - (k + \frac{1}{2}\Delta k)x]} + e^{-i[(\omega - \frac{1}{2}\Delta\omega)t - (k - \frac{1}{2}\Delta k)x]} \tag{54}$$

Let us concentrate on the real part of this wavefunction (the imaginary part can be handled by a calculation similar to that for the real part). According to Eq. (32),

$$\mathrm{Re}\,\Psi = \cos\left[(\omega + \tfrac{1}{2}\Delta\omega)t - (k + \tfrac{1}{2}\Delta k)x\right]$$
$$+ \cos\left[(\omega - \tfrac{1}{2}\Delta\omega)t - (k - \tfrac{1}{2}\Delta k)x\right] \tag{55}$$

As we already saw in Section 1.6, two such cosine waves can be combined by means of the trigonometric identity for the sum of two cosines, with the result

$$\mathrm{Re}\,\Psi = 2\cos\left[(\tfrac{1}{2}\Delta\omega)t - (\tfrac{1}{2}\Delta k)x\right]\cos(\omega t - kx) \tag{56}$$

If $\Delta\omega$ and Δk are small, then the factor $\cos(\omega t - kx)$ in the expression (56) represents a wave very similar to each of the original waves. The factor $\cos\left[(\tfrac{1}{2}\Delta\omega)t - (\tfrac{1}{2}\Delta k)x\right]$ represents a wave of much smaller frequency and much longer wavelength (or smaller wave number) which can be regarded as a slowly varying amplitude that modulates the wave $\cos(\omega t - kx)$. Figure 5.20 shows a plot of the expression (56) at one instant of time. The dashed curve forming the envelope indicates the amplitude modulation.

Obviously, each of the regions of large amplitude has the shape of a wave packet. Thus, by superposition of two waves of slightly different frequency and wavelength we have constructed a sequence of wave packets. This is the best we can do with two harmonic waves; if we want to construct a *single* wave packet, we must add

Re Ψ

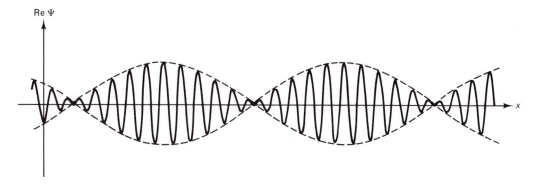

x

Fig. 5.20 The superposition of two harmonic waves of slightly different frequencies.

many more harmonic waves to our superposition so that all the wave packets except one, say, the central one in Fig. 5.20, are canceled. As a crude estimate of the width Δx of the resulting single wave packet we can take one-half the wavelength of the modulation in Fig. 5.20,

$$\Delta x \sim \frac{2\pi}{\Delta k} \tag{57}$$

and as a crude estimate of the uncertainty in the momentum we can take*

$$\Delta p_x = \hbar \, \Delta k \tag{58}$$

The product of (57) and (58) gives

$$\Delta x \, \Delta p_x \sim \frac{2\pi}{\Delta k} \times \hbar \, \Delta k \sim h \tag{59}$$

which is what we expect from the Heisenberg relation.

Our simple example of the superposition of two waves also yields some instructive results concerning the velocity of waves and of wave packets. As we know from Section 1.6, the velocity of a harmonic wave, called the phase velocity, can be expressed as the ratio of angular frequency to wave number. Thus, the phase velocity of the wave (52) is

$$v_p = \frac{\omega}{k} \tag{60}$$

---

* This is only a crude, order-of-magnitude estimate since the addition of extra waves to the superposition will make the uncertainty in the wave number larger than the value Δk given in Eqs. (52) and (53).

Since $\omega = E/\hbar = \frac{1}{2}mv^2/\hbar$ and $k = p/\hbar = mv/\hbar$, this phase velocity is

$$v_p = \frac{\frac{1}{2}mv^2/\hbar}{mv/\hbar} = \frac{1}{2}v \tag{61}$$

i.e., the phase velocity of the wave is one-half the particle velocity. This puzzling disagreement between the phase velocity and the particle velocity need not worry us because if we want to time the motion of a particle over some given distance we must describe the particle by a wave packet. Thus the relevant velocity is the velocity of the envelope of the wave packet, or the group velocity. From Section 1.6 we know that the group velocity can be expressed as

$$v_g = \frac{d\omega}{dk} \tag{62}$$

Since

$$d\omega = d\left(\frac{\frac{1}{2}mv^2}{\hbar}\right) = \frac{mv\,dv}{\hbar} \tag{63}$$

and

$$dk = d\left(\frac{mv}{\hbar}\right) = \frac{m\,dv}{\hbar} \tag{64}$$

we find that

$$v_g = \frac{mv\,dv/\hbar}{m\,dv/\hbar} = v \tag{65}$$

Thus, the group velocity coincides with the particle velocity. This is very reasonable since a traveling wave packet is the wave-mechanical version of a traveling particle.

For a wave packet we can recast the Heisenberg relation in an instructive form involving energy and time. If the uncertainty in the momentum is Δp, then

$$\Delta E = \Delta\left(\frac{p^2}{2m}\right) = \frac{p}{m}\Delta p = v\,\Delta p \tag{66}$$

If the width of the packet is Δx, then the time that the packet takes to travel a distance equal to the width is

$$\Delta t = \frac{\Delta x}{v} \tag{67}$$

This can be regarded as the **characteristic time** for the evolution of the packet—it is the time required for the packet to move a clearly noticeable distance. The product of ΔE and Δt is

$$\Delta E\,\Delta t = v\,\Delta p\,\frac{\Delta x}{v} = \Delta p\,\Delta x \qquad (68)$$

and, by Heisenberg's uncertainty relation,

$$\Delta E\,\Delta t \gtrsim h \qquad (69)$$

This is the **energy–time uncertainty relation.*** Although for a wave packet this uncertainty relation is a direct consequence of the position-momentum uncertainty relation, for other quantum-mechanical systems it is an independent relation. The time Δt appearing in this relation is always a characteristic time required for an appreciable change to occur in the system.

Example 4 The **lifetime** of a typical excited state in an atom is about 10^{-8} s, i.e., the electron typically takes about 10^{-8} s to emit a photon and complete the transition to a lower state. What is the energy uncertainty of such an excited state?

Solution By the energy–time uncertainty relation, the energy uncertainty of the excited state is

$$\Delta E \gtrsim h/\Delta t = 6.6 \times 10^{-34}\,\text{J}\cdot\text{s}/10^{-8}\,\text{s}$$

$$\gtrsim 6.6 \times 10^{-26}\,\text{J} = 4.1 \times 10^{-7}\,\text{eV} \quad \blacksquare$$

Chapter 5: SUMMARY

de Broglie wavelength:

$$\lambda = \frac{h}{p}$$

Probability interpretation of wavefunction:

$$\begin{pmatrix} \text{probability for electron} \\ \text{to be found between} \\ y \text{ and } y + dy \end{pmatrix} = |\Psi(y,t)|^2\,dy$$

Superposition Principle: If the amplitude due to one source is Ψ_1, and the amplitude due to a second source is Ψ_2, then the total amplitude for both sources acting together is $\Psi_1 + \Psi_2$.

---

* The exact form of the energy–time uncertainty relation is $\Delta E\,\Delta t \geqslant \hbar/2$.

Heisenberg uncertainty relation:

$$\Delta x \, \Delta p_x \geqslant \hbar/2$$

Duality: Quantum-mechanical particles (wavicles) sometimes behave like classical particles, and sometimes like classical waves.

Complementarity: An experiment that measures particle properties will not detect wave properties, and conversely.

Wavefunctions for free particle:

$$e^{-i(\omega t - kx)}$$

$$e^{-i(\omega t + kx)}$$

Schrödinger's equation for free particle:

$$-\frac{\hbar^2}{2m}\frac{\partial^2 \Psi}{\partial x^2} = i\hbar \frac{\partial \Psi}{\partial t}$$

Energy–time uncertainty relation:

$$\Delta E \, \Delta t \geqslant \hbar/2$$

Chapter 5:   PROBLEMS

1. Show that the de Broglie wavelength of an electron of energy E can be expressed as

$$\lambda = \frac{12.26}{\sqrt{E}}$$

where λ is measured in angstroms and E in electron volts.

2. What is the de Broglie wavelength of electrons of an energy of 20,000 eV in the beam of a TV tube? Is classical mechanics likely to be a satisfactory approximation for the operation of TV tubes?

3. As described in Example 1.2, one of the oil droplets used by Millikan had a mass of 8.4×10^{-14} kg and a speed of 2.0×10^{-5} m/s. What is the de Broglie wavelength for such an oil droplet?

4. The electron-microscope picture of Fig. 5.7 was made with an electron beam of energy 40 keV. What is the de Broglie wavelength of such electrons?

5. The protons accelerated in the main ring at Fermilab attain a momentum of 5.3×10^{-16} kg · m/s. What is the de Broglie wavelength of these protons? [Note that Eq. (2) remains valid for relativistic particles.]

6. According to classical statistical mechanics, the mean thermal kinetic energy of a particle at temperature $T = 300$ K is $\frac{3}{2}kT = 6.21 \times 10^{-21}$ J. What is the de Broglie wavelength of an electron ($m = 9.11 \times 10^{-31}$ kg), a grain of dust ($m = $

10^{-17} kg), a tennis ball ($m = 6.0 \times 10^{-2}$ kg), and an automobile ($m = 1500$ kg) of this energy?

7. Rutherford used alpha particles of energy 7.7 MeV to probe the interior of the atom. What is the de Broglie wavelength of such alpha particles? Is it adequate to treat the motion of these alpha particles by classical mechanics, down to distances of 10^{-14} m from the nucleus, as Rutherford did?

8. The "thermal" neutrons in a nuclear reactor may be regarded as a gas at a temperature of 300°C.
   (a) What is the average kinetic energy and the speed of such neutrons?
   (b) What is their de Broglie wavelength?

9. For what speed of an electron is the de Broglie wavelength equal to the Compton wavelength $h/m_e c$? [Note: You must use the relativistic formula

$$p = mv/\sqrt{1 - v^2/c^2}$$

   in Eq. (2).]

10. In order to examine the interior structure of the proton and the neutron, experimenters at the Stanford Linear Accelerator Center (SLAC) used a beam of electrons of energy 20 GeV as a probe. What is the de Broglie wavelength of such electrons? What is the size of the smallest structural detail that can be resolved by such electrons? [*Hint:* Equation (2) remains valid for relativistic electrons, but the momentum must be calculated from the relativistic formula for momentum, $c^2 p^2 = E^2 - m^2 c^4$.]

11. At high temperatures and/or low densities, the behavior of a gas can be described by classical statistical mechanics because, to a good approximation, the atoms move along classical paths. However, when the temperature of the gas is low and/or the density high, the classical approximation fails and quantum effects become important. The classical approximation breaks down completely when the de Broglie wavelength of a typical atom becomes comparable to the average interatomic distance (defined as $n^{-1/3}$, where n is the density of atoms).
    (a) Show that this happens when $(h^2/mkT)^{1/2} \sim n^{-1/3}$.
    (b) Consider helium gas at room temperature. At what density and pressure does the classical approximation break down?
    (c) Consider the gas of conduction electrons in a typical metal. Assume that there is one conduction electron per atom. According to the above criterion, is the classical approximation valid?

12. Since the wavelength of visible light is ~ 5000 Å, a light microscope cannot resolve any details smaller than this length. What must be the minimum energy of the electrons in an electron microscope if it is to attain a resolution at least as good as a light microscope? (In practice, energies used in electron microsopes are usually more than 10 keV, and their resolution is not limited by the wave properties of the electrons.)

13. A diffraction grating used in optics has 20,000 rulings per cm. Suppose we use such a grating for an electron interference experiment. What would be the angular separation between the (principal) maxima if the electron energy is 100 eV? If the

maxima on the fluorescent screen (see Fig. 5.15) are to be separated by 1 mm, how large must we make the distance from the grating to the fluorescent screen?

14. When scattering a beam of monoenergetic electrons on one of the faces of a crystal of nickel, Davisson and Germer found a strongly reflected beam at an angle of $\phi = 50.0°$ from the incident beam (see Fig. 5.21).
(a) The spacing between the relevant Bragg planes in the nickel crystal is 0.91 Å. From the Bragg condition, with $n = 1$, calculate the wavelength of the electrons that gave the strong reflected beam at $\phi = 50.0°$.
(b) The energy of these electrons was 54.0 eV. Calculate the wavelength from de Broglie's relation. Can you explain the (small) discrepancy between the results (a) and (b)? (*Hint:* Think of the work function of nickel.)

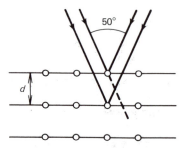

Fig. 5.21 Electron beam incident on a crystal.

15. Nuclear reactors produce a copious flux of neutrons, and these will form a beam if they are allowed to escape through a hole in the wall of the reactor. The neutrons in such a beam have mixed energies. We can use Bragg reflection by a crystal to separate this beam of neutrons of mixed energies into monoenergetic beams emerging at different angles, just as we can use an optical "diffraction" grating to separate a beam of light of mixed colors into monochromatic beams. Suppose we use a beryllium crystal with a Bragg-plane spacing of 1.80 Å to intercept a beam of neutrons of mixed energies. At what angle (with respect to the incident beam) will neutrons of energy 0.020 eV emerge from this crystal?

16. A beam of electrons of energy 64.0 eV is incident on an aluminum foil containing many microcrystals. The interference pattern produced by the scattered beams is observed on a fluorescent screen placed 15.0 cm beyond the foil. The pattern consists of several concentric circles, such as shown in Fig. 5.4. Calculate the radii of the circles that result from Bragg reflections by a set of planes with a spacing of 4.04 Å.

17. Show that if electrons are to display Bragg reflection, they must have an energy of at least $h^2/(8m_e d^2)$. What is this minimum energy (in eV) for electrons reflected by Bragg planes in nickel with $d = 0.91$ Å?

18. The first measurement of the de Broglie wavelength of atoms was performed by O. Stern. He aimed a beam of helium atoms, from an oven, at the surface of a lithium-fluoride crystal. The regular rows of atoms with a spacing of 2.07 Å on the surface of the crystal act like the rulings on the surface of an optical grating, at oblique incidence. It is known from optics that if a beam is incident on a reflection grating at an angle ϕ (see Fig. 5.22), then beams of maximum intensity emerge at angles θ satisfying the equation

$$d \sin \theta - d \sin \phi = n\lambda$$

where d is the spacing between the rulings or rows of the grating (for $n = 0$ this gives $\theta = \phi$, which corresponds to ordinary reflection). Suppose that helium atoms of an energy equal to the mean thermal energy at $T = 300 \text{ K}$ are incident with an angle $\phi = 20°$. Calculate the angle θ at which the first-order ($n = 1$) beam will emerge.

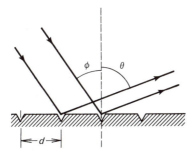

Fig. 5.22 Electron beam incident on a grating. Note that the angles are measured with respect to the normal to the grating.

19. In an experiment performed by E. Rupp soon after that of Davisson and Germer, the wave character of electrons was demonstrated with an optical grating. In this experiment, a low-energy beam of electrons was aimed at the grating at oblique incidence, with an angle ϕ very near 90° (this angle is measured with respect to the normal, as in Fig. 5.22). For oblique incidence, the angles of the emerging beams of maximum intensity satisfy the equation

$$d \sin \theta - d \sin \phi = n\lambda$$

where d is the spacing between the rulings of the grating. Suppose that an electron beam of energy 2.0 eV is incident on a grating with 10,000 rulings per centimeter at an angle $\phi = 89.0°$. At what angle will we observe the first-order ($n = -1$) emerging beam? Could we observe this beam if the angle of incidence were $\phi = 0$?

20. Figure 5.23 shows a neutron interferometer* consisting of a single crystal of silicon that has been cut away to leave three parallel ridges. By Bragg reflections at the first two ridges, a neutron beam incident from the left is split into two

---

* R. Colella et al., *Phys. Rev. Lett.* **34**, 1472 (1975).

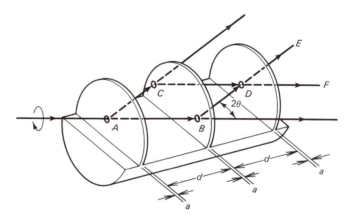

Fig. 5.23 Neutron interferometer. [From R. Colella et al., *Phys. Rev. Lett.* **34**, 1472 (1975).]

beams (*ABD* and *ACD*) which superpose at the third ridge (*D*). A final Bragg reflection at the third ridge yields the emerging beams (*DE* and *DF*).

(a) The emerging beam *DF* is a superposition of the two beams *ABDF* and *ACDF*. Taking into account that all the Bragg reflections are identical (since they occur in the same crystal), show that these two beams are of equal intensities and of equal phases, giving constructive interference.

(b) The emerging beam *DE* is a superposition of the two beams *ABDE* and *ACDE*. Taking into account that neutron beams suffer a phase change during reflection, are these two beams of equal intensities and phases?

21. The interference pattern shown in Fig. 5.10 has the intensity distribution

$$|\Psi(\theta)|^2 = \cos^2\left(\frac{\pi d}{\lambda}\sin\theta\right)$$

Calculate the ratio of the probability for finding the electron in the interval $0 \leqslant \theta \leqslant \sin^{-1}(\lambda/d)$ to the probability for finding the electron in the interval $\sin^{-1}(\lambda/d) \leqslant \theta \leqslant \sin^{-1}(2\lambda/d)$. For the sake of simplicity, assume that $\lambda/d \ll 1$.

22. A photon passes through a horizontal slit of width 5×10^{-6} m. What uncertainty in the vertical position will this photon have as it emerges from the slit? What minimum uncertainty in vertical momentum?

23. What is the minimum uncertainty in the velocity of an automobile of mass 1500 kg, given that the position of its center of mass has been measured to within 1.0×10^{-12} m.

24. Suppose that the *x*-component of the velocity of an electron has been measured to an accuracy of 10^{-2} m/s. What is the minimum uncertainty in its position along the *x*-axis? Along the *y*-axis? Solve the same problem if the particle is a proton.

25. Suppose that the uncertainty in the instantaneous position of an electron moving along the x-axis is $10\,\text{Å}$. The velocity of the electron is $10^6\,\text{m/s}$. What is the minimum uncertainty in momentum? In velocity? In kinetic energy?

26. According to de Broglie's naive picture of closed waves traveling around the orbit of an electron in a hydrogen atom, what would be the wavelength of such a wave in the ground state? According to the Uncertainty Principle, what would be the uncertainty in the wavelength? What conclusion can you draw?

†27. At any given instant of time, the position and the momentum of the Earth are somewhat uncertain because of quantum effects. This implies that the year (defined as the time it takes the Earth to complete one orbit) is somewhat uncertain. Give a *rough* estimate for the latter uncertainty. Express your answer in percent.

28. For a relativistic particle, Eqs. (1) and (2) for the frequency and the de Broglie wavelength become*

$$v = \frac{E}{h} = \frac{mc^2/h}{\sqrt{1 - v^2/c^2}}$$

$$\lambda = \frac{p}{h} = \frac{mv/h}{\sqrt{1 - v^2/c^2}}$$

Show that the phase velocity and the group velocity of the de Broglie waves for such a particle are

$$v_p = \frac{c^2}{v} \quad \text{and} \quad v_g = v$$

29. Consider a wave packet whose width is initially as small as possible, i.e., $\Delta x \simeq h/\Delta p_x$.
(a) Show that in a time $\Delta t \simeq m(\Delta x)^2/h$ the width of this wave packet will double (because the high-momentum components of the packet will move ahead of the low-momentum components).
(b) Show that this doubling time Δt and the energy uncertainty ΔE in the rest frame of the packet are related by $\Delta E\,\Delta t \simeq h$.

30. Consider a radio wave in the form of a pulse lasting $0.001\,\text{s}$. This pulse then has a length of $0.001\,\text{s} \times c = 3 \times 10^5\,\text{m}$. Since an individual photon of this radio wave can be anywhere within this pulse, the uncertainty in the position of the photon is $\Delta x = 3 \times 10^5\,\text{m}$ along the direction of propagation.
(a) According the Heisenberg's relation, what is the corresponding uncertainty in the momentum of the photon?
(b) What is the uncertainty in the frequency of the photon?

31. The lifetime of a typical excited state in an atom is $10^{-8}\,\text{s}$. Suppose that the atom emits a photon of wavelength about $6000\,\text{Å}$. What is the energy uncertainty of this photon? What is the wavelength uncertainty of this photon?

---

* The rest-mass energy is included in the energy E. Because of this, the frequency for a particle at rest is not zero, but mc^2/h. This quantity plays the role of an additive constant in the frequency; it has no observable consequences.

32. The extremely unstable particles, or "resonances," discovered by high-energy physicists have such a short lifetime that it is not accessible to direct measurement. Instead, the lifetime must be estimated from the energy–time uncertainty relation, using the observed uncertainty in the rest-mass energy of these particles. For instance, the unstable particle ϕ (1680) has an energy uncertainty of 150 MeV, i.e., some of the ϕ particles are produced with an energy of 150 MeV more (or less) than the energy of some other ϕ particles. What lifetime does this energy uncertainty imply?

Wave Mechanics II
—Particles in Potentials

In this chapter we will deal with the behavior of quantum-mechanical particles in the presence of a force represented by some given potential-energy function. This behavior is determined by the general Schrödinger equation, which includes the potential.* The general Schrödinger equation is the basic equation of wave mechanics. It plays the role of an equation of motion for the wavefunction, permitting us to calculate the wavefunction at any later time from its initial condition at some initial time.

Among the solutions of the Schrödinger equation, the standing-wave solutions are of the greatest importance. Wave mechanics explains the quantization of energy in terms of standing waves. The discrete frequencies of a wave-mechanical system—such as an atom—are quite analogous to the discrete frequencies of a musical instrument—such as a violin or a flute—, which are likewise due to standing waves. In these musical instruments, the possible standing waves must have an integral number of wavelengths or half-wavelengths within the length of the string or the air column. This requirement selects a discrete set of possible standing waves, and therefore a discrete, quantized set of frequencies. The discrete frequencies and energies of a wave-mechanical system arise in

---

* In quantum mechanics, the words *potential energy* and *potential* are often used interchangeably. This is in contrast to the practice in electromagnetism, where *potential* is reserved for potential energy per unit charge.

much the same way. As we will see in this chapter, if a quantum-mechanical particle is bound within a system by some attractive force, the standing wave describing this particle must have an integral number of wavelengths or half-wavelengths within the boundaries of the system. This requirement selects a discrete, quantized set of frequencies and of energies (in a wave-mechanical system, discrete frequencies imply discrete energies, since $E = \hbar\omega$). Of course, the analogy between musical instruments and atoms is not perfect: the wave equation for an electron is more complicated than the wave equation for a string, and the harmonics of the atomic music are more complicated than the harmonics of the violin string.

6.1 Particle in a Box

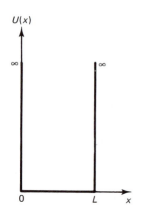

$U(x)$

∞ ∞

0   L x

Fig. 6.1 Potential energy for a particle in a one-dimensional box. The potential energy is infinite at $x = 0$ and $x = L$.

The simplest example of a quantum-mechanical system that exhibits standing waves is a particle in a one-dimensional "box." The walls of the box are assumed to be perfectly rigid and elastic, and the particle is assumed to move freely between these walls. Figure 6.1 shows the box extending from $x = 0$ to $x = L$ on the x-axis. The walls are represented by an infinitely large positive potential energy at $x = 0$ and $x = L$; this potential energy may be regarded as representing an infinitely deep potential well.

Classically, a particle confined in such a box would simply bounce back and forth and, of course, it would not be subject to any quantization of energy. Quantum mechanically, we can describe the motion of the particle by using a suitable superposition of the free-particle wavefunctions known from the preceding chapter. If the particle has an energy E and momentum $p = \sqrt{2mE}$, then the harmonic waves traveling to the right and the left along the x-axis are, respectively,

$$\mathrm{e}^{-i(Et/\hbar - \sqrt{2mE}\,x/\hbar)} \tag{1}$$

and

$$\mathrm{e}^{-i(Et/\hbar + \sqrt{2mE}\,x/\hbar)} \tag{2}$$

As we know from Section 1.6, the superposition of two such traveling waves of equal amplitudes gives a standing wave. For the sake of generality, let us multiply the two waves (1) and (2) by some arbitrary complex coefficients A and B and construct the superposition

$$\Psi(x,t) = Ae^{-i(Et/\hbar - \sqrt{2mE}\, x/\hbar)} + Be^{-i(Et/\hbar + \sqrt{2mE}\, x/\hbar)} \qquad (3)$$

This superposition will be a standing wave provided that the magnitudes of the coefficients A and B are equal ($|A| = |B|$).* Roughly, this equality of magnitudes means that the wavefunction has equal amounts of waves traveling toward the right and the left, and thus has no net traveling motion.

At $x = 0$ and at $x = L$ the wavefunction must be zero at all times,

$$\Psi(0, t) = 0 \qquad \text{at } x = 0 \qquad (4)$$

and

$$\Psi(L, t) = 0 \qquad \text{at } x = L \qquad (5)$$

These are called **boundary conditions.** To understand how these boundary conditions come about, note that for $x < 0$ and for $x > L$, the wavefunction must be identically zero because the probability for finding the particle outside of the box is identically zero. By continuity, if $\Psi = 0$ for $x < 0$ and for $x > L$, then $\Psi = 0$ at $x = 0$ and at $x = L$.† Note that the boundary conditions for our quantum-mechanical wavefunction are exactly the same as the boundary conditions for a vibrating string, fixed at both ends.

With our wavefunction (3), the boundary conditions (4) and (5) read

$$Ae^{-iEt/\hbar} + Be^{-iEt/\hbar} = 0 \qquad \text{at } x = 0 \qquad (6)$$

$$Ae^{-i(Et/\hbar - \sqrt{2mE}\, L/\hbar)} + Be^{-i(Et/\hbar + \sqrt{2mE}\, L/\hbar)} = 0 \qquad \text{at } x = L \qquad (7)$$

The first of these equations immediately implies $B = -A$. Consequently, the second equation reduces to

$$Ae^{-iEt/\hbar}(e^{i\sqrt{2mE}\, L/\hbar} - e^{-i\sqrt{2mE}\, L/\hbar}) = 0 \qquad (8)$$

which is equivalent to

$$2iA \sin(\sqrt{2mE}\, L/\hbar) = 0 \qquad (9)$$

---

* For a *real* function $f(x,t)$, a standing wave has the form $f(x,t) \propto \cos(\omega t + \phi_1) \times \cos(kx + \phi_2)$; the peaks of this wave remain at fixed positions x, and they grow and decrease in unison. For our *complex* function $\Psi(x,t)$, a standing wave has the form $\Psi(x,t) \propto e^{-i\omega t + i\phi_1} \cos(kx + \phi_2)$; the peaks of this wave remain at fixed positions x, and the complex phase factor $e^{-i\omega t}$ changes in unison. It is a simple exercise to check that with $|A| = |B|$ the wave (3) is indeed a standing wave.

† This argument presupposes that Ψ is a continuous function. For now, we will accept this without proof.

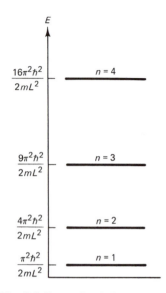

Fig. 6.2 Energy-level diagram for a particle in a box.

The vanishing of the sine function requires that the argument is an integer multiple of π,

$$\sqrt{2mE}\, L/\hbar = n\pi \tag{10}$$

or

$$E_n = \frac{n^2 \pi^2 \hbar^2}{2mL^2} \qquad n = 1, 2, 3, \ldots \tag{11}$$

This equation tells us how the energies of the particle in a box are quantized. Note that although $n = 0$ gives a valid solution for Eq. (9), it must be excluded because it makes the wavefunction identically zero everywhere, which is unacceptable. Thus the lowest value of n is $n = 1$ and the energy of the ground state is $\pi^2 \hbar^2/(2mL^2)$. The energy of the first excited state is four times as large; the energy of the second excited state is nine times as large, etc. Figure 6.2 shows the energy-level diagram for our particle in a box. In wave mechanics, the quantized energies of a physical system are customarily called the **energy eigenvalues***. Thus, Eq. (11) tells us the energy eigenvalues for the particle in a box.

For the energy eigenvalues (11), our wavefunction becomes

$$\Psi_n(x, t) = 2iAe^{-in^2\pi^2\hbar t/(2mL^2)} \sin(n\pi x/L) \qquad n = 1, 2, 3, \ldots \tag{12}$$

The states of a system that correspond to definite values of the quantized energy are called **energy eigenstates.** Thus, the wavefunctions Ψ_n are the wavefunctions for the energy eigenstates. The dependence of Ψ_n on x is entirely contained in the functions $\sin(n\pi x/L)$. Figure 6.3 shows plots of these functions for $n = 1, 2,$ and 3. For $n = 1$, the box contains one-half wavelength (one antinode); for $n = 2$ it contains one wavelength (two antinodes); for $n = 3$ it contains three-halves wavelength (three antinodes), etc.

The probability distribution is

$$|\Psi_n(x, t)|^2 = |2iA|^2 \sin^2(n\pi x/L) \tag{13}$$

This is constant in time. In wave mechanics, a system is said to be in a **stationary state** if the probability distribution $|\Psi_n|^2$ is constant in time. Thus, the particle in a box is in a stationary state whenever its wavefunction is one or another of the wavefunctions listed in Eq. (12).

In our calculation the value of the constant A has been left undetermined. This value must be adjusted to satisfy the normalization condition (5.49),

$$\int_{-\infty}^{\infty} |\Psi_n(x, t)|^2\, dx = 1 \tag{14}$$

---

* This is a mongrelization of the German word "Eigenwert," which means "proper value."

Since $\Psi_n = 0$ for $x \leq 0$ and $x \geq L$, the integration can be restricted to the range $0 < x < L$. With Eq. (13) we obtain

$$\int_0^L |2iA|^2 \sin^2\left(\frac{n\pi x}{L}\right) dx = 1 \tag{15}$$

The integral of $\sin^2(n\pi x/L)$ can be readily evaluated by the substitution

$$\sin^2(n\pi x/L) = \tfrac{1}{2}[1 - \cos(2n\pi x/L)]$$

which immediately yields*

$$\int_0^L \sin^2\left(\frac{n\pi x}{L}\right) dx = \tfrac{1}{2}L \tag{16}$$

Hence Eq. (15) reduces to

$$|2iA|^2 \times \tfrac{1}{2} \times L = 1 \tag{17}$$

or

$$|2iA| = \sqrt{\frac{2}{L}} \tag{18}$$

This fixes the magnitude of the complex number iA. The phase of this number is still undetermined; we will arbitrarily assign a phase so that iA is real and positive. The final expression for our wavefunction then becomes

$$\Psi_n(x, t) = \sqrt{\frac{2}{L}}\, e^{-in^2\pi^2\hbar t/(2mL^2)} \sin(n\pi x/L) \tag{19}$$

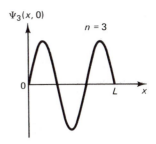

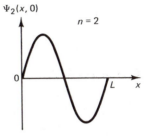

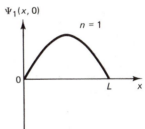

Fig. 6.3 Space dependence of the wavefunctions Ψ_n.

Example 1   A proton or a neutron in a nucleus can be roughly regarded as a particle in a box; the surface of the nucleus plays the role of the wall of the box, and the proton moves more or less freely between these walls. What is the energy released when a proton makes a transition from the first excited state to the ground state of a box of nuclear size, say 1.0×10^{-14} m?

Solution   With $L = 1 \times 10^{-14}$ m and $m = m_p = 1.67 \times 10^{-27}$ kg, the energy of the ground state is

$$E_1 = \frac{\pi^2 \hbar^2}{2m_p L^2} = \frac{\pi^2 \times (1.05 \times 10^{-34}\,\text{J·s})^2}{2 \times 1.67 \times 10^{-27}\,\text{kg} \times (1.0 \times 10^{-14}\,\text{m})^2}$$

$$= 3.3 \times 10^{-13}\,\text{J} \tag{20}$$

---

* This result can be easily remembered by noting that the average value of the sine squared over any number of full wavelengths or half wavelengths is $\tfrac{1}{2}$. Thus

$$\int_0^L \sin^2\left(\frac{n\pi x}{L}\right) dx = (\text{average value}) \times (\text{range of integration}) = \frac{1}{2} \times L$$

The energy of the first excited state is four times as large:

$$E_2 = 4E_1 = 13.2 \times 10^{-13}\,\text{J} \tag{21}$$

Hence the energy released is the difference

$$E_2 - E_1 = 13.2 \times 10^{-13}\,\text{J} - 3.3 \times 10^{-13}\,\text{J} = 9.9 \times 10^{-13}\,\text{J} \tag{22}$$

Expressed in MeV, this energy difference equals 6.2 MeV. In fact, the observed energy differences between stationary states in a nucleus are typically a few MeV, in rough agreement with our simple calculation. ■

Example 2   According to the Superposition Principle, any superposition of wavefunctions is a possible wavefunction. Suppose that a particle in a box is in a superposition of the ground state and the first excited state, with an amplitude of $\frac{1}{2}$ for the former and $\sqrt{3}/2$ for the latter. The wavefunction is

$$\Psi(x, t) = \frac{1}{2}\,\Psi_1(x, t) + \frac{\sqrt{3}}{2}\,\Psi_2(x, t)$$

$$= \frac{1}{2}\sqrt{\frac{2}{L}}\,e^{-iE_1 t/\hbar} \sin(\pi x/L) + \frac{\sqrt{3}}{2}\sqrt{\frac{2}{L}}\,e^{-iE_2 t/\hbar} \sin(2\pi x/L) \tag{23}$$

(This implies that the probability for the ground state is $\frac{1}{4}$ and the probability for the excited state is $\frac{3}{4}$.) Find the probability distribution for this wavefunction.

Solution   We must evaluate the square of the absolute value of $\Psi(x, t)$. Since the square of the absolute value of any complex number equals the product of the number and its conjugate,

$$|\Psi(x, t)|^2 = \Psi(x, t)\Psi^*(x, t)$$

$$= \left[\frac{1}{2}\sqrt{\frac{2}{L}}\,e^{-iE_1 t/\hbar} \sin(\pi x/L) + \frac{\sqrt{3}}{2}\sqrt{\frac{2}{L}}\,e^{-iE_2 t/\hbar} \sin(2\pi x/L)\right]$$

$$\times \left[\frac{1}{2}\sqrt{\frac{2}{L}}\,e^{iE_1 t/\hbar} \sin(\pi x/L) + \frac{\sqrt{3}}{2}\sqrt{\frac{2}{L}}\,e^{iE_2 t/\hbar} \sin(2\pi x/L)\right]$$

$$= \frac{1}{4}\frac{2}{L} \sin^2(\pi x/L) + \frac{3}{4}\frac{2}{L} \sin^2(2\pi x/L)$$

$$+ \frac{\sqrt{3}}{4}\frac{2}{L} (e^{-iE_1 t/\hbar + iE_2 t/\hbar} + e^{-iE_2 t/\hbar + iE_1 t/\hbar}) \sin(\pi x/L) \sin(2\pi x/L)\,.$$

$$= \frac{1}{2L} \sin^2(\pi x/L) + \frac{3}{2L} \sin^2(2\pi x/L)$$

$$+ \frac{\sqrt{3}}{L} \cos[(E_2 - E_1)t/\hbar] \times \sin(\pi x/L) \sin(2\pi x/L) \tag{24}$$

Note that the first two terms on the right side are time independent, but the last term oscillates with a frequency $\omega = (E_2 - E_1)/\hbar$. Thus the superposition of two stationary states is *not* a stationary state. The oscillating term in the probability distribution plays an important role in the emission of electromagnetic waves. If the particle is charged, then the oscillating term represents an oscillating charge distribution, with an oscillating electric dipole moment. Such a charge distribution will radiate electromagnetic waves with a frequency equal to the frequency of oscillation, $\omega = (E_2 - E_1)/\hbar$. This is, of course, exactly the transition frequency postulated by Bohr. Thus, when a particle makes a quantum jump between two stationary states, the emission of radiation is due to oscillating terms in the probability distribution calculated from a superposition of the two stationary states. ■

Incidentally, we can obtain a rough estimate of the ground-state energy of the particle in a box by appealing to Heisenberg's uncertainty relation,

$$\Delta p_x \, \Delta x \gtrsim h \tag{25}$$

Since the particle is known to be within the box, $\Delta x \leqslant L$. Hence

$$\Delta p_x \gtrsim \frac{h}{L} \tag{26}$$

For the ground state, we expect that the energy is as small as possible, consistent with the condition (26). This leads us to the estimate

$$E = \frac{(\Delta p_x)^2}{2m} \simeq \frac{h^2}{2mL^2} \tag{27}$$

which is in reasonable agreement with the exact value $\pi^2 \hbar^2/(2mL^2)$. Note that the energy (27) does not represent an energy uncertainty. The ground state has a sharply defined energy, with no uncertainty ($\Delta E = 0$). Only the momentum is uncertain because it has equal probability for positive and for negative values.

6.2 Schrödinger's Equation with Potential

In Section 5.4 we found the Schrödinger equation for a free particle. Now we must look for the Schrödinger equation for a particle subject to a force or, more generally, an interaction. Since at the microscopic level the forces are usually conservative, we can represent the interaction by the potential energy. For one-dimensional motion along the x axis, the potential energy will be some function $U(x)$. To see how this potential energy is to be included in the Schrödinger equation, we find it instructive to examine the case of a potential energy that is constant within some region, say, $U(x) = U_0$. In the presence of such a constant potential energy the particle experiences no force, and the motion proceeds with a constant momentum, as though the particle were free. The only change lies in the relation between momentum and energy: the kinetic energy $p^2/2m$ equals the difference $E - U_0$ of total energy and potential energy,

$$\frac{p^2}{2m} = E - U_0 \tag{28}$$

or

$$p = \sqrt{2m(E - U_0)} \tag{29}$$

The harmonic wave $e^{-i(Et/\hbar - px/\hbar)}$ now takes the form

$$\Psi(x, t) = e^{-i(Et/\hbar - \sqrt{2m(E-U_0)}x/\hbar)} \tag{30}$$

What differential equation does this wavefunction satisfy? As in Section 5.4, we can discover this by taking the first derivative of (30) with respect to time and the second derivative with respect to space. Upon comparing these derivatives, we immediately find

$$-\frac{\hbar^2}{2m}\frac{\partial^2}{\partial x^2}\Psi(x, t) + U_0\Psi(x, t) = i\hbar\frac{\partial}{\partial t}\Psi(x, t)$$

This is the correct equation for a *constant* potential.

We now take a step which is plausible, but which we cannot justify rigorously: we assume that the differential equation valid for a constant potential energy U_0 is also valid for a varying potential energy $U(x)$. Thus

$$-\frac{\hbar^2}{2m}\frac{\partial^2}{\partial x^2}\Psi(x, t) + U(x)\Psi(x, t) = i\hbar\frac{\partial}{\partial t}\Psi(x, t) \tag{31}$$

This is **Schrödinger's equation with potential.** It can be justified more fully by a somewhat more sophisticated argument based on the Correspondence Principle,* but we will not attempt to do this here. The Schrödinger equation is for wave mechanics what Newton's Second Law is for classical mechanics. Given a potential energy $U(x)$ and an initial wavefunction $\Psi(x, t)$ at an initial time, say, $t = 0$, the Schrödinger equation permits us to calculate the wavefunction at any later time. Hence the time evolution of the

---

* See, e.g., D. S. Saxon, *Elementary Quantum Mechanics*, Section 5.3.

wavefunction is deterministic. What is not deterministic about wave mechanics is the connection between the wavefunction and the observed particle positions—the probability interpretation of the wavefunction leaves an element of chance in this connection.

The solution of the wave equation (31) often poses a rather difficult mathematical problem. Sometimes the best procedure is numerical integration, conveniently done on a fast computer (for an example for this, see the sequence of pictures in Section 4).

If the potential energy $U(x)$ corresponds to an attractive force, capable of confining the particle within some finite range, then we expect that the solution of the wave equation should yield a discrete set of standing waves, with quantized eigenvalues of the energy. As in the case of the particle in the box, the standing waves must be adjusted to "fit" the potential energy. However, in general the standing-wave solutions of Eq. (31) are complicated functions of position—if the potential is not constant, then they are not simple harmonic waves with a well-defined wavelength. We run into trouble with the wavelength as soon as we try to replace the constant potential U_0 in Eq. (29) by a variable potential energy $U(x)$. When we blindly make this replacement, Eq. (29) leads us to a wavelength

$$p = \frac{h}{\lambda} = \sqrt{2m[E - U(x)]} \qquad (32)$$

If $U(x)$ is a very gradually varying function, the wavelength given by (32) is also a gradually varying function; under these conditions the concept of wavelength retains some meaning.* But if $U(x)$ is a strongly varying function that changes by a substantial amount in a short distance, then the concept of wavelength loses all meaning. Under these conditions, the standing wave is a complicated function of position which has nodes and antinodes but bears little resemblance to a harmonic wave; its shape must be calculated from the Schrödinger equation.

Even though the space dependence of the standing wave is complicated, the time dependence remains simple. If the particle is in an energy eigenstate, all parts of the standing wave oscillate in unison, with the same frequency. Thus, we expect that the time dependence of the standing wave is of the form $e^{-i\omega t}$ or $e^{-iEt/\hbar}$,

---

* Such a gradually varying wavelength occurs in some problems in optics, for instance, the propagation of sunlight through the layers of the Earth's atmosphere. The index of refraction of the atmosphere is a gradually varying function because it depends on the density of air; this gives rise to a gradual variation in wavelength and a gradual deflection, or refraction, of the rays of sunlight, a phenomenon dramatically illustrated by the weird distortions sometimes seen in the image of the Sun at sunset.

as it is for the simple case of the particle in the box.* Taking into account this time dependence, we can write the wavefunction $\Psi(x, t)$ as the product of a function of time and a function of space,

$$\Psi(x, t) = e^{-iEt/\hbar}\psi(x) \tag{33}$$

The function $\psi(x)$ describes the shape of the standing wave. Upon substituting this expression for $\Psi(x, t)$ into the Schrödinger equation, we obtain

$$-\frac{\hbar^2}{2m}\frac{d^2}{dx^2}\psi(x)e^{-iEt/\hbar} + U(x)\psi(x)e^{-iEt/\hbar} = E\psi(x)e^{-iEt/\hbar} \tag{34}$$

and canceling the overall factor $e^{-iEt/\hbar}$,

$$-\frac{\hbar^2}{2m}\frac{d^2}{dx^2}\psi(x) + U(x)\psi(x) = E\psi(x) \tag{35}$$

This is the **time-independent Schrödinger equation.** The solution of this equation give us the standing waves and the energy eigenvalues.

Note that if the wavefunction $\Psi(x, t)$ has the form given in Eq. (33), with the time dependence entirely contained in the complex exponential, then the probability distribution is constant in time,

$$|\Psi(x, t)|^2 = |e^{-iEt/\hbar}\psi(x)|^2 = |\psi(x)|^2 \tag{36}$$

As in the case of the particle in a box, the energy eigenstates are stationary states.

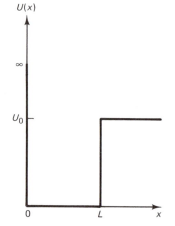

Fig. 6.4 Finite potential well. The height of the potential energy is U_0 on the right side and infinite on the left side.

6.3 **Particle in a Finite Potential Well**

Now that we have available the Schrödinger equation (35), we can solve the problem of a particle in a finite potential well. Figure 6.4 shows the potential energy as a function of x. Note that this potential energy rises to a finite height U_0 on the right side, but to an infinite height on the left side. Hence the left side coincides with the infinite potential well of the particle in a box of Section 6.1. Because of this coincidence, the solution for our finite well re-

---

* This conclusion can be made rigorous by appealing to a theorem of advanced wave mechanics: in an energy eigenstate, the time dependence of the wavefunction is necessarily of the form $e^{-iEt/\hbar}$.

tains some of the simple features of the solution for the infinite well. (A potential well with finite height on both sides is slightly more difficult to treat.) In a later chapter, we will see that the potential well illustrated in Fig. 6.4 provides a rough description of the "strong" nuclear binding force, especially in the case of the deuteron.

Within the region $0 < x < L$, the standing wave is simply a superposition of two waves traveling toward the right and the left. Since the boundary condition at $x = 0$ is $\psi = 0$, exactly the same as for the infinite well in Section 6.1, the superposition will also be exactly the same as in Section 1 [see Eq. (3) with $B = -A$],

$$\psi(x) = 2iA \sin\left(\frac{\sqrt{2mE}\,x}{\hbar}\right) \qquad 0 \leqslant x \leqslant L \qquad (37)$$

For convenience, we rewrite this as

$$\psi(x) = 2iA \sin(kx) \qquad 0 \leqslant x \leqslant L \qquad (38)$$

where $k = \sqrt{2mE}/\hbar$. Note that here we omitted the time dependence $e^{-iEt/\hbar}$. We will have to insert this extra factor after we complete the rest of the calculation [see Eq. (33)].

So far we have not made explicit use of the Schrödinger equation [we can, of course, check that in the range $0 < x < L$ the function (38) is a solution of Schrödinger's equations with $U(x) = 0$, as it should be]. But we have to make explicit use of Schrödinger's equation to find the wavefunction in the region $x > L$; in this region the finite well differs from the infinite well. With $U(x) = U_0$, the Schrödinger equation is

$$-\frac{\hbar^2}{2m}\frac{d^2}{dx^2}\psi(x) + U_0\psi(x) = E\psi(x) \qquad (39)$$

or

$$\frac{d^2}{dx^2}\psi(x) = \frac{2m}{\hbar^2}(U_0 - E)\psi(x) \qquad (40)$$

For a bound state, $U_0 > E$; hence the factor $(U_0 - E)$ appearing on the right side of Eq. (40) is *positive*. If we introduce the abbreviation

$$\kappa = \sqrt{2m(U_0 - E)}/\hbar \qquad (41)$$

we can write the differential equation (40) as

$$\frac{d^2}{dx^2}\psi(x) = \kappa^2\psi(x) \qquad (42)$$

It is easy to verify that the two solutions of this differential equation are

$$\psi(x) = Ce^{-\kappa x} \qquad x \geqslant L \qquad (43)$$

and

$$\psi(x) = De^{\kappa x} \qquad x \geqslant L \qquad (44)$$

where C and D are arbitrary constants. The first of these solutions is a decreasing exponential, the second an increasing exponential. Obviously, the latter is unacceptable since it would make the wavefunction infinitely large at $x = \infty$. The correct solution for $x \geqslant L$ is therefore (43).

Now, let us compare the solutions (38) and (43) at $x = L$. The point $x = L$ is a boundary point for each of the two solutions (38) and (43). At this point, the solutions must satisfy the following **boundary conditions:**

$$d\psi/dx \text{ is continuous} \qquad (45)$$

$$\psi \text{ is continuous} \qquad (46)$$

We can understand these conditions as follows: If $d\psi/dx$ were not continuous the graph of $\psi(x)$ would have a kink at $x = L$ (see Fig. 6.5a), and the graph of $d\psi/dx$ would have a jump, or discontinuity (see Fig. 6.5b). It would then follow that $d^2\psi/dx^2$ (the derivative of $d\psi/dx$) would be infinite at $x = L$. But this is impossible because the Schrödinger equation tells us that $d^2\psi/dx^2 = -(2m/\hbar^2)(E - U)\psi$, which is finite whenever U and ψ are finite. Likewise, if ψ were not continuous, then $d\psi/dx$ would be infinite, and $d^2\psi/dx^2$ would be even more infinite, again contradicting the Schrödinger equation.

The continuity of ψ requires that the two solutions (38) and (43) must be equal at $x = L$,

$$2iA \sin kL = Ce^{-\kappa L} \qquad (47)$$

and the continuity of $d\psi/dx$ requires that their derivatives must also be equal,

$$2ikA \cos kL = -\kappa Ce^{-\kappa L} \qquad (48)$$

If we divide Eq. (48) by Eq. (47), we obtain

$$k \cot kL = -\kappa \qquad (49)$$

Since both k and κ depend on E, this equation can only be satisfied if E has some special value, i.e., this equation is the quantization

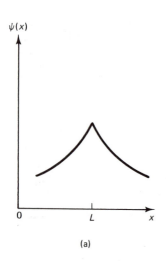

(a)

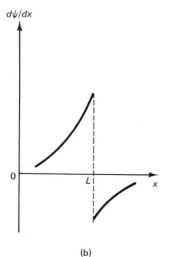

(b)

Fig. 6.5 (a) Hypothetical function with a kink at $x = L$. (b) The derivative of this function has a discontinuity at $x = L$. The second derivative is infinite (does not exist) at $x = L$.

condition for E. Unfortunately, the equation cannot be solved explicitly for E; the best we can do is to construct a graphical solution. For this purpose we adopt the following method.* We square both sides of Eq. (49),

$$k^2 \cot^2 kL = \kappa^2 \tag{50}$$

With the identity $\cot^2 kL = 1/(\sin^2 kL) - 1$, this becomes

$$\frac{k^2}{\sin^2 kL} - k^2 = \kappa^2 \tag{51}$$

or

$$\sin^2 kL = \frac{k^2}{\kappa^2 + k^2} = \frac{k^2}{2m(U_0 - E)/\hbar^2 + 2mE/\hbar^2}$$
$$= \frac{\hbar^2 k^2}{2mU_0} \tag{52}$$

Taking the square root of both sides, we finally obtain

$$\sin kL = \pm \sqrt{\frac{\hbar^2}{2mU_0}}\, k \tag{53}$$

We accomplish the graphical solution of this equation by plotting $\sin kL$ vs. k, and $\pm \sqrt{\hbar^2/(2mU_0)}\, k$ vs. k. The solutions of Eq. (53) correspond to the intersections of these two plots (see Fig. 6.6). However, not all of these graphical solutions are solutions of Eq. (49). When we squared this equation, we eliminated a minus sign; consequently, we must select from the solutions of Eq. (53) those that correspond to $\cot kL < 0$. This condition rules out the intervals $0 < kL < \pi/2$, $\pi < kL < 3\pi/2$, etc; these excluded intervals have been shaded in Fig. 6.6. All the intersections outside the shaded areas are valid solutions.

Note that if U_0 is small, the slopes of the straight lines $\pm \sqrt{\hbar^2/(2mU_0)}\, k$ vs. k will be steep (see Fig. 6.7), and it may happen that there is not even a single valid solution. This means that if U_0 is small, there is no bound state in the finite well. Conversely, if U_0 is large, the slopes of the straight lines are nearly horizontal and there are many intersections, i.e., many bound states.

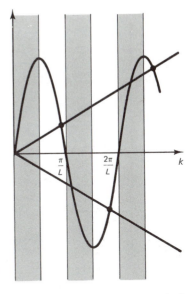

Fig. 6.6 Plots of $\sin kL$ vs. k and of $\pm \sqrt{\hbar^2/(2mU_0)}\, k$ vs. k. The intersections of these two plots give solutions of Eq. (49). The intersections in the shaded intervals must be excluded.

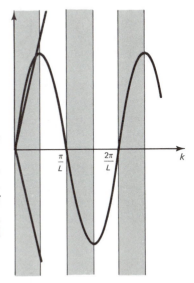

Fig. 6.7 If U_0 is small, there are no valid intersections.

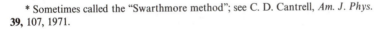

* Sometimes called the "Swarthmore method"; see C. D. Cantrell, *Am. J. Phys.* **39**, 107, 1971.

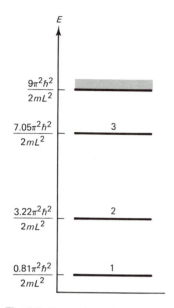

Fig. 6.8 Energy-level diagram for a finite well with $U_0 = 9\pi^2\hbar^2/(2mL^2)$. This well has three energy levels.

Figure 6.8 shows an energy-level diagram for a finite well with $U_0 = 9\pi^2\hbar^2/(2mL^2)$. For this depth, the well has three energy eigenstates. Figure 6.9 shows the wavefunctions of these eigenstates.

One remarkable feature of these wavefunctions is that the particle has a finite probability for being found in the region $x > L$. Classically, the point $x = L$ is a turning point for the motion of the particle, and $x > L$ is a forbidden region for the particle. This can be readily seen by examining the kinetic energy; in this region

$$K = E - U_0 \tag{54}$$

which is *negative* since $E < U_0$. Thus, it would seem that if we ever find the particle in this region, we will find it with a negative kinetic energy! However, it turns out that Heisenberg's uncertainty relation saves us from this disaster. To see what role uncertainties play in the measurement, note that according to Eq. (43), the wavefunction decreases by a factor e^{-1} when the distance x increases by $1/\kappa$ (the characteristic decay length for the exponential function is $1/\kappa$); because of this rapid decrease of the wavefunction, we can regard $1/\kappa$ as the typical penetration distance of the particle into the forbidden region. If we want to detect the particle in this region, we must therefore measure its position with an uncertainty less than $1/\kappa$,

$$\Delta x < 1/\kappa \tag{55}$$

By the Heisenberg relation, the uncertainty in the momentum will then be at least

$$\Delta p_x \gtrsim h/\Delta x \gtrsim h\kappa \tag{56}$$

The uncertainty of kinetic energy associated with this uncertainty in momentum is

$$\Delta K = \frac{(\Delta p_x)^2}{2m} \gtrsim \frac{h^2\kappa^2}{2m} \tag{57}$$

or, with Eq. (41),

$$\Delta K > (2\pi)^2(U_0 - E) \tag{58}$$

Comparing this with (54), we see that after the position measurement, the uncertainty in the kinetic energy is larger than the magnitude of the nominal negative kinetic energy. Hence the negative kinetic energy is hidden by the uncertainty; the negative kinetic energy is an unobservable, virtual kinetic energy.

6.4 Barrier Penetration

A potential barrier is a region of high potential energy. Figure 6.10 shows a square potential barrier of height U_0 extending from $x = a$ to $x = b$ on the x-axis. The dashed horizontal line represents the kinetic energy of a particle incident on this barrier from the left. Classically, the point $x = a$ is a turning point for the motion of this particle, and the region $a < x < b$ is forbidden. However, as we saw in the preceding section, a quantum-mechanical particle can move beyond the classical turning point and penetrate the forbidden region. After traversing this region, it can then emerge on the far side. This is sometimes called tunneling through a barrier. Figure 6.11 shows the wavefunction for a particle tunneling through a barrier. The average energy of the particle is one-half of the height of the barrier. The initial wavefunction is a wavepacket incident from the left (Fig. 6.11a). Upon impact on the barrier, most of the wavepacket is reflected and returns toward the left (Fig. 6.11i). But a small part penetrates the barrier (see Fig. 6.11e) and a very small part ultimately leaks out on the far side (in Fig. 6.11i, the amount of wave leaking out on the right side is hardly noticeable). Thus there is a small probability for transmission through the barrier.

The calculations displayed in Fig. 6.11 for barrier penetration by a wavepacket were performed numerically by computer. The calculations for barrier penetration by a harmonic wave are much simpler; they can be done by hand. On the left side of the

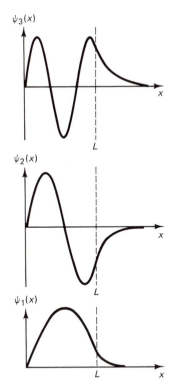

Fig. 6.9 Space dependence of the wavefunctions for the finite well with $U_0 = 9\pi^2\hbar^2/(2mL^2)$.

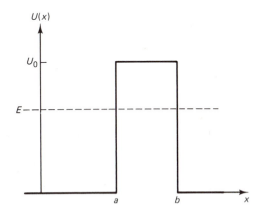

Fig. 6.10 A square potential barrier. A particle is incident from the left; its kinetic energy is indicated by the dashed line.

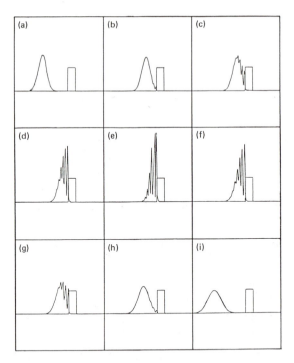

Fig. 6.11 Wave packet incident on a square barrier. The wave packet is partially transmitted and partially reflected. [From A. Goldberg, H. M. Schey, and J. L. Schwartz, *Am. J. Phys.* **35**, 177 (1967).]

barrier the wavefunction is a superposition of the incident and the reflected waves, i.e., $\psi(x)$ is a superposition of e^{ikx} and e^{-ikx}.* In the classically forbidden region $a < x < b$, the wavefunction is a superposition of $e^{-\kappa x}$ and $e^{\kappa x}$ [see Eqs. (43) and (44)]. Finally, to the right of the barrier, the wavefunction consists of the transmitted wave e^{ikx}, multiplied by some amplitude. This amplitude, and the other amplitudes in the superpositions, can be determined by writing down the boundary conditions (45) and (46) at each of the two boundaries $x = a$ and $x = b$. The calculation is slightly messy, and we will leave it for Example 3. Instead, we will try to make a rough estimate of the probability for transmission through a fairly thick or fairly high barrier. For such a barrier, the contribution from the increasing exponential $e^{\kappa x}$ in the forbidden region is unimportant (note that this exponential is completely absent in the case of the infinitely wide barrier that constitutes the right side of the potential

---

* Note that we are here dealing with the time-independent wavefunction $\psi(x)$. As usual, the time-dependent wavefunction is $\Psi(x,t) = e^{-iEt/\hbar}\psi(x)$.

of the preceding section; see Fig. 6.4). This exponential arises from a reflection of the wavefunction at the inner right edge of the barrier; in the case of a thick, high barrier the amount of wave that penetrates the barrier is quite small, and the amount that remains after reflection at the far end of the barrier is even smaller. Thus, the wavefunction in the forbidden region is approximately $\sim e^{-\kappa x}$. At $x = a$, this exponential function is $e^{-\kappa a}$ and at $x = b$ it is $e^{-\kappa b}$; hence the probability decreases by a factor $e^{-2\kappa b}/e^{-2\kappa a}$ or $e^{-2\kappa(b-a)}$ between one end of the barrier and the other. From this we see that the probability for transmission through the barrier is

$$P \simeq e^{-2\kappa(b-a)} \tag{59}$$

We must keep in mind that this is an approximate result. In our comparison of the probabilities at one end of the barrier and the other, we have not only neglected the contribution of the extra wavefunction $e^{\kappa x}$ in the forbidden region, but also the contribution of the reflected wave at the left of the barrier. A more careful calculation shows that the probability (59) contains an extra factor with an extra dependence on energy. However, in most practical applications the rough estimate (59) is correct as to order of magnitude, and it gives the main dependence on energy.

Example 3   For a wave incident on the square potential barrier from the left, calculate the amplitude of the transmitted wave emerging on the right and calculate the probability for transmission.

Solution   We will assume that the amplitude of the incident wave is $A = 1$. According to the qualitative discussion given above, for $x \leqslant a$ the wavefunction is then

$$\psi = e^{ikx} + Re^{-ikx} \qquad x \leqslant a \tag{60}$$

For $a \leqslant x \leqslant b$ the wavefunction is

$$\psi = Ce^{-\kappa x} + De^{\kappa x} \qquad a \leqslant x \leqslant b \tag{61}$$

and for $x \geqslant b$ it is

$$\psi = Te^{ikx} \qquad x \geqslant b \tag{62}$$

Here R, C, D, and T are unknown amplitudes which we will have to evaluate by means of the boundary conditions (45) and (46). At $x = a$, these boundary conditions read

$$ike^{ika} - ikRe^{-ika} = -\kappa Ce^{-\kappa a} + \kappa De^{\kappa a} \tag{63}$$

$$e^{ika} + Re^{-ika} = Ce^{-\kappa a} + De^{\kappa a} \tag{64}$$

and at $x = b$

$$-\kappa Ce^{-\kappa b} + \kappa De^{\kappa b} = ikTe^{ikb} \tag{65}$$

$$Ce^{-\kappa b} + De^{\kappa b} = Te^{ikb} \tag{66}$$

Equations (63)–(66) are four equations for the four unknown amplitudes R, C, D, and T. To solve these equations, multiply (64) by κ and add the result to (63); this yields an expression for D:

$$D = \frac{e^{-\kappa a}}{2\kappa} \left[(\kappa + ik)e^{ika} + R(\kappa - ik)e^{-ika} \right] \tag{67}$$

Multiply (64) by κ and subtract the result from (63); this gives an expression for C:

$$C = \frac{e^{\kappa a}}{2\kappa} \left[(\kappa - ik)e^{ika} + R(\kappa + ik)e^{-ika} \right] \tag{68}$$

Upon substituting these expressions for C and D into Eqs. (65) and (66), we obtain two equations for the two unknowns R and T. From this we can solve for T, the amplitude of the transmitted wave:

$$T = \frac{-4ik\kappa e^{-\kappa(b-a)}e^{-ik(b-a)}}{(\kappa - ik)^2 - (\kappa + ik)^2 e^{-2\kappa(b-a)}} \tag{69}$$

By some further algebra with Eqs. (63)–(66) we can solve for the other amplitudes. Figure 6.12 shows the real part of the wavefunction (Re Ψ) at one instant of time. Note that the amplitude within the barrier decreases from left to right, i.e., the decreasing exponential $e^{-\kappa x}$ dominates over the increasing exponential $e^{\kappa x}$.

The probability for finding a particle in an interval dx is $|e^{-ikx}|^2 \, dx = 1 \times dx$ for the incident wave, and $|Te^{-ikx}|^2 \, dx = |T|^2 \, dx$ for the transmitted wave. Hence $|T|^2$ tells us the factor by which the probability for the transmitted wave is reduced compared to the incident wave, i.e., $|T|^2$ is the probability for transmission:

$$P = |T|^2 = \frac{16k^2\kappa^2}{\left| (\kappa - ik)^2 - (\kappa + ik)^2 e^{-2\kappa(b-a)} \right|^2} \, e^{-2\kappa(b-a)} \tag{70}$$

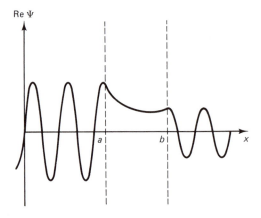

Fig. 6.12 Harmonic wave tunneling through a square barrier.

If the barrier is fairly thick, then $e^{-2\kappa(b-a)} \ll 1$ and the second term in the denominator can be neglected compared to the first, so that

$$P \simeq \frac{16k^2\kappa^2}{|\kappa - ik|^4}e^{-2\kappa(b-a)} = \frac{16k^2\kappa^2}{(k^2 + \kappa^2)^2}e^{-2\kappa(b-a)} \qquad (71)$$

with $k = \sqrt{2mE}/\hbar$ and $\kappa = \sqrt{2m(U_0 - E)}/\hbar$, this reduces to

$$P \simeq \frac{16E(U_0 - E)}{(E + U_0 - E)^2}e^{-2\kappa(b-a)}$$

$$\simeq 16\frac{E}{U_0}\left(1 - \frac{E}{U_0}\right)e^{-2\kappa(b-a)} \qquad (72)$$

Provided that E is *not* near zero, nor near U_0, the factor $(E/U_0)(1 - E/U_0)$ is of order of magnitude ~ 1. The probability for transmission is then roughly $P \sim 16e^{-2\kappa(b-a)}$; except for the factor of 16, this is in agreement with our rough estimate. The factor of 16 is of little importance because the usual comparison of observed probabilities and energies relies on a plot of $\ln P$ vs. energy; in such a plot the factor of 16 merely adds an uninteresting constant increment of $\ln 16 \simeq 2.8$ to the ordinate. ∎

From our result for a square barrier we can deduce the probability for transmission through a barrier of arbitrary shape. We simply approximate such a barrier as a succession of thin square barriers (see Fig. 6.13). For a thin barrier, we can approximate $e^{-2\kappa(b-a)} \simeq 1$ in the *denominator* of Eq. (70). This denominator then reduces to

$$|(\kappa - ik)^2 - (\kappa + ik)^2|^2 = 16k^2\kappa^2$$

and Eq. (70) reduces to $P \simeq e^{-2\kappa(b-a)}$. Thus, the transmission probabilities for the individual thin barriers in Fig. 6.13 are $P_i \simeq e^{-2\kappa\Delta x_i}$. According to the usual rule for combining the probabilities of successive independent events, the overall probability for transmission through the entire barrier is then

$$P = P_1 \cdot P_2 \cdot P_3 \cdots \simeq e^{-2\sum_i \kappa \Delta x_i} \qquad (73)$$

If the intervals Δx_i are small, we can replace the sum by an integral, and we obtain

$$P \simeq e^{-2\int_a^b \kappa \, dx} \qquad (74)$$

With $\kappa = \sqrt{2m[U(x) - E]}/\hbar$, this becomes

$$P \simeq e^{-2\int_a^b \sqrt{2m[U(x) - E]}/\hbar \, dx} \qquad (75)$$

Again, this is an approximation, but it is quite adequate for many problems.

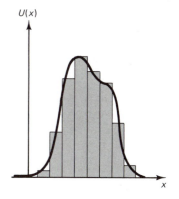

$U(x)$

Fig. 6.13 A barrier of arbitrary shape can be approximated as a succession of thin square barriers.

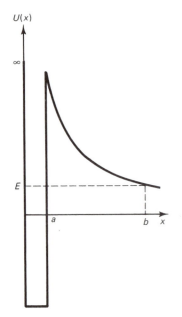

U(x)

∞

E

a   *b*   *x*

Fig. 6.14 The Coulomb potential barrier for an alpha particle. In the interior of the nucleus the potential energy is very low, because of the strongly attractive nuclear force. The alpha particle is incident from the left; its kinetic energy is indicated by the horizontal dashed line.

Equation (75) finds immediate application in an interesting problem in nuclear physics: the theory of alpha decay. It is well known that many heavy elements—radium, uranium, thorium, etc.—emit alpha particles. This is a tunneling process. The alpha particle is initially confined within the nucleus by a high potential barrier, but after repeated impacts on the barrier, it ultimately succeeds in penetrating through the barrier and making good its escape. Figure 6.14 shows the barrier holding the alpha particle in the nucleus. The sloping right side of this barrier is simply the Coulomb potential energy

$$U(x) = \frac{(2e)(Ze)}{4\pi\varepsilon_0 x} \tag{76}$$

corresponding to the repulsive electric force between the alpha particle (charge $2e$) and the nucleus (charge Ze)*. The steep left side of the barrier arises from the strongly attractive nuclear binding force. The dashed line shows a typical energy for an alpha particle. Obviously, the left side of the barrier ($x = a$) is a classical turning point. To escape from the nucleus, the alpha particle must tunnel through the barrier, with a probability given by Eq. (75). For convenience, we take the natural logarithm of P and then substitute for the potential,

$$\ln P \simeq -\frac{2}{\hbar} \int_a^b \sqrt{2m_\alpha\left(\frac{2e^2 Z}{4\pi\varepsilon_0 x} - E\right)}\, dx \tag{77}$$

At the point b, the Coulomb potential energy is equal to the total energy,

$$\frac{2e^2 Z}{4\pi\varepsilon_0 b} = E$$

This determines the upper limit of integration,

$$b = \frac{2e^2 Z}{4\pi\varepsilon_0 E} \tag{78}$$

In most cases of practical interest, the energy E is much smaller than the height of the barrier. This implies that b is very large; more precisely, $b \gg a$. It is then possible to make the approximation $a = 0$ in the integral on the right side of Eq. (77),

$$\ln P \simeq -\frac{2}{\hbar} \int_0^b \sqrt{2m_\alpha\left(\frac{2e^2 Z}{4\pi\varepsilon_0 x} - E\right)}\, dx \tag{79}$$

---

* For alpha emission by, say, uranium, the value of Z is 90, corresponding to the 90 positive charges remaining in the nucleus *after* the alpha particle has been emitted.

If $b \gg a$, this is a good approximation because the integral (79) receives most of its contribution from the large interval $a \leqslant x \leqslant b$, and only a very small contribution from the small interval $0 \leqslant x \leqslant a$.

To evaluate (79), we introduce a new variable

$$\xi = \frac{x}{b} = \frac{x}{2e^2 Z/4\pi\varepsilon_0 E} \tag{80}$$

so that

$$\ln P \simeq -\frac{2\sqrt{2m_\alpha E}\, b}{\hbar} \int_0^1 \sqrt{\frac{1}{\xi} - 1}\, d\xi \tag{81}$$

With the further substitution $\xi = \sin^2 \theta$ this becomes

$$\ln P \simeq -\frac{2\sqrt{2m_\alpha E}\, b}{\hbar} \int_0^{\pi/2} \sqrt{\frac{1 - \sin^2 \theta}{\sin^2 \theta}}\, 2\sin\theta \cos\theta\, d\theta \tag{82}$$

$$\simeq -\frac{4\sqrt{2m_\alpha E}\, b}{\hbar} \int_0^{\pi/2} \cos^2 \theta\, d\theta \tag{83}$$

The integral of $\cos^2 \theta$ around a quarter of a circle is $\pi/4$. Hence

$$\ln P = -\frac{4\sqrt{2m_\alpha E}\, b}{\hbar} \frac{\pi}{4} \tag{84}$$

or

$$\ln P \simeq -\frac{2\pi}{\hbar} \frac{Ze^2}{4\pi\varepsilon_0} \sqrt{\frac{2m_\alpha}{E}} \tag{85}$$

In terms of the final velocity v_α of the alpha particle, the energy is $E = \frac{1}{2}m_\alpha v_\alpha^2$, so that

$$\ln P \simeq -\frac{4\pi}{\hbar} \frac{Ze^2}{4\pi\varepsilon_0} \frac{1}{v_\alpha} \tag{86}$$

Thus the probability for tunneling through the nuclear barrier is

$$P \simeq e^{-(4\pi/h)(Ze^2/4\pi\varepsilon_0)(1/v_\alpha)} \tag{87}$$

This is called the **Gamow factor.*** We will deal with some numerical examples of this factor when we study radioactive decay in Chapter 10.

---

* **George Gamow,** 1904–1968, Russian and later American theoretical physicist, professor at George Washington University and at the University of Colorado. He worked on theories of α decay, β decay, and fission, and also on the theory of element formation in the early universe and on the genetic code in DNA. Gamow is widely known as the author of several charming books on science and mathematics for the general public.

6.5 The Harmonic Oscillator

The potential energy for a harmonic oscillator with a spring constant k is

$$U(x) = \tfrac{1}{2}kx^2 \tag{88}$$

In order to prevent confusion between the spring constant k and the wave number k of a de Broglie wave, it is convenient to express the potential energy in terms of the classical angular frequency of oscillation, $\omega_0 = \sqrt{k/m}$, so that

$$U(x) = \tfrac{1}{2}m\omega_0^2 x^2 \tag{89}$$

This function is plotted in Fig. 6.15.

The time-independent Schrödinger equation for this potential energy is

$$-\frac{\hbar^2}{2m}\frac{d^2}{dx^2}\psi(x) + \tfrac{1}{2}m\omega_0^2\psi(x) = E\psi(x) \tag{90}$$

The acceptable standing waves must satisfy the boundary condition that the wavefunction is $\psi(x) = 0$ at $x = \pm\infty$. The mathematical investigation of the solutions of the differential equation (90) shows that, as in the case of the square potential well, acceptable standing waves exist only for certain discrete eigenvalues of the energy. We will not deal with the rigorous mathematical investigation of this problem; instead we will try to find the eigenvalues by a simple approximation based on the expression (32) for the position-dependent wavelength

$$\frac{1}{\lambda(x)} = \frac{\sqrt{2m[E - U(x)]}}{h} \tag{91}$$

In our present case, this becomes

$$\frac{1}{\lambda(x)} = \frac{\sqrt{2m[E - \tfrac{1}{2}m\omega_0^2 x^2]}}{h} \tag{92}$$

If we neglect the penetration of the wavefunction into the barrier, the classical turning points are (approximately) nodes. Then the condition for a standing wave is that the number of wavelengths between the turning points equals $\tfrac{1}{2}$, or 1, or $\tfrac{3}{2}$, or etc. This condition is an obvious generalization of the results of Section 6.1, where we found that for a particle in a box, the number of wavelengths between the turning points is $\tfrac{1}{2}$, or 1, or $\tfrac{3}{2}$, or etc. If the wavelength

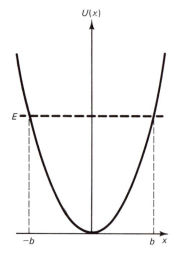

Fig. 6.15 Potential energy for the harmonic oscillator. The horizontal dashed line indicates the energy. The classical turning points are at $\pm b$.

is constant—as in Section 1—, then the number of wavelengths is simply the distance divided by the wavelength. If the wavelength is position-dependent—as in Eq. (91)—, then the number of wavelengths is the integral of $1/\lambda(x)$. Thus, the condition for a standing wave is

$$\int_a^b \frac{\sqrt{2m[E - U(x)]}}{h}\, dx = \tfrac{1}{2}, 1, \tfrac{3}{2}, \ldots \tag{93}$$

or

$$2 \int_a^b \sqrt{2m[E - U(x)]}\, dx = nh \qquad n = 1, 2, 3, \ldots \tag{94}$$

This formula is called the **WKB approximation** for the energy eigenvalues.* Although, strictly speaking, this approximation is valid only when the potential has a very gradual variation and when the wavelength is short, it turns out that it gives reasonable answers for the harmonic-oscillator potential and for the Coulomb potential, even when these potentials have a strong variation. This is a lucky break, and we will make the most of it.

For the potential energy $U(x) = \tfrac{1}{2}m\omega_0^2 x^2$, the classical turning points are symmetrically located with respect to the origin so that $a = -b$ and Eq. (94) then becomes

$$2 \int_{-b}^b \sqrt{2m[E - \tfrac{1}{2}m\omega_0^2 x^2]}\, dx = nh \tag{95}$$

The turning point is given by

$$\tfrac{1}{2}m\omega_0^2 b^2 = E \tag{96}$$

or

$$b = \sqrt{\frac{2E}{m\omega_0^2}} \tag{97}$$

To evaluate the integral on the left side of equation (95), we introduce a new variable θ with

$$\sin \theta = \sqrt{\frac{m\omega_0^2}{2E}}\, x \tag{98}$$

The integral then becomes

$$2 \int_{-\pi/2}^{\pi/2} \sqrt{2mE}\, \sqrt{1 - \sin^2 \theta}\, \sqrt{\frac{2E}{m\omega_0^2}}\, \cos \theta\, d\theta \tag{99}$$

---

* Named after G. Wentzel, H. A. Kramers, and L. Brillouin.

or

$$\frac{4E}{\omega_0} \int_{-\pi/2}^{\pi/2} \cos^2\theta \, d\theta$$

Since the integral of $\cos^2\theta$ around half a circle is $\pi/2$, this expression equals $2\pi E/\omega_0$ and our quantization condition is

$$\frac{2\pi}{\omega_0} E = nh \qquad (100)$$

or

$$E_n = n\hbar\omega_0 \qquad n = 1, 2, 3, \ldots \qquad (101)$$

This agrees almost exactly with Planck's quantization postulate,

$$E_n = n\hbar\omega_0 \qquad n = 0, 1, 2, 3 \qquad (102)$$

The only difference lies in the energy of the ground state. According to our WKB approximation, the energy of the ground state is $\hbar\omega_0$; according to Planck it is 0. The rigorous mathematical solution of the Schrödinger equation shows that neither of these is quite right; the exact result is

$$E_n = (n + \tfrac{1}{2})\hbar\omega_0 \qquad n = 0, 1, 2, \ldots \qquad (103)$$

Thus, the energy of the ground state is $\frac{1}{2}\hbar\omega_0$. Figure 6.16 displays the energy levels of the harmonic oscillator according to Eq. (103). Note that the deviations between Eqs. (101), (102), and (103) are of no consequence for the evaluation of the energy *differences* between the states of the harmonic oscillator—according to all these equations the energy difference between adjacent states is $\hbar\omega_0$.

The rigorous mathematical solution of the Schrödinger equation shows that the wavefunctions of the energy eigenstates are polynomials in x multiplied by $e^{-m\omega_0 x^2/(2\hbar)}$. For instance, the wavefunctions for the ground state and for the first two excited states are*

$$\psi_1(x) = \left(\frac{m\omega_0}{\hbar\pi}\right)^{1/4} e^{-m\omega_0 x^2/(2\hbar)} \qquad (104)$$

$$\psi_2(x) = \sqrt{2}\left(\frac{m\omega_0}{\hbar\pi}\right)^{1/4}\left(\frac{m\omega_0}{\hbar}\right)^{1/2} x\, e^{-m\omega_0 x^2/(2\hbar)} \qquad (105)$$

$$\psi_3(x) = \frac{1}{\sqrt{2}}\left(\frac{m\omega_0}{\hbar\pi}\right)^{1/4}\left(2\frac{m\omega_0}{\hbar} x^2 - 1\right) e^{-m\omega_0 x^2/(2\hbar)} \qquad (106)$$

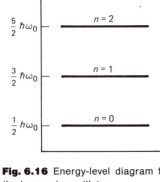

Fig. 6.16 Energy-level diagram for the harmonic oscillator.

---

* For a derivation of these wavefunctions, see, e.g., D. S. Saxon, *Elementary Quantum Mechanics*.

Figure 6.17 shows plots of these wavefunctions. Note that, as expected, they all look like standing waves. The wavefunction $\psi_1(x)$ has a single antinode; $\psi_2(x)$ has two antinodes; $\psi_3(x)$ has three; etc. The time-dependent wavefunctions are

$$\Psi_1(x, t) = e^{-i\omega_0 t/2}\psi_1(x) \tag{107}$$

$$\Psi_2(x, t) = e^{-3i\omega_0 t/2}\psi_2(x) \tag{108}$$

$$\Psi_3(x, t) = e^{-5i\omega_0 t/2}\psi_3(x) \tag{109}$$

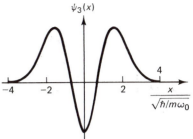

6.6 The Hydrogen Atom

The wave-mechanical treatment of the hydrogen atom poses a rather difficult mathematical problem because, in contrast to the simple systems we have examined so far, the hydrogen atom is a three-dimensional system. This means we need the three-dimensional version of the Schrödinger equation. Since the one-dimensional equation contains a second derivative with respect to x, the three-dimensional equation must contain corresponding derivatives with respect to y and z,

$$-\frac{\hbar^2}{2m}\left(\frac{\partial^2\psi}{\partial x^2} + \frac{\partial^2\psi}{\partial y^2} + \frac{\partial^2\psi}{\partial z^2}\right) + U(x, y, z)\psi = E\psi \tag{110}$$

For the hydrogen atom, m is the electron mass m_e and U is the Coulomb potential energy

$$U = -\frac{e^2}{4\pi\varepsilon_0 r} \tag{111}$$

where $r = \sqrt{x^2 + y^2 + z^2}$. Thus,

$$-\frac{\hbar^2}{2m_e}\left(\frac{\partial^2\psi}{\partial x^2} + \frac{\partial^2\psi}{\partial y^2} + \frac{\partial^2\psi}{\partial z^2}\right) - \frac{e^2}{4\pi\varepsilon_0 r}\psi = E\psi \tag{112}$$

This is a **partial differential equation,** with three variables x, y, and z; the solution is a function $\psi(x, y, z)$ of these three variables. Partial differential equations are hard to solve because it is necessary to deal with all the variables simultaneously. The easiest way to find the solution of Eq. (112) is to make a transformation of variables, replacing the rectangular coordinates by spherical coordinates; the partial differential equation can then be separated into three independent, ordinary differential equations, one for each of the

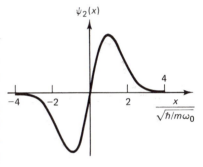

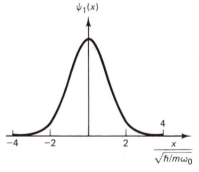

Fig. 6.17 Space dependence of the wavefunctions for the lowest three states of the harmonic oscillator.

spherical coordinates. We will not pursue the exact solution of Schrödinger's equation. Instead, we will again rely on the WKB approximation. But first we need to look at the quantization of the orbital angular momentum of the electron.

In Section 5.1 we saw how de Broglie's simple argument with periodic waves traveling around a circular orbit gave a rough explanation for Bohr's quantization condition of the angular momentum, $L = n\hbar$. The mathematical solution of the three-dimensional Schrödinger equation confirms this quantization, but with some modifications and refinements. In three dimensions, the angular momentum **L** is a vector with three components L_x, L_y, L_z. As shown in textbooks of quantum mechanics,* the magnitude of this angular momentum vector is quantized according to the rule

$$|\mathbf{L}| = \sqrt{l(l+1)}\,\hbar \qquad l = 0, 1, 2, 3, \ldots \qquad (113)$$

Thus, the magnitudes of the angular momentum are

$$|\mathbf{L}| = 0, \sqrt{2}\,\hbar, \sqrt{6}\,\hbar, \sqrt{12}\,\hbar, \ldots \qquad (114)$$

Furthermore, *one* of the components L_x, L_y, L_z is also quantized. Which of these components we select for preferential treatment is a matter of choice. If we choose L_z, then

$$L_z = m_l \hbar \qquad m_l = 0, \pm 1, \pm 2, \ldots, \pm l \qquad (115)$$

Note that, for a given value of l, there are $2l + 1$ possible values of L_z. For instance, if $l = 2$, then the possible values of the z-component of the angular momentum are

$$L_z = -2\hbar, -\hbar, 0, \hbar, 2\hbar \qquad (116)$$

Figure 6.18 shows the corresponding possible directions of the angular momentum vector. If L_z has one of the quantized values specified in Eq. (115), then L_x and L_y do not have well-defined values. The reason for this weird behavior of the components of angular momentum is that these components are subject to uncertainty relations, analogous to the uncertainty relation for Δx and Δp_x. If one of the components has no uncertainty ($\Delta L_z = 0$), then the others necessarily have large uncertainties.[†]

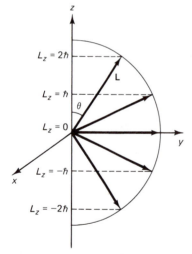

Fig. 6.18 Possible values of the z-component of angular momentum for the case $l = 2$.

---

* See, e.g., D. S. Saxon, *Elementary Quantum Mechanics.*

[†] However, in the exceptional case $|\mathbf{L}| = 0$, all three components are zero ($L_x = L_y = L_z = 0$), and all are free of uncertainties.

The simultaneous quantization of $|\mathbf{L}|$ and L_z can be described graphically by means of the **vector model**. In this model, we are asked to think of the vector $\mathbf{L}$ as quickly precessing around the z-axis (see Fig. 6.19). The angle between the vector $\mathbf{L}$ and the z-axis remains constant; according to Fig. 6.19,

$$\cos\theta = \frac{L_z}{|\mathbf{L}|} = \frac{m_l\hbar}{\sqrt{l(l+1)}\hbar^2} = \frac{m_l}{\sqrt{l(l+1)}} \qquad (117)$$

The smallest possible angle occurs for $m_l = l$, which gives $\cos\theta = l/\sqrt{l(l+1)}$; the largest angle occurs for $m_l = 0$, which gives $\cos\theta = 0$ and $\theta = 90°$. The average values of L_x and L_y are zero, but the instantaneous values of L_x and L_y oscillate around zero. Because of the precession, the instantaneous values of L_x and L_y change quickly—successive measurements give different results, and the values of L_x and L_y are not well defined. Only the average values of L_x and L_y are well defined—both averages are zero.

Finally, let us compare the quantization conditions derived from wave mechanics with Bohr's simple quantization condition $L = \hbar, 2\hbar, 3\hbar$, etc. Obviously, Bohr's condition agrees approximately with Eq. (113) if l is large ($l \gg 1$); and it also agrees with Eq. (115) if $m_l = l$. Hence, Bohr's condition is a special limiting case of the exact quantization condition.

With this brief survey of the quantization of angular momentum in wave mechanics, we are ready to discuss the quantization of energy in the hydrogen atom. The energy of the electron is

$$E = \frac{p^2}{2m_e} + U = \frac{p^2}{2m_e} - \frac{e^2}{4\pi\varepsilon_0 r} \qquad (118)$$

Here p is the magnitude of the three-dimensional momentum vector. This vector can be resolved into two components: one component $\mathbf{p}_r$ along the radial direction, and one component $\mathbf{p}_\perp$ perpendicular to the radial direction (see Fig. 6.20). By definition, the orbital angular momentum is $L = rp_\perp$, and therefore

$$p^2 = p_r^2 + p_\perp^2 = p_r^2 + \frac{L^2}{r^2} \qquad (119)$$

Taking into account the quantization of the angular momentum, we obtain

$$p^2 = p_r^2 + \frac{l(l+1)\hbar^2}{r^2} \qquad (120)$$

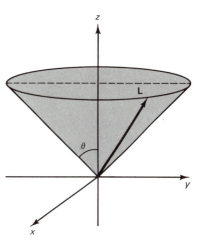

Fig. 6.19 According to the vector model, the angular-momentum vector precesses around the z-axis, tracing out a cone.

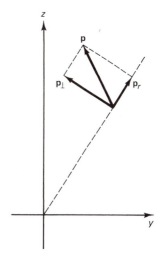

Fig. 6.20 The momentum has a radial component $\mathbf{p}$, and a perpendicular, or transverse, component $\mathbf{p}_\perp$. The vectors $\mathbf{p}$, $\mathbf{p}_r$, and $\mathbf{p}_\perp$ all lie in the same plane, but this plane need not coincide with the plane of the page.

Hence, we can express the energy as

$$E = \frac{1}{2m_e}\left(p_r^2 + \frac{l(l+1)\hbar^2}{r^2}\right) - \frac{e^2}{4\pi\varepsilon_0 r} \qquad (121)$$

The momentum in the radial direction is therefore

$$p_r = \sqrt{2m_e\left(E + \frac{e^2}{4\pi\varepsilon_0 r}\right) - \frac{l(l+1)\hbar^2}{r^2}} \qquad (122)$$

From this we deduce a position-dependent wavelength

$$\frac{1}{\lambda(r)} = \frac{1}{h}\sqrt{2m_e\left(E + \frac{e^2}{4\pi\varepsilon_0 r}\right) - \frac{l(l+1)\hbar^2}{r^2}} \qquad (123)$$

As in Section 5, we can now formulate the WKB condition for a standing wave in the radial direction,

$$2\int_a^b \sqrt{2m_e\left(E + \frac{e^2}{4\pi\varepsilon_0 r}\right) - \frac{l(l+1)\hbar^2}{r^2}}\, dr = nh \qquad (124)$$

This equation determines the energy eigenvalues. For the sake of simplicity, we will only deal with the case $l = 0$ [the integral (124) cannot be evaluated by elementary means if $l > 0$]. With $l = 0$, our equation reduces to

$$2\int_a^b \sqrt{2m_e\left(E + \frac{e^2}{4\pi\varepsilon_0 r}\right)}\, dr = nh \qquad (125)$$

Since the energy E is negative, corresponding to a bound state, we write $E = -|E|$ so that

$$2\int_a^b \sqrt{2m_e\left(\frac{e^2}{4\pi\varepsilon_0 r} - |E|\right)}\, dr = nh \qquad (126)$$

Figure 6.21 shows the potential energy $U = -e^2/(4\pi\varepsilon_0 r)$ as a function of the radial coordinate r; the dashed line indicates the (negative) energy E. Since the particle must always have a positive radial coordinate, we can regard the potential as infinite for $r < 0$.* Obviously, one classical turning point is $a = 0$. The other classical turning point is determined by

$$|E| = \frac{e^2}{4\pi\varepsilon_0 b} \qquad (127)$$

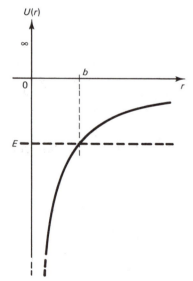

Fig. 6.21 The Coulomb potential energy. The horizontal dashed line indicates the energy of the electron.

---

* This does not mean that the particle cannot pass through the origin in three dimensions. It merely means that in the diagram in Fig. 6.21, such a particle has to have positive values of r both before and after it reaches $r = 0$; thus in this diagram the particle is reflected at $r = 0$.

or

$$b = \frac{4\pi\varepsilon_0}{e^2} |E| \tag{128}$$

At this stage we can save some labor by noting that, with $a = 0$, the integral appearing in Eq. (126) is essentially the same as the integral appearing in Eq. (79)—to make them identical we merely have to alter some of the constants. With these alterations, we can make use of our earlier result (85),

$$2 \int_0^b \sqrt{2m_e \left(\frac{e^2}{4\pi\varepsilon_0 r} - |E| \right)} \, dr = \pi \frac{e^2}{4\pi\varepsilon_0} \sqrt{\frac{2m_e}{|E|}} \tag{129}$$

With this, our condition (125) for standing waves becomes

$$\pi \frac{e^2}{4\pi\varepsilon_0} \sqrt{\frac{2m_e}{|E|}} = nh \tag{130}$$

Taking into account that E is negative, this yields

$$E_n = -\frac{e^4 m_e}{2(4\pi\varepsilon_0)^2 \hbar^2} \frac{1}{n^2} \qquad n = 1, 2, 3, \ldots \tag{131}$$

These energy eigenvalues are exactly the same as those obtained from Bohr's theory. However, there is a difference: all these energy eigenvalues belong to states of zero angular momentum, whereas in Bohr's theory zero angular moment was not permitted.

What energy eigenvalues does wave mechanics give if the angular momentum is not zero? The general answer found by solving the three-dimensional wave equation is*

$$E = -\frac{e^4 m_e}{2(4\pi\varepsilon_0)^2 \hbar^2} \frac{1}{(n' + l)^2} \qquad n' = 1, 2, 3, \ldots \tag{132}$$

Thus, for $l = 0$, the state of lowest energy has $E = -13.6 \, \text{eV}$; but for $l = 1$, the state of lowest energy has $E = -13.6/4$; and for $l = 2$ it has $E = -13.6/9$, etc. Figure 6.22 shows the energy levels for different values of l. In atomic physics, it is standard practice to label a state with $l = 0$ with the letter s; $l = 1$ with p; $l = 2$ with d, etc. These labels appear in Fig. 6.22. Note that the energies of states

---

* Incidentally: for $l > 0$, the WKB approximation does not quite give the right result. Instead of the factor $(n' + l)^2$ in the denominator of Eq. (132), it gives $[n' + \sqrt{l(l + 1)}]^2$.

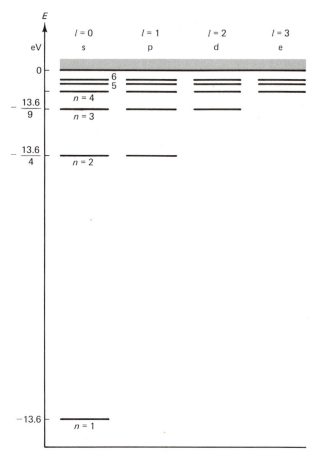

Fig. 6.22 Energy-level diagram for hydrogen. Different columns in this diagram correspond to different values of *l*. The letters s, p, d, etc. stand for *l* = 0, 1, 2, etc.

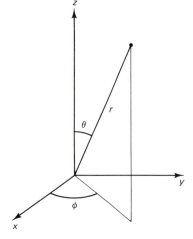

Fig. 6.23 The spherical polar coordinates r, θ, and ϕ.

with different l coincide. Such a coincidence among the energies of two or more distinct states is called **degeneracy.** All the states of the hydrogen atom, except the ground state, are degenerate.

The wavefunctions obtained by solving the partial differential equation are rather complicated functions of x, y, z or of the spherical coordinates r, θ, ϕ (see Fig. 6.23). Only the ground-state wavefunction is fairly simple. The following equations give the wavefunctions for some of the eigenstates of lowest energy. These wavefunctions are designated by ψ_{nlm_l} where the subscripts n, l, and m_l indicate, respectively, the energy according to Eq. (131), the orbital angular momentum according to Eq. (113), and the z-

component of the orbital angular momentum according to Eq. (115):*

$$\psi_{100}(r) = \frac{1}{\sqrt{4\pi}} \frac{2}{a_0^{3/2}} e^{-r/a_0} \tag{133}$$

$$\psi_{200}(r) = \frac{1}{\sqrt{4\pi}} \frac{2}{(2a_0)^{3/2}} \left(1 - \frac{r}{2a_0}\right) e^{-r/(2a_0)} \tag{134}$$

$$\psi_{211}(r) = -\frac{1}{\sqrt{8\pi}} \frac{1}{(2a_0)^{3/2}} \sin\theta \, e^{i\phi} \frac{r}{a_0} e^{-r/(2a_0)} \tag{135}$$

$$\psi_{210}(r) = \frac{1}{\sqrt{4\pi}} \frac{1}{(2a_0)^{3/2}} \cos\theta \, \frac{r}{a_0} e^{-r/(2a_0)} \tag{136}$$

$$\psi_{300}(r) = \frac{1}{\sqrt{4\pi}} \frac{2}{(3a_0)^{3/2}} \left(1 - \frac{2}{3}\frac{r}{a_0} + \frac{2}{27}\frac{r^2}{a_0^2}\right) e^{-r/(3a_0)} \tag{137}$$

Here a_0 is the Bohr radius familiar from Chapter 4,

$$a_0 = \frac{4\pi\varepsilon_0}{e^2} \frac{\hbar^2}{m_e} \tag{138}$$

In three dimensions, the wavefunction is related to the probability for finding the electron in a small volume $dx\,dy\,dz$,

$$\begin{pmatrix} \text{probability for} \\ \text{finding electron} \\ \text{in volume } dx\,dy\,dz \end{pmatrix} = |\psi|^2 \, dx\,dy\,dz \tag{139}$$

Hence the units of ψ must be $(\text{meter})^{-3/2}$, in agreement with Eqs. (133)–(137).

Figure 6.24 gives plots of the radial dependence of the wavefunctions above. Since separate plots of the radial and the angular dependence do not give a clear impression of the shape of the wavefunction, it is preferable to represent the wavefunctions graphically as probability clouds, shown in Fig. 6.25. The density of the portions of these clouds has been adjusted to be proportional to $|\psi|^2$. These figures actually show cross sections through the clouds; the complete cloud is obtained by rotating the figures about the vertical axis. The figures show that for $n = 1$, $l = 0$ the probability cloud is dense in a spherical region, with a maximum at the origin; for $n = 2$, $l = 0$ it is dense in spherical region and in a surrounding concentric shell; for $n = 2$, $l = 1$, $m_l = 0$ it is dense in a spherical region with a hole (a torus), etc.

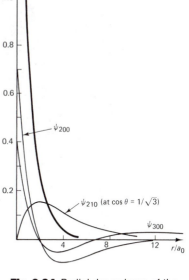

$\psi(r) \times \sqrt{4\pi a_0^3}$

Fig. 6.24 Radial dependence of the wavefunctions ψ_{nlm_l}.

* For a derivation of these wavefunctions, see D. S. Saxon, *Elementary Quantum Mechanics*.

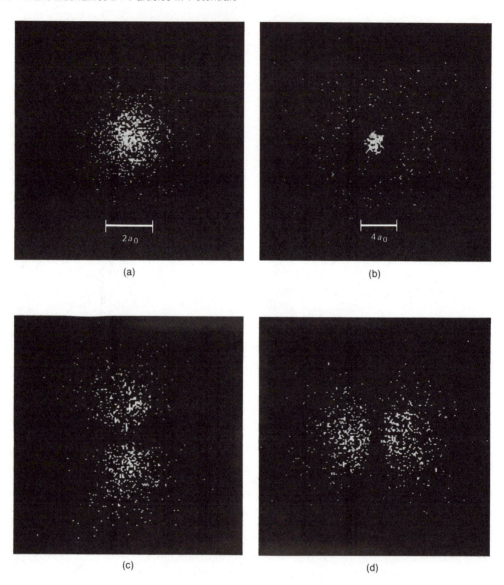

(a)

(b)

(c)

(d)

Fig. 6.25 Graphical representation of the wavefunctions ψ_{nlm_l} by means of probability clouds. The density of points in the cloud is proportional to $|\psi_{nlm_l}|^2$. (a) $n = 1$, $l = 0$, $m_l = 0$. (b) $n = 2$, $l = 0$, $m_l = 0$. (c) $n = 2$, $l = 1$, $m_l = 0$. (d) $n = 2$, $l = 1$, $m_l = \pm 1$.

Example 4   Calculate the mean value of r for an electron in the ground state of the hydrogen atom.

Solution   The probability for finding the electron in a small volume dV is $|\psi_{100}|^2\, dV$. Consider a volume in the shape of a thin spherical shell of radius r and thickness dr (see Fig. 6.26). For this volume, $dV = 4\pi r^2\, dr$ and hence

$$\begin{pmatrix}\text{probability for} \\ \text{finding electron} \\ \text{in interval from} \\ r \text{ to } r + dr\end{pmatrix} = 4\pi|\psi_{100}|^2 r^2\, dr \tag{140}$$

This can also be interpreted as the probability for finding the value of r in the interval dr. To obtain the mean value of r, we must multiply this probability by r and integrate over all values of r:

$$\bar{r} = \int_0^\infty 4\pi|\psi_{100}|^2 r^3\, dr \tag{141}$$

$$= \int_0^\infty \frac{4}{a_0^3}\, e^{-2r/a_0} r^3\, dr \tag{142}$$

This kind of integral is related to the factorial function; in general,

$$\int_0^\infty e^{-\beta r} r^n\, dr = \frac{n!}{\beta^{n+1}} \qquad \text{for } n = 0, 1, 2, \ldots \tag{143}$$

Hence

$$\bar{r} = \frac{4}{a_0^3}\frac{3!}{(2/a_0)^4} = \frac{3}{2} a_0 \quad \blacksquare \tag{144}$$

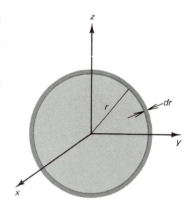

Fig. 6.26 A thin spherical shell of radius r and thickness dr.

Chapter 6:   SUMMARY

Energies of stationary states of particle in a box:

$$E_n = \frac{n^2\pi^2\hbar^2}{2mL^2} \qquad n = 1, 2, 3, \ldots$$

Wavefunctions of stationary states:

$$\Psi_n(x, t) = \sqrt{\frac{2}{L}}\, e^{-in^2\pi^2\hbar t/(2mL^2)} \sin(n\pi x/L)$$

Schrödinger's equation with potential:

$$-\frac{\hbar^2}{2m}\frac{\partial^2\Psi}{\partial x^2} + U(x)\Psi = i\hbar\frac{\partial\Psi}{\partial t}$$

Time dependence of stationary state:

$$\Psi(x, t) = e^{-iEt/\hbar}\psi(x)$$

Time-independent Schrödinger equation:

$$-\frac{\hbar^2}{2m}\frac{d^2\psi}{dx^2} + U(x)\psi = E\psi$$

Wavefunction in constant potential:

$$E \geqslant U_0: \quad \psi(x) \propto e^{\pm ikx}, \qquad k = \sqrt{2m(E - U_0)}/\hbar$$

$$E \leqslant U_0: \quad \psi(x) \propto e^{\pm \kappa x}, \qquad \kappa = \sqrt{2m(U_0 - E)}/\hbar$$

Barrier penetration probability:

$$P \simeq e^{-2\int_a^b \sqrt{2m[U(x) - E]}/\hbar\, dx}$$

Gamow factor:

$$P \simeq e^{-(4\pi/\hbar)(Ze^2/4\pi\varepsilon_0)(1/v_\alpha)}$$

WKB approximation:

$$2\int_a^b \sqrt{2m[E - U(x)]}\, dx = nh$$

Energy of stationary states of harmonic oscillator:

$$E_n = (n + \tfrac{1}{2})\hbar\omega_0$$

Quantization of angular-momentum vector:

$$|\mathbf{L}| = \sqrt{l(l + 1)}\,\hbar \qquad l = 0, 1, 2, \ldots$$

$$L_z = m_l\hbar \qquad m_l = 0, \pm 1, \pm 2, \ldots, \pm l$$

Energies of stationary states of hydrogen:

$$E = -\frac{e^4 m_e}{2(4\pi\varepsilon_0)^2\hbar^2}\frac{1}{(n' + l)^2} \qquad \begin{array}{l} l = 0, 1, 2, \ldots \\ n' = 1, 2, 3, \ldots \end{array}$$

Wavefunction of ground state of hydrogen:

$$\Psi_{100}(x, t) = \frac{1}{\sqrt{4\pi}}\frac{2}{a_0^{3/2}}e^{-r/a_0}e^{-iE_1 t/\hbar}$$

Chapter 6: PROBLEMS

1. Show that if $|A| = |B|$, then the superposition of two traveling waves given in Eq. (3) is a standing wave, as described in the footnote on p. 217.

2. According to some early (erroneus) speculations on nuclear structure, the nucleus was supposed to contain a number of electrons. If we treat such an electron as a particle in a (one-dimensional) box of nuclear size, about 1.0×10^{-14} m, what

would be the ground-state energy of the electron? What would be the energy difference between the ground state and the first excited state? Typical nuclear energy differences are of the order of ~ 1 MeV. Is this consistent with the existence of electrons in the nucleus?

3. An oxygen molecule confined in a container can be regarded as a particle in a box. Suppose that the size of this box is 10 cm and treat the motion of the oxygen molecule as one-dimensional.
(a) What is the ground-state energy?
(b) If the molecule has an energy equal to the mean (one-dimensional) thermal energy $\frac{1}{2}kT$ at $T = 300$ K, what is the quantum number n? What is the energy difference between the nth state and the $(n + 1)$th state?

4. A particle in a box is in the first excited state, described by the wavefunction of Eq. (19) with $n = 2$. What is the probability for finding the particle in the (almost) infinitesimal interval $dx = 0.001L$ near $x = \frac{1}{4}L$? In a similar interval near $x = \frac{1}{3}L$?

5. Suppose that a particle in a box is in the first excited state. What is the probability for finding the particle in the interval $0 \leq x \leq L/4$?

6. With the probability distribution given in Eq. (13), calculate the mean value of x and the mean value of x^2 for a particle in a box in a given energy eigenstate.

7. Consider a particle in a box in the superposition of the ground state and the first excited state given by Eq. (23). Calculate the mean value of x for this particle. Is this value time independent?

8. A particle in a box is in the energy eigenstate of eigenvalue E_n. Calculate the force that the particle exerts on each wall. (*Hint: $F = -dE_n/dL$. Why?*)

9. A given particle in a box has a 40% chance of being found in the ground state and a 60% chance of being found in the first excited state.
(a) What is the wavefunction of this particle at time $t = 0$? Assume that at this time the two contributions to the wavefunction are in phase near $x = 0$.
(b) What is the wavefunction at time $t > 0$?
(c) What is the mean value of the energy?

10. Prove that if $\psi_n(x, t)$ and $\psi_{n'}(x, t)$ are two wavefunctions of a particle in a box corresponding to different energy eigenvalues, then

$$\int_0^L \psi_n^*(x, t)\psi_{n'}(x, t)\, dx = 0$$

This is called the **orthonormality condition**; it is generally valid for any two wavefunctions belonging to different energy eigenvalues in any quantum-mechanical system.

11. Consider a particle in a three-dimensional box extending from $x = 0$ to $x = L$ along the x-axis, from $y = 0$ to $y = L$ along the y-axis, and from $z = 0$ to $z = L$ along the z-axis. Taking into account that x, y, and z motions of the particle are dynamically and probabilistically independent, find the wavefunction $\Psi(x, y, z, t)$ describing an arbitrary energy eigenstate. What are the energy eigenvalues?

12. For a particle in a three-dimensional box measuring $L \times L \times L$, the energy eigenvalues are

$$E_{n_1 n_2 n_3} = \frac{\pi^2 h^2}{2mL^2} (n_1^2 + n_2^2 + n_3^2)$$

where n_1, n_2, and n_3 are the quantum numbers for the motions in the x-, y-, and z-directions. What are the energy eigenvalues and the quantum numbers for the four lowest eigenstates? Draw the energy-level diagram.

13. Use the uncertainty principle to estimate the ground-state energy for a particle of mass m moving in the potential well

$$U(x) = k|x|$$

where $k > 0$.

14. Show that whenever $\psi_1(x, t)$ and $\psi_2(x, t)$ are solutions of the Schrödinger equation for a given potential, then $\psi_1(x, t) + \psi_2(x, t)$ is also a solution of the Schrödinger equation.

15. Consider a particle of mass m, energy E moving in the potential $U = \infty$ for $x < 0$, and $U = 0$ for $x > 0$ (see Fig. 6.27). Find the solution of the Schrödinger equation. Is the energy quantized?

16. A particle of mass m, energy E is moving in the potential shown in Fig. 6.28. The energy E, indicated by the dashed line in Fig. 6.28, is less than the height U_0 of the potential step. Find the solution of the Schrödinger equation in the regions $x < 0$ and $x > 0$.

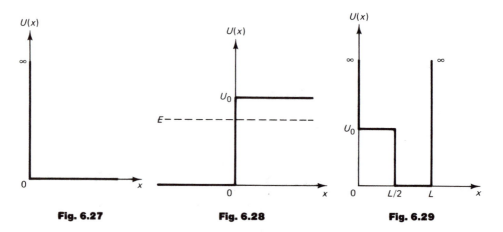

Fig. 6.27   **Fig. 6.28**   **Fig. 6.29**

17. A particle of mass m moves in one dimension in a potential (see Fig. 6.29)

$$U(x) = \begin{cases} \infty & \text{for } x \leqslant 0 \\ U_0 & \text{for } 0 < x < L/2 \\ 0 & \text{for } L/2 < x < L \\ \infty & \text{for } x \geqslant L \end{cases}$$

Find the equation for the energy eigenvalues of the particle confined in this potential. Assume that $E > U_0$. Your answer should be a formula involving E, U_0, L, and m.

18. Prepare a plot similar to Fig. 6.6 for a finite well with $U_0 = 4\hbar^2\pi^2/(2mL^2)$. How many bound states are there for this value of the potential? Roughly, what are their energies? Draw the energy-level diagram.

19. Show that the square well described in Section 6.3 has no bound states if $U_0 < \hbar^2\pi^2/(8mL^2)$.

20. By careful inspection of diagrams similar to Fig. 6.6, show that the finite square well described in Section 6.3 has N bound states if U_0 lies in the range

$$\frac{(2N - 1)^2\hbar^2\pi^2}{8mL^2} < U_0 < \frac{(2N + 1)^2\hbar^2\pi^2}{8mL^2}$$

21. Find the equation for the energy eigenvalues of a square potential well finite on *both* sides (see Fig. 6.30) by the following procedure: (a) First, assume that the wavefunction in the region $-a < x < a$ is of the form $A \sin kx$. Show that this leads to exactly the same equation for the energy eigenvalues as in Section 3. (b) Second, assume that the wavefunction in the region $-a < x < a$ is of the form $A \cos kx$. Find the new equation for the energy eigenvalues.

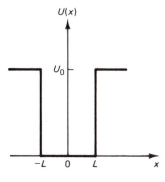

Fig. 6.30

22. Assume that electrons of energy 2.0 eV are incident on a square potential barrier of height 6.0 eV and width 8.0 Å. What fraction of the electrons will penetrate through this barrier?

23. A particle of energy E and mass m is incident (from the left) on the square potential barrier shown in Fig. 6.10. Suppose that the energy is such that $E > U_0$.
(a) Write down the solution of Schrödinger's equation (including the time dependence) for each of the regions $x < 0$, $0 < x < a$, and $x > a$. The solution should contain several unknown constants. Be sure to express all the parameters of the solution in terms of E, m, and U_0.
(b) State the boundary conditions that apply to this problem and use them to obtain a sufficient number of equations so that all the unknown constants can be solved in principle.

24. From the equations given in Example 3, solve for the coefficient R representing the amplitude of the reflected wave. Verify that $|R|^2 + |T|^2 = 1$. What is the physical interpretation of this equation?

25. Calculate the transmission probability for a particle of mass m and energy E incident from the left on the potential step shown in Fig. 6.31. The energy E, indicated by the dashed line is above the height U_0 of the potential step. Do this calculation by first finding the reflection probability; then use (transmission probability) = 1 − (reflection probability). (You cannot calculate the transmission probability directly by comparing the wave amplitudes on the right and the left because the slower speed of propagation of the waves on the right piles up their amplitude—in much the same way as surf piles up when ocean waves slow down in the shallows of a beach—and gives a misleading impression of a high probability of transmission.)

26. Repeat Problem 25 for a particle incident on the potential step of Fig. 6.31 from the *right*.

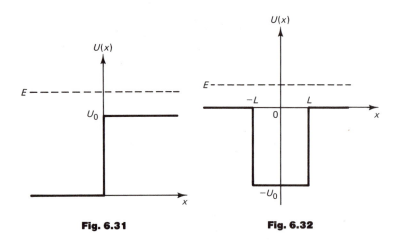

Fig. 6.31   **Fig. 6.32**

27. A particle of *positive* energy approaches the square well illustrated in Fig. 6.32 from the left. Some of the wave will be transmitted, and some will be reflected. Calculate the amplitude of the transmitted wave emerging on the right. What is the transmission probability if the energy of the particle is much larger than U_0?

28. According to the Gamow factor, what is the probability of penetration of an alpha particle of energy 8.0 MeV approaching a uranium nucleus?

29. Show that according to the approximation (75), the tunneling probability for a particle approaching a barrier of arbitrary shape is independent of whether the particle approaches from the right or the left. Would you expect this to be true even when the potential is such that the approximation (75) is not valid (assuming, of course, that the potential is at the same level on both sides of the barrier)?

30. Use the approximate formula given in Section 6.4 to calculate the transmission probability for a particle of energy E incident on the potential barrier shown

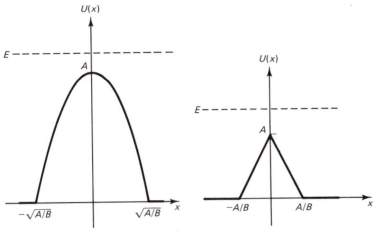

Fig. 6.33   **Fig. 6.34**

in Fig. 6.33:

$$U(x) = \begin{cases} 0 & \text{for } |x| > \sqrt{A/B} \\ A - Bx^2 & \text{for } |x| < \sqrt{A/B} \end{cases}$$

Assume that $E < A$.

31. With the approximate formula given in Section 6.4, calculate the transmission probability for a particle of energy E incident on the potential barrier (see Fig. 6.34)

$$U(x) = \begin{cases} 0 & \text{for } |x| > A/B \\ A - B|x| & \text{for } |x| < A/B \end{cases}$$

Assume that $E < A$.

32. Use the uncertainty principle to estimate the energy of the ground state of the harmonic oscillator. (*Hint:* Begin by substituting $x \simeq \Delta x$ and $p \simeq \Delta p \simeq h/\Delta x$ in the expression $\frac{1}{2}kx^2 + p^2/2m$ for the energy. Then find the value of Δx that makes this energy minimum.)

33. The eigenvalues of the harmonic oscillator can also be estimated by the following, somewhat crude recipe. For a given value of the energy E, calculate the position of the two classical turning points. Then pretend that the potential is an infinitely deep well of a width equal to the distance between the two turning points. Then use the equation for the energy eigenvalues of such a well [Eq. (11)] to calculate the energy eigenvalues. Show that this recipe gives $E_n = n\hbar\omega_0\pi/4$. Compare these values with the exact result $E_n = (n + \frac{1}{2})\hbar\omega_0$.

34. Verify that Eqs. (104) and (105) are normalized correctly.

35. A pendulum consists of a bob of mass 100 g suspended from a (massless) string of length 100 cm. The pendulum swings with an amplitude of 12°.
(a) Treating this pendulum as a harmonic oscillator, estimate the quantum number n. What is the energy difference between the nth state and the $(n + 1)$th state?

(b) Show that the swinging pendulum is *not* in an energy eigenstate, but rather in a superposition of several energy eigenstates. (*Hint:* If the pendulum were in an energy eigenstate, the probability distribution would be time independent.)

36. Suppose that a harmonic oscillator is in some arbitrary superposition of the ground state and several excited states. Show that the superposition is necessarily a periodic function of time, with a period $2\pi/\omega$.

37. Verify that Eqs. (107), (108), and (109) are solutions of the time-dependent Schrödinger equation for the harmonic oscillator.

38. The vibrations of a H_2 molecule are mathematically equivalent to the vibrations of a simple harmonic oscillator of spring constant $k = 1.13 \times 10^3$ N/m and mass 1.67×10^{-27} kg. What are the energy eigenvalues (in eV) for the vibrations of the molecule? What is the energy (in eV) and the wavelength (in Å) of the photons emitted when the molecule makes a transition from an excited state to the next lower excited state?

39. According to a theoretical model, the potential describing the force acting between the two atoms in a hydrogen molecule is

$$U(x) = U_0(e^{-2(x-x_0)/b} - 2e^{-(x-x_0)/b})$$

where U_0 and b are constants, and x_0 is the (classical) equilibrium separation between the atoms.
(a) By performing a Taylor-series expansion about $x = x_0$, show that for small vibrations about the equilibrium point, this potential can be approximated by a harmonic-oscillator potential.
(b) Find the energy eigenvalues for such small vibrational motions of the molecule; your answer should be an expression involving U_0 and b.

40. The wavefunction for a particle in the ground state of a harmonic oscillator is given by Eq. (104). From the corresponding probability distribution, calculate the mean value of x^2 for this particle.

41. A particle is in the ground state of the harmonic oscillator, with the wavefunction given by Eq. (104). Using the corresponding probability distribution, calculate the mean value of the potential energy. From this, deduce the mean value of the kinetic energy.

42. A particle of mass m moves in one dimension in the following potential (see Fig. 6.35):

$$U(x) = \begin{cases} \infty & \text{for } x \leqslant 0 \\ Ax & \text{for } 0 \leqslant x \end{cases}$$

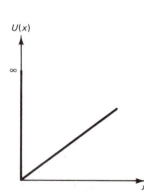

Fig. 6.35

Use the WKB approximation to find the energy eigenvalues.

43. By means of the WKB approximation, find the energy eigenvalues for a particle of mass m moving in the potential well

$$U(x) = k|x|$$

where $k > 0$.

44. Consider an electron in the second excited state ($n = 3$) in a hydrogen atom. List all the possible values of l and m_l for this electron.

45. Suppose that an electron in a hydrogen atom is in a state characterized by the angular momentum quantum numbers l and m_l. Show that although in such a state neither L_x nor L_y is well defined, $L_x^2 + L_y^2$ is well defined. Express the value of the latter quantity in terms of l and m_l.

46. Use the uncertainty principle to estimate the ground-state energy of the hydrogen atom. (*Hint:* See Problem 32.)

47. Verify that Eq. (133) is normalized correctly.

48. An electron is in the ground state of the hydrogen atom, with the wavefunction given by Eq. (133). What is the probability that the electron be found somewhere within the volume of the sphere $r \leqslant a_0$?

49. Calculate the mean value of r^2 for an electron in the ground state of the hydrogen atom, with the wavefunction given by Eq. (133).

50. Consider an electron in the ground state of the hydrogen atom, with the wavefunction given by Eq. (133). Using the corresponding probability distribution, calculate the mean value of the Coulomb potential energy. From this, calculate the mean value of the kinetic energy, and calculate the rms value of the speed of the electron.

51. Suppose that at time $t = 0$ the electron in a hydrogen atom is in the state described by the wavefunction

$$\psi(r) = \frac{1}{2}\psi_{100}(r) + \frac{\sqrt{3}}{2}\psi_{200}(r)$$

(a) What will be the wavefunction at time $t > 0$?
(b) What is the probability for finding the electron in the ground state? In the first excited state?
(c) What is the mean value of the energy of this electron?

Spin and the

Exclusion Principle

In our study of the wave mechanics we have treated the electron as a quantum-mechanical particle characterized by mass and charge, but devoid of any other features. Now we will see that the electron has some extra features: it has a **spin**, or intrinsic angular momentum, and it has a magnetic moment. From a classical point of view, we are tempted to picture the electron as a very small billiard ball, and to associate the spin angular momentum with a rotational motion of this billiard ball about its axis. But such a naive classical picture is totally misleading. It is known from high-energy collision experiments that the size of the electron is less than 10^{-18} m. A billiard ball of this size would have a quite small moment of inertia, and it would have to rotate with an impossibly large equatorial speed (much faster than the speed of light) if its rotational angular momentum were to match the spin of the electron.

To achieve an understanding of spin, we must appeal to an advanced formulation of wave mechanics including relativity. The picture of spin presented by relativistic wave mechanics shows that the spin is much more akin to the rotational angular momentum found within a circularly polarized light wave (see Section 1.6) than to any kind of rotation of a rigid body. In the relativistic electron wave, as in a circularly polarized light wave, there is a circulating flow of energy and momentum around the direction of motion of the wave; this gives the wave an angular momentum.* Thus the existence of spin hinges on the wave aspects of the electron. In this chapter we will discuss the spin phenomenologically, without inquiring into the underlying mechanism that causes it. However, we should always keep in mind that the spin is an essentially quantum-mechanical property of the electron.

We will also discuss another important quantum-mechanical property of electrons: their complete indistinguishability and the implications of this for a system containing several electrons. We will see that electrons obey an Exclusion Principle that forbids the presence of more than one electron in a given orbital and spin state.

7.1 **The Spin of the Electron**

In Section 6.6 we found that wave mechanics gives the same energy levels for the hydrogen atom as the Bohr theory, in excellent agreement with the spectroscopic observational data. However, the agreement with observational data is not perfect. For instance, high-resolution spectroscopy reveals that the first Balmer line consists of two separate spectral lines differing in wavelength by about 0.1 Å. Such a pair of nearly identical spectral lines is called a **doublet.** Careful examination of the light emitted in other transitions in the hydrogen atom reveals other doublets and quadruplets of lines. The same holds true for the light emitted in transitions in other atoms. The wavelength difference between the members of a multiplet is often much larger than in hydrogen; for instance, the yellowish light of sodium is due to a pair of spectral lines D_1, D_2 (the famous "sodium doublet") with a wavelength difference of about 6 Å. In the spectra of all atoms, multiplets are the rule rather than the exception.

This multiplet structure, or **fine structure,** of spectral lines posed a puzzle for quantum physics. It is clear that the fine structure of spectral lines indicates a fine structure of the energy levels; for instance, in the hydrogen atom all the energy levels except the s levels are closely spaced pairs of levels (see Fig. 7.1). The puzzle is: what

---

* H. C. Ohanian, *Am. J. Phys.*, June 1986.

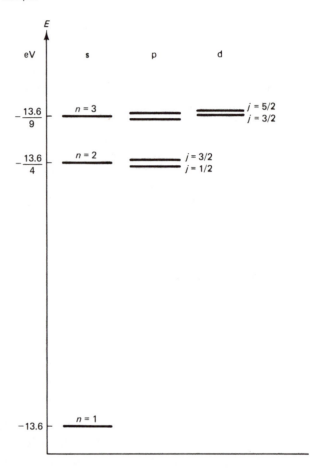

Fig. 7.1 Fine structure of the lowest energy levels of hydrogen. The splitting between the doublets is typically ∼10^{-5} eV. In the energy-level diagram, the splitting has been exaggerated for the sake of clarity.

causes this splitting of the energy levels? A first step toward a solution was taken in 1924 by W. Pauli,* who established that the doublet energy levels could be classified by a new quantum number

* **Wolfgang Pauli,** 1900–1958, Austrian, and later Swiss theoretical physicist, professor at Zürich. Pauli studied under Sommerfeld and under Bohr. His investigations of atomic spectra and atomic structure led him to introduce a new quantum number, and to postulate his Exclusion Principle. The latter earned him the Nobel Prize in 1945. He made an important contribution to the theory of β decay, proposing that the emission of the β particle is accompanied by the emission of a massless neutral particle (later called the neutrino), which carries away some of the energy released in the decay.

j with the two values

$$j = l + \tfrac{1}{2} \quad \text{and} \quad j = l - \tfrac{1}{2} \tag{1}$$

for the upper and lower levels of the doublet, respectively.* This quantum number gives the magnitude of the total angular momentum **J** of the atom, according to the usual formula

$$|\mathbf{J}| = \sqrt{j(j+1)}\,\hbar \tag{2}$$

But Pauli offered no physical explanation of the difference between the orbital angular momentum (characterized by *l*) and the total angular momentum (characterized by *j*).

The explanation for this difference was finally supplied in 1925 by S. Goudsmit and G. Uhlenbeck,[†] who proposed that the electron is endowed with spin **s**, or intrinsic angular momentum, of magnitude

$$|\mathbf{s}| = \sqrt{\tfrac{3}{4}}\,\hbar \tag{3}$$

or

$$|\mathbf{s}| = \sqrt{\tfrac{1}{2}(\tfrac{1}{2} + 1)}\,\hbar \tag{4}$$

Thus, the spin quantum number of the electron is $s = \tfrac{1}{2}$, and the electron is called a spin-$\tfrac{1}{2}$ particle. The *z*-component of the spin is quantized in the usual way,

$$s_z = m_s\hbar = \pm\tfrac{1}{2}\hbar \tag{5}$$

If $s_z = +\tfrac{1}{2}\hbar$, the electron is said to have spin "up"; if $s_z = -\tfrac{1}{2}\hbar$, spin "down."

This intrinsic, nonorbital angular momentum of the electron emerges in a natural way from a relativistic wave equation, first formulated by P. A. M. Dirac[‡] in 1928. As we mentioned in the introduction, the spin is an angular momentum stored within the quantum-mechanical wave. But the relativistic wave equation is mathematically quite complicated, and we cannot deal with the details in this book.

---

* For $l = 0$, the quantum number *j* has only a single value, $j = \tfrac{1}{2}$.

[†] **Samuel A. Goudsmit,** 1902– , Dutch and later American physicist, professor at the University of Michigan, chairman at Brookhaven National Laboratory, and editor of *The Physical Review*. **George E. Uhlenbeck,** 1900– , Dutch and later American physicist, professor at the University of Michigan and at Utrecht.

[‡] **Paul Adrien Maurice Dirac,** 1902–1984, British theoretical physicist, professor at Cambridge. He predicted the existence of antielectrons on the basis of his relativistic wave equation for the electron, and received the Nobel Prize in 1933 when antielectrons were discovered experimentally.

The total angular momentum of the hydrogen atom is the vector sum of the orbital and the spin vectors,

$$\mathbf{J} = \mathbf{L} + \mathbf{s} \tag{6}$$

If we were dealing with classical angular momentum vectors $\mathbf{L}$ and $\mathbf{s}$, then their vector sum could have any magnitude in the range

$$\sqrt{l(l+1)}\,\hbar + \sqrt{s(s+1)}\,\hbar \quad \text{to} \quad \sqrt{l(l+1)}\,\hbar - \sqrt{s(s+1)}\,\hbar$$

(these extremes correspond to the addition of parallel vectors and of antiparallel vectors, respectively). However, the addition of quantum-mechanical angular-momentum vectors is subject to a special restriction that arises from the quantization conditions. We state this restriction without proof:* the vector sum has the magnitude $\sqrt{j(j+1)}\,\hbar$ where j must be one or another of the following values:

$$j = l + s, \quad l + s - 1, \quad \ldots, \quad l - s + 1, \quad \text{or} \quad l - s \tag{7}$$

In this statement of the rule for addition of angular momentum we have assumed that $l > s$, but l and s are otherwise arbitrary angular-momentum quantum numbers.[†] Figure 7.2 shows how the vectors $\mathbf{L}$ and $\mathbf{S}$ add to give the resultant vector $\mathbf{J}$. Since the magnitudes of all three vectors are quantized, the angles between the vectors can only assume a discrete set of values—in the case $l = 1$, $s = \frac{1}{2}$ illustrated in Fig. 7.2, only two sets of values. For the electron in the hydrogen atom, $s = \frac{1}{2}$ and Eq. (7) reduces to

$$j = l + \tfrac{1}{2} \quad \text{or} \quad l - \tfrac{1}{2} \tag{8}$$

This is in agreement with Pauli's values as given above.

The energy difference between the $j = l + \frac{1}{2}$ level and the $j = l - \frac{1}{2}$ level in a doublet arises from the intrinsic magnetic moment of the electron. According to classical physics, we expect that a spinning ball with electric charge has a magnetic moment because each of the charge elements on the ball moves around a circle forming an electric current loop (see Fig. 7.3); if the electric charge is negative, the direction of the magnetic moment is opposite to the direction of the spin. Although this naive classical picture does not provide a correct description of the wave structure of the

---

* If $l < s$, then the possible values of j are

$$j = s + l, \quad s + l - 1, \quad \ldots, \quad s - l + 1, \quad s - l$$

For $l = 0$, this automatically yields $j = \frac{1}{2}$ as the *only* possible value.

[†] For a proof, see D. S. Saxon, *Elementary Quantum Mechanics*.

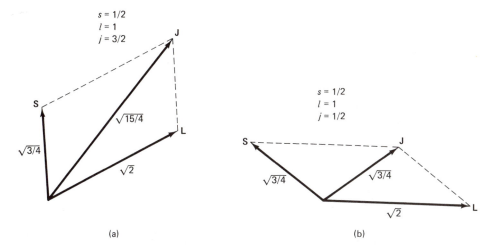

Fig. 7.2 Addition of the orbital angular momentum vector (with $l = 1$) and the spin angular momentum vector (with $s = \frac{1}{2}$). The resultant is the total angular momentum vector (with either $j = \frac{1}{2}$ or $j = \frac{3}{2}$).

electron, the theoretical analysis of the relativistic properties of quantum-mechanical particles confirms that an electron endowed with both spin and electric charge must necessarily have a magnetic moment. The magnetic moment is proportional to the spin,*

$$\boldsymbol{\mu}_e = -\frac{e}{m_e}\mathbf{s} \tag{9}$$

The minus sign in this equation indicates that the direction of the magnetic moment is opposite to the direction of the spin. According to Eq. (3), the magnitude of the magnetic moment is

$$\mu_e = \frac{e\hbar}{m_e}\sqrt{\frac{3}{4}} = \frac{\sqrt{3}}{2}\frac{e\hbar}{m_e} \tag{10}$$

and according to Eq. (5), the z-component of the magnetic moment has the values

$$\mu_{e,z} = -\frac{e}{m_e}m_s\hbar = \mp\frac{e}{m_e}\frac{\hbar}{2} \tag{11}$$

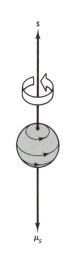

Fig. 7.3 A naive classical picture of the electron as a spinning, charged ball. Since the charge is negative, the resulting magnetic moment vector is opposite to the spin vector.

---

* Some further theoretical refinements show that the factor of proportionality in Eq. (9) is not exactly e/m_e, but $1.001160\ldots \times e/m_e$. The first experimental measurement of this anomaly in the factor of proportionality was performed by **Polykarp Kusch,** 1911– , American physicist, professor at Columbia, who thereby gained the 1955 Nobel Prize.

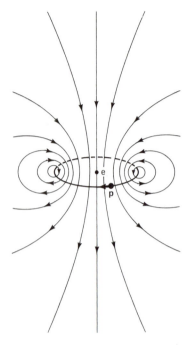

Fig. 7.4 In the reference frame of the electron, the proton moves around a circle in a clockwise direction, producing the magnetic field shown.

The quantity $e\hbar/(2m_e) = 9.27408 \times 10^{-24}$ J/T is called the **Bohr magneton.** Thus, the z-component of the electron's magnetic moment is ± 1 Bohr magneton.

The explanation of the fine structure of the energy levels of the hydrogen atom hinges on the interaction of the magnetic moment of the electron with the magnetic field generated by the orbital motion; this is called the **spin-orbit interaction.** The magnetic moment $\boldsymbol{\mu}_e$ placed in a magnetic field **B** has a potential energy*

$$U = -\boldsymbol{\mu}_e \cdot \mathbf{B} = \frac{e}{m_e} \mathbf{s} \cdot \mathbf{B} \tag{12}$$

i.e., the potential energy is positive if **s** is parallel to **B** and negative if it is antiparallel. In the calculation of this potential energy, the magnetic field must be evaluated in the rest frame of the electron. To see how the orbital motion generates a magnetic field in the rest frame, let us appeal to the Bohr model, with the electron moving in a clockwise circular orbit around the proton. In the reference frame of the electron, the proton will then be moving in a clockwise circular orbit around the electron. This motion of the proton is equivalent to a current loop with a clockwise positive current. The magnetic field of this current at the position of the electron is then in the downward direction, as shown in Fig. 7.4. Hence the potential energy (12) is negative if $s_z = \frac{1}{2}\hbar$ and positive if $s_z = -\frac{1}{2}\hbar$. Since the orbital angular momentum of the electron in a clockwise orbit is downward, the potential energy is negative for the $l - \frac{1}{2}$ state and positive for the $l + \frac{1}{2}$ state, in agreement with the energy-level diagram in Fig. 7.1. This accounts for the energy splitting of the doublet levels. The quantitative calculation confirms this qualitative discussion.

An experiment first performed in 1922 by O. Stern and W. Gerlach[†] gives direct empirical evidence for the spin of the electron and for the quantization of the angular momentum. This experiment was originally performed with silver atoms, but it was later repeated with hydrogen atoms. Figure 7.5 is a sketch of the appa-

---

* The expression for the potential energy of a magnetic moment in a magnetic field is derived in many textbooks on classical electromagnetism, e.g., H. C. Ohanian, *Physics*, Chapter 31.

[†] **Otto Stern,** 1888–1969, German and later American experimental physicist, professor at Hamburg and at the Carnegie Technical Institute. He developed the molecular-beam method and used this for the experiment on silver atoms, in collaboration with **Walther Gerlach,** 1899– , German physicist. Later, Stern used his method to measure the magnetic moment of the proton, and discovered its anomalous value. For this, he received the Nobel Prize in 1943.

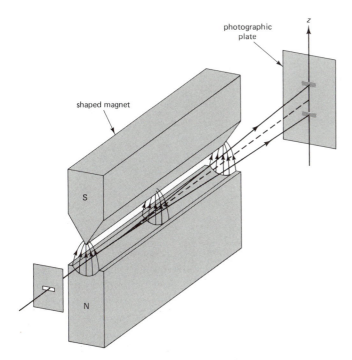

Fig. 7.5 The apparatus for the Stern–Gerlach experiment. The pole pieces of the magnet are shaped to produce a very inhomogeneous magnetic field, increasing in strength from bottom to top.

ratus. A beam of atoms travels along the gap between the poles of a magnet and strikes a photographic plate. The magnet has carefully shaped pole pieces, so that the magnetic field is very inhomogeneous (in Fig. 7.5, the magnetic field increases in a vertical direction). In such an inhomogeneous magnetic field, a magnetic moment experiences a vertical force depending on the orientation of the magnetic moment: according to Eq. (12) the potential energy of a magnetic moment in the vertical magnetic field in the median plane of the magnet is $U = -\mu_z B_z$ and therefore the force is

$$F_z = -\frac{\partial U}{\partial z} = \mu_z \frac{\partial B_z}{\partial z} \tag{13}$$

If μ_z is positive, then the force is positive; if μ_z is negative, then the force is negative. Atoms with magnetic moments will therefore be deflected up or down by an amount proportional to the z component of the spin. In the incident beam, the distribution of directions of the magnetic moments of the atoms is random. Classically, this means that the distribution of magnetic moments of these

(a)

(b)

Fig. 7.6 Photographs obtained by Stern and Gerlach with a beam of silver atoms. (a) When the magnetic field is zero, all the atoms strike in a single impact zone. (b) When the magnetic field is nonzero, the atoms strike in an upper and a lower zone. [From O. Stern and W. Gerlach, *Zeitschr. f. Physik* **9**, 349 (1922).]

TABLE 7.1   The Spins of Some Particles

Particle	Spin Quantum Number, s
Electron	1/2
Proton	1/2
Neutron	1/2
Photon	1
Neutrino	1/2
Muon	1/2
Pion (π^+, π^-, π^0)	0
Kaon (K^+, K^-, K°)	0
$J/\psi\,(3100)$	1
$\Sigma\,(2030)$	7/2
$\Lambda\,(1520)$	3/2

atoms spans a continuous range of values of μ_z, and the deflection of these magnetic moments by the magnetic field should result in a continuous range of impact zones on the photographic plate. Experimentally, what we find is quite different: we find only a discrete set of impact zones—in the case of silver or of hydrogen, only two impact zones (see Fig. 7.6). This experimental result establishes that μ_z has only two permitted, discrete values, i.e., μ_z is quantized. Since the magnetic moment is associated with the rotational motion of electric charge and is proportional to the angular momentum of this motion, the quantization of the magnetic moment implies the quantization of the angular momentum. In the case of hydrogen, the absence of orbital angular momentum in the atom allows us to draw the conclusion that the quantized magnetic moment must be associated with some kind of quantized nonorbital angular momentum. This conclusion is consistent with the notion of spin of the electron. Besides the electron, many other "elementary" quantum-mechanical particles have spin. Table 7.1 lists the spin quantum number s for some of these particles; the table includes a few particles without spin ($s = 0$).

7.2   The Total Angular Momentum; *L-S* Coupling

We saw in the preceding section that the energy eigenstates in a multiplet in the hydrogen atom can be characterized by the quantum numbers l (for orbital angular momentum), s (for spin), and j (for total angular momentum). For a more complete characterization of the energy eigenstates, we also need the quantum number

m_j (for the z-component of total angular momentum), which distinguishes between states with different orientations of the total angular momentum.* All of these are **"good" quantum numbers**—their values remain fixed as long as the electron remains in the given energy eigenstate. In classical mechanics, these good quantum numbers correspond to constants of the motion. In wave mechanics, we cannot measure, say, the orbital angular momentum in one atom as a function of time and verify that its magnitude remains constant; but we can measure this angular momentum in each of several atoms in the same energy eigenstate and verify that this yields the same result on each occasion, with zero uncertainty.

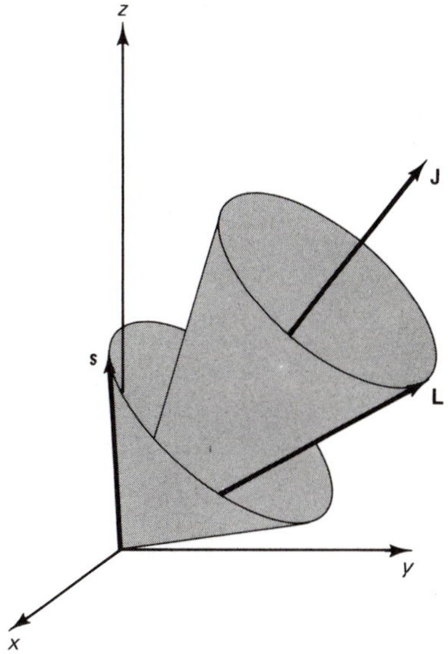

On the other hand, the quantum numbers m_l and m_s (for the z-components of the orbital angular momentum and the spin) are not good quantum numbers—their values do not remain fixed. In classical mechanics, the variation of m_l and m_s as a function of time is a simple consequence of the spin-orbit coupling given in Eq. (12). The orbital motion exerts a torque on the spin, and the spin exerts a corresponding torque on the orbit; consequently, both the orbital angular momentum and the spin experience a precession, and only the total angular momentum maintains a fixed direction (see Fig. 7.7). In wave mechanics, the mathematical treatment of the spin-orbit interaction is somewhat more complicated, but the broad picture remains the same. The variation of m_l or m_s shows up whenever we measure the z-components of the orbital angular momentum or the spin in each of several hydrogen atoms in the same energy eigenstate. In such measurements, we sometimes find one value of m_l or m_s, sometimes another, i.e., the result of the measurement is uncertain.[†]

The energy levels of most other atoms, with more than one electron, can be characterized by a similar set of quantum numbers. The net orbital angular momentum of the electrons of such an atom is the sum of the individual angular momenta of the

Fig. 7.7 The total angular momentum **J** remains constant while **L** and **S** precess around it, changing their direction but not their magnitude.

---

* States with different values of m_j have not been mentioned explicitly in the energy-level diagram (Fig. 7.1) because—for given quantum numbers l, s, and j within a multiplet—all the states with different values of m_j have the same energy; they are degenerate. If we were to place the atom in an external magnetic field, states with different values of m_j would acquire different energies, and the energy-level diagram would have to be altered.

[†] Since the orbital angular momentum and spin obey the usual quantization conditions, the result of these measurements is always one or another of the values $m_l = 0, \pm 1, \pm 2, \ldots$ or $m_s = \pm \frac{1}{2}$. But which of these values emerges on any particular occasion is uncertain.

electrons,

$$L = \sum_i L_i \tag{14}$$

Likewise, the net spin is

$$S = \sum_i s_i \tag{15}$$

and the total angular momentum is

$$J = L + S = \sum_i L_i + \sum_i s_i \tag{16}$$

All of these angular momenta obey the usual quantization rules. Thus

$$|L| = \sqrt{L(L + 1)}\,\hbar \tag{17}$$

$$|S| = \sqrt{S(S + 1)}\,\hbar \tag{18}$$

and

$$|J| = \sqrt{J(J + 1)}\,\hbar \tag{19}$$

Note that here the quantum numbers have been designated by uppercase letters L, S, and J, which must not be confused with the magnitudes $|L|$, $|S|$, and $|J|$. This usage of uppercase letters for the net angular momentum of an atom, in contrast to lowercase letters for the angular momentum of a single electron, is somewhat silly, but sanctified by tradition.* In Eq. (17), the net orbital quantum number L is an integer, $L = 0, 1, 2, \ldots$. In Eq. (18), the net spin quantum number S is an integer if the number of electrons in the atom is even, $S = 0, 1, 2, \ldots$; but S is a half-integer if the number of electrons is odd, $S = \frac{1}{2}, \frac{3}{2}, \frac{5}{2}, \ldots$. The same rule holds for the total angular momentum quantum number J.

As in the case of the hydrogen atom, the quantum numbers L, S, and J are all good quantum numbers, and so is the quantum number M_J. To understand why these are good quantum numbers, and why they are the *only* good quantum numbers, we must examine the interactions among the electrons. These interactions arise directly and indirectly from the repulsive Coulomb forces among all the electrons. In a direct way, these forces perturb the orbits of the electrons, changing their orbital angular momenta in

---

* For the hydrogen atom, with only a single electron, uppercase letters and lowercase letters are equally appropriate. We will opt for uppercase letters hereafter.

magnitude and in direction. Thus, the individual orbital angular momenta are not constants of the motion; only the net orbital angular momentum is constant. In an indirect way, the Coulomb forces bring about an interaction of the spins of the electrons. This spin interaction involves a subtle quantum mechanism, whose qualitative explanation we will leave to Section 7.4. The result is that the spins exert torques on each other, changing their magnitudes and directions. Consequently, the individual spins are not constants of the motion; only the net spin is constant. We therefore have two good quantum numbers, L and S. The remaining interaction we have not yet taken into account is the spin-orbit coupling. The situation is now exactly as in the hydrogen atom: the spin-orbit coupling causes **L** and **S** to precess about each other, so that only the total angular momentum **J** maintains a fixed direction. Thus neither m_L nor m_S is a good quantum number, but m_J is a good quantum number. The strong coupling of the individual orbital angular momenta into a net orbital angular momentum **L** and a net spin **S**, on which is superimposed a weaker coupling of **L** and **S** that causes these vectors to precess about each other without changing their magnitudes is called *L-S*, or **Russell–Saunders coupling.** This coupling scheme is approximately valid for most atoms.*

When listing the angular-momentum quantum numbers for some state of an atom, it is customary to indicate the value of L by an uppercase letter, according to the scheme

$$L = \quad 0 \quad 1 \quad 2 \quad 3 \quad 4 \quad 5 \ldots$$

$$\text{Letter} \quad S \quad P \quad D \quad F \quad G \quad H \ldots$$

The value of J is written as a subscript after this letter, and the value $2S + 1$ is written as a superscript before this letter, as follows:

$$^{2S+1}L_J \tag{20}$$

This is the **spectroscopic-term** notation for the angular-momentum quantum numbers of a state. For example, the ground state of

---

* Some atoms of large Z, such as lead, display a different coupling scheme called *j-j* **coupling.** This consists of a strong coupling of the individual orbital angular momentum and spin of each electron into an individual angular momentum **J**, on which is superimposed a weaker coupling of the individual **J**'s that causes these vectors to precess about each other.

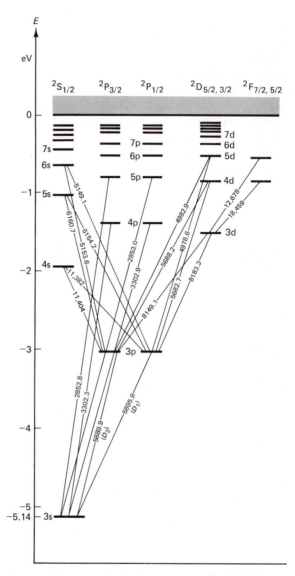

Fig. 7.8 Energy-level diagram of sodium. The slanting lines indicate transitions; the numbers give the emitted wavelengths in air, in Å. (From H. G. Kuhn, *Atomic Spectra*. Reproduced by permission of Longman Group Ltd.)

hydrogen has $L = 0$, $J = \frac{1}{2}$, and $S = \frac{1}{2}$ which we write as*

$$^2S_{1/2} \tag{21}$$

In energy-level diagrams of atoms, the angular-momentum quantum numbers of the ground state and of the excited states are usually indicated by this spectroscopic term notation. Figure 7.8 shows the energy-level diagram of sodium. Note that the ground state is $^2S_{1/2}$, i.e., it has the same net angular momentum quantum numbers as hydrogen; the lowest excited state is $^2P_{1/2}$; the next is $^2P_{3/2}$, etc. The transitions indicated by the slanting lines in Fig. 7.8 are the strongest transitions; these obey the selection rules

$$\Delta J = 0, \pm 1 \qquad \text{(but } J = 0 \text{ to } J = 0 \text{ is excluded)} \tag{22}$$

$$\Delta m_J = 0, \pm 1 \qquad \begin{array}{l}\text{(but } m_J = 0 \text{ to } m_J = 0 \\ \text{is excluded if } \Delta J = 0)\end{array} \tag{23}$$

$$\Delta L = 0, \pm 1 \tag{24}$$

$$\Delta S = 0 \tag{25}$$

and

$$\Delta l = \pm 1 \tag{26}$$

where l is the orbital angular-momentum quantum number of the electron making the transition. These selection rules apply not only to sodium, but to all atoms. Transitions obeying these selection rules are called **allowed;** transitions not obeying these selection rules are called **forbidden.** The latter transitions are not absolutely forbidden, but they occur only very rarely, and they therefore produce only very faint spectral lines.

The first two selection rules [Eqs. (22) and (23)] have a simple physical explanation: they express the conservation of angular momentum. As we know from Table 7.1, the photon has a spin angular momentum of one unit ($s = 1$). An atom that emits a photon must supply this angular momentum, and hence the net angular momentum vector of the atom must change by one unit, either by a change of magnitude or else by a change of direction. The other selection rules [Eqs. (24)–(26)] can be understood by an examination of the mechanism for the emission of radiation. We mentioned

---

* Don't confuse the letter S (for $L = 0$) with the letter S for the spin quantum number!

in Example 6.2 that this mechanism hinges on the existence of an oscillating electric dipole moment in the atom during the transition. The selection rules (24)–(26) ensure that this dipole moment is nonzero. If the transition violates these selection rules, then the oscillating electric dipole moment is zero, and the atom cannot radiate by means of the dipole mechanism (the atom can then still radiate by means of an oscillating octupole moment or a magnetic moment, but this kind of "forbidden" radiation tends to be weak).

7.3 The Zeeman Effect

The net magnetic moment of an atom with one or several electrons is the vector sum of all the magnetic moments associated with the spins plus all the magnetic moments associated with the orbital motion. By appealing, once more, to the Bohr model, we can understand how the orbital motion of an electron generates a magnetic moment. Imagine an electron of speed v in a circular orbit of radius r (see Fig. 7.9). The moving electron is equivalent to a circular current loop with a current

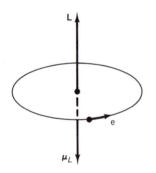

$$I = \frac{\text{(charge)}}{\text{(time)}} = \frac{e}{2\pi r/v} = \frac{ev}{2\pi r} \tag{27}$$

The magnetic moment of such a current loop is

$$\mu = \text{(area)} \times \text{(current)} = \pi r^2 \frac{ev}{2\pi r} = \frac{evr}{2} \tag{28}$$

We can rewrite this as

$$\mu = \frac{e}{2m_e} L \tag{29}$$

Fig. 7.9 Electron in a circular orbit. The orbital magnetic moment is opposite to the orbital angular momentum.

where $L = m_e vr$ is the orbital angular momentum of the electron. Taking into account that the direction of the magnetic moment is related by the right-hand rule to the direction of the electric current (which is opposite to the direction of motion of the electron), we can express (29) as a vector equation as follows:

$$\boldsymbol{\mu}_L = -\frac{e}{2m_e} \mathbf{L} \tag{30}$$

Although our derivation of this result is questionable, it is confirmed by a more rigorous quantum-mechanical treatment.* Note that since Eq. (30) is true for *each* electron, it is also true for *all* the electrons jointly, i.e., the net orbital magnetic moment of the atom is equal to $-e/(2m_e)$ multiplied by the net orbital angular momentum. Likewise according to Eq. (9), the net spin magnetic moment of the atom is equal to $-e/m_e$ multiplied by the net spin angular momentum. Thus, the net magnetic moment of the atom is the vector sum of these orbital and spin contributions,

$$\boldsymbol{\mu} = -\frac{e}{2m_e}\mathbf{L} - \frac{e}{m_e}\mathbf{S} \tag{31}$$

$$= -\frac{e}{2m_e}(\mathbf{L} + 2\mathbf{S}) \tag{32}$$

To find the magnitude $|\boldsymbol{\mu}|$ of this net magnetic moment and the component μ_z along the z-axis, we need to take into account the angular orientation of $\mathbf{L}$ and of $\mathbf{S}$. For a state with given values of the total angular momentum quantum numbers J and m_J, the relevant angles can be calculated from a vector diagram, similar to Fig. 7.2. The calculation is somewhat messy,† and we will merely state the final result for μ_z,

$$\mu_z = -gm_J\frac{e\hbar}{2m_e} \qquad m_J = J, J - 1, \ldots, 1 - J, -J \tag{33}$$

where the factor g, called the **Landé g-factor**,‡ is

$$g = 1 + \frac{J(J + 1) + S(S + 1) - L(L + 1)}{2J(J + 1)} \tag{34}$$

Here the quantum numbers L, S, and J are related in the usual way to the magnitudes of the net orbital angular momentum, the net spin angular momentum, and the total angular momentum. Note that the total angular momentum depends on the state of the atom. For instance, sodium in the ground state ($L = 0$, $S = \frac{1}{2}$, $J = \frac{1}{2}$) has $g = 2$ and $\mu_z = e\hbar/m_e$ or $-e\hbar/m_e$; but sodium in the

---

* See, e.g., J. L. Powell and B. Craseman, *Quantum Mechanics*, Section 10–5.

† See, e.g., M. Born, *Atomic Physics*.

‡ **Alfred Landé,** 1888– , German and later American physicist, professor at Tübingen and at Ohio State University.

lowest excited P state ($L = 1$, $S = \frac{1}{2}$, $J = \frac{1}{2}$) has $g = \frac{2}{3}$ and $\mu_z = \frac{1}{3}e\hbar/m_e$ or $-\frac{1}{3}e\hbar/m_e$.

If we immerse an atom in an external magnetic field, the inter-action between the atomic magnetic moment and the magnetic field will shift and split the energy levels of the atom. This leads to corresponding splittings of the spectral lines of the light emitted by the atom—all the spectral lines split into multiplets. Thus, in addition to the fine structure generated by the internal magnetic field of the atom, we now get an extra structure generated by the external magnetic field acting on the atom. The splitting of spectral lines by an external magnetic field is called the **Zeeman effect.** Since it depends on the quantum numbers of the energy levels, it is a very useful tool in the identification of these quantum numbers. The Zeeman effect is also a very useful tool in astronomy since it permits the detection of the magnetic fields on stars; for instance, light emitted in sunspot regions on the Sun displays Zeeman split-tings indicating magnetic fields of about 0.3 T.

We can calculate the energy shift of an energy level in a mag-netic field from an equation similar to Eq. (12). Let us assume that the magnetic field lies along the z-axis. Then

$$U = -\boldsymbol{\mu} \cdot \mathbf{B} = -\mu_z B$$

or

$$U = gm_J \frac{e\hbar B}{2m_e} \tag{35}$$

Note that this energy shift removes the degeneracy of states with different values of m_J (and equal values of the other quantum numbers): in the presence of a magnetic field, states with different values of m_J have different energies. Figure 7.10a shows the Zeeman shifts of the ground state and the first excited states of sodium. The doublet spectral line D_1, D_2 of sodium splits into a quadruplet plus a sextuplet—ten spectral lines altogether (see Fig. 7.10b).

Example 1   In the spectrum of sodium, the spectral line D_1 generated by the transition from the first excited state $^2P_{1/2}$ to the ground state $^2S_{1/2}$ has a wavelength of 5897.6 Å.* When sodium is placed in a magnetic field, this spectral line splits into four spectral lines (see Fig. 7.10b). Cal-culate the wavelength difference between the shortest and the longest wavelengths in this quadruplet in a magnetic field of 0.2 T.

---

* This is the wavelength in vacuum. The wavelength in air is 5895.9 Å.

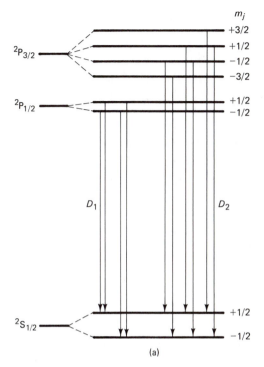

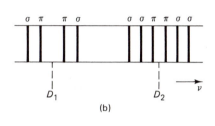

m_j

+3/2
+1/2
−1/2
−3/2

+1/2
−1/2

+1/2

−1/2

$^2P_{3/2}$

$^2P_{1/2}$

$^2S_{1/2}$

D_1

D_2

(a)

(b)

D_1 D_2

Fig. 7.10 (a) Zeeman energy shifts of the ground state ($^2S_{1/2}$) and of the first excited states ($^2P_{1/2}$ and $^2P_{3/2}$) of sodium. The possible transitions obey the selection rules stated in Eqs. (22)–(26). (b) Splitting of the doublet spectral lines D_1, D_2 of sodium. Each of the resulting ten spectral lines is polarized; the labels π and σ indicate the polarization. When observed from a direction perpendicular to the magnetic field, spectral lines marked with π are polarized parallel to the magnetic field, and lines marked with σ are polarized perpendicular to the magnetic field.

Solution As stated in the discussion following Eq. (34), the $^2S_{1/2}$ state has $\mu_z = \pm e\hbar/m_e$ and the $^2P_{1/2}$ state has $\mu_z = \pm\frac{1}{3}e\hbar/m_e$. Thus the shifts of the energy levels are, respectively,

$$U = \pm\frac{e\hbar}{m_e} B \tag{36}$$

and

$$U' = \pm\frac{1}{3}\frac{e\hbar}{m_e} B \tag{37}$$

Figure 7.10a shows these energy shifts. This figure also shows the transitions that produce the longest and the shortest wavelengths. The photons of longest wavelength have a net energy shift of

$$-\frac{e\hbar}{m_e} B - \frac{1}{3}\frac{e\hbar}{m_e} B = -\frac{4}{3}\frac{e\hbar}{m_e} B \tag{38}$$

and those of shortest wavelength have a net energy shift of

$$\frac{e\hbar}{m_e} B + \frac{1}{3}\frac{e\hbar}{m_e} B = +\frac{4}{3}\frac{e\hbar}{m_e} B \tag{39}$$

The difference of energy between these two kinds of photons is

$$\Delta E = -\frac{8}{3}\frac{e\hbar}{m_e} B \tag{40}$$

which implies a wavelength difference

$$\Delta\lambda = \Delta\left(\frac{hc}{E}\right) \simeq -\frac{hc}{E^2}\Delta E \tag{41}$$

$$\simeq -\frac{\lambda^2}{hc}\Delta E$$

$$\simeq \frac{\lambda^2}{hc}\frac{8}{3}\frac{e\hbar}{m_e} B \tag{42}$$

With $\lambda = 5898$ Å and $B = 0.2$ T this yields

$$\Delta\lambda = \frac{(5.898 \times 10^{-7}\,\text{m})^2 \times 8 \times 1.6 \times 10^{-19}\,\text{C} \times 0.2\,\text{T}}{3.0 \times 10^8\,\text{m/s} \times 3 \times 9.1 \times 10^{-31}\,\text{kg} \times 2\pi}$$

$$= 1.7 \times 10^{-11}\,\text{m} = 0.17\,\text{Å} \quad \blacksquare \tag{43}$$

The complicated splitting Zeeman effect illustrated in Fig. 7.10 is sometimes called the **"anomalous" Zeeman effect.** If the external magnetic field is very strong, of the order of 10 or more tesla, then the splitting becomes much simpler; it is then called the **"normal" Zeeman effect.*** This simplification is due to a breakdown of the *L-S* coupling. A strong magnetic field exerts strong torques on the orbital and the spin magnetic moments of the atom, causing them to precess rapidly about the direction of the magnetic field. As we saw at the beginning of this section, the *L-S* coupling is due to an interaction between the orbital and the spin magnetic

---

* The gradual changeover from the anomalous Zeeman effect to the normal Zeeman effect as the magnetic field increases from weak to very strong is called the **Paschen-Back effect.** What is meant by "very strong" depends on the atom and the spectral line under discussion; roughly, a magnetic field is to be regarded as very strong if it produces Zeeman splittings considerably larger than the zero-field multiplet splittings.

moments. In a very strong external magnetic field, this interaction between the orbital and the spin magnetic moments is much weaker than the interaction of each moment with the external magnetic field. It is then a good approximation to ignore the interaction between the two moments, and only take into account the interaction of each with the magnetic field. The torque exerted by the magnetic field on the orbital magnetic moment causes **L** to precess around the direction of the magnetic field, or the z-axis. Likewise the torque on the spin magnetic moment causes **S** to precess around this axis, independently of **L**. During the precessional motion m_L and m_S remain constant, i.e., they are now good quantum numbers. We can then use Eq. (32) directly to calculate μ_z,

$$\mu_z = -\frac{e\hbar}{2m_e}(m_L + 2m_S) \tag{44}$$

so that the energy shift is

$$U = -\mu_z B = (m_L + 2m_S)\frac{e\hbar B}{2m_e} \tag{45}$$

Figure 7.11a shows these energy shifts for the ground state and the first excited states of sodium. Although there are six possible transitions, there are only three distinct spectral lines (see Fig. 7.11b) because the energy differences for several of the six transitions coincide.

If we immerse an atom in an external *electric* field, the interaction of the atomic electrons with this electric field will also shift the energy levels and consequently split the spectral lines into multiplets. This is the **Stark effect.*** It is rather more difficult to treat mathematically than the Zeeman effect because it is a second-order effect which depends on an induced electric dipole moment. An undisturbed atom in a stationary state usually does not possess an electric dipole moment; the electric field must first induce an electric dipole moment by distorting the motion of the electrons, and then interact with this induced dipole moment to produce the energy shift. Thus, the energy shift is proportional to the square of the electric field strength.

---

* **Johannes Stark,** 1874–1957, German physicist, professor at Aachen, and president of the Physikalische-Technische Reichsanstalt. He received the Nobel Prize in 1919 for his discoveries of the Doppler effect of light emitted by moving ions and of the Stark effect.

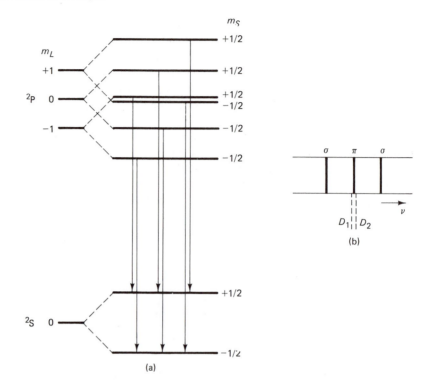

Fig. 7.11 (a) "Normal" Zeeman energy shifts of the ground state and the first excited state of sodium in a strong magnetic field. (b) Zeeman splitting of the spectral lines D_1, D_2 of sodium in a strong magnetic field. The labels π and σ indicate the polarization, as in Fig. 7.10b.

7.4 Pauli's Exclusion Principle

All electrons are identical—they all have exactly the same mass, electric charge, spin, magnetic moment, and every other physical attribute. But this does not, by itself, imply that electrons are indistinguishable. In classical physics, we can distinguish between two particles by continually observing them as they move along their separate trajectories. Even if their paths were to cross, and the two particles were to collide, we could still tell which is which because classical physics allows us to predict the trajectories after the collision. But in quantum physics the situation is drastically different: the identity of electrons implies their indistinguishability because the quantum uncertainties prevent us from "tracking" electrons. Whenever two electrons collide or approach closely, so that their wavefunctions overlap, we cannot be sure which is which after the collision.

The indistinguishability of electrons must be taken into account in the construction of the wavefunction for a system of several electrons. For instance, the wavefunction describing a system of two electrons, with coordinates x and x', is a function of two variables,*

$$\psi = \psi(x, x') \tag{46}$$

The probability for finding the first electron in the interval dx and the second in dx' is

$$|\psi(x, x')|^2 \, dx \, dx' \tag{47}$$

and the probability for finding the first electron in the interval dx' and the second in dx is

$$|\psi(x', x)|^2 \, dx \, dx' \tag{48}$$

Since the electrons are indistinguishable, it makes no difference which electron we regard as "first" and which as "second." Thus, the two expressions (47) and (48) must be equal—each merely gives the probability of finding *some* electron in the interval dx and *some* electron in dx'. From the equality

$$|\psi(x, x')|^2 = |\psi(x', x)|^2 \tag{49}$$

we deduce that the two wavefunctions differ at most by a phase factor:

$$\psi(x, x') = e^{i\phi}\psi(x', x) \tag{50}$$

The phase ϕ must be a universal constant, i.e., it must be independent of the details of the wavefunction. We recognize the necessity for this universality if we contemplate the superposition of two wavefunctions, $\psi(x, x') + \chi(x, x')$. Unless both ψ and χ acquire the same phase factor upon the exchange of x and x', the superposition would change by *more* than a phase factor, in violation of Eq. (49). We can now evaluate the phase factor by a trick: exchange x and x' once again on the right side of Eq. (50); this yields once again the same phase factor:

$$\psi(x, x') = e^{i\phi}\psi(x', x) = e^{i\phi}[e^{i\phi}\psi(x, x')] \tag{51}$$

From this we see that $e^{2i\phi} = 1$ and that

$$e^{i\phi} = +1 \quad \text{or} \quad -1 \tag{52}$$

---

* For the sake of simplicity, we deal only with the time-independent wavefunction corresponding to some eigenstate of energy; and we deal only with one-dimensional motion along the x-axis.

Thus, the wavefunctions of indistinguishable particles are either symmetric functions under exchange of their variables,

$$\psi(x, x') = \psi(x', x) \tag{53}$$

or else antisymmetric functions,

$$\psi(x, x') = -\psi(x', x) \tag{54}$$

Particles whose wavefunctions are symmetric are called **bosons;** particles whose wavefunctions are antisymmetric are called **fermions.*** Experimental observations have established that electrons and all other particles of spin $\frac{1}{2}$, or $\frac{3}{2}$, or $\frac{5}{2}$, ... are fermions, whereas photons and all particles of spin 0, or 1, or 2, ... are bosons.[†]

In the application of Eqs. (53) and (54), we must keep in mind that the complete specification of the state of, say, an electron includes not only the spatial dependence, but also the spin. Thus, when we exchanged the variables x and x' in these equations, it was understood that we also exchanged the spins. To make this explicit, we must include the spin variables m_s and $m_{s'}$ in the wavefunction. We will indicate the spin state of the electron by the symbol $|m_s = \frac{1}{2}\rangle$ for spin up, and $|m_s = -\frac{1}{2}\rangle$ for spin down. The symbol $|m_s = \pm\frac{1}{2}\rangle$ is called a **ket** (it is the rear half of a brac*ket*). If we include the spin variables explicitly, the antisymmetry requirement (54) becomes

$$\psi(x, x')|m_s = a\rangle |m_{s'} = b\rangle = -\psi(x', x)|m_s = b\rangle |m_{s'} = a\rangle \tag{55}$$

We will see how to apply this equation in the following example.

Example 2   Suppose that two electrons are confined to a (one-dimensional) box, as described in Section 6.1. If both electrons are in the ground state, what is their wavefunction? For the purposes of this example, ignore the electric repulsion between the electrons, so that each acts as a free particle except when in contact with the walls.

Solution   For a single electron in the ground state, the wavefunction is

$$\psi_1(x)|m_s = \frac{1}{2}\rangle \quad \text{or} \quad \psi_1(x)|m_s = -\frac{1}{2}\rangle \tag{56}$$

Since we are assuming that the second electron does not exert any forces on the first, the dependence of the wavefunction on x is not altered by

---

* **Enrico Fermi,** 1901–1954, Italian and later American physicist (for more on Fermi, see p. 363).

[†] This observational fact is in agreement with a theorem of advanced, relativistic quantum mechanics: the **spin-statistics theorem.**

the presence of the second electron. Likewise, the dependence of the wavefunction on x' is not altered by the presence of the first electron. Thus, the wavefunction $\psi(m_s, m_{s'}; x, x')$ for both electrons together must be a product of the form

$$\psi_1(x)|m_s = \pm\tfrac{1}{2}\rangle \psi_1(x')|m_{s'} = \pm\tfrac{1}{2}\rangle \tag{57}$$

or a superposition of several such products. In view of the antisymmetry requirement (55), the only acceptable superposition is

$$\psi(m_s, m_{s'}; x, x') = \psi_1(x)\psi_1(x')|m_s = \tfrac{1}{2}\rangle|m_{s'} = -\tfrac{1}{2}\rangle$$
$$-\psi_1(x)\psi_1(x')|m_s = -\tfrac{1}{2}\rangle|m_{s'} = \tfrac{1}{2}\rangle \tag{58}$$

or some multiple of this.* Note that according to this wavefunction, the spin of one electron is up, and the spin of the other is down (since the electrons are indistinguishable, it is meaningless to ask *which* electron has spin up, and *which* has spin down). If we were to make both spins equal, say, both spins up, then the resulting wavefunction would be identically zero, which is no wavefunction at all. We therefore see that whenever two electrons are in the ground state, the z-components of their spins must be opposite. Furthermore, it can be shown that for the wavefunction (58) the x- and y-components of the spin are also opposite, and thus all the components of the net spin are zero (the cancellation of the z-components of the spins is immediately obvious from Eq. (58), but the cancellation of the x- and y-components is not so obvious; the proof of this cancellation requires some acquaintance with the mathematical rules for the addition of spins in quantum mechanics). ■

The argument of the preceding example applies to any orbital state in any system: whenever two electrons are in the same orbital state, their spins must be opposite. If we include the spin as part of the specification of a state, then we can rephrase this as follows:

No more than one electron can be in any given orbital and spin state of a system.

This statement is called the **Exclusion Principle.** It was discovered by Pauli in 1925 during his theoretical investigations of the orbital arrangement of the electrons in atoms. In the form stated above, the Exclusion Principle is valid not only for a system with two electrons, but also for a system with more than two electrons. As we will see in the next section, the Exclusion Principle has far-reaching consequences for the electron structure of atoms.

---

* To achieve the correct normalization, we must multiply (58) by $1/\sqrt{2}$. This is not hard to check, but we must take into account that the normalization condition involves integrations over x and x', and summations over s and s'.

If two or more electrons are in *different* orbital states, the anti-symmetry requirement for the wavefunction does not impose restrictions on the orientations of the spins; such electrons can have opposite spins or parallel spins.

Example 3   Suppose that one of the two electrons in a box is in the ground state, and the other in the first excited state. What is the wavefunction if both spins are up? As in the preceding example, ignore the electric repulsion between the electrons.

Solution   The wavefunction for both electrons together must be a product of the ground-state wavefunction and the excited-state wavefunction, or a superposition of several such products. Taking into account the anti-symmetry requirement, we see that the only acceptable superposition is

$$\psi(m_s, m_{s'}; x, x') = \psi_1(x)\psi_2(x')|m_s = \tfrac{1}{2}\rangle|m_{s'} = \tfrac{1}{2}\rangle$$

$$-\psi_2(x)\psi_1(x')|m_s = \tfrac{1}{2}\rangle|m_{s'} = \tfrac{1}{2}\rangle \quad (59)$$

or some multiple of this.   ■

Note that for $x = x'$, the wavefunction (59) is zero; thus, the probability for finding the two electrons at the same place is zero. This is a characteristic property of the wavefunction for electrons with parallel spins, and it holds in general, even in systems more complicated than electrons in a box. We can establish this property in general by recognizing that for two electrons with parallel spins the antisymmetry of the wavefunction cannot be due to the spins—the exchange of equal spins has no effect on the wavefunction. Hence the antisymmetry must be entirely due to the spatial variables x and x'; for $x = x'$, Eq. (54) therefore becomes

$$\psi(m_s, m_{s'}; x, x) = -\psi(m_s, m_{s'}; x, x) \quad (60)$$

which implies that $\psi = 0$.

The zero probability for finding two electrons at the same place can be interpreted to mean that electrons of parallel spins tend to avoid one another. In contrast, electrons of opposite spins have no such tendency (their wavefunction usually does not vanish for $x = x'$). This correlation between the spin orientation and the probability distribution has an important effect on the Coulomb interaction between the electrons. If the electrons tend to avoid one another, then their mutual Coulomb energy will tend to be low. Thus, we see that electrons of parallel spins tend to have a lower Coulomb energy than electrons of opposite spins. This difference in Coulomb energy is effectively equivalent to a spin-spin interaction which tends to align the spins of the electrons. This effective spin-spin interaction (produced indirectly by the Coulomb interaction) plays an important role in atoms, where it is much stronger than the ordinary magnetic interaction of the spins (produced directly by the interaction of the spin magnetic moments).

7.5 **The Periodic Table**

For an atom with two or more electrons, say, Z electrons, the calculation of the wavefunction and of the energy eigenvalues is an extremely difficult mathematical problem. The complications arising from the Coulomb interactions among the electrons prevent us from finding exact solutions; instead, we must rely on numerical solutions constructed by a variety of clever approximation schemes. In one such scheme, each electron is regarded as moving under the influence of the central Coulomb field of the nucleus plus the *average* Coulomb field of all the other $Z - 1$ electrons. This means that all the other electrons are treated as a spherically symmetric cloud of negative electric charge which is concentric with the pointlike positive electric charge of the nucleus (see Fig. 7.12). The energy eigenstates of an electron moving in such a cloud are somewhat similar to those of hydrogen. They can be characterized by the usual quantum numbers n, l, m_l, and m_s (corresponding to the radial motion, the magnitude of the orbital angular momentum, the z-component of this angular momentum, and z-component of spin). As a next step in this approximation scheme, we must take into account the deviation between the average Coulomb field and the true Coulomb field, and also the interactions among spins and orbital angular momenta (L-S coupling). In consequence of these residual interactions, none of the quantum numbers n, l, m_l, m_s are good quantum numbers. Nevertheless, we can use these numbers in conjunction with a simple counting procedure based on the Exclusion Principle to gain some qualitative insights into the arrangement of the electrons in atoms.

The Exclusion Principle tells us that for every electron in an atom there must be one occupied state. By counting the number of such occupied states we can therefore discover the net orbital angular momentum, the net spin, and the total angular momentum of the atom. When performing this counting procedure, we can pretend that the residual interaction among the electrons is absent because this interaction does not affect the number of available

Fig. 7.12 Spherically symmetric average charge distribution produced by $Z - 1$ electrons and the nucleus.

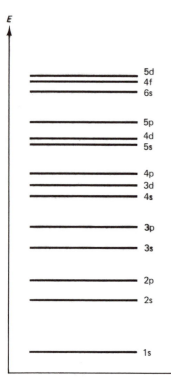

E

5d
4f
6s

5p
4d
5s

4p
3d
4s

3p

3s

2p

2s

1s

Fig. 7.13 Sequential order of energy levels of the subshells for the outermost electron of an atom. This diagram gives the energy of each subshell when all the subshells below it are filled and all the subshells above it are empty. The energy scale in this diagram is only qualitative; exact level separations depend on Z.

states of given orbital angular momentum and spin; it affects only the shape of the wavefunctions of these states.

For given values of n and l, there are $2l + 1$ possible values of m_l, and for each of these there are two possible values of m_s. This means that for given values of n and l, there are $(2l + 1) \times 2$ available states. These states are said to form a **subshell**.* If $l = 0$, the subshell has 2 available states; if $l = 1$, it has 6; if $l = 2$ it has 10; if $l = 3$ it has 14; etc.

In the ground state of an atom, the electrons will occupy the available states of lowest energy, one electron for each available orbital and spin state. We can therefore build up the ground-state configuration of the atoms in the Periodic Table by the following construction: begin with the hydrogen configuration and add one positive charge to the nucleus and one electron, then add another positive charge and another electron, etc., always placing the additional electron in the unoccupied state of lowest energy. We will use this construction to determine the ground-state configuration of the first few atoms in the Periodic Table. But before we start, we need to know the sequential order of the energy levels of the subshells. The order is (see Fig. 7.13)[†]

$$1s, 2s, 2p, 3s, 3p, 4s, 3d, 4p, 5s, 4d, 5p, 6s, 4f, 5d, \ldots \quad (61)$$

This order can be deduced from calculations of the energy eigenvalues of electrons moving under the influence of the average Coulomb field of other electrons, according to the approximation scheme described above.

To determine the angular-momentum quantum numbers of the ground state, we also need to know the arrangement of lowest energy for the angular momenta of several electrons sharing a

---

* They are given this name to distinguish them from the K, L, M,... **shells** we considered in Section 4.7. Each of these shells contains one or more subshells. For instance, the K shell contains an $l = 0$ subshell; the L shell contains an $l = 0$ and an $l = 1$ subshell; etc.

[†] A simple mnemonic for this sequential order of energy levels is as follows: write the subshells for given shells in rows, and read downward along successive diagonals:

1s
2s 2p
3s 3p 3d
4s 4p 4d 4f
5s 5p 5d 5f 5g
6s 6p 6d ...

subshell ("equivalent electrons"). This arrangement is given by **Hund's rule:** the arrangement of lowest energy is that with the largest value of S; and among arrangements of equal values of S, the lowest is that with the largest value of L. This rule can be shown to be a consequence of the spin and orbit interactions mentioned in Section 1.

We are now ready to construct the ground-state configurations of the atoms:

Hydrogen; Z = 1. The orbital configuration is 1s. The quantum numbers for the orbital angular momentum, the spin, and the total angular momentum are $L = 0$, $S = \frac{1}{2}$, and $J = \frac{1}{2}$. The spectroscopic term notation for these values is $^2S_{1/2}$.

Helium; Z = 2. We must add one electron to hydrogen. Since the 1s subshell has room for two electrons, we can place both electrons in this subshell, with opposite spins. The orbital configuration is 1s 1s.* The quantum numbers for the net orbital angular momentum, the net spin, and the total angular momentum are $L = 0$, $S = 0$, $J = 0$, giving a spectroscopic term $^1S_0$.

Lithium; Z = 3. There is no more room in the 1s subshell; hence we must place the next electron in the 2s subshell. This yields an orbital configuration 1s 1s 2s. The angular-momentum quantum numbers are $L = 0$, $S = \frac{1}{2}$, $J = \frac{1}{2}$. Note that the net spin is entirely due to the unpaired 2s electron. The spectroscopic term is $^2S_{1/2}$, the same as for hydrogen.

Beryllium; Z = 4. We must add the next electron to the 2s subshell, with spin opposite to that of the electron already in this shell. The configuration is 1s 1s 2s 2s. The angular-momentum quantum numbers are $L = 0$, $S = 0$, $J = 0$, giving a spectroscopic term $^1S_0$.

Boron; Z = 5. Now both the 1s and the 2s subshells are full; we must place the next electron in the 2p subshell, so that the configuration is 1s 1s 2s 2s 2p. The orbital and spin angular momentum is entirely due to the unpaired 2p electron, which gives $L = 1$, $S = \frac{1}{2}$. Since the spin-orbit interaction (see Section 1) results in a lower energy for $J = \frac{1}{2}$ than for $J = \frac{3}{2}$, the ground state has $J = \frac{1}{2}$, and the spectroscopic term is $^2P_{1/2}$.

---

* This is usually abbreviated as $(1s)^2$ or simply $1s^2$. We will not use this abbreviation here since it is somewhat confusing.

Carbon; Z = 6. We add the next electron to the 2p subshell and obtain the configuration 1s 1s 2s 2s 2p 2p. Since the p subshell has 5 separate orbital states (with different m_l), there is no need to pair the two 2p electrons in the same orbital state, with opposite spins. It is better to place them in different orbital states since Hund's rule demands parallel spins for these electrons, i.e., $S = 1$. Also, this rule demands the largest possible L. Since the electrons do not have equal m_l, they cannot attain $L = 2$; the next largest value is $L = 1$, which leads to $J = 0$. Thus, the spectroscopic term is $^3P_0$.

Nitrogen; Z = 7. We add another electron to the 2p subshell. The configuration is then 1s 1s 2s 2s 2p 2p 2p. Hund's rule demands parallel spins for the three 2p electrons, i.e., $S = \frac{3}{2}$. Since the three 2p electrons are distributed over the states $m_l = 1, 0, -1$, they span all possible directions of the orbital angular momentum. This implies that the net orbital angular is a sum of vectors distributed over all possible directions; such a sum is zero, i.e., $L = 0$. The spectroscopic term is $^4S_{3/2}$.

Oxygen; Z = 8. We again add an electron to the 2p subshell, obtaining a configuration 1s 1s 2s 2s 2p 2p 2p 2p. The orbital angular momentum is entirely due to this added electron, i.e., $L = 1$. The added electron has to share the orbital state of one of the electrons already in the subshell. Since such paired electrons always have opposite spins, they do not contribute to the net spin. Only the other two, unpaired, electrons contribute, giving $S = 1$. On the basis of the spin-orbit interaction, we would then expect a ground state with $J = 0$. Actually, the ground state has $J = 2$. (This is an instance of inversion of the energy order of a multiplet. The explanation of this inversion lies in advanced quantum mechanics; suffice it to say that it tends to happen whenever a shell is more than half full.) The spectroscopic term is $^3P_2$.

Fluorine; Z = 9. Adding one more electron to the 2p subshell, we obtain the configuration 1s 1s 2s 2s 2p 2p 2p 2p 2p. To determine the angular-momentum quantum numbers, we can take advantage of a clever trick: imagine that we add one more electron to fluorine, completely filling the 2p subshell. The net orbital angular momentum will then be zero because, as in the case of nitrogen, the electrons are then equally distributed over the states $m_l = 1, 0, -1$. Furthermore, the net spin will then be zero because all the individual spins are paired so that they cancel. Hence the configuration for 10 electrons has $L = 0$, $S = 0$, and $J = 0$. It follows that in the configuration for 9 electrons, the angular momentum must be entirely due to the *absence* of one electron from the full 2p shell. Since one such electron has $L = 1$ and $S = \frac{1}{2}$, these must be the quantum numbers for fluorine. The total angular-momentum quantum number is $J = \frac{3}{2}$, corresponding to an inverted multiplet, as in the case of oxygen. The spectroscopic term is $^2P_{3/2}$.

Neon; $Z = 10$. The tenth electron fills the 2p subshell, giving a configuration 1s 1s 2s 2s 2p 2p 2p 2p 2p 2p. As discussed above, the full 2p subshell has zero orbital angular momentum and zero spin, i.e., $L = 0$, $S = 0$ and $J = 0$, with a spectroscopic term $^1S_0$.

Sodium; $Z = 11$. We must add the next electron in the 3s subshell, so that the configuration is 1s 1s 2s 2s 2p 2p 2p 2p 2p 2p 3s. The net angular momentum is that of this added electron, i.e., $L = 0$, $S = \frac{1}{2}$, $J = \frac{1}{2}$, with a spectroscopic term $^2S_{1/2}$.

We can continue with this procedure and construct atoms of larger values of Z. Table 7.2 gives the ground-state configurations and the spectroscopic terms for all the atoms. Inspection of this table reveals certain repetitive, periodic patterns: H, Li, Na, K, Rb, ... all have the spectroscopic term $^2S_{1/2}$; He, Ne, A, Kr, Xe, ... all have $^1S_0$; F, Cl, Br, I, ... all have $^2P_{3/2}$, etc. These atoms belong to the columns IA, 0, and VIIA of the Periodic Table (see Fig. 7.14). The agreement of their spectroscopic terms explains the similarities in their spectra. Furthermore, the atoms in column IA all have an outermost single electron in an s state; the atoms in column 0 all have their outermost electrons in a completely filled p shell; the atoms in column VIIA have their outermost electrons in a p shell that is one electron short of being full. These similarities in the arrangement of the outermost electrons (valence electrons) explain the similarities in the chemical behavior of these atoms: the atoms in column IA are alkalis, those in column 0 are noble gases, those in column VIIA are halogens.

We therefore see that the simple counting procedure based on the Pauli Principle can explain the broad features of the Periodic Table. Note that as we move along the second and third row of the Table, we are progressively filling an s and a p subshell. This requires 8 electrons, and explains why these rows have 8 columns. The next four rows of the Table display a different pattern because we are filling not only an s and a p subshell, but also an additional subshell of higher angular momentum. For instance, in the third row we are filling the 4s, 3d, and 4p subshells. The filling of the 3d subshell gives rise to "transition elements" ($Z = 21$ to $Z = 30$; see the columns labeled with B in Fig. 7.14).

TABLE 7.2 Electron Configurations and Spectroscopic Terms for the Ground States of the Atoms[a]

Atom	Z	K	L		M			N				O				Term	Ionization Potential
		1s	2s	2p	3s	3p	3d	4s	4p	4d	4f	5s	5p	5d	5f		
H	1	1														$^2S_{1/2}$	13.5981 eV
He	2	2														$^1S_0$	24.5868
Li	3	2	1													$^2S_{1/2}$	5.3916
Be	4	2	2													$^1S_0$	9.322
B	5	2	2	1												$^2P_{1/2}$	8.298
C	6	2	2	2												$^3P_0$	11.260
N	7	2	2	3												$^4S_{3/2}$	14.534
O	8	2	2	4												$^3P_2$	13.618
F	9	2	2	5												$^2P_{3/2}$	17.422
Ne	10	2	2	6												$^1S_0$	21.564
Na	11	2	2	6	1											$^2S_{1/2}$	5.139
Mg	12	2	2	6	2											$^1S_0$	7.646
Al	13	2	2	6	2	1										$^2P_{1/2}$	5.986
Si	14	2	2	6	2	2										$^3P_0$	8.151
P	15	2	2	6	2	3										$^4S_{3/2}$	10.486
S	16	2	2	6	2	4										$^3P_2$	10.360
Cl	17	2	2	6	2	5										$^2P_{3/2}$	12.967
Ar	18	2	2	6	2	6										$^1S_0$	15.759
K	19	2	2	6	2	6		1								$^2S_{1/2}$	4.341
Ca	20	2	2	6	2	6		2								$^1S_0$	6.113
Sc	21	2	2	6	2	6	1	2								$^2D_{3/2}$	6.54
Ti	22	2	2	6	2	6	2	2								$^3F_2$	6.82
V	23	2	2	6	2	6	3	2								$^4F_{3/2}$	6.74
Cr	24	2	2	6	2	6	5	1								$^7S_3$	6.765
Mn	25	2	2	6	2	6	5	2								$^6S_{5/2}$	7.432
Fe	26	2	2	6	2	6	6	2								$^5D_4$	7.870
Co	27	2	2	6	2	6	7	2								$^4F_{9/2}$	7.86
Ni	28	2	2	6	2	6	8	2								$^3F_4$	7.635
Cu	29	2	2	6	2	6	10	1								$^2S_{1/2}$	7.726
Zn	30	2	2	6	2	6	10	2								$^1S_0$	9.394
Ga	31	2	2	6	2	6	10	2	1							$^2P_{1/2}$	5.999
Ge	32	2	2	6	2	6	10	2	2							$^3P_0$	7.899
As	33	2	2	6	2	6	10	2	3							$^4S_{3/2}$	9.81
Se	34	2	2	6	2	6	10	2	4							$^3P_2$	9.752
Br	35	2	2	6	2	6	10	2	5							$^2P_{3/2}$	11.814
Kr	36	2	2	6	2	6	10	2	6							$^1S_0$	13.999
Rb	37	2	2	6	2	6	10	2	6			1				$^2S_{1/2}$	4.177
Sr	38	2	2	6	2	6	10	2	6			2				$^1S_0$	5.693
Y	39	2	2	6	2	6	10	2	6	1		2				$^2D_{3/2}$	6.38
Zr	40	2	2	6	2	6	10	2	6	2		2				$^3F_2$	6.84
Nb	41	2	2	6	2	6	10	2	6	4		1				$^6D_{1/2}$	6.88
Mo	42	2	2	6	2	6	10	2	6	5		1				$^7S_3$	7.10
Tc	43	2	2	6	2	6	10	2	6	5		2				$^6S_{5/2}$	7.28
Ru	44	2	2	6	2	6	10	2	6	7		1				$^5F_5$	7.366
Rh	45	2	2	6	2	6	10	2	6	8		1				$^4F_{9/2}$	7.46
Pd	46	2	2	6	2	6	10	2	6	10						$^1S_0$	8.33
Ag	47	2	2	6	2	6	10	2	6	10		1				$^2S_{1/2}$	7.576
Cd	48	2	2	6	2	6	10	2	6	10		2				$^1S_0$	8.993
In	49	2	2	6	2	6	10	2	6	10		2	1			$^2P_{1/2}$	5.786
Sn	50	2	2	6	2	6	10	2	6	10		2	2			$^3P_0$	7.344
Sb	51	2	2	6	2	6	10	2	6	10		2	3			$^4S_{3/2}$	8.641
Te	52	2	2	6	2	6	10	2	6	10		2	4			$^3P_2$	9.01
I	53	2	2	6	2	6	10	2	6	10		2	5			$^2P_{3/2}$	10.457
Xe	54	2	2	6	2	6	10	2	6	10		2	6			$^1S_0$	12.130

[a] Numbers in parentheses are doubtful.

TABLE 7.2 (continued)

Atom	Z	K	L	M	4f	5s	5p	5d	5f	6s	6p	6d	7s	Term	Ionization Potential (eV)
Cs	55	Same				2	6			1				$^2S_{1/2}$	3.894 eV
Ba	56					2	6			2				$^1S_0$	5.211
La	57					2	6	1		2				$^2D_{3/2}$	5.5770
Ce	58				1	2	6	1		2				$^1G_4$	5.466
Pr	59				3	2	6			2				$^4I_{9/2}$	5.422
Nd	60				4	2	6			2				$^5I_4$	5.489
Pm	61				5	2	6			2				$^6H_{5/2}$	5.554
Sm	62				6	2	6			2				$^7F_0$	5.631
Eu	63				7	2	6			2				$^8S_{7/2}$	5.666
Gd	64				7	2	6	1		2				$^9D_2$	6.141
Tb	65				(8)	2	6	(1)		(2)				$(^8G_{13/2})$	5.852
Dy	66				10	2	6			2				$^5I_8$	5.927
Ho	67				11	2	6			2				$^4I_{15/2}$	6.018
Er	68				12	2	6			2				$^3H_6$	6.101
Tm	69				13	2	6			2				$^2F_{7/2}$	6.184
Yb	70				14	2	6			2				$^1S_0$	6.254
Lu	71				14	2	6	1		2				$^2D_{3/2}$	5.426
Hf	72				14	2	6	2		2				$^3F_2$	6.865
Ta	73				14	2	6	3		2				$^4F_{3/2}$	7.88
W	74				14	2	6	4		2				$^5D_0$	7.98
Re	75				14	2	6	5		2				$^6S_{5/2}$	7.87
Os	76				14	2	6	6		2				$^5D_4$	8.5
Ir	77				14	2	6	7		2				$^4F_{9/2}$	9.1
Pt	78				14	2	6	9		1				$^3D_3$	9.0
Au	79				14	2	6	10		1				$^2S_{1/2}$	9.22
Hg	80				14	2	6	10		2				$^1S_0$	10.43
Tl	81				14	2	6	10		2	1			$^2P_{1/2}$	6.108
Pb	82				14	2	6	10		2	2			$^3P_0$	7.417
Bi	83				14	2	6	10		2	3			$^4S_{3/2}$	7.289
Po	84				14	2	6	10		2	4			$^3P_2$	8.43
At	85				14	2	6	10		2	5			$^2P_{3/2}$	
Rn	86				14	2	6	10		2	6			$^1S_0$	10.749
Fr	87				14	2	6	10		2	6		(1)		
Ra	88				14	2	6	10		2	6		2	$^1S_0$	5.278
Ac	89				14	2	6	10		2	6	1	2	$^2D_{3/2}$	5.17
Th	90				14	2	6	10		2	6	2	2	$^3F_2$	6.08
Pa	91				14	2	6	10	2	2	6	1	2	$^4K_{11/2}$	5.89
U	92				14	2	6	10	3	2	6	1	2	$^5L_6$	6.05
Np	93				14	2	6	10	4	2	6	1	2	$^6L_{11/2}$	6.19
Pu	94				14	2	6	10	6	2	6		2	$^7F_0$	6.06
Am	95				14	2	6	10	7	2	6		2	$^8S_{7/2}$	5.993
Cm	96				14	2	6	10	7	2	6	1	2	$^9D_2$	6.02
Bk	97				14	2	6	10	(9)	2	6	(0)	(2)	$^6H_{5/2}$	6.23
Cf	98				14	2	6	10	(10)	2	6	(0)	(2)	$^5I_8$	6.30
Es	99				14	2	6	10	(11)	2	6	(0)	(2)	$^4I_{15/2}$	6.42
Fm	100				14	2	6	10	(12)	2	6	(0)	(2)	$^3H_6$	6.50
Md	101				14	2	6	10	(13)	2	6	(0)	(2)	$^2F_{7/2}$	6.58
No	102				14	2	6	10	(14)	2	6	(0)	(2)	$^1S_0$	6.65
Lw	103				14	2	6	10	(14)	2	6	(1)	(2)		

The Periodic Table of Chemical Elements

IA	IIA	IIIB	IVB	VB	VIB	VIIB	VIII	VIII	VIII	IB	IIB	IIIA	IVA	VA	VIA	VIIA	0
1 **H** 1.0079																	2 **He** 4.00260
3 **Li** 6.94	4 **Be** 9.01218											5 **B** 10.81	6 **C** 12.011	7 **N** 14.0067	8 **O** 15.9994	9 **F** 18.998403	10 **Ne** 20.179
11 **Na** 22.98977	12 **Mg** 24.305											13 **Al** 26.98154	14 **Si** 28.0855	15 **P** 30.97376	16 **S** 32.06	17 **Cl** 35.453	18 **Ar** 39.948
19 **K** 39.0983	20 **Ca** 40.08	21 **Sc** 44.9559	22 **Ti** 47.88	23 **V** 50.9415	24 **Cr** 51.996	25 **Mn** 54.9380	26 **Fe** 55.847	27 **Co** 58.9332	28 **Ni** 58.69	29 **Cu** 63.546	30 **Zn** 65.39	31 **Ga** 69.72	32 **Ge** 72.59	33 **As** 74.9216	34 **Se** 78.96	35 **Br** 79.904	36 **Kr** 83.80
37 **Rb** 85.4678	38 **Sr** 87.62	39 **Y** 88.9059	40 **Zr** 91.22	41 **Nb** 92.9064	42 **Mo** 95.94	43 **Tc** 98.9062	44 **Ru** 101.07	45 **Rh** 102.9055	46 **Pd** 106.42	47 **Ag** 107.8682	48 **Cd** 112.41	49 **In** 114.82	50 **Sn** 118.71	51 **Sb** 121.75	52 **Te** 127.60	53 **I** 126.9045	54 **Xe** 131.29
55 **Cs** 132.9054	56 **Ba** 137.33	57–71 Rare Earths	72 **Hf** 178.49	73 **Ta** 180.9479	74 **W** 183.85	75 **Re** 186.207	76 **Os** 190.2	77 **Ir** 192.22	78 **Pt** 195.08	79 **Au** 196.9665	80 **Hg** 200.59	81 **Tl** 204.383	82 **Pb** 207.2	83 **Bi** 208.9804	84 **Po** (209)	85 **At** (210)	86 **Rn** (222)
87 **Fr** (223)	88 **Ra** 226.0254	89–103 Acti-nides	104 **Rf** (260)	105 **Ha** (260)	106 (263)	107 (263)		109 (266)									

Rare Earths (Lanthanides)

57 **La** 138.9055	58 **Ce** 140.12	59 **Pr** 140.9077	60 **Nd** 144.24	61 **Pm** (145)	62 **Sm** 150.36	63 **Eu** 151.96	64 **Gd** 157.25	65 **Tb** 158.9254	66 **Dy** 162.50	67 **Ho** 164.9304	68 **Er** 167.26	69 **Tm** 168.9342	70 **Yb** 173.04	71 **Lu** 174.967

Actinides

89 **Ac** (227)	90 **Th** 232.0381	91 **Pa** 231.0359	92 **U** 238.029	93 **Np** 237.0482	94 **Pu** (244)	95 **Am** (243)	96 **Cm** (247)	97 **Bk** (247)	98 **Cf** (251)	99 **Es** (254)	100 **Fm** (257)	101 **Md** (258)	102 **No** (259)	103 **Lr** (260)

Fig. 7.14 The Periodic Table of Chemical Elements. In each box, the upper number is the *atomic number*. The lower number is the *atomic mass,* i.e., the mass (in grams) of one mole or, alternatively, the mass (in u) of one atom. Numbers in parentheses denote the atomic masses of the most stable or best-known isotope of the element; all other numbers represent the average masses of a mixture of several isotopes as found in naturally occurring samples of the element.

7.6 Molecular Energy Levels

In a molecule consisting of two or more atoms, the energy depends not only on the electronic state of each atom, but also on the vibrational and the rotational state of the molecule as a whole. Crudely, we can think of a molecule as an arrangement of masses held together by nearly massless springs (see Fig. 7.15). The masses represent the atomic nuclei, and the springs represent the atomic electrons whose interactions are mainly responsible for the interatomic forces. For a fixed electronic state of each atom, the spring constant then has some fixed value. Apart from the fixed electronic energy, the energy of the molecule is the sum of vibrational energy and rotational energy.

The energy eigenvalues associated with the vibrational motion are simply those of a harmonic oscillator. If the frequency of oscillation of the masses shown in Fig. 7.15 is ω_0, the vibrational energy is quantized according to the familiar formula

$$E_{\text{vib}} = (n + \tfrac{1}{2})\hbar\omega_0 \tag{62}$$

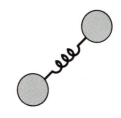

Fig. 7.15 Crude picture of a vibrating diatomic molecule.

We can make a rough estimate of the frequency ω_0 for a molecule by the following argument: Imagine we squeeze the atoms together, so that the distance between them is reduced by a factor of 2. Since the typical radius of each atom is the Bohr radius a_0, the initial separation is $\sim 2a_0$ and the displacement is $\sim a_0$. The energy stored in the spring is therefore $\sim \tfrac{1}{2}ka_0^2 \sim \tfrac{1}{2}M\omega_0^2 a_0^2$, where M is the mass of one atom. This spring energy is actually an energy of deformation of the electronic state of the atoms. A large deformation, such as contemplated above, is roughly equivalent to a transition from one electronic state to the next; typically, the energy difference is then of the order of magnitude of the energy differences found in hydrogen, $E_0 \sim (e^2/4\pi\varepsilon_0)^2 m_e/\hbar^2$. This is to be identified with the spring energy:

$$\tfrac{1}{2}M\omega_0^2 a_0^2 \sim E_0 \tag{63}$$

or

$$\sqrt{M}\,\omega_0 a_0 \sim \sqrt{E_0} \tag{64}$$

Solving this for ω_0 and inserting the expression for the Bohr radius, we find

$$\hbar\omega_0 \sim \frac{e^2}{4\pi\varepsilon_0}\frac{m_e}{\hbar}\frac{\sqrt{E_0}}{\sqrt{M}} \sim \sqrt{\frac{m_e}{M}}\,E_0 \tag{65}$$

For most atoms m_e/M is 10^{-4} or 10^{-5}. Hence the typical vibrational energy is about $10^{-2} E_0$. The typical wavelength emitted in a purely vibrational transition is therefore about 100 times larger than the typical wavelength emitted in an electronic transition—the typical wavelength lies in the infrared.

The energy eigenvalues for the rotational motion emerge from the quantization of angular momentum. The molecule rotates about an axis passing through the center of mass (see Fig. 7.16). If the moment of inertia about this axis is I, and the angular momentum $\mathbf{J}$, then

$$E_{\text{rot}} = \frac{|\mathbf{J}|^2}{2I} \tag{66}$$

The angular momentum must obey the usual quantization rule, $|\mathbf{J}^2| = J(J+1)\hbar^2$. Hence

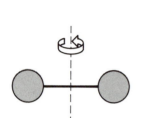

Fig. 7.16 Crude picture of a rotating diatomic molecule.

$$E_{\text{rot}} = \frac{J(J+1)\hbar^2}{2I} \qquad J = 0, 1, 2, \ldots \tag{67}$$

For a rough estimate, we can assume that the radial distance of each mass from the center of mass is $\sim a_0$, so that the moment of inertia of both masses together is $\sim 2Ma_0^2$ and

$$E_{\text{rot}} \sim \frac{J(J+1)\hbar^2}{2Ma_0^2} \tag{68}$$

In a typical transition, J will be of the order of magnitude of 1 and the change of rotational energy will therefore be of the order of magnitude of $\hbar^2/(2Ma_0^2)$,

$$\frac{\hbar^2}{2Ma_0^2} \sim \frac{\hbar^2}{2M} \left(\frac{e^2}{4\pi\varepsilon_0}\right)^2 \frac{m_e^2}{\hbar^4} \sim \frac{m_e}{M} E_0 \tag{69}$$

With $m_e/M \sim 10^{-4}$, this shows that the typical rotational energy is about $10^{-4} E_0$. The typical wavelength emitted in a purely rotational transition is therefore about 10^4 times larger than the wavelength emitted in an electronic transition—the typical wavelength lies in the far infrared, close to the microwave region.

In our calculation we have assumed that the molecule rotates about the transverse axis (see Fig. 7.16). Of course, the molecule can also rotate about the longitudinal axis. However, the moment of inertia about this axis is quite small (remember that most of the mass is concentrated in the very small nucleus); according to Eq. (67), the energy of the first excited state is therefore rather large. Hence, at ordinary temperatures, all the molecules remain in the ground state, and the rotation about the longitudinal axis remains absent.

Figure 7.17 shows the energy-level diagram for vibrational excited states of a diatomic molecule.* The spacing between these energy levels is constant. (However, for very large n the potential energy of the interatomic "springs" deviates from the harmonic-oscillator potential, and the spacing between the energy levels does not stay constant.)

Figure 7.18 shows the energy-level diagram for rotational excited states of the molecule. Note that the spacing between the energy levels increases. It is easy to show that the spacing is directly proportional to J:

$$E_{\text{rot},J} - E_{\text{rot},J-1} \propto J(J+1) - (J-1)J \propto 2J \qquad (70)$$

If a molecule is initially in a state of high J, it will make a sequence of transitions from J to $J-1$ to $J-2$ to $J-3$, etc.† The light emitted by a sample of molecules therefore displays a **band spectrum,** containing a sequence of wavelengths.

Although the wavelengths of the purely vibrational or rotational transitions are far from the visible region, the wavelengths of electronic transitions in the atoms accompanied by simultaneous vibrational or rotational transitions in the molecule often fall into the visible region or the ultraviolet region. For instance, Fig. 7.19 shows two electronic energy levels in a molecule and their modification by vibrational and rotational excitations. The arrows indicate a set of electronic-vibrational-rotational transitions; the transitions in this set involve a change in the electronic state and simultaneous changes in the vibrational and rotational states. This set of transitions gives rise to a spectral band, consisting of a sequence of closely spaced spectral lines. Figure 7.20 shows several such spectral bands seen in the spectrum of the NO molecule.

The simultaneous electronic-vibrational-rotational transitions are somewhat complicated to describe mathematically because both the vibrational frequency and the moment of inertia of the molecule depend on the electronic configuration. Thus, the spacings between the vibrational levels and the spacings between the rotational levels are different for each electronic state. Furthermore, it is necessary to take into account the couplings among the spins of the electrons, the orbital angular momenta of the electrons,

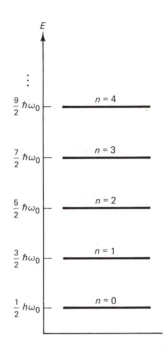

Fig. 7.17 Energy-level diagram for the vibrational excitations of a diatomic molecule.

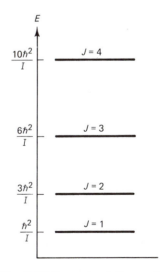

Fig. 7.18 Energy-level diagram for the rotational excitations of a diatomic molecule.

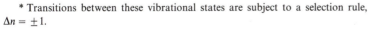

* Transitions between these vibrational states are subject to a selection rule, $\Delta n = \pm 1$.

† The transitions are subject to the selection rule $\Delta J = \pm 1$.

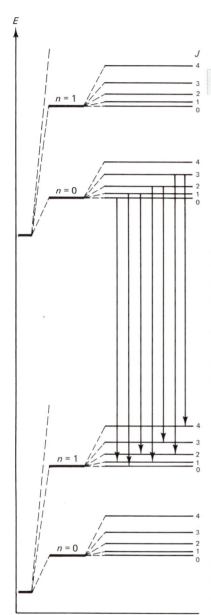

Fig. 7.19 Modification of electronic energy levels by vibrational and rotational excitations. For the sake of clarity, the separations between the vibrational levels ($n = 0, 1$) have been exaggerated; the separations between the rotational levels ($J = 0, 1, 2$, etc.) have been exaggerated even more. The transitions shown give a rotational band.

Fig. 7.20 Band spectrum of NO. (From G. Herzberg, *Molecular Spectra and Molecular Structure*. Courtesy G. Herzberg.)

and the orbital angular momenta of the nuclei. For the sake of simplicity, we will only deal with the case of vibrational-rotational transitions, i.e., transitions in which vibrational and rotational states change simultaneously, while the electronic state remains fixed. Such transitions give rise to absorption lines in the infrared, which can be readily observed with an absorption spectrometer that measures intensity as a function of wavelength for light transmitted through a sample of gas. At ordinary temperatures, the most likely initial vibrational state of the molecules in the sample of gas is $n = 0$. However, the molecules in the sample are usually distributed over a wide range of rotational states. A vibrational transition $n = 0$ to $n = 1$ accompanied by a simultaneous rotational transition J to $J - 1$ leads to the absorption of a photon of energy

$$E = \Delta E_{\text{vib}} + \Delta E_{\text{rot}} \tag{71}$$

$$= \hbar\omega_0 - 2J\frac{\hbar^2}{2I} \qquad J = 1, 2, 3, \ldots \tag{72}$$

Likewise, for the rotational transition J to $J + 1$,

$$E = \hbar\omega_0 + 2(J + 1)\frac{\hbar^2}{2I} \qquad J = 0, 1, 2, \ldots \tag{73}$$

Each of the two equations (72) and (73) describes a set of possible photon energies. Since the molecules in a sample of gas are distributed over a wide range of different values of J, each of these sets of energies gives rise to an **absorption band spectrum** with a sequence of wavelengths. Equation 72 yields an ascending sequence of wavelengths, and Eq. (73) a descending sequence of wavelengths.

A careful measurement of the wavelengths in band spectra permits a precise experimental determination of the vibration frequency ω_0 and of the moment of inertia I. Thus, the spectroscopy of bands is a valuable tool in the investigation of the molecular structure.

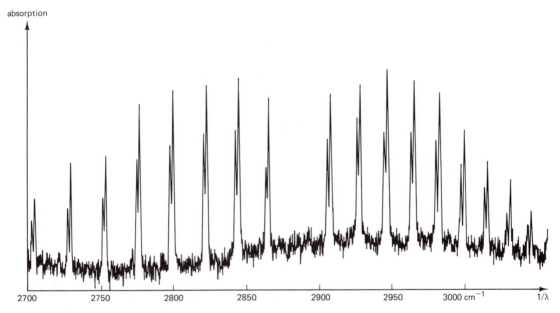

absorption

2700   2750   2800   2850   2900   2950   3000 cm^{-1} 1/λ

Fig. 7.21 Absorption band spectrum of HCl. Note that each spectral line consists of two closely spaced peaks; the higher peak is due to the molecules containing the isotope ^{35}Cl, the lower to molecules containing ^{37}Cl. This spectrum was obtained with a commercial infrared spectrometer. (Courtesy J. Anderson, Union College.)

Example 4   In the absorption band spectrum of the HCl molecule (see Fig. 7.21), the wavelength differences between adjacent spectral lines correspond to photon energy differences of 2.62×10^{-3} eV. What is the moment of inertia of the molecule? What is the center-to-center distance between the atoms? Assume that the Cl atom is the isotope ^{35}Cl.

Solution   The energies of the absorbed photons are given by Eq. (72) or (73). According to these equations, the energy difference between the photons belonging to two adjacent lines in the ascending or descending band is a constant. Evaluating this energy difference for, say $J = 1$ and $J = 2$ in Eq. (72), we find

$$\Delta E = 2 \times 2\frac{\hbar^2}{2I} - 2\frac{\hbar^2}{2I} = \frac{\hbar^2}{I} \tag{74}$$

Consequently,

$$I = \frac{\hbar^2}{\Delta E} = \frac{\hbar^2}{2.62 \times 10^{-3}\,\text{eV}} \tag{75}$$

$$= \frac{(1.055 \times 10^{-34}\,\text{J} \cdot \text{s})^2}{4.19 \times 10^{-22}\,\text{J}} = 2.66 \times 10^{-47}\,\text{kg} \cdot \text{m}^2 \tag{76}$$

From this we can deduce the center-to-center distance between the H and the Cl atoms. The distances of the hydrogen atom and of the chlorine atom from the center of mass are, respectively,

$$r_1 = \frac{rm_2}{m_1 + m_2} \quad \text{and} \quad r_2 = \frac{rm_1}{m_1 + m_2}$$

where r is the center-to-center distance, m_1 the mass of hydrogen, and m_2 the mass of chlorine. Hence

$$I = m_1 r_1^2 + m_2 r_2^2 = m_1 \left(\frac{rm_2}{m_1 + m_2} \right)^2 + m_2 \left(\frac{rm_1}{m_1 + m_2} \right)^2 \tag{77}$$

$$= r^2 \frac{m_1 m_2}{m_1 + m_2} \tag{78}$$

and

$$r = \sqrt{\frac{(m_1 + m_2)I}{m_1 m_2}} \tag{79}$$

$$= \sqrt{\frac{(1 + 35) \times 1.66 \times 10^{-27} \text{ kg} \times 2.66 \times 10^{-47} \text{ kg} \cdot \text{m}^2}{(1.66 \times 10^{-27} \text{ kg}) \times (35 \times 1.66 \times 10^{-27} \text{ kg})}}$$

$$= 1.28 \times 10^{-10} \text{ m} = 1.28 \text{Å} \quad \blacksquare \tag{80}$$

Example 5   Although a molecule will preferentially (resonantly) absorb a photon of an energy equal to the energy difference ΔE between the initial state and a final state, the molecule will sometimes absorb a photon of larger energy. Whenever this happens, the transition of the molecule from the initial to the final state will be accompanied by the emission of a new photon, which carries away the energy excess. The frequency of this new photon is given by

$$h v' = h v - \Delta E$$

where v is the frequency of the incident photon. The new frequency will be smaller than the original frequency if the molecule makes a transition to a higher excited state ($\Delta E > 0$), but it will be larger than the original frequency if the molecule is initially in an excited state and makes a transition to a lower state ($\Delta E < 0$). The net result is that some of the light incident on a sample of molecules emerges with a smaller or a larger frequency, i.e., the light is scattered with a changed frequency. This is called the **Raman effect.*** The intensity of the scattered light produced by this effect is much less than the intensity produced by ordinary, or

---

* **Chandrasekhara Venkata Raman,** 1888–1970, Indian physicist, professor at Calcutta, founder of the *Indian Journal of Physics*, and director of the Raman Research Institute at Bangalore. He received the Nobel Prize in Physics in 1930 for his discovery of the Raman effect.

Rayleigh, scattering (without change of frequency); however, the Raman effect is a very useful tool for the experimental investigation of rotational and vibrational energy levels of molecules.

Suppose that light of frequency v from a laser is incident on a sample of molecules of some given moment of inertia. Calculate what frequencies will be present in the scattered light. Take into account only rotational transitions, and use the selection rule $\Delta J = \pm 2$, which applies for rotational transitions in the Raman effect.

Solution   With $\Delta J = -2$, Eq. (67) yields

$$\Delta E = (J-2)(J-1)\frac{\hbar^2}{2I} - J(J+1)\frac{\hbar^2}{2I} = -(2J-1)\frac{\hbar^2}{I}$$

Hence

$$v' = v - \frac{\Delta E}{h} = v + (2J-1)\frac{\hbar}{2\pi I} \qquad J = 2, 3, 4, \ldots$$

Likewise, with $\Delta J = +2$,

$$v' = v - \frac{\Delta E}{h} = v - (2J+3)\frac{\hbar}{2\pi I} \qquad J = 0, 1, 2, \ldots$$

The scattered light will contain components at these shifted frequencies produced by the Raman effect, and also a component at the original frequency produced by ordinary scattering (the latter component will be the most intense).   ■

7.7   Stimulated Emission; the Laser

Under ordinary conditions, an atom in an excited state will spontaneously make a transition to a lower state emitting light. This process is called **spontaneous emission** because it does not require any outside stimulus. The process is essentially probabilistic, much like radioactive decay: in some given time interval the atom has some probability for making the transition. The time interval that corresponds to a 50% probability of transition is called the half-life of the excited state. The probability and the half-life can be calculated by the methods of advanced quantum mechanics. Typically, the half-life for an atomic transition in the visible region is about 10^{-8} second. However, if the transition is "forbidden" by a selection rule, then the lifetime is much longer.

By a simple theoretical argument (see below) we can establish that atoms emit light not only by spontaneous emission, but also by **stimulated emission.** The process of stimulated emission involves the following: an electromagnetic wave, generated somewhere else, is incident on an atom and disturbs its electrons; if the atom is in an excited state, the disturbance caused by the wave triggers a transition from the excited state to a lower state, i.e., the wave provides a stimulus for the transition. The process of stimulated emission is closely related to **stimulated absorption.** In the latter process, an electromagnetic wave is incident on an atom in the lower state and triggers a transition to the excited state. Note that in stimulated absorption, the atom absorbs a photon from the wave, whereas in stimulated emission the atom ejects a photon into the wave, thereby increasing the intensity of the wave. In either case, the energy of the photon must be equal to the energy difference $E_2 - E_1$ between the states, i.e.,

$$hv = E_2 - E_1 \tag{81}$$

Thus, stimulated absorption and stimulated emission are resonance processes—they will occur only if the frequency of the wave has the particular value given by Eq. (81). Figure 7.22 shows transitions involving stimulated absorption and stimulated emission in energy-level diagrams.

The argument that establishes the existence of stimulated emission is as follows: Consider a sample of atoms immersed in blackbody radiation, with which they are in thermal equilibrium. Atoms in the lower state (E_1) will occasionally absorb radiation and make a transition to the upper state (E_2). Atoms in the upper state will occasionally emit radiation and make a transition to the lower state. In thermal equilibrium, the numbers of atoms in the lower and the upper states are fixed by the Boltzmann factor; to keep these numbers constant, the rates of upward and downward transitions must be equal. The crux of our argument is that these two transition rates cannot be equal unless there is stimulated emission. The rate of upward transitions depends on the intensity of the blackbody radiation—more intense radiation is a stronger stimulus and therefore leads to more absorption—and hence the rate of downward transitions must also depend on the intensity of the radiation, which means there must be stimulated emission.

By means of a calculation first carried out by Einstein in 1916, we can obtain a quantitative relation between spontaneous emission, stimulated emission, and stimulated absorption. The rate of stimulated absorptions, and therefore the rate of upward transitions, is proportional to the spectral energy density u_v of the blackbody radiation, and also proportional to the number n_1 of atoms

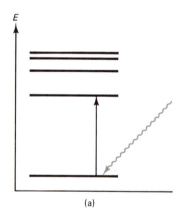

(a)

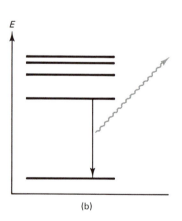

(b)

Fig. 7.22 (a) Stimulated absorption; (b) stimulated emission.

in the lower state,

$$\begin{pmatrix} \text{rate of upward} \\ \text{stimulated transitions} \end{pmatrix} = B_{12} n_1 u_v \qquad (82)$$

Here B_{12} is a constant of proportionality, called Einstein's B coefficient. Likewise, the rate of stimulated downward transitions is proportional to the energy density of the radiation, and to the number n_2 of atoms in the upper state:

$$\begin{pmatrix} \text{rate of downward} \\ \text{stimulated transitions} \end{pmatrix} = B_{21} n_2 u_v \qquad (83)$$

where B_{21} is another constant. The rate of spontaneous downward transitions is independent of u_v, but it is still proportional to n_2:

$$\begin{pmatrix} \text{rate of downward} \\ \text{spontaneous transition} \end{pmatrix} = A_{21} n_2 \qquad (84)$$

Here A_{21} is yet another constant, called Einstein's A coefficient. Equating the upward rate to the sum of the downward rates, we obtain

$$n_1 B_{12} u_v = n_2 (B_{21} u_v + A_{21}) \qquad (85)$$

In thermal equilibrium, the numbers n_1 and n_2 are proportional to Boltzmann factors $e^{-E_1/kT}$ and $e^{-E_2/kT}$, respectively. Hence

$$e^{-E_1/kT} B_{12} u_v = e^{-E_2/kT} (B_{21} u_v + A_{21}) \qquad (86)$$

If we take the limit $T \to \infty$ in this equation, the Boltzmann factors will become equal, and the energy density u_v will become very large [u_v increases with temperature; see Eqs. (3.6) and (3.7)], so that A_{21} can be neglected compared to $B_{21} u_v$ (the Einstein coefficients are independent of temperature). Hence, Eq. (86) reduces to

$$B_{12} = B_{21} \qquad (87)$$

This means that the probabilities for stimulated emission and for stimulated absorption are equal—an atom in the excited state is just as likely to make a downward stimulated transition as an atom in the lower state is to make an upward stimulated transition.

Note that if we solve Eq. (86) for u_v, and take into account Eqs. (81) and (87), we obtain

$$u_v = \frac{A_{21}}{B_{12} e^{-(E_1 - E_2)/kT} - B_{21}} = \frac{A_{21}/B_{21}}{e^{hv/kT} - 1} \qquad (88)$$

This expression must agree with the Planck formula, Eq. (3.7):

$$u_\nu = \frac{4}{c} S_\nu = \frac{8\pi\nu^2}{c^3} \frac{h\nu}{e^{h\nu/kT} - 1} \tag{89}$$

The agreement demands*

$$\frac{A_{21}}{B_{21}} = \frac{8\pi h\nu^3}{c^3} \tag{90}$$

Hence the probabilities for stimulated emission and for spontaneous emission are proportional—an atom highly susceptible to spontaneous emission is also highly susceptible to stimulated emission.

A remarkable feature of the photon emitted by stimulated emission is that it always has the same polarization and the same phase as the incident wave. Thus, the photon enhances the amplitude of the incident wave. This suggests that electromagnetic waves of extremely high amplitude could be generated by the combined stimulated emissions from a large sample of atoms. However, if the atoms are in thermal equilibrium, the incident wave loses more energy by absorption than it gains by emission since, according to the Boltzmann factor, the lower states of the atoms are more abundantly populated than the upper states ($n_1 > n_2$), and therefore the rate of stimulated absorption exceeds the rate of stimulated emission. To achieve a net gain in the energy of the wave, we need a sample of atoms with an excess population in the upper states ($n_2 > n_1$). Such an abnormal condition, which deviates drastically from thermal equilibrium, is called a **population inversion.** It can only be established by artificial means—we must somehow "pump" an excess of atoms into the upper state by supplying energy from some external source.

Stimulated emission from an inverted population of atoms or molecules is the basic mechanism exploited in the operation of masers and lasers.[†] The first maser was built by C. H. Townes in

---

* Einstein regarded his calculation leading to Eq. (88) as a derivation of Planck's law. He derived Eq. (90) not by comparison with Planck's law, but by comparison with the classical blackbody-radiation formula obtained by Rayleigh, $u_\nu = 8\pi\nu^3 h/c^3$, with which Eq. (88) must coincide in the high-temperature limit.

[†] These neologisms originated as acronyms standing for *m*icrowave *a*mplification by *s*timulated *e*mission of *r*adiation, and for *l*ight *a*mplification by *s*timulated *e*mission of *r*adiation.

1954 and the first laser by T. H. Maiman in 1960.* Figure 7.23 is a schematic diagram of Maiman's laser. It consists of a rod of ruby crystal surrounded by a flash lamp. The red color of the ruby crystal is due to the presence of chromium impurities. In the ruby laser, only the chromium atoms participate in stimulated emission. The flash lamp supplies the energy for pumping chromium atoms into the excited state. Figure 7.24 shows the relevant energy levels of chromium. The intense flash of light from the flash lamp kicks a large number of chromium atoms into the highest energy level 3. Some of these atoms make spontaneous transitions to the intermediate level 2. This level is **metastable,** i.e., it has a low probability for spontaneous decay. Thus a large population accumulates in this level, and the result is a population inversion between levels 1 and 2. Note that the population inversion can only be achieved by this indirect route, via level 3. If we were to use a flash lamp to produce direct stimulated transitions from level 1 to level 2, we would not achieve a population inversion because such upward stimulated transitions would be accompanied by (premature) downward stimulated transitions, and the population in level 2 would not accumulate to the extent required for our purposes.

The excited chromium atoms in level 2 are now ready to begin the lasing action. The first chromium atom that makes a spontaneous downward transition to level 1 radiates a light wave, which triggers stimulated emission by the other excited chromium atoms. In order to ensure that all, or almost all, of the excited chromium atoms participate in stimulated emission, it is necessary to make the light wave travel through the ruby rod several times so that any excited atoms not triggered on the first pass are triggered on a later pass. For this purpose, the ends of the rod are carefully polished and silvered, making them into mirrors. One end is only partially silvered, so that the light can ultimately escape. The back-and-forth reflections strongly enhance the lasing action for those light waves that are emitted exactly parallel to the axis of the rod and that have a wavelength fitting into the length of the rod

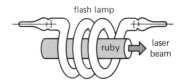

Fig. 7.23 Ruby laser of T. H. Maiman.

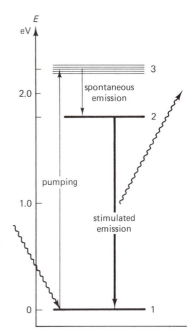

Fig. 7.24 Energy levels of chromium that participate in the action of the laser.

---

* **Charles H. Townes,** 1915– , American physicist, professor at Columbia University, the Massachussetts Institute of Technology, and Berkeley. For his invention of the maser, he shared the Nobel Prize in 1964 with **Nikolai G. Basov,** 1922– , and **Alexander M. Prochorov,** 1916– , Soviet physicists, who made the same invention independently. Townes and A. Schawlow (see p. 302) formulated the theoretical principles for the operation of the laser, and **Theodore Harold Maiman,** 1927– , American physicist, built the first laser while working at the Hughes Research Laboratories.

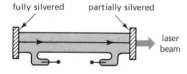

Fig. 7.25 He-Ne laser.

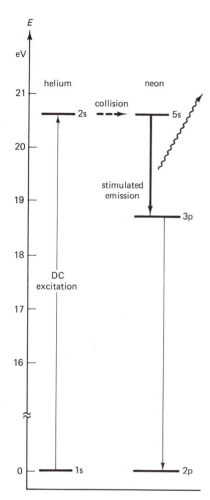

Fig. 7.26 Energy levels of helium and of neon.

exactly an integral number of times. For such light waves, the rod acts as a resonant cavity oscillating in one of its normal modes. In this cavity, the light wave is a standing wave, and the stimulated emissions by the excited atoms increase the amplitude of this standing wave. Ultimately the wave leaks out of the half-silvered end of the ruby rod, forming the external, usable beam of the laser.

The wave emerging from the laser is **coherent**: it is a single plane wave in which all of the individual waves emitted by the individual atoms have combined in phase. The net amplitude of the wave is proportional to the number of atoms that contributed to it, and the net intensity is proportional to the square of the number of atoms. By contrast, ordinary light emitted by, say, a neon tube is **incoherent**: it contains a superposition of many waves of random phases. The net intensity of such incoherent light is proportional to the number of atoms, and not to the square of the number of atoms. Since the number of atoms in a light source is very large, coherent emission yields an enormously higher intensity than incoherent emission. Furthermore, the light emerging from the laser is extremely well collimated (unidirectional) because it originated from a light wave traveling exactly parallel to the axis of the rod. Typically, the beam of a laser has an angular spread of no more than a minute of arc.

The beam from a ruby laser emerges as a pulse that lasts only as long as excited atoms remain available. When the excited atoms become depleted, the lasing action ends, and must be started again by another flash from the flash lamp. Some other kinds of lasers are capable of continuous operation. The most popular laser of this kind is the helium–neon laser. It consists of a glass tube containing a mixture of helium and neon at low pressure (see Fig. 7.25). One end of this tube is silvered, the other partially silvered. Electrodes near the ends are connected to a source of high voltage. The stimulated emission is due entirely to the neon; the helium merely serves to pump the neon. The source of high voltage pushes a current of electrons through the tube. These electrons collide with the helium atoms and excite them. The excited helium atoms in turn collide with the neon atoms. In such a collision, the energy of the helium atom is likely to be transferred to the neon atom, which has an excited level of roughly the same energy as that of helium (see the level 5s in Fig. 7.26). An inverted population accumulates in this level of the neon atoms. The lasing action is due to stimulated transitions from level 5s to level 3p. From the latter level, the neon atom makes a quick, spontaneous transition to the ground state. It is then ready to be pumped again.

Lasers have a wide range of practical and scientific applications. For instance, the development of **stabilized lasers** has recently led to a new definition of the meter, our unit of length. These lasers contain a feedback mechanism that continually compares the wavelength emitted by the laser with the wavelength of a reference spectral line, usually a spectral line of iodine. Whenever there is an incipient deviation between these wavelengths, the feedback mechanism makes a slight adjustment to the length of the laser cavity, to restore the laser wavelength. Stabilized lasers developed by the National Bureau of Standards maintain a constant wavelength and frequency to within 2 parts in 10^{12}. These lasers were used to measure the speed of light with unprecedented precision: the frequency of the laser was measured by comparison with the cesium atomic clock (the standard of time), and the wavelength was measured by comparison with the ^{86}Kr wavelength (the standard of length); from this, the speed of light can be calculated as $c = \lambda v$. The result of these measurements was

$$c = 299792458 \, \text{m/s} \tag{91}$$

with an uncertainty of ± 1 m/s. Most of this uncertainty is due to the intrinsic lack of sharpness of the ^{86}Kr spectral line. Like all atomic spectral lines, this line has a finite width, arising from the wave-mechanical energy uncertainty of the photon.* To circumvent the limitations set by the lack of sharpness of the ^{86}Kr spectral line, the General Conference on Weights and Measures in 1983 adopted a new definition of the meter: *One meter is the distance traveled by light in vacuum in a time interval of 1/299792458 of a second.* This definition takes the speed of light as a standard and adjusts the definition of the meter so that the speed of light always retains the standard value given by Eq. (91). Any new measurements of the speed of light will therefore have to be regarded as new determinations of the meter. Note that the new definition of the meter does not specify the reference frame used for fixing the standard speed of light. This is in accord with the theory of special relativity, according to which the speed of light is independent of the choice of (inertial) reference frame.

In practice, the new definition of the meter is implemented by means of stabilized lasers. The frequency of such a laser is measured by comparison with the cesium standard clock, and the

---

* The energy–time uncertainty relation implies that $\Delta E \sim \hbar / \Delta t$, where Δt is the half-life of the excited state.

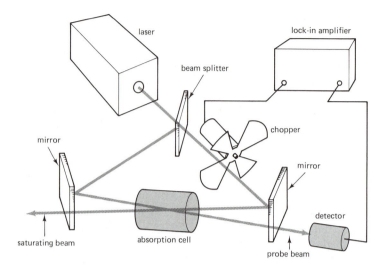

Fig. 7.27 Saturation spectrometer of Hänsch, Shahin, and Schawlow. The absorption cell contains a sample of the atoms under investigation. Both the saturating beam and the probe beam pass through this sample, but in opposite directions. The lock-in amplifier picks up the portion of the probe signal that is modulated at the chopper frequency. (After A. L. Schawlow, *Physics Today*, December 1982.)

wavelength is then *calculated* using $\lambda = c/v$ and the standard speed of light. This gives the wavelength of the laser in meters. Any other wavelength, or any other length, can then be measured by an interferometric comparison with the laser wavelength.

Lasers have an important scientific application in spectroscopy, where they have led to spectacular achievements in recent years.* The lasers used for this purpose are **tunable lasers,** which operate with molecules of organic dyes. These molecules have a very large number of closely spaced, overlapping spectral lines and they therefore are capable of emitting light over a continuous range of wavelengths. By means of a selective device attached to the laser, the light can be sharply tuned to any chosen wavelength in this range. When such a laser is used to illuminate a sample of atoms, the light will be resonantly absorbed if the wavelength matches that of one of the possible atomic transitions. This resonant absorption is exploited in **Doppler-free saturation spectroscopy,** a clever technique for eliminating the blurring or broadening of spectral lines caused by the random motion of the atoms in the sample. Figure 7.27

---

* **Arthur L. Schawlow,** 1921– , and **Nicholaas Bloembergen,** 1920– , American physicists, professors at Stanford and at Harvard, respectively, were awarded the Nobel Prize in 1981 for their contributions to the development of laser spectroscopy.

shows the experimental arrangement. The light from the tunable laser is split into two beams, the saturating beam and the probe beam, which pass through the sample of atoms in opposite directions. The first of these beams is periodically interrupted by a chopper. When this saturating beam is off, the probe beam is absorbed by the sample. When the saturating beam is on, it saturates the sample, i.e., it triggers transitions from the ground state to the excited state and thereby depletes the population of atoms in the ground state. The probe beam then encounters no atoms in the ground state and it passes through the sample without absorption. Thus, the intensity of the probe beam is modulated at the chopper frequency. If we adjust the wavelength of the tunable laser so as to maximize the modulation, we achieve resonance for both the saturating beam and the probe beam. But the only atoms that can be in resonance with two beams of light incident from opposite directions are those that have no velocity component along the beams—if an atom has a velocity component along the beams, then the wavelengths incident from opposite directions are Doppler shifted by opposite amounts in the rest frame of the atom, and the beams cannot both be in resonance. Hence, the experimental procedure selects atoms that are instantaneously free of Doppler shifts, and it therefore makes the spectral lines extremely sharp.

Saturation spectroscopy has been used to good advantage with hydrogen atoms, whose small mass and consequent large Doppler broadening severely limit the resolution attainable by ordinary spectroscopy. This has led to an improved value of the Rydberg constant. It has also revealed directly for the first time the fine structure and the hyperfine structure of the spectral lines of hydrogen. For instance, Fig. 7.28 displays the structure of the Balmer α line (6565 Å), as obtained by a saturation spectrometer at Stanford University. The fine structure is due to spin-orbit coupling, and the hyperfine structure is due to the combined effects of nuclear spin, relativistic corrections in the motion of the electron, and quantum fluctuation effects. The Lamb shift* displayed in Fig. 7.28 arises from these latter effects, which can be calculated from the relativistic quantum theory of electromagnetic interactions, or quantum electrodynamics (QED). To gain some appreciation for the amazing resolution attained in Fig. 7.28, note that on a scale of wavelengths, the Lamb shift is only 0.015 Å.

---

* This shift was first measured with a method relying on the resonant absorption of radio waves, by **Willis Eugene Lamb,** 1913–   , American physicist, professor at Columbia, Stanford, and Yale, who thereby gained the 1955 Nobel Prize.

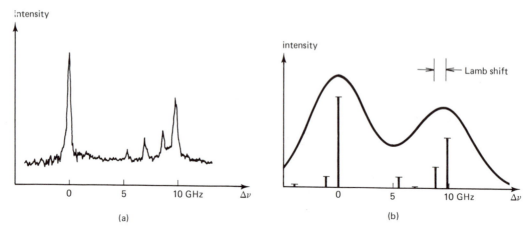

Fig. 7.28 (a) Fine structure and hyperfine structure of the Balmer α line, as obtained with the saturation spectrometer. (b) Spectrum calculated from theory. The heights of the lines indicate the intensities. (Courtesy T. W. Hänsch, Stanford University.)

Chapter 7: SUMMARY

Spin of electron:

$$|\mathbf{s}| = \sqrt{\tfrac{3}{4}}\,\hbar$$

$$s_z = \pm\tfrac{1}{2}\hbar$$

Addition of orbital and spin angular momentum of electron:

$$\mathbf{J} = \mathbf{L} + \mathbf{s}$$

$$j = l + \tfrac{1}{2} \quad \text{or} \quad l - \tfrac{1}{2} \quad (j = \tfrac{1}{2} \text{ if } l = 0)$$

Spin magnetic moment of electron:

$$\boldsymbol{\mu}_{\mathrm{e}} = -\frac{e}{m_{\mathrm{e}}}\,\mathbf{s}$$

$$\mu_{\mathrm{e},z} = \mp\frac{e}{m_{\mathrm{e}}}\frac{\hbar}{2}$$

Russell–Saunders coupling:

$$\mathbf{L} = \sum_i \mathbf{L}_i, \qquad |\mathbf{L}| = \sqrt{L(L+1)}\,\hbar$$

$$\mathbf{S} = \sum_i \mathbf{s}_i, \qquad |\mathbf{S}| = \sqrt{S(S+1)}\,\hbar$$

$$\mathbf{J} = \mathbf{L} + \mathbf{S}, \qquad |\mathbf{J}| = \sqrt{J(J+1)}\,\hbar$$

Spectroscopic-term notation:

$$^{2S+1}L_J$$

Orbital magnetic moment of electron:

$$\boldsymbol{\mu}_L = -\frac{e}{2m_e}\mathbf{L}$$

Net magnetic moment of atom (z-component):

$$\mu_z = -gm_J\frac{e\hbar}{2m_e} \qquad m_J = J, J-1, \ldots, 1-J, -J$$

Landé g-factor:

$$g = 1 + \frac{J(J+1) + S(S+1) - L(L+1)}{2J(J+1)}$$

Energy shift for "anomalous" Zeeman effect:

$$U = -\mu_z B = gm_J\frac{e\hbar B}{2m_e}$$

Energy shift for "normal" Zeeman effect:

$$U = -\mu_z B = (m_L + 2m_S)\frac{e\hbar B}{2m_e}$$

Pauli's Exclusion Principle: No more than one electron can be in any given orbital and spin state of a system.

Molecular energy levels:

$$\text{Vibrational:} \quad E_{\text{vib}} = (n + \tfrac{1}{2})\hbar\omega_0$$

$$\text{Rotational:} \quad E_{\text{rot}} = \frac{J(J+1)\hbar^2}{2I}$$

Raman effect:

$$h\nu' = h\nu - \Delta E$$

Chapter 7: PROBLEMS

1. Using a naive classical model, pretend that the electron is a rigid sphere of radius 10^{-18} m with a uniform mass density. From the known mass and the known spin angular momentum $s_z = \hbar/2$, calculate the rotational speed at the equator.

2. If we use a beam of carbon atoms in a Stern–Gerlach experiment, how many distinct beams will emerge from the magnet? If we use a beam of nitrogen atoms? If we use a beam of oxygen atoms?

3. The separation between the upper and the lower beams of atoms emerging from the magnet in a Stern–Gerlach experiment depends on the inhomogeneity of the magnetic field and on the length of the magnet. What value of $\partial B_z/\partial z$ is required to produce a separation of 1 mm between the upper and the lower beams

of hydrogen atoms emerging from a magnet 12 cm long if the speed of the atoms is 600 m/s?

4. Suppose that an electron in an atom is in a state with $l = 3$. What is the magnitude of the vector **L**? What are the possible quantum numbers j and the possible magnitudes of the vector $J = L + s$?

5. An electron in an atom is in a state with $l = 1$ and $j = \frac{3}{2}$ (see Fig. 7.2).
(a) Calculate the angle between the vectors **L** and **s**.
(b) Calculate the magnitude of the magnetic moment vector $\mu = -(e/2m_e) \times (L + 2s)$.

6. Give the spectroscopic term notation for each of the lowest eight excited states of hydrogen indicated in the energy-level diagram of Fig. 7.1.

7. The Ω^- particle has a negative charge $-e$ and a spin $s = \frac{3}{2}$. If such a particle is in an orbit of orbital angular momentum quantum number l around a heavy atomic nucleus, what are the possible values of the quantum number j? Which of the states of different values of j has the lowest energy? What is the spectroscopic term for the ground state?

8. According to Example 6.2, if a charged particle is in a superposition of two different states in a box, the particle creates an oscillating charge distribution. If the charge on the particle is q, the charge density of the charge distribution is $q|\psi(x, t)|^2$. For this charge distribution, the electric dipole moment is defined as $\int xq|\psi(x, t)|^2 \, dx$.
(a) Consider a particle in the superposition of the ground and the first excited states described in Example 6.2. Evaluate the dipole moment.
(b) Consider a particle in a superposition of the ground and the *second* excited state. Evaluate the dipole moment for this superposition and show that this dipole moment *does not* oscillate. Consequently, is the transition from the second excited state to the ground state allowed or forbidden?

9. Consider an atom with $L = 1$, $J = \frac{3}{2}$, and $m_j = +\frac{3}{2}$. On the basis of Figure 7.2, calculate the angle between **L** and **S**. Calculate the magnitude of the magnetic moment μ. Calculate the angle between μ and the z-axis. Calculate μ_z. Does your value for μ_z agree with the value obtained from the Landé formula?

10. Calculate the magnetic moments of the atoms of the He, Li, Be, and B in their ground states (see Section 7.5 for their spectroscopic terms).

11. The energy-level diagram for sodium in Fig. 7.10a shows six different transitions from the $^2P_{3/2}$ state into the ground state. For each of these transitions, list the values of ΔJ, ΔL, ΔS, and Δl. Do these values satisfy the selection rules given in Section 7.2?

12. If sodium is in a magnetic field, the spectral line at 5897.6 Å splits into a quadruplet (see Example 1). Calculate the wavelength shift, relative to the undisturbed line, for each member of this quadruplet, if the magnetic field is 0.2 T. Express your answer in Å.

13. Prepare a diagram similar to Fig. 7.10a showing the Zeeman energy shifts for a $^2D_{3/2}$ level of sodium. Calculate the energy shift (relative to the undisturbed energy level) for each of the new energy levels in a magnetic field of 0.1 T. Express your answers in eV.

14. Sodium has several excited energy levels of the type $^2P_{3/2}$. If sodium is placed in a magnetic field, each such excited level splits into several levels. For a magnetic field of 0.2 T, calculate the energy shifts of these new levels relative to the original undisturbed level. Express the energies in eV.

15. The spectral line $\lambda = 584.4$ Å in the spectrum of helium is due to a transition from a $^1P_1$ level to a $^1S_0$ level. If we place the sample of helium atoms in a magnetic field of 0.4 T, this spectral line splits into a multiplet of lines. Calculate the wavelengths shifts of the members of this multiplet relative to the original spectral line.

16. The yellow sodium doublet ($\lambda = 5895.9$ Å and $\lambda = 5889.9$ Å in air) consists of two spectral lines separated by a fine-structure splitting of about 6 Å. If the sodium atom is placed in an external magnetic field, roughly what field strength is required to achieve a Zeeman splitting comparable to the fine-structure splitting?

17. If sodium is placed in a magnetic field of 15 T, the spectral line D_1 ($\lambda = 5897.6$ Å) splits into a "normal" Zeeman multiplet. Calculate the wavelengths of the members of this multiplet.

18. Suppose that two electrons are in a box, one in the ground state and one in the first excited state. If the z-components of the spins of the two electrons are opposite, the possible wavefunctions are

$$\psi_1(x)\psi_2(x')|m_s=\tfrac{1}{2}\rangle\,|m_{s'}=-\tfrac{1}{2}\rangle + \psi_2(x)\psi_1(x')|m_s=\tfrac{1}{2}\rangle\,|m_{s'}=-\tfrac{1}{2}\rangle$$
$$-\psi_1(x)\psi_2(x')|m_s=-\tfrac{1}{2}\rangle\,|m_{s'}=\tfrac{1}{2}\rangle - \psi_2(x)\psi_1(x')|m_s=-\tfrac{1}{2}\rangle\,|m_{s'}=\tfrac{1}{2}\rangle$$

and

$$\psi_1(x)\psi_2(x')|m_s=\tfrac{1}{2}\rangle\,|m_{s'}=-\tfrac{1}{2}\rangle - \psi_2(x)\psi_1(x')|m_s=\tfrac{1}{2}\rangle\,|m_{s'}=-\tfrac{1}{2}\rangle$$
$$+\psi_1(x)\psi_2(x')|m_s=-\tfrac{1}{2}\rangle\,|m_{s'}=\tfrac{1}{2}\rangle - \psi_2(x)\psi_1(x')|m_s=-\tfrac{1}{2}\rangle\,|m_{s'}=\tfrac{1}{2}\rangle$$

(a superposition of these two wavefunctions is also a possible wavefunction).
(a) Check that each of these wavefunctions satisfies the antisymmetry requirement.
(b) Which of these wavefunctions is zero for $x = x'$?

19. Three electrons are confined in a one-dimensional box of length L. Find the lowest energy of this system; ignore the electric repulsion between the electrons. Find the net spin of the state of lowest energy.

20. Two neutral pions (π^0) are in the ground state of a one-dimensional box of length L, between $x = 0$ and $x = L$. Pretend that the pions do not interact.
(a) What is the wavefunction for this system of two pions? Include the correct normalization for the wavefunction.
(b) What is the probability for finding the two pions in the interval $L/2 - dx/2$ to $L/2 + dx/2$?
(c) What is the probability for finding one pion in this interval, regardless of the location of the other pion?

21. Consider two electrons in different p states of an atom. Prepare a list of all the possible spectroscopic terms that can result from combining the orbital and spin angular momenta of these electrons in diverse ways.

22. The spectroscopic terms for the lowest excited states of the He atom are $^3S_1$, $^1S_0$, $^3P_0$, and $^1P_1$. What are the corresponding configurations of the two electrons? Assume that only one electron is raised into an excited state.

23. The spectroscopic terms for some of the lowest excited states of the Li atom are $^2P_{1/2}$, $^2S_{1/2}$, and $^2D_{3/2}$. What are the corresponding configurations of the three electrons? Assume that only the outermost electron is raised into an excited state.

24. What is the spectroscopic term for the ground state of the Cl^- ion?

25. By means of arguments similar to those used in Section 7.5, determine the electronic configurations and the spectroscopic terms for the ground states of magnesium, aluminum, and silicon.

26. For each of the columns of the Periodic Table, list the typical, or most popular, spectroscopic term of the ground state (e.g., $^1S_0$ for the noble gases, $^2S_{1/2}$ for the alkalis, etc.).

27. Give rough numerical estimates for the energies of the first excited vibrational and the first excited rotational states of a H_2 molecule.

28. From spectroscopic observations, the energy difference between the ground state and the second excited rotational state of the N_2 molecule is known to be $14.9 \times 10^{-4}\,eV$. Deduce the moment of inertia of the molecule. Deduce the center-to-center distance between the N atoms.

29. Figure 7.29 shows the dimensions of the H_2O molecule. Calculate the moment of inertia of this molecule about the axis of symmetry (indicated by the dashed line). Calculate the energy of the first excited rotational state; express your answer in eV.

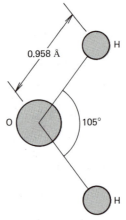

0.958 Å

O   105°   H   H

Fig. 7.29

30. Suppose that a carbon atom rotates rigidly about an axis through its center. Estimate the energy of the first excited rotational state; express your answer in eV. (Assume that the electrons in carbon are uniformly distributed over a sphere of radius equal to the Bohr radius, and treat the atom as a spherical mass distribution; ignore the nucleus.)

31. Monochromatic light of wavelength 4358 Å from a mercury lamp is incident on a sample of HCl molecules. Calculate the wavelengths of the four lines in the rotational Raman spectrum that are closest to the wavelength of the incident light.

32. When light of wavelength 6328 Å from a laser is incident on a sample of HCl molecules, two spectral lines of 5321 Å and 7805 Å are observed in the scattered light. These spectral lines arise from the Raman effect in vibrational transitions, with the selection rule $\Delta n = \pm 1$. What value of the vibrational frequency of the HCl molecule can you deduce from this?

Electrons in Solids

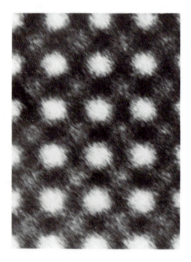

Fig. 8.1 Regular arrangement of atoms in a thin crystal of cubic barium titanate. The large spots are titanate ions and the small spots barium ions. This picture was taken with a new, extremely powerful electron microscope. (Courtesy Lawrence Berkeley Laboratory.)

In all solids the building blocks, whether molecules or atoms, are permanently and almost rigidly locked into their positions. Crystalline solids, such as metals and minerals, are distinguished from amorphous solids, such as glasses and plastics, by the regularity of the arrangement of their building blocks. Their atomic or molecular building blocks form an orderly, repetitive lattice (see Fig. 8.1). Thus, we can regard metals and other crystalline solids as giant molecules in which the atoms are held in specific places according to a well-defined plan, just as in an ordinary molecule. These giant molecules are polymers, but they are three-dimensional polymers rather than ordinary, stringlike, one-dimensional polymers.

The bonds that hold the atoms in their places within a crystalline solid are essentially the same as the bonds that hold the atoms within an ordinary molecule. Ultimately, these bonds arise from the Coulomb interactions among the negatively charged electrons and the positively charged nuclei. Whenever two atoms are in close contact, the electron distribution of each is disturbed by the presence of the other. This redistribution of the electrons usually leads to a binding force between the atoms. Depending on just how the electrons are redistributed, we speak of ionic bonds, covalent bonds, van der Waals bonds, hydrogen bonds, or metallic bonds. All of these kinds of bonds, with the exception of metallic bonds, occur in ordinary molecules as well as in solids.

The characteristic feature of metals is the presence of an abundance of free electrons. When the atoms of a metal join in a crystal, the outermost, or valence, electrons become detached from their atoms. These electrons are then more or less free to roam through the entire volume of the metal—they form a "gas" of free electrons which fills the entire volume of the metal. As we will see, this gas is not only responsible for the bonds that hold the ions in their places in the lattice, but it is also responsible for the electrical properties of metals. A similar gas of free electrons is responsible for the remarkable electrical properties of semiconductors, of such crucial importance in our electronic technology.

8.1 Bonds Between Atoms

In this section we will briefly discuss the kinds of bonds that hold the atoms in a solid. We will keep the discussion qualitative. A quantitative treatment of the theory of bonds requires the full machinery of wave mechanics and is quite difficult. Also, we had better keep in mind that in some crystals two of the bonding mechanisms may be present simultaneously. For instance, in AgCl the bonds are of an intermediate kind, partially ionic and partially covalent.

Ionic Bonds. Some atoms, such as Na, tend to lose electrons easily; some others, such as Cl, tend to acquire extra electrons. When atoms of these two kinds interact, the rearrangement of the electron distribution is very simple: an electron from an atom of the first kind becomes attached to an atom of the second kind. Thus the atoms become positive and negative ions, such as Na^+ and Cl^-. The bond between these ions is then simply due to the Coulomb attraction between the positive and the negative charges. Since both Na^+ and Cl^- have the spherically symmetric electron configuration of the atoms of noble gases, their Coulomb interaction is that of two point charges, with a potential energy $U = -e^2/(4\pi\varepsilon_0 r)$ corresponding to an attractive, inverse-square force. This Coulomb interaction is responsible for holding the Na^+ and Cl^- ions in the NaCl lattice (see Fig. 8.2).

However, if the distance between the atoms is very small, so that the wavefunctions of the inner electrons of the two atoms overlap, then the potential energy is modified by a repulsive effect

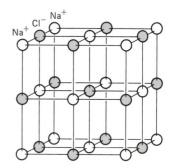

Fig. 8.2 Arrangement of Na^+ and Cl^- ions in the NaCl lattice.

arising from the Exclusion Principle. This can be most easily understood by considering the extreme case of complete overlap between the atoms, so that their nuclei coincide and the two atoms form a single, combined atom. In this case it is obviously impossible for both atoms to keep their electron configurations unchanged since this would mean that any state occupied in each of the separate atoms would be doubly occupied in the combined atom. Consequently, the Exclusion Principle requires that whenever atoms overlap, electrons must be forced into formerly unoccupied states, of higher energy. This effect occurs not only in the case of complete overlap, but also in the case of partial overlap (the latter case is rather more complicated because the wavefunctions of atoms in intimate contact are strongly distorted).

Figure 8.3 shows a plot of the mutual potential energy of a Na^+ ion and a Cl^- ion as a function of their separation (this potential energy has been calculated from the quantum-mechanical wave function). The sharp increase of the potential energy at distances shorter than 2 Å is due to electrons that are forced into states of higher energy. At a distance larger than about 3 Å, the

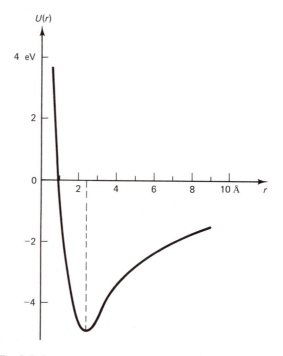

Fig. 8.3 Potential energy of interacting Na^+ and Cl^- as a function of distance.

exclusion effect disappears, and the Coulomb interaction takes over. From this plot we see that at large distances the force between the ions is attractive, but at short distances it is repulsive. At a distance of 2.4 Å these attractive and repulsive forces cancel, giving a (classical) equilibrium point.

The repulsive contribution to the force at short distances is called the **repulsive core.** This repulsive core is a general feature of the interaction of atoms—it is found in all atoms and it is essentially the same for all the different kinds of bonds. Thus we see that the Exclusion Principle is responsible for the resistance that all atoms offer to interpenetration.

Covalent Bonds. Atoms that are not very willing to give up electrons or to accept electrons may be willing to share electrons with their neighbors. The shared electrons tend to concentrate in the region between the atoms. Usually the shared electrons occur in pairs of opposite spins, each atom contributing one electron. Such a redistribution of the electrons gives rise to a bond between the atoms because the shared electrons exert a Coulomb attraction on the positive charges in the two atoms.

Covalent bonds exhibit strong directional preferences. For instance, the carbon atom likes to form four covalent bonds directed toward the corners of a tetrahedron centered on the atom (see Fig. 8.4). The bonding of carbon atoms with other carbon atoms in such a tetrahedral geometry creates the lattice of diamond, the hardest of all crystals (see Fig. 8.5).

Van der Waals Bonds. The atoms of the noble gases and most molecules do not have available any electrons to give up or to share. The interaction between two such atoms will then not lead to any large-scale redistribution of electrons, but merely to a correlation of the motions of the electrons in the two atoms. We can roughly understand this by means of the simple classical Bohr model. Imagine an atom with electrons in circular orbits. On the average, the electric dipole moment of the atom is zero, since each electron is just as likely to be found on one side of the nucleus as on the other. However, at each instant of time, the dipole moment has some nonzero value, depending on the instantaneous positions of the electrons (see Fig. 8.6). This fluctuating dipole moment generates an electric field which tends to polarize the charge distribution in neighboring atoms, inducing a correlated, aligned dipole moment. As we know from electromagnetism, the force between aligned dipole moments is attractive. (In terms of the motions of the electrons, we can say that the phases of the orbits in neighboring

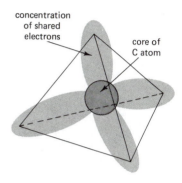

Fig. 8.4 Tetrahedral bonds of the carbon atom.

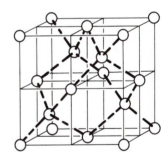

Fig. 8.5 Arrangement of carbon atoms in the diamond lattice.

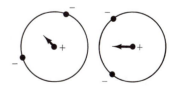

Fig. 8.6 Instantaneous positions of the electrons in two adjacent helium atoms. Each atom has an instantaneous dipole moment, indicated by an arrow.

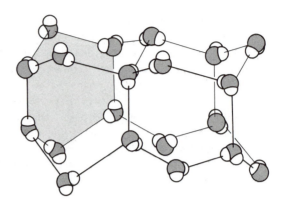

Fig. 8.7 Arrangment of water molecules in an ice crystal. Adjacent oxygen atoms are joined by hydrogen bonds.

atoms tend to be correlated in such a way that the electrons avoid coming near each other, and therefore the net force between the charge distributions of the two atoms is attractive.) This force between neighboring atoms is called the van der Waals force.

The bonds produced by the van der Waals force are much weaker than ionic or covalent bonds. Hence the noble gases only crystallize at very low temperatures or at very high pressures. Note that since the mechanism underlying the van der Waals forces is quite general, these forces act between all kinds of atoms and molecules, but they are only important when the ionic or covalent forces are absent. The cohesive forces in liquids are van der Waals forces.

Hydrogen Bonds. The hydrogen atom can lose its electron to neighboring atoms, leaving the hydrogen with a positive charge. This positive charge placed between two adjacent negative ions produces a bond between the ions. Obviously, this bond is somewhat similar to a covalent bond, with the positively charged hydrogen playing the role of the electron pair. Hydrogen bonds are responsible for joining the water molecules in ice crystals (see Fig. 8.7). They are also responsible for joining the two strands of the double helix that makes up the molecule of DNA.

Metallic Bonds. The electron redistribution that creates the bonds between the atoms in a metal is somewhat similar to that found in covalent bonds: the atoms share electrons. But instead of merely sharing one electron with a neighbor, each atom shares one (or two) of its electrons with all of the other atoms in the crystal. The shared electrons are free to move through the entire volume of the crystal; they form a gas of free electrons. Thus the metal consists of a lattice of positive ions embedded in an interpenetrating sea of negative charges. The Coulomb attraction between these positive and negative charges holds the ions in their places.

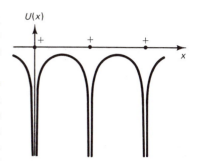

Fig. 8.8 Potential energy of a "free" electron moving through a metal along a straight line (x-axis).

Although the "free" electrons in a metal are not attached to any particular atom, they are not completely free of forces. When an electron passes near an ion, it experiences a strong, attractive Coulomb force. The net effect of all the ions on an electron can be described by a periodically repeating potential-energy function (see Fig. 8.8). Classically, an electron moving in such a periodic potential would simply speed up near each ion, and slow down in the spaces between. Wavemechanically, the motion of an electron is rather more complicated—as we will see in Section 8.3, for certain ranges of wavelength the motion is actually impossible (forbidden energy bands). However, for an electron of long wavelength, or low momentum, the motion is that of a free particle, i.e., the periodic potential produces no effect on the motion. The reason is that, in general, a wave is insensitive to any disturbance whose size is small compared to a wavelength; for example, ocean waves of long wavelength (swells) are unaffected by a small island in their path. An electron wave only senses the potential averaged over roughly one wavelength. If the wavelength is long, so that it includes several of peaks and valleys shown in Fig. 8.8, then the average potential is a constant, which is equivalent to zero force. Under these conditions the "free" electrons in the metal will behave like free particles.

Table 8.1 is a summary of different types of crystals, classified according to their bonds. This table also lists the typical binding energies per atom. This binding energy is largest for covalent bonds (these are the strongest bonds) and it is least for van der Waals bonds (these are the weakest bonds).

TABLE 8.1   Kinds of Bonds in Crystals

Bond	Arrangement of Electrons	Typical Binding Energy
Ionic	Excess electrons at − ions, deficit at + ions	5 eV/atom
Covalent	Shared electron pairs	10
van der Waals	Correlation of orbital motions	0.1
Hydrogen	Deficit of electrons at H^+ ions	0.5
Metallic	Free-electron gas	3

8.2 The Free-Electron Gas

We can readily explain several of the important physical properties of metals in terms of the free-electron gas. For instance, the motion of the free electrons is responsible for the high electric conductivity of metals. Accordingly, the free electrons are often called **conduction electrons.** One of the first calculations of the conductivity of a free-electron gas was made by P. Drude* around 1900. Drude treated the electrons in a piece of metal as classical particles, with large thermal velocities of random directions. These thermal velocities of the electrons do not bring about a net transport of electric charge, since, on the average, for every electron moving in some given direction there is another electron moving in the opposite direction. However, when the ends of the piece of metal are connected to the terminals of a battery or a generator, an electric field **E** will be produced within the piece of metal and, under the influence of the electric force, the electrons will accelerate in the direction of −**E**, acquiring an additional nonrandom velocity in this direction. This additional velocity of the electrons will give rise to a transport of charge. The magnitude of the additional velocity, called the drift speed, is limited by the decelerations that the electrons suffer in collisions with the ions of the lattice. These collisions play the role of a frictional force that slows the electrons down. The drift speed is always quite small compared to the random speeds. Under typical conditions, the electron gas in a metallic wire carrying a current has a drift speed of perhaps 10^{-2} m/s, but the individual electrons have random speeds of about 10^6 m/s. Thus, the motion of an electron consists of rapid zigzags on which is superimposed a much slower drift motion along the wire; qualitatively, the motion resembles that shown in Fig. 8.9, but the amount of drift per zigzag is much less than shown in this figure. The number of collisions per second that an electron makes with the ions of the lattice is proportional to its speed. Since the drift speed is negligible compared to the random speed, the drift speed has next to no effect on the collision rate. This means we can regard the collision rate as a constant, independent of the drift speed and of the electric field. We will designate the collision rate by $1/\tau$. The quantity τ is called the **relaxation time;** roughly, it is the time interval between successive collisions of an electron.

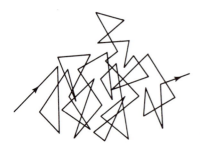

Fig. 8.9 Path of an electron in a conductor. The electron zigzags and gradually drifts from left to right.

---

* **Paul Karl Ludwig Drude,** 1863–1906, German physicist, editor of *Annalen der Physik.*

TABLE 8.2   Free-Electron Densities and Approximate
Relaxation Times in Some Metals[a]

Metal	n	τ
Li	$4.7 \times 10^{28}/\text{m}^3$	$0.83 \times 10^{-14}\,\text{s}$
Na	2.6	3.0
K	1.3	4.5
Cu	8.5	2.5
Ag	5.8	3.8

[a] The electron density assumes one free electron per atom; the relaxation time is calculated from the observed conductivity by means of Eq. (8).

Typically, the relaxation time is of the order of $10^{-14}\,\text{s}$ (See Table 8.2).

The drift motion of the electrons determines the net, macroscopic current carried by the wire. We can calculate the drift speed by examining the losses and gains of momentum of this electron. If the drift speed of an electron is v_d, then the momentum is $m_e v_d$. We expect that, on the average, a collision will absorb all of this momentum, i.e., a collision will destroy the forward drift motion and leave the electron with only the random thermal motion. This means that, in each collision, the electron loses a momentum $m_e v_d$. The average rate at which the electron loses momentum in collisions is therefore

$$\left(\frac{\Delta p}{\Delta t}\right)_{\text{loss}} = \frac{m_e v_d}{\tau} \qquad (1)$$

On the other hand, the rate at which the electron gains momentum by the action of the electric field is

$$\left(\frac{\Delta p}{\Delta t}\right)_{\text{gain}} = -eE \qquad (2)$$

Under steady-state conditions, the rate of loss of momentum must match the rate of gain. By setting the right sides of Equations (1) and (2) equal, we immediately obtain

$$v_d = -\frac{eE\tau}{m_e} \qquad (3)$$

This is the speed with which the electron gas flows along the wire. The negative sign in Eq. (3) indicates that the direction of flow is opposite to the direction of the electric field.

To find the electric current carried by this flow of the electron gas, we need to take into account the number of free electrons. Suppose that the metal of the wire has n free electrons per unit volume. In metals, the number of free electrons is usually one per atom; typically, this amounts to 10^{28} or 10^{29} electrons per m^3 (see Table 8.2). If the wire has a cross-sectional area A and a length l, then its total number of free electrons is $n \times$ (volume) $= nAl$ and the total charge associated with these electrons is

$$\Delta q = -enAl \qquad (4)$$

Within a time interval $\Delta t = l/|v_d|$ all of these electrons emerge at one end of the wire. Hence the current in the wire is

$$I = \frac{|\Delta q|}{\Delta t} = \frac{enAl}{l/|v_d|} = \frac{e^2 n\tau}{m_e} AE \qquad (5)$$

If we express this in terms of the potential difference $\Delta V = El$ across the ends of the wire, we obtain

$$I = \left(\frac{e^2 n\tau}{m_e} \frac{A}{l} \right) \Delta V \qquad (6)$$

This is **Ohm's law:** the current is proportional to the potential difference. According to Eq. (6), the resistance of the wire is

$$R = \frac{\Delta V}{I} = \frac{m_e}{e^2 n\tau} \frac{l}{A} \qquad (7)$$

and the conductivity is

$$\sigma = \frac{1}{R} \frac{l}{A} = \frac{e^2 n\tau}{m_e} \qquad (8)$$

Example 1   The conductivity of silver is $6.2 \times 10^7/\Omega \cdot$ m and the number of free electrons per unit volume is $5.8 \times 10^{28}/$m^3. Calculate the relaxation time. Calculate the drift speed of the electrons in a silver wire of length 2.0 m subjected to a potential difference of 40 V.

Solution   From Eq. (8),

$$\tau = \frac{\sigma m_e}{e^2 n} = \frac{(6.2 \times 10^7/\Omega \cdot \text{m}) \times 9.1 \times 10^{-31}\,\text{kg}}{(1.6 \times 10^{-19}\,\text{C})^2 \times 5.8 \times 10^{28}/\text{m}^3}$$

$$= 3.8 \times 10^{-14}\,\text{s} \qquad (9)$$

The values of the relaxation times for the other metals listed in Table 8.2 were calculated similarly, from their measured conductivities.

The electric field in the wire is $E = \Delta V/d = 40\,\text{V}/2.0\,\text{m} = 20\,\text{V/m}$. Hence, from Eq. (3),

$$v_d = -\frac{eE\tau}{m_e} = -\frac{1.6 \times 10^{-19}\,\text{C} \times 20\,\text{V/m} \times 3.8 \times 10^{-14}\,\text{s}}{9.1 \times 10^{-31}\,\text{kg}}$$

$$= -0.13\,\text{m/s}$$

Such a low drift speed is typical for the motion of the free electrons in metals. (In contrast, the *random* speed of these free electrons is much larger; as we will see below, the typical value of the random speed is $10^6\,\text{m/s}$ or more.) ■

Although Drude's simple classical calculation leads to the correct form of Ohm's law, the attempts at determining the relaxation time τ from the atomic parameters and the classical theory of random thermal motions were a complete failure—the value of the relaxation time calculated from classical theory disagrees with experiment, and also fails to reproduce the observed dependence of the conductivity on temperature. Furthermore, according to classical theory we would expect that the free-electron gas has a heat capacity of its own amounting to $3\,\text{cal/K} \cdot \text{mole}$,* which has to be added to the heat capacity of the crystal lattice. But experiment shows that the heat capacity of metals is not noticeably different from that of other crystals. All these difficulties of the classical theory of the free-electron gas disappear when we treat the electrons by wave mechanics and take into account the Exclusion Principle. As we will see in this section, the Exclusion Principle plays a crucial role in determining the properties of the free-electron gas because it forces electrons into states of very high kinetic energy, much higher than the classical thermal kinetic energy $\frac{3}{2}kT$.

A gas of quantum-mechanical particles obeying the Exclusion Principle—electrons or other particles of half-integer spin—is called a **Fermi gas.** The distinction between the classical gas and the Fermi gas shows up drastically if we consider the simple case of zero temperature. Classically, the particles of a gas at zero temperature ($T = 0\,\text{K}$) all have zero kinetic energy—they all are at rest. Quantum mechanically, electrons at zero temperature cannot all have zero kinetic energy because this would mean that they all are in the ground state, which would contradict the Exclusion Principle. Instead, at zero temperature the electrons will occupy

---

* This is the usual heat capacity at constant volume for a gas of classical particles.

the available states of lowest energy, two electrons with opposite spins in each orbital state.

We can roughly regard the free electrons in a metal as free particles in a box. The available states are then described by standing waves (see Section 6.1), and the energies of the available states are given by the following three-dimensional generalization of Eq. (6.11):

$$E = (n_x^2 + n_y^2 + n_z^2) \frac{\pi^2 \hbar^2}{2m_e L^2} \tag{10}$$

Here, we have assumed that the box is a cube, measuring $L \times L \times L$; this assumption is convenient, but not necessary. The possible values of n_x, n_y, and n_z can be represented as points in a three-dimensional space labeled with n_x-, n_y-, and n_z-axes. Figure 8.10 shows some of these points; note that the x-, y-, and z-intervals between adjacent points are one unit, and that all the points are in the first octant ($n_x > 0$, $n_y > 0$, $n_z > 0$). We can use this figure to count the number of available states up to some given energy E. The radial distance between a given point n_x, n_y, n_z and the origin is

$$R = \sqrt{n_x^2 + n_y^2 + n_z^2} \tag{11}$$

We can express the energy in terms of this radius,

$$E = R^2 \frac{\pi^2 \hbar^2}{2m_e L^2} \tag{12}$$

and, conversely, we can express the radius in terms of the given energy,

$$R = \sqrt{\frac{2m_e L^2}{\pi^2 \hbar^2} E} \tag{13}$$

The number of available orbital states of energy less than E is simply the number of points within the radius R. Each point is associated with a volume element of one unit; these volume elements can be regarded as infinitesimal provided R is a large number. The number of points within the radius R is therefore simply the volume within the first octant, i.e., $(\frac{1}{8})(4\pi/3)R^3$. Taking into account the two spin states per orbital state, we find that the number of states of energy less than E is

$$N = 2 \times \frac{1}{8} \frac{4\pi}{3} \left(\frac{2m_e L^2}{\pi^2 \hbar^2} \right)^{3/2} E^{3/2} \tag{14}$$

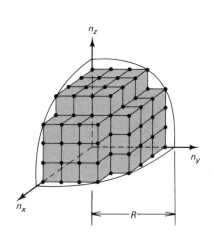

Fig. 8.10 Each point in this plot represents one possible value of the quantum numbers n_x, n_y, n_z. Note that all points are in the first octant. Those points within a given radius R corresponds to quantum states of an energy less than E.

Since the volume of our box is $V = L^3$, this reduces to

$$N = \frac{1}{3}(2m_e)^{3/2} \frac{V}{\pi^2 \hbar^3} E^{3/2} \qquad (15)$$

In a crystal at zero temperature, the electrons fill the available states of lowest energy. Hence if we set the number of states in Eq. (15) equal to the number of electrons, the energy E will equal the energy of the highest occupied state, or **Fermi level.** Solving for this energy E, we obtain

$$E_F = (3\pi^2)^{2/3} \frac{\hbar^2}{2m_e} \left(\frac{N}{V}\right)^{2/3} \qquad (16)$$

This is called the **Fermi energy** of the electron gas. Note that this energy depends only on the density of the electron gas.*

Example 2   What is the Fermi energy for the free-electron gas in silver? What is the speed of an electron with this energy?

Solution   With $N/V = 5.8 \times 10^{28}/\text{m}^3$, Eq. (16) becomes

$$E_F = (3\pi^2)^{2/3} \frac{(1.05 \times 10^{-34}\,\text{J} \cdot \text{s})^2}{2 \times 9.1 \times 10^{-31}\,\text{kg}} (5.8 \times 10^{28}/\text{m}^3)^{2/3} \qquad (17)$$

$$= 8.7 \times 10^{-19}\,\text{J} = 5.4\,\text{eV} \qquad (18)$$

The speed of an electron with this energy is

$$v_F = \sqrt{\frac{2E_F}{m_e}} = 1.4 \times 10^6\,\text{m/s} \quad \blacksquare \qquad (19)$$

Thus the typical speeds of the electrons in the Fermi gas are quite large. For comparison, note that if we wanted to give the molecules of a classical gas typical speeds of this order of magnitude, we would have to heat the gas to a temperature of $6 \times 10^4\,\text{K}$! This is a striking illustration of the difference between the classical gas and the Fermi gas.

---

* Equation (16) gives the Fermi energy for zero temperature. The Fermi energy for nonzero temperature is defined as the energy of the level that has a probability of $\frac{1}{2}$ for being occupied; it depends on the temperature and the density.

The number of available states per unit energy interval is dN/dE, obtained by differentiating Eq. (15),

$$\frac{dN}{dE} = \frac{1}{2} (2m_e)^{3/2} \frac{V}{\pi^2 \hbar^3} E^{1/2} \qquad (20)$$

Figure 8.11 is a plot of dN/dE, with the energy plotted along the vertical axis, as customary in an energy-level diagram. The shaded area indicates those states occupied at zero temperature.

With the expression (20) we can calculate the total energy of the Fermi gas:

$$E_{\text{total}} = \int_0^{E_F} E\, dN = \int_0^{E_F} E\, \frac{1}{2} (2m_e)^{3/2} \frac{V}{\pi^2 \hbar^3} E^{1/2}\, dE$$

$$= \frac{1}{5} (2m_e)^{3/2} \frac{V}{\pi^2 \hbar^3} E_F^{5/2} \qquad (21)$$

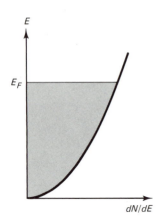

Fig. 8.11 Number of states per unit energy as a function of energy. The energy is plotted vertically, as in an energy-level diagram. The shaded area shows the energy levels that are filled at zero temperature.

or, according to Eq. (16),

$$E_{\text{total}} = \tfrac{3}{5} N E_F \qquad (22)$$

Thus, at zero temperature the average energy per electron is $E_{\text{total}}/N = \tfrac{3}{5} E_F$.

Next, we must examine the changes that occur in the state of the free-electron gas when we heat it from zero temperature to some finite temperature. Obviously, some of the electrons will then make transitions to states of higher energy, but the transitions are severely limited by the Exclusion Principle, which forbids transitions to states already occupied. Since the final state of thermal equilibrium is independent of the mechanism of heat transfer, let us make the convenient assumption that the electrons acquire their thermal energy by collisions with the ions of the lattice. At a temperature T, the energy of the ions is of the order of kT. Hence a collision can at most transfer that much energy to the electron. At room temperature, $kT \sim 0.03$ eV. For most of the electrons, 0.03 eV is not enough energy to permit a transition to an unoccupied state; hence most electrons are unable to accept this energy and they will remain locked in their original state. Only those electrons that are within 0.03 eV of an unoccupied state—or within 0.03 eV of the zero-temperature Fermi level—can accept this energy and jump to a higher state. Thus, a few states just below the zero-temperature Fermi level become partially depleted, and a few states just above become partially filled (see Fig. 8.12), but most of the states remain as they were at zero temperature. This means that

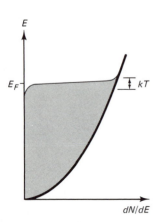

Fig. 8.12 The shaded area shows the energy levels that are filled at room temperature. Note that the width of the region of partially full, partially empty levels is $\sim kT$. At room temperature kT is very small compared to E_F; hence this width is actually much smaller than drawn in the figure.

the electron gas in a metal at room temperature is pretty much
the same as an electron gas at zero temperature. Even in a metal
near its melting temperature, the electron gas is nearly unchanged.
The total energy and other properties of the electron gas are nearly
independent of temperature, over a very wide range of temperatures.
This explains why the electron gas makes only a small contribution
to the specific heat of the metal—the electron gas absorbs only a
small amount of thermal energy.

We can estimate the specific heat of the electron gas by the
following simple argument. As stated above, the Exclusion Principle prevents most electrons from absorbing thermal energy because the adjacent energy levels are full. Only the electrons within
a distance $\sim kT$ of the Fermi level can absorb thermal energy.
The fraction of these electrons is roughly $\sim (kT/E_F)$. Thus, we
expect that the specific heat of the Fermi gas is smaller than that
of a classical gas by a factor kT/E_F. Since the specific heat per
mole of a classical gas is $\frac{3}{2}N_0 k$, the specific heat per mole of the
Fermi gas must be roughly*

$$C_V \sim \frac{kT}{E_F} \frac{3}{2} N_0 k \tag{23}$$

For the free-electron gas in a metal, E_F is typically a few eV (see
Example 2), whereas $kT \sim 0.03\,\text{eV}$ at room temperature. This
means that kT/E_F is of the order of magnitude of 10^{-2}, and therefore the specific heat of the free-electron gas is only $\sim 1\%$ of the
value expected for a classical gas.

We might suppose that the Exclusion Principle also inhibits the
conduction of electric current by the free-electron gas. When acted
upon by the electric field, electrons try to absorb energy and momentum from this field, making transitions to nearby states of
slightly higher energy and momentum. If all such states are occupied, how can the electrons make the transitions? The answer is
that the electric field acts also on the electrons in the higher states;
it increases the momentum of *all* the electrons at the same rate.
The electric field removes electrons from the higher states, lifting
them into even higher, unoccupied states, and creates vacancies at
just the rate needed to accommodate the electrons arriving from
the lower states. Hence the Exclusion Principle does not interfere
with the acceleration process, which proceeds pretty much as in the
classical case. Only in the deceleration process does the Exclusion

---

* The exact result is $C_V = \frac{1}{2}\pi^2 (kT/E_F)N_0 k$.

Principle make a difference—most collisions are forbidden because the final state is occupied. The only collisions that occur are those that reverse the directions of motion of the electrons that have attained maximum speed in the direction of the current. These electrons make transitions to states of maximum or nearly maximum speed in the direction opposite to the current*. They are then ready to be again accelerated by the electric field.

The wave-mechanical result for the conductivity therefore has the same form as Drude's classical result [Eq. (8)]. However, the theoretical expression for the collision rate $1/\tau$ is quite different. We will not deal with the derivation of this theoretical expression; suffice it to say that for metals at reasonably high temperature (room temperature or above), the theoretical expression for $1/\tau$ is proportional to T; hence the conductivity is proportional to $1/T$, in agreement with experimental measurements. In essence, this decrease of conductivity with temperature is due to an increase in the rate of collisions between the electrons and the ions of the lattice. The ions of the lattice are endowed with thermal motions consisting of oscillations about their equilibrium positions; the amplitudes of these oscillations increase with temperature. These oscillations increase the likelihood of collision for an electron trying to pass through the lattice, i.e., they increase the effective friction force opposing the motion of the electron.

8.3 The Band Theory of Solids; Conductors, Semiconductors, and Insulators

The free-electron model of metals discussed in the preceding section rests on the rather crude approximation that the interactions between the electrons and the lattice ions can be ignored. A more refined treatment of the motion of electrons in a solid must take these interactions into account—they determine the spectrum of energy levels of the electrons in the solid. We will see in this section that the energy levels in a solid occur in **bands,** analogous to the shells or subshells found in isolated atoms. The electrical properties of the solid hinge on the extent to which these bands are filled. Partially filled bands permit electron motion and make the

---

* Such a transition is called an *Umklapp* process, from the German for flipping over.

solid into a conductor, whereas completely filled bands restrict electron motion and make the solid into an insulator.

To understand why a solid has energy bands, let us examine the simple problem of an electron moving in one dimension in a periodic potential consisting of square wells separated by a distance a (see Fig. 8.13). The potential wells roughly represent the attraction between the electron and the positive ions of the lattice. We will pretend that this periodic potential continues indefinitely in both directions, i.e., the crystal is infinite (the boundaries of the crystal add messy, but inessential, complications to the problem). The wavefunction for a free electron traveling to the right or toward the left is e^{ikx} or e^{-ikx}. The energy of such a free electron is proportional to the square of the wave number,

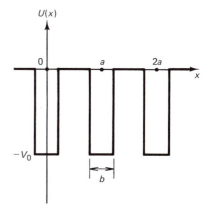

Fig. 8.13 Periodic potential consisting of square wells of width b separated by a distance a.

$$E = \frac{\hbar^2}{2m_e} k^2 \qquad (24)$$

The interaction with the periodic potential modifies the wavefunctions,* and modifies the expression for the energy. The mathematical details of this modification emerge from the solution of the Schrödinger equation, and are rather complicated; but even without entering into these details, we can recognize that, for small values of k (or large values of $\lambda = 2\pi/k$), the energy of the electron must be of the form

$$E = E_0 + Bk^2 \qquad (25)$$

where E_0 and B are constants. The justification of this expression is simply that, for small values of k, the energy $E(k)$ can be expressed as a Taylor series $E \simeq E_0 + Ak + Bk^2$, and the first-order term must be absent ($A = 0$) because the energy cannot depend on the sign of k if the potential is symmetric. It is customary to write Eq. (25) in a form resembling the kinetic energy of a free electron:

$$E = E_0 + \frac{\hbar^2}{2m^*} k^2 \qquad (26)$$

where m^* is a new constant, called the **effective mass.** In most metals, the effective mass of an electron is somewhat larger than

---

* It can be shown that the wavefunctions in the presence of the periodic potential are of the form $e^{\pm ikx}u(x)$, where $u(x)$ is a periodic function with the same period as the potential; this is **Bloch's theorem** (see, e.g., C. Kittel, *Introduction to Solid State Physics*).

TABLE 8.3 Effective Mass of Electron

Crystal	Effective mass, m^*
Li	$2.2m_e$
Na	$1.3m_e$
Al	$1.5m_e$
Cu	$1.4m_e$
Ag	$1.0m_e$
Mg	$1.3m_e$
Pb	$2.0m_e$

the ordinary mass. This means that, in consequence of its inter-action with the ions, the electron is somewhat more sluggish to re-spond to external disturbances, i.e., the electron has larger inertia. Table 8.3 lists values of the effective mass for a few metals.

Although Eq. (26) is a good approximation at small values of k, it begins to fail when k becomes comparable to the reciprocal of the distance a between the potential wells, i.e., when $k \simeq 1/a$. And it fails completely when $k = \pi/a$. At this value of k, and at integer multiples of this value, the energy has discontinuities (see Fig. 8.14). The reason for these discontinuities is that for these values of k, reflections of the waves become very important. Each ion reflects a small portion of the wave and, if all these reflected portions are

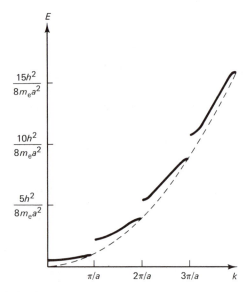

Fig. 8.14 Energy as a function of k for an electron moving in the peri-odic potential of square wells. (For the special case $b \to 0$ and $V_0 \to \infty$.)

in phase, the amplitude of the net reflected wave traveling backward will become equal to the amplitude of the wave traveling forward. This will happen if the distance between adjacent ions equals one-half wavelength, or a multiple of one-half wavelength, i.e.

$$2a = n\lambda \qquad n = 1, 2, 3, \ldots \qquad (27)$$

or

$$k = \frac{n\pi}{a} \qquad (28)$$

Note that Eq. (27) is simply the condition for Bragg reflection, with $\theta = 90°$ [see Eq. (1.153)]. The superposition of the traveling wave and the reflected traveling wave yields a standing wave. For instance, if $k = \pi/a$, the possible superpositions are

$$\psi_1 = e^{i\pi x/a} - e^{-i\pi x/a} = 2i\sin(\pi x/a) \qquad (29)$$

and

$$\psi_2 = e^{i\pi x/a} + e^{-i\pi x/a} = 2\cos(\pi x/a) \qquad (30)$$

Figure 8.15 shows these two standing waves. According to the first of these waves, the electron probability is small (zero) at the location of each ion; according to the second wave, it is large (maximum) at these locations. The second wave has a lower energy because it places the electron near the position of an ion, where the potential energy is very low. Thus, the two waves (29) and (30) have different energies even though they have the same wavenumbers, i.e., as a function of k, the energy has a discontinuity at $k = \pm\pi/a$. Similar discontinuities occur at $k = \pm n\pi/a$. The ranges of k within which the function $E(k)$ is continuous are called **Brillouin zones.** For example, the first Brillouin zone consists of the interval $-\pi/a < k < \pi/a$; the second Brillouin zone consists of the intervals $-2\pi/a < k < -\pi/a$ and $\pi/a < k < 2\pi/a$, etc. For each Brillouin zone, there is a permitted range of energies, called an **energy band.** Adjacent energy bands are separated by a forbidden range of energies, called a **gap,** or a forbidden band. Figure 8.16 shows the bands and the gaps associated with the function $E(k)$ of Fig. 8.14.

In the above discussion we pretended that the crystal has infinite extent and an infinite number of atoms. Each energy band then consists of a continuous range of energies, i.e., an infinite number of energy levels. In the more realistic case of a crystal with finite boundaries and a finite number of atoms, each energy band consists of a discrete set of closely spaced energy levels. The number of (orbital) energy levels per band is equal to the number of atoms in the crystal. To see how this comes about, let us examine what

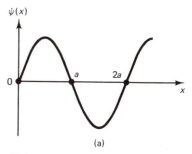

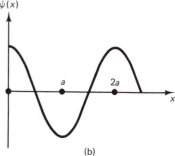

Fig. 8.15 Standing waves in the periodic potential: (a) $\psi_1 \propto \sin(\pi x/a)$; (b) $\psi_2 \propto \cos(\pi x/a)$.

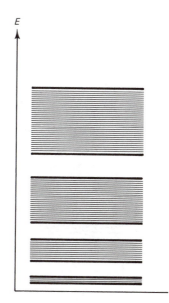

Fig. 8.16 Energy levels for the periodic potential. The permitted energies occur in bands separated by forbidden gaps.

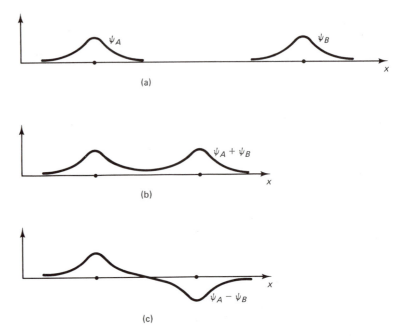

Fig. 8.17 Wavefunction of two hydrogen atoms: (a) wide separation; (b) small separation: isolated wavefunctions combined with the same sign; (c) small separation: isolated wavefunction combined with opposite signs.

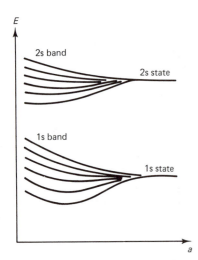

Fig. 8.18 Energy levels as a function of separation for six hydrogen atoms arranged on a straight line.

happens to the energy levels of isolated atoms when they are brought together. Consider, for instance, two hydrogen atoms, each in the ground state. When the two atoms are widely separated, the wavefunction of each electron is simply that of an isolated atom (see Figure 8.17a). When the atoms are in contact, these wavefunctions overlap and each electron orbits around both nuclei The wavefunction of each electron is, roughly, a combination of the wavefunctions of the isolated atoms. Figure 8.17b and c show two possible wavefunctions: in the first the two isolated wavefunctions are combined with the same signs, in the second with opposite signs. The first wavefunction corresponds to a lower energy than the second because the electron is more likely to be found midway between the nuclei, which leads to a lower Coulomb energy (expressed another way: an electron with the first wavefunction creates a covalent bond). Thus, whenever two hydrogen atoms are brought close together, the ground-state energy level splits into two separate energy levels. More generally, when N atoms are brought together, each energy level splits into N energy levels, forming a band. For instance, Fig. 8.18 illustrates the formation of such bands when six hydrogen atoms are brought together. Thus, we see how bands in a solid arise from the distortion of the subshells of the isolated atoms.

In a solid at zero temperature, the electrons settle into the available states of lowest energy. The lower energy bands will therefore be completely filled, and the uppermost energy band will be either filled or partially filled, depending on the number of electrons and on the number of available states. The difference between a conductor and an insulator arises from a partially filled or a completely filled uppermost energy band. A conductor, such as copper, has a partially filled band (in copper, the uppermost band is only half filled, see Fig. 8.19). When subjected to an external electric field, the electrons in this band absorb energy from the field and make transitions to available, unoccupied states in the same band; this permits the electrons to move and to carry an electric current. An insulator, such as diamond, has a completely filled band (see Fig. 8.20). The electrons in this band are locked into their states because there are no available, unoccupied states. The nearest unoccupied states are in the next band, but this is separated from the filled band by an energy gap of about 6 eV (see Fig. 8.20). Since it is very difficult for an electron to absorb an energy of 6 eV from an electric field of ordinary strength, most electrons refuse to respond to the electric field. Thus, the electrons in diamond refuse to carry an electric current.

A semiconductor, such as silicon, has a completely filled band, i.e., it resembles an insulator. However, the gap between this filled band and the next band is small, about 1 eV or less (see Fig. 8.21). Hence electrons can fairly easily make the transition from one band to the next, and then carry an electric current. The filled band of an insulator or a semiconductor is called the **valence band,** and the nearby empty band is called the **conduction band.** The conductivity of a semiconductor increases with temperature because thermal disturbances excite some of the electrons into the conduction band, helping the semiconductor to carry a current. This is in sharp contrast to the behavior of metals, whose conductivity always decreases with temperature, due to the increase of "friction" with temperature. In a semiconductor, this increase of friction is more than compensated by the increase of the number of electrons participating in the conduction of the current.

A semiconductor carries a current not only in the conduction band, but also in the valence band. The electrons that move from the valence band to the conduction band leave behind unoccupied states, and the electrons remaining in the valence band can then take advantage of these unoccupied states to carry a current. The unoccupied states are called **holes.** Since such a hole represents a deficit of negative charge, it can be regarded as an effective positive

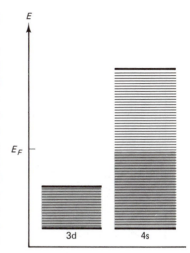

Fig. 8.19 Energy bands of a copper crystal. The 4s and 3d bands overlap to some extent. The band containing the electrons of highest energy is the 4s band. The electrons fill this band only partially.

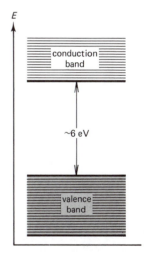

Fig. 8.20 Energy bands of a diamond crystal. The electrons fill the lower band completely.

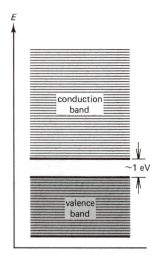

Fig. 8.21 Energy bands of a silicon crystal. The electrons fill the lower band completely, but the gap between this band and the next is small.

charge. Thus, the current carried by the electrons in the valence band can be associated with the motion of holes. Roughly, the electrons in the valence band play a game of musical chairs: the electrons move in one direction, by jumping into available holes, and the holes therefore move in the opposite direction. Mathematically, it is easier to describe the current as a flow of holes than a flow of electrons because there are few holes but very many electrons in the valence band.

For practical applications, semiconductors are usually doped (contaminated) with carefully controlled amounts of other atoms. Such a contaminated semiconductor is called an **impurity semiconductor.** The impurity atoms can be electron donors or electron acceptors. Arsenic atoms in silicon act as donors. The arsenic atom has one more electron in its outer shell than silicon; if an arsenic atom is substituted for a silicon atom in a silicon crystal, it will yield up an extra free electron to the crystal. The donated electrons enhance the conductivity of the silicon. On the other hand, gallium impurities in silicon act as acceptors. The gallium atom has one less electron in its outer shell than silicon; if a gallium atom is substituted for a silicon atom, it will trap a free electron, and thereby create a hole. Such holes also enhance the conductivity of silicon. Impurity semiconductors with donors are called *n*-type; those with acceptors are called *p*-type.

A mere few parts per million of impurity atoms in a semiconductor crystal suffice to modify the conductivity by several orders of magnitude. This permits us to tailor the electrical properties of semiconductors so as to suit a wide range of technological applications.

8.4 **Semiconductor Devices**

Some of the most interesting properties of semiconductors emerge when we join two pieces of semiconductors of different types (*n*-type and *p*-type). The behavior of electrons at the interface between such semiconductors underlies the operation of semiconductor devices, such as rectifiers, transistors, solar cells, light-emitting diodes, etc. In the following, we will briefly describe the operation of these devices.

Rectifier (Diode). If we join a piece of *n*-type and a piece of *p*-type semiconductor, some of the free electrons will diffuse from the *n* region into the *p* region, and some of the free holes will dif-

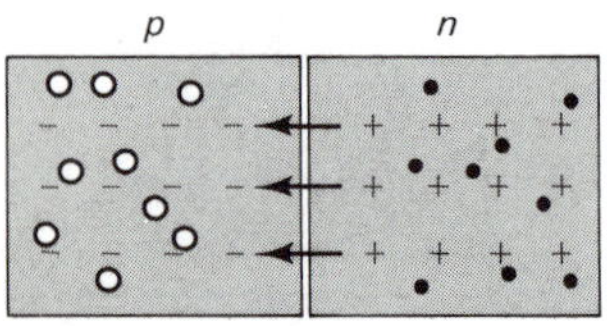

Fig. 8.22 Charge distribution in two semiconductors of different types joined together. The + and − signs indicate the ions of the lattice; the black dots indicate electrons, and the white dots holes. The layers of charge near the interface generate an electric field.

fuse from the *p* region into the *n* region. These diffused electrons and holes meet and annihilate—the electrons jump into the holes, and both disappear. This leaves the ions near the junction bare and results in a separation of electric charge: the *n* region acquires a layer of positive charge and the *p* region a layer of negative charge. The electric field associated with this charge separation opposes the diffusion. Equilibrium is achieved when the electric field is just strong enough to stop the diffusion (see Fig. 8.22).

If we connect this *p-n* junction to a battery or some other source of emf, it will permit the flow of current in one direction, but not in the other. Figure 8.23 shows the *p-n* junction connected so that the *n* region of semiconductor is at high potential and the *p* region at low, a configuration called "reverse bias." Free electrons in the *n* region flow away through the wire, and free holes in the *p* region flow away through the other wire. Thus each region is depleted of its charge carriers, and the current stops almost immediately. Occasionally, thermal fluctuations will create an electron-hole pair at the junction (by exciting an electron from the valence band to the conduction band); but the rate of creation of such pairs is low, and the resulting residual current is quite small, and is completely independent of the applied emf. Figure 8.24 shows the *p-n* junction connected so that the *n* region is at low potential and the *p* region at high, a configuration called "forward bias." Now the source of emf pumps a steady flow of electrons into the *n* region, and pumps holes into the *p* region (i.e., it removes electrons from the *p* region). The electrons and the holes meet at the junction and annihilate. Thus, the *p-n* junction permits a flow of current in the forward direction, but not in the reverse direction. Under forward bias, the current increases quite steeply with the applied emf (see Fig. 8.25).

Because of its directional property, the *p-n* junction can be used as a rectifier. When connected to a source of alternating emf, the junction will only pass current during the "forward" half of the cycle. Such solid-state rectifiers find wide application in the so-called alternators, which have taken the place of conventional mechanical commutators in the electric generators used in automobiles.

Transistor (Triode).   A simple junction transistor consists of a thin piece of semiconductor of one type sandwiched between two pieces of the opposite type. Figure 8.26 shows a *n-p-n* junction transistor; the piece in the middle is called the base, and the pieces at the ends are called the emitter and the collector, respectively. The three terminals of this transistor are connected to two sources of emf, so that the emitter–base junction has a forward bias and

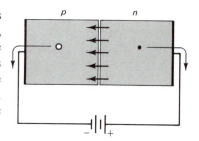

Fig. 8.23 A *p-n* junction connected to a source of emf. The *n* region is at high potential (reverse bias).

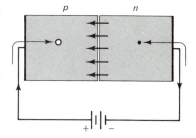

Fig. 8.24 A *p-n* junction connected to a source of emf. The *n* region is at low potential (forward bias).

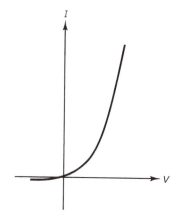

Fig. 8.25 Current as a function of voltage for a *p-n junction rectifier.*

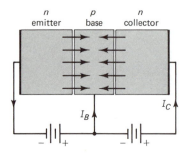

| n | p | n |
| emitter | base | collector |

I_B

I_C

Fig. 8.26 A *n-p-n* transistor. The *n* region on the left is the emitter, the *p* region in the center is the base, and the *n* region on the right is the collector.

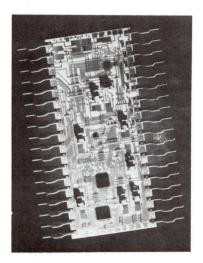

Fig. 8.27 An integrated circuit. (Courtesy AT&T Bell Laboratories.)

the base–collector junction a reverse bias. Under these conditions, the emitter–base junction permits a flow of electrons from the emitter into the base. If the *p* region in the base were fairly thick, then the electrons that enter the *p* region would annihilate with holes, as in an ordinary *p-n* junction rectifier. However, the *p* region in the transistor is very thin and it is doped much less heavily than the emitter so that the density of holes is relatively low. Hence almost all the electrons wander across the width of this region without meeting a hole. At the base–collector junction these electrons feel the electric field associated with the charge separation at this junction. This field pulls the electrons into the collector, and they then continue around the circuit forming the current I_C. Only a small fraction of the electrons entering the base leave via the terminal connected to the base. This fraction forms a current I_B. The ratio of I_C to I_B depends on the geometry of the transistor and on the characteristics of the semiconductor materials. For a given transistor, this ratio is a constant,

$$\frac{I_C}{I_B} = \text{const.} \tag{31}$$

For typical transistors the ratio is between 20 and 200. Whenever we change the base current I_B, we will also change the collector current I_C. Hence the transistor acts as a current amplifier— whenever we change the base current by a small amount, we change the collector current by a large amount.

In practice, transistors are not manufactured by joining separate slices of semiconductor, but by diffusing the required concentrations of acceptors or donors into a single crystal of silicon or germanium. The crystal is immersed in a gas of acceptor or donor atoms, which are absorbed by the crystal and then gradually migrate into its interior. The shape of the *n* and *p* regions is controlled by masking off suitable areas of the crystal. Many transistors, rectifiers, or capacitors can be built on a single chip of silicon. This diffusion technique is also used in the manufacture of integrated circuits consisting of many very small transistors, rectifiers, and capacitors embedded in a chip of silicon just a few millimeters in size (see Fig. 8.27).

Light-Emitting Diode (LED). We have seen that when a *p-n* junction is operated in the forward bias configuration, electrons arriving from the *n* region and holes from the *p* region meet at the junction. These electrons and holes annihilate each other when they meet at the interface. The annihilation is simply a transition of an electron from the conduction band to an empty state (hole) in the

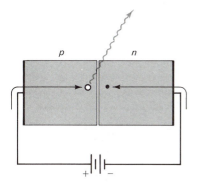

Fig. 8.28 (a) Schematic diagram of a solid-state laser. (b) Photograph of a solid-state laser held in the eye of an ordinary needle. (Courtesy AT&T Bell Laboratories.)

valence band, a transition that releases energy. In some materials, such as gallium arsenide, the energy is released in the form of a photon, i.e., the *p-n* junction emitts light whenever an electric current passes through it.

Such light-emitting diodes find widespread application in luminous digital displays for clocks, radios, and all kinds of measuring instruments. Light-emitting diodes can also be used as lasers. Whenever there are electrons in the conduction band and a corresponding number of holes in the valence band, we have a population inversion. The emission of one initial photon can then lead to the stimulated emission of many other, coherent, photons. Figure 8.28 shows a schematic diagram and a photograph of a solid-state laser. This kind of laser, which can be of a very small size, is used to generate the light pulses that carry information in some of the new telephone lines consisting of optical fibers.

Solar Cell. In principle, a solar cell is nothing but a light-emitting diode operating in reverse. Sunlight is absorbed at the *p-n* interface and excites electrons into the conduction band, i.e., sunlight creates electron–hole pairs. The electric field at the junction pulls the electrons toward the *n* region and the holes toward the *p* region. Consequently, in the external circuit an electric current flows from the *p* terminal to the *n* terminal (see Fig. 8.29). The semiconductor material in a solar cell must be very thin so that it is transparent to sunlight. This is usually achieved by depositing a thin layer of *p*-type silicon on a wafer of *n*-type silicon (see Fig. 8.30).

The emf of a silicon solar cell is about 0.6 V, but the current that it delivers is fairly small. For example, a typical solar cell of

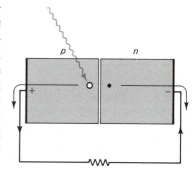

Fig. 8.29 A *p-n* junction operating as a solar cell.

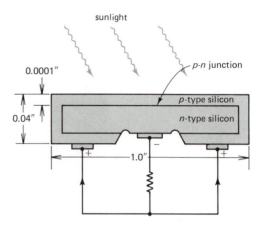

Fig. 8.30 Design of a solar cell.

area $5\,\text{cm}^2$ exposed to full sunlight delivers only 0.1 A; this amounts to roughly 10% conversion of the energy of sunlight into electric energy.

8.5 **Superconductivity**

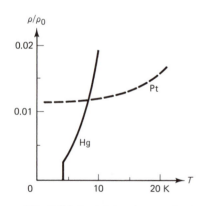

Fig. 8.31 Resistivity of a sample of mercury as a function of temperature. The vertical axis gives the resistivity relative to the resistivity at 273 K. The resistivity vanishes at 4.15 K, when mercury becomes superconducting. For comparison, the dashed curve represents the resistivity of platinum, which remains a normal conductor.

Superconductivity was discovered by K. Onnes* in 1911 during electrical experiments with a sample of frozen mercury. Figure 8.31 shows a plot of the measured values of the resistivity of mercury as a function of temperature. At a temperature of 4.15 K, the resistivity drops sharply; and below this critical temperature, the resistivity is zero. Besides mercury, many other metals exhibit superconductivity at low temperatures. Table 8.4 lists some superconducting metals and their transition temperatures. Furthermore, a large variety of compounds and alloys exhibit superconductivity. The highest known transition temperature is found in an alloy containing niobium and germanium (Nb_3Ge), whose transition temperature is 23.2 K.

---

* **Heike Kammerlingh Onnes,** 1853–1926, Dutch experimental physicist, professor at Leiden. He was the first to succeed in the liquefaction of helium. The availability of liquid helium initiated the modern era in low-temperature physics. By immersion in liquid helium at reduced pressure, Onnes cooled samples of materials to within 0.8 degree of absolute zero. He received the Nobel Prize in 1913 for his investigations of the behavior of matter at low temperatures.

TABLE 8.4   Some Superconductors

Element	T_c
Aluminum	1.19 K
Zinc	0.546
Gallium	1.09
Niobium	9.46
Indium	3.40
Tin	3.722
Osmium	0.66
Tungsten	0.012
Mercury	4.153
Lead	7.18

The superconducting state must be counted as a new state of matter, in addition to the solid, liquid, gas, and plasma states. The transition from ordinary conductor to superconductor is a thermodynamic phase transition, which involves an increase of order (decrease of entropy), just as does the transition from, say, liquid to solid. However, the increased order in a superconductor is not due to a rearrangement of the atoms, but rather a rearrangement of the conduction electrons, which condense into a highly ordered orbital configuration, with strong correlations among their motions.

A superconductor is a perfect conductor—its resistance is truly zero. Even the most precise experiments have not been able to detect any residual resistance in a superconductor. Once a current is started in a closed loop of superconducting wire, this current will continue to keep flowing around the loop of its own accord, as long as the wire is kept cold. Such a steady current that flows without any resistive loss is called a **persistent current.** Persistent currents have been observed to flow for several years with undiminished strengths. The experimental evidence shows that the decay time of these currents exceeds 10^5 years. Theoretical calculations indicate that the decay time is actually of the order of $10^{4 \times 10^7}$ years.

Superconducting loops with persistent currents produce magnetic fields and therefore can serve as magnets. Such a superconducting magnet does not require any source of emf to maintain its current and its magnetic field. Superconducting magnets are used to produce intense magnetic fields for a variety of scientific and engineering applications. Unfortunately, the maximum intensity of the magnetic field that can be attained by a superconducting magnet is subject to a severe limitation: intense magnetic fields destroy superconductivity. For instance, a magnetic field of 0.04 T will destroy the superconductivity of mercury if the temperature is near absolute zero. An even smaller magnetic field suffices to destroy the superconductivity if the temperature is a few degrees above absolute zero. The minimum magnetic field that will quench the superconductivity of a material is called the **critical magnetic field;** the strength of this field is a function of temperature. Figure 8.32 is a plot of the critical field strength for mercury as a function of temperature. Superconductors of the second kind, which are made of alloys (see below), have much higher values of the critical field. Some of these superconductors tolerate magnetic fields a thousand times stronger than those tolerated by mercury. Hence superconductors of the second kind are preferred for the construction of magnets.

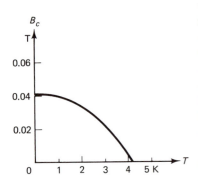

Fig. 8.32 Critical magnetic field for mercury as a function of temperature.

In an ordinary conductor, a steady current requires an electric field to overcome the resistance. The electric field within a conductor carrying a given current is directly proportional to the resistance (Ohm's law). A superconductor, with zero resistance, always has zero electric field in its interior. From this, we can conclude that the rate of change of magnetic field in a superconductor must always be zero; if it were not, then the changing magnetic flux would induce an electric field, in contradiction with the requirement that the electric field remain zero. For example, Fig. 8.33a shows what happens if we transport a superconducting cylinder into a magnetic field; the superconductor pushes the magnetic field lines aside so that none of these penetrate its interior. Wherever the superconductor first touches the magnetic field, eddy currents are induced on its surface, and the magnetic field of these currents produces a deformation of the original magnetic field lines to prevent their penetration into the volume of the supercon-

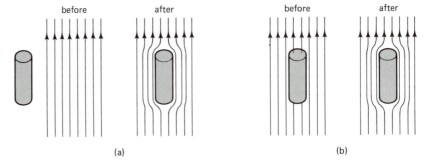

Fig. 8.33 (a) A superconductor initially outside of the magnetic field pushes the field lines aside when transported into the magnetic field. (b) A normal conductor ($T > T_c$) initially in the magnetic field expels the field lines when it becomes a superconductor ($T < T_c$).

ductor. This behavior is characteristic of a perfect conductor, but a superconductor is more than just a perfect conductor. A superconductor not only prevents the penetration of magnetic field lines that are initially outside of the superconducting material, but it also *expels* any magnetic field lines that were initially inside the material, before it became superconducting (see Fig. 8.33b). This behavior differs from that of a mere perfect conductor. For instance, if we convert a ball of gas into a plasma (by ionizing it), the magnetic field lines are trapped (or frozen) in the plasma; when we switch the external magnet off, the magnetic field lines inside the plasma remain unchanged, the induced currents in the plasma generating just enough magnetic field to keep the flux constant. In contrast, the superconductor expels the magnetic field lines instead of trapping them. This expulsion of magnetic flux during the transition from the normal to the superconducting state is called the **Meissner effect.*** It means that the superconductor is not only a perfect conductor, but also a perfect diamagnet. The expulsion of the magnetic field from the volume of the superconductor is brought about by eddy currents that flow along the surface; the magnetic field of these currents produces a cancellation of magnetic field in the volume of the superconductor. These currents actually flow in a thin surface layer on the superconductor, about 10^{-6} cm deep. Within this depth, called the **London penetration depth,**† the strength of the magnetic field decreases exponentially from its external value to zero.

The ideal Meissner effect, with complete expulsion of the magnetic flux from the entire volume of the metal, only takes place if the metal has the shape of a very long cylinder (a wire), oriented parallel to the magnetic field. For other shapes, the extent of expulsion of magnetic flux depends on the geometry. In the extreme case of a large slab of metal (a large plate) perpendicular to the magnetic field, there is no expulsion of flux at all. This is an immediate consequence of the requirement that magnetic field lines cannot stop or start anywhere (no sources or sinks of field lines). If a large superconducting plate is located between the poles of a magnet (see Fig. 8.34), the field lines have no choice but to penetrate through the plate. Under these conditions, the volume of the plate

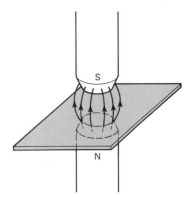

Fig. 8.34 Large flat superconducting plate between poles of a magnet.

---

* **Karl Wilhelm Meissner,** 1891–1959, German and later American physicist, professor at Purdue.

† **Fritz London,** 1900–1954, German and later American physicist, professor at Duke University collaborated with **Heinz London,** 1907–   , German and later British physicist, Chief Scientist at the Atomic Energy Research Establishment at Harwell.

Fig. 8.35 Superconducting domains (white) and normal domains (black) seen on the surface of a plate of pure tantalum placed in a magnetic field. The magnification is 50X. (Courtesy U. Essmann, Max-Planck-Institut f. Metallforschung.)

in the magnetic field splits into domains of superconducting material and of normal material. These domains are thin, parallel sheets (laminae) folded in a convoluted pattern around the direction of the magnetic field lines. Figure 8.35 shows these sheets seen end-on where they intersect the surface of the plate. The magnetic field is zero in the superconducting domains, but it is not zero in the normal domains. Thus, a metal with such intermingled domains will not expel the magnetic flux from its entire volume. In a strong magnetic field, the size of the normal domains increases at the expense of the superconducting domains and, when the field reaches the critical value, the entire volume of metal becomes normal.

In pure metals, such as those listed in Table 8.4, the expulsion of magnetic flux from each of the superconducting domains in the metal is an all-or-nothing affair: if the metal is held at a fixed temperature and immersed in a magnetic field, it will prevent the penetration of the magnetic flux as long as the magnetic field is weaker than the critical value; but the metal will suddenly cease to be a superconductor when the magnetic field becomes stronger than the critical value, and it will then freely permit the penetration of the magnetic flux.

In alloys, the expulsion of magnetic flux from a superconducting domain is a much more complicated affair. The alloy will permit a partial penetration of flux if the magnetic field is of intermediate strength; and it will finally cease to be a superconductor and permit complete penetration of flux when the magnetic field becomes strong enough. Thus, an alloy has two critical values of magnetic field strength: at a value B_{c_1}, the flux begins to penetrate and at a value B_{c_2} the flux penetrates completely and superconductivity breaks down. For example, at a temperature of 4.2 K, the niobium–tin alloy Nb_3Sn has critical values $B_{c_1} = 0.019$ T and $B_{c_2} = 22$ T.

Superconductors, such as Nb_3Sn, that permit a partial penetration of the magnetic flux when immersed in a magnetic field of intermediate strength are called **superconductors of the second kind.** In a magnetic field of intermediate strength, such a superconductor is in a mixed state: the bulk of the material is superconducting, but it is threaded by very thin filaments of normal material; these filaments are oriented parallel to the external magnetic field and they serve as conduits for the penetrating lines of this external magnetic field. A current circulates around the perimeter of each filament; this current shields the bulk of the superconductor from the magnetic field in the filament. The flow of this current has the character of a vortex; because of this, the filaments are called **vortex lines.**

Theoretical considerations (see below), confirmed by experiments, show that the amount of flux associated with each vortex line has a fixed value

$$\Phi_0 = \frac{h}{2e} = 2.06785 \times 10^{-15} \, \text{T} \cdot \text{m}^2 \tag{32}$$

In a superconductor of the second kind, an increase of the strength of the external magnetic field will not cause an increase of the flux in each vortex line; instead, it will cause an increase in the number of vortex lines threading the superconductor. The stronger the external magnetic field, the more densely will the vortex lines be packed. Figure 8.36 shows the vortex lines (view end-on) in a sample of niobium; the vortex lines are packed in a regular triangular pattern.

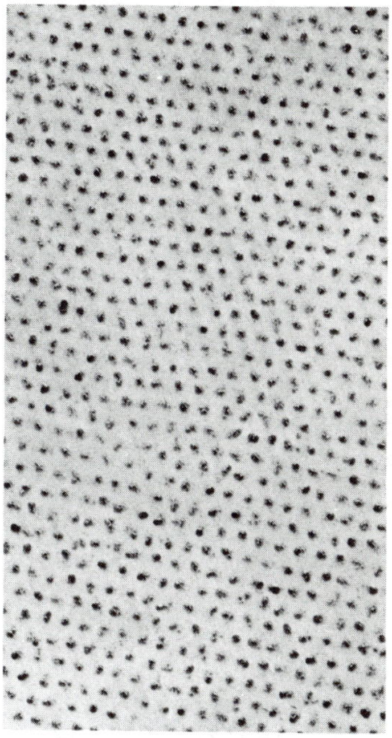

The details of the mechanism underlying superconductivity were finally spelled out in the **Bardeen–Cooper–Schrieffer (BCS)* theory** of superconductivity, some fifty years after the first experiments. The key to this mechanism is an interaction between the free electrons in the metal, an interaction that occurs via the lattice. Roughly, this involves the following: the negative charge of each free electron exerts an attractive force on the positive charges of the ions of the lattice; consequently, the nearby ions contract slightly toward the electron. This slight concentration of positive charge, in turn, attracts other electrons. The net effect is that a free electron exerts a small attractive force on another free electron.

Although the attractive electron–electron force is too small to be of any consequence at normal temperatures, it is strong enough to permanently bind two electrons into a pair, called a **Cooper pair,** when the temperature is within a few degrees of absolute zero, where the thermal disturbances nearly disappear. In a superconducting material in electrostatic equilibrium (no current), each Cooper pair consists of two electrons of opposite momenta. Obviously such a configuration makes no sense from a classical point of view: if two particles have opposite and constant momenta, they will travel away from each other in opposite directions; they will then cease to interact and they cannot remain

Fig. 8.36 Triangular pattern of vortex lines seen on a niobium disc placed in a magnetic field of 0.098 T. The magnification is 15400X. (Courtesy U. Essmann, Max-Planck-Institut f. Metallforschung.)

---

* **John Bardeen,** 1908– , **Leon N. Cooper,** 1930– , and **John Robert Schrieffer,** 1931– , American physicists. Bardeen's work on semiconductors at Bell Laboratories led to the invention of transistor, for which he received a Nobel Prize in 1956 (jointly with William Shockley and Walter H. Brattain). He moved to the University of Illinois, where he collaborated with Cooper and Schrieffer on the theory of superconductivity, for which the three shared the Nobel Prize in 1972.

bound. However, the configuration makes sense from a quantum-mechanical point of view, where each particle is described by a wave: even if two waves have opposite directions of motion they can continue to overlap for a long time and they can continue to interact.

Since the interaction of the electrons with the lattice plays a crucial role in the mechanism of superconductivity, it is not surprising that the best of the normal conductors, such as silver, copper, and gold, fail to display superconductivity. In these conductors the interaction of the electrons with the lattice is fairly small; this permits the electrons to move with fairly small friction, but it also prevents the electrons from forming the Cooper pairs required for superconductivity. A simple experimental observation confirms that the lattice plays some role in superconductivity: the critical transition temperature of a superconducting material depends on the mass of the ions of the lattice—if we replace the atoms by heavier atoms (replace isotopes by heavier isotopes), then the transition temperature decreases somewhat. This **isotope effect** gave the earliest indication that the lattice is intimately involved in the mechanism of superconductivity.

The wavefunction of a Cooper pair extends over a fairly large volume, and overlaps the wavefunctions of other Cooper pairs. In a typical superconductor, the volume of a given pair encompasses as many as 10^6 other pairs. This dense overlap gives rise to strong correlations among the motions of all the pairs. Hence the superconducting state is a **collective state,** in which all the conduction electrons act cooperatively. The wavefunction describing the superconducting state extends coherently over the entire volume of the superconductor. Thus, the electrons in superconductors display quantum effects on a macroscopic scale, in contrast to the electrons in individual atoms or molecules, which display quantum effects only on a microscopic scale. In the words of Schrieffer: the conduction electrons in a superconductor are condensed into a single "macromolecule," which extends over the entire volume of the superconducting system and is capable of motion as a whole.

The characteristic properties of superconductors—zero resistivity, Meissner effect—can be explained in terms of the energy spectrum of the excited states of the "macromolecule." The energy spectrum has a gap of the order of 10^{-3} eV between the highest occupied state and the first excited state (see Fig. 8.37).* For a

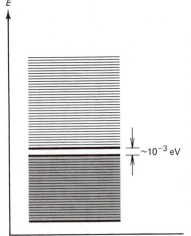

Fig. 8.37 Energy bands of a superconductor. The electrons completely fill the band below the energy gap.

$\sim 10^{-3}$ eV

---

* This gap is reminiscent of the gap found in insulators; but whereas in insulators the fully occupied band below the gap cannot conduct current, in superconductors this band *does* conduct.

superconductor at zero temperature, the width of the gap is directly proportional to the magnitude of the critical transition temperature T_c,

$$E_g = 3.52kT_c$$

Essentially, this gap represents the energy needed to break up one of the Cooper pairs.

Because of this energy gap, the "macromolecule" in the superconducting ground state resists perturbations, unless the energy of the perturbation exceeds the gap energy. This means that the wavefunction of the ground state of the "macromolecule" has a certain stiffness or rigidity—it resists changes of state. This rigidity explains the zero resistivity of the superconductor. An electric current along the superconductor involves an overall rigid translational motion of the "macromolecule." Such a translational motion requires only a small amount of energy—it involves no change of the correlations among the electrons or transition to an excited state above the energy gap. The motion proceeds without friction because the random scattering of an electron by an irregularity in the lattice would affect the correlations of the electrons, and require a transition to an excited state, above the energy gap; thus, the scattering is inhibited. The rigidity of the wave function also explains the Meissner effect. When the superconductor is immersed in a magnetic field, the wavefunction of the "macromolecule" does not change, i.e., the orbital configuration of the electrons does not change. However, the electrons, or the Cooper pairs, change speed while remaining in the same orbital configuration. These changes of speed of the charge carriers amount to an induced current, which expels the magnetic field from the volume of the superconductor. Such a change of speed without a change of orbital configuration is analogous to what happens in the Bohr model if an electron in a circular orbit is gradually immersed in a magnetic field. As is well known,* the electric field induced by the increase of magnetic flux accelerates the electron, changing its speed without changing its orbital radius; this change of speed gives rise to the diamagnetism of atoms. The analogous change of speed of the Cooper pairs in the "macromolecule" gives rise to the perfect diamagnetism of superconductors.

The large-scale coherence of the wavefunction in the superconductor leads to several remarkable phenomena, such as magnetic

---

* See, e.g., H. C. Ohanian, *Physics*, Section 33.4.

flux quantization and the Josephson effect. We will conclude this chapter with a breif discussion of these two phenomena.

Flux Quantization. Consider a body with a hole, such as a ring, placed in a magnetic field (Fig. 8.38a). If we reduce the temperature of the ring and make it superconducting, the magnetic flux in the hole will be trapped—the magnetic field lines cannot pass through the surrounding superconducting material, and hence must remain where they are. The magnetic field lines will remain in the hole even if we now reduce the original external magnetic field to zero; this changes the pattern of the field lines into that of a ring of current (see Fig. 8.38b), but does not change the number of lines within the hole. Long before the days of BCS theory, F. London predicted that the trapped flux must be quantized, i.e., the flux must be an integer multiple of a fundamental quantum of flux. London showed that this quantization of flux is a consequence of the phase relationships of the wavefunction of the electrons in the presence of the magnetic field; his argument relied on an analysis of the effect of the magnetic vector potential on the wavefunction; here we will give a simplified, nonrigorous version of the argument, without invoking the vector potential.

As a first step in our argument, we need to know the effect of a magnetic field on the phase of a wavefunction. This effect is given by a simple mathematical rule: if a particle of charge q moves around a closed path encircling some magnetic field, then the phase change δ contributed by the magnetic field is proportional to the magnetic

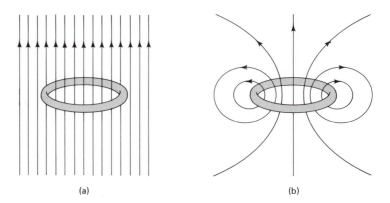

(a)   (b)

Fig. 8.38 (a) A superconducting ring placed in a magnetic field intercepts magnetic flux. (b) If the external magnetic field is removed, the magnetic flux will remain trapped, and a current will flow around the superconducting ring.

flux Φ intercepted by the area within the path,

$$\delta = \frac{q}{h}\Phi \qquad (33)$$

Note that in this context, the phase δ is the difference between the phase changes for the same path with and without the magnetic field.

To justify this equation for the phase change, let us examine a simple case of deflection of the charged particle by a magnetic field. Figure 8.39 shows a narrow region of width Δy in which there is a uniform magnetic field. A charged particle of speed v passes through this region and emerges with a (small) transverse deflection. We can calculate this deflection according to classical mechanics, from the magnetic force. Assuming that the region of magnetic field is sufficiently narrow, we can use an impulse approximation. The time the particle spends in the magnetic field is approximately $\Delta y/v$. Since the tranverse magnetic force is qvB, the transverse momentum the particle acquires is $p_x = -qvB \times \Delta y/v = -qB\,\Delta y$, and the deflection angle is

$$\theta = \frac{|p_x|}{p} = \frac{qB\,\Delta y}{p} \qquad (34)$$

In this calculation we have adopted a particle picture. But we ought to be able to deduce the same deflection by adopting a wave picture. The deflection is then due to a change in the direction of the wavefronts, brought about by transverse variation in the wavelength.* Figure 8.40 shows the instantaneous wave fronts, one wavelength apart. Since the direction of advance of the waves is supposed to swing toward the left, the wavelength in this figure must decrease from right to left along each wave front in the magnetic field. The deflection angle can be expressed as

$$\theta = \frac{\Delta l}{\Delta x} \qquad (35)$$

where Δl is the difference between the heights of the right and left edges of the group of wave fronts in the magnetic field (see Fig. 8.40).

---

* Since the magnitude of the momentum is unchanged when a particle enters a magnetic field, the de Broglie relation $\lambda = h/p$ suggests that the wavelength is also unchanged. But the de Broglie relation is *not* valid in a magnetic field; the relationship between λ and p is more complicated in a magnetic field (it involves the magnetic vector potential).

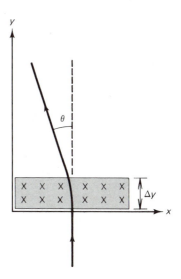

Fig. 8.39 Uniform magnetic field in a narrow region of width Δy in the x-y plane. The magnetic field is perpendicular to this plane; the crosses show the tails of the magnetic field vectors. A positively charged particle passing through this magnetic field is deflected toward the left.

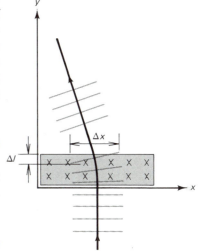

Fig. 8.40 Instantaneous wavefronts of the particle passing through the magnetic field.

Comparing this with Eq. (34), we find that

$$\frac{\Delta l}{\Delta x} = \frac{q\,\Delta y\,B}{p} \tag{36}$$

or

$$\Delta l = \frac{q}{p}\,\Delta x\,\Delta y\,B \tag{37}$$

The phase difference between two adjacent, roughly parallel paths in the magnetic field (see Fig. 8.41) is then

$$\delta = 2\pi\,\frac{\Delta l}{\lambda} = \frac{2\pi q}{\lambda p}\,\Delta x\,\Delta y\,B \tag{38}$$

But λp is h, by de Broglie's relation; and $B\,\Delta x\,\Delta y$ is the magnetic flux. Hence

$$\delta = \frac{2\pi q}{h}\,\Phi \tag{39}$$

This phase difference between the two paths is equal to the net phase change produced by the action of the magnetic field on a particle that travels up along one path and down along the other, completing a roundtrip (we assume that the paths merge at some distance above and below the edge of Fig. 8.41; exactly how this is brought about is irrelevant—it does not affect the phase contributed by the magnetic field).* Thus, the analysis of the phase relationships for a particle moving through a uniform magnetic field leads us to Eq. (33).

Although we have obtained Eq. (33) in the context of a rather special and simple example, the result is valid in general, for any configuration of path and magnetic field. Note that the phase depends on the magnetic field only indirectly, through the magnetic flux. This has the surprising consequence that the wave suffers a phase shift even if it never comes into direct contact with the magnetic field, for instance, if the magnetic field is confined to the interior

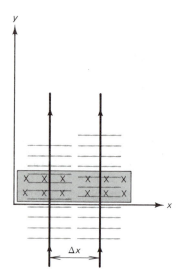

Fig. 8.41 Paths of two charged particles passing through the magnetic field. The paths are approximately parallel (the deflections of the particles have been ignored in drawing this figure). The particle on the left has the shorter wavelength in the magnetic field.

---

* Incidentally: cursory examination of the deflection of a particle traveling in the downward direction suggests that $\Delta\lambda/\Delta x$ has to be negative, to obtain the deflection toward the right demanded by a classical calculation for a particle traveling in the downward direction. This would be in conflict with Eq. (35). Actually, there is no such conflict because, for a particle traveling in the downward direction, the wavelength is effectively negative, or, more precisely, the wavenumber k_z is negative. Taking into account the sign of k_z, it is then easy to check that the values of $\Delta k_z/\Delta x$ are the same for the upward motion ($k_z > 0$) and the downward motion ($k_z < 0$).

of a solenoid and the particle moves along some path *outside* the solenoid. This is another illustration of the drastic differences between classical mechanics and wave mechanics.

We are now ready to apply Eq. (33) to the case of the superconducting ring. The "particles" moving around this ring are Cooper pairs of charge $q = -2e$. Their wavefunction is coherent around the entire circumference of the right, i.e., it is everywhere well defined. Thus, the phase must change by an integer multiple of 2π for a closed circular path around the ring:

$$2\pi n = \frac{2e}{h} \Phi \tag{40}$$

This yields

$$\Phi = \frac{h}{2e} n \tag{41}$$

where n is a positive or negative integer, or zero. According to this quantization condition, the flux is an integer multiple of the fundamental **magnetic flux quantum** $\Phi_0 = h/(2e) = 2.07 \times 10^{-15}\,\text{T}\cdot\text{m}^2$. This quantization has been experimentally confirmed in delicate experiments, first performed by Deaver and Fairbank.*

The filaments of normal material in a superconductor of the second kind may be regarded as "holes" in the superconductor. As in the case of a superconducting ring, the flux in the hole must therefore be quantized. Indeed, the value of the flux quoted in Eq. (32) corresponds exactly to one quantum of flux.

Josephson Effect. If a thin layer of insulator is sandwiched between two pieces of superconductor (Fig. 8.42), the Cooper pairs can tunnel through the insulator, giving a current from one superconductor into the other. The tunneling of a current through an insulating layer was discovered experimentally by I. Giaever. The behavior of the current through an insulating layer sandwiched between two superconductors was first analyzed theoretically by B. D. Josephson,[†] and the arrangement of insulator and two superconductors shown in Fig. 8.42 is called a **Josephson junction.** In practice, such a junction is manufactured by depositing a thin layer

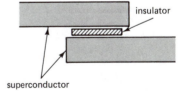

Fig. 8.42 A Josephson junction.

---

* B. S. Deaver and W. M. Fairbank, *Phys. Rev. Lett.* **7,** 34 (1961).

[†] **Ivar Giaever,** 1929– , American physicist at the General Electric Company and professor at Rensselaer Polytechnic Institute and **Brian D. Josephson,** 1940– , English physicist, professor at Cambridge, received the Nobel Prize in 1973.

of an oxide, only 10 to 20 Å thick, on the surface of one superconductor, and then placing the second superconductor on top of this.

The layer of insulator acts as a potential barrier for the Cooper pairs. A fraction of the current approaching this barrier from the left or the right manages to tunnel through the barrier and continue on the other side, without any net loss of energy. Thus, the junction permits the flow of a current even if the potential difference across it is zero. This is called the **DC Josephson effect.** Since it results from the familiar tunneling phenomenon of quantum mechanics, it will not seem very surprising. But what is surprising is that if we apply a DC voltage across the junction, the result is an AC current—the constant voltage generates an oscillating current! The frequency of this oscillating current is directly proportional to the voltage:

$$v = \frac{2e\,\Delta V}{h} \tag{42}$$

This is called the **AC Josephson effect.** It is caused by the coupling between the wave in one superconductor and the wave in the other. Crudely, it is analogous to the classical beat phenomenon seen in the superposition of two waves of different frequencies. The wavefunctions of a Cooper pair on the two sides of the junction differ in energy by $2e\,\Delta V$, and hence differ in frequency by $2e\,\Delta V/h$. Since the charge density and the current only depend on $|\psi|^2$, the individual wave frequencies are not observable. However, when the waves are coupled, the net amplitude is modulated by the beat frequency, which equals the frequency difference. This modulation shows up in the current.

According to Eq. (42), a voltage of 1×10^{-6} V generates an AC current of a frequency of 483.6 MHz. Measurement of the frequencies of AC Josephson currents can be used as a very precise and convenient method for the measurement of voltages. Potentiometers based on this method attain a precision of about 1 part in 10^8.

Chapter 8:   SUMMARY

Drift velocity of conduction elections:

$$v_d = -\frac{eE\tau}{m_e}$$

Conductivity:

$$\sigma = \frac{e^2 n \tau}{m_e}$$

Fermi energy:

$$E_F = (3\pi^2)^{2/3} \frac{\hbar^2}{2m_e} \left(\frac{N}{V}\right)^{2/3}$$

Average energy per electron:

$$\tfrac{3}{5} E_F$$

Effective mass:

$$E = E_0 + \frac{\hbar^2}{2m^*} k^2$$

Energy gap in insulator:

Several eV

Energy gap in semiconductor:

$$\sim 1 \text{ eV}$$

Energy gap in superconductor:

$$E_g = 3.52 k T_c \sim 10^{-3} \text{ eV}$$

Phase change due to magnetic field:

$$\delta = \frac{q}{\hbar} \Phi$$

Flux quantization:

$$\Phi = \frac{h}{2e} n$$

AC Josephson effect:

$$\nu = \frac{2e}{h} \Delta V$$

Chapter 8: PROBLEMS

1. Consider the crystals of HF, Na, Hg, Ne, KCl, Si, SiC, and LiF. What kinds of bonds do you expect to find in each of these crystals? Explain why.

2. In a crystal of NaCl, the ions of Na^+ and Cl^- are arranged on a cubic lattice. Given that the density of NaCl is 2.16 g/cm^3, calculate the distance between the centers of adjacent ions (atomic masses are listed in Appendix 3).

3. The crystal of KCl consists of ions of K^+ and Cl^- arranged on a cubic lattice with a distance of 3.15 Å between the centers of adjacent ions. Estimate the electrostatic binding energy per ion. For the sake simplicity, pretend that each ion interacts only with its nearest positive and negative neighbors.

4. In copper, each atom contributes one electron to the free-electron gas. Calculate the number of free electrons per unit volume. The density of copper is $8.94 \times 10^3 \, kg/m^3$.

5. Given that the conductivity of potassium metal is $1.39 \times 10^7 /\Omega \cdot m$ and that the density of free electrons is $1.40 \times 10^{28}/m^3$, calculate the relaxation time.

6. A copper wire of diameter 1.0 mm and length 2.0 m carries a DC current of 5.0 A. What is the drift speed of the electrons in the copper? How long will it take an electron to travel from one end of the wire to the other? The conductivity of copper is $5.88 \times 10^7 /\Omega \cdot m$.

7. In silver, the relaxation time is $3.8 \times 10^{-14} \, s$. Using the speed calculated in Eq. (19), estimate the mean distance between collisions, and compare this distance with the interatomic distance.

8. A conducting strip of width d is placed perpendicularly and face-on to a uniform magnetic field **B**. If a current, consisting of electrons of drift speed v_d, flows along this strip, the magnetic force will deflect the electrons in the transverse direction, and cause them to concentrate at one edge of the strip. Corresponding to this excess of electrons at one edge of the strip, there will be a deficit of electrons at the opposite edge (see Fig. 8.43). This charge separation generates an electric field and an electric potential across the strip, a phenomenon called the **Hall effect**. Under equilibrium conditions, the transverse force exerted by the electric field on an electron must match the transverse force exerted by the magnetic field.

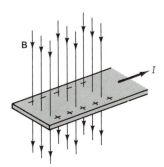

Fig. 8.43 Experimental arrangement for the Hall effect.

(a) Show that the electric field required for equilibrium is $E = v_d B$. Show that the corresponding potential difference across the strip is $\Delta V = v_d B d$.
(b) On a diagram, indicate which edge of the strip is at positive potential and which at negative.

(c) If, instead of electrons, the current carriers were holes, how would the answer to part (b) change?

(d) Evaluate the Hall potential difference numerically for a copper strip of width 1.0 cm and thickness 0.10 mm carrying a current of 120 A in a magnetic field of 2.0 T. The conductivity of copper is $5.88 \times 10^7/\Omega \cdot m$.

9. Calculate the Fermi energy for the free electrons in copper, taking into account the effective mass of these electrons.

10. What is the total kinetic energy of all the free electrons in one mole of silver?

11. In Section 8.2 we derived an equation [Eq. (20)] for the number of standing electron waves per unit energy interval in a box of volume V. Carry out a similar derivation for the number of standing light waves per unit frequency interval in a cavity of volume V. Your result should agree with the result quoted in Section 3.2.

12. Starting with Eq. (20), obtain formulas for the average speed and the rms speed of the electrons in the Fermi gas.

13. The compressibility of a material is defined as $(V \, \partial P/\partial V)^{-1}$, where ∂V is the change of volume produced by a change of pressure ∂P. Assuming that the compressibility of a "soft" metal, such as lithium or sodium, is entirely due to the compressibility of the free-electron gas, calculate the compressibilities of lithium and of sodium, and compare your results with the measured values, $8.3 \times 10^{-11} \, m^2/N$ and $14.7 \times 10^{-11} \, m^2/N$, respectively. (*Hint:* The pressure can be expressed as the derivative of the energy,

$$P = \frac{\partial E_{total}}{\partial V} = \frac{3}{5} N \frac{\partial E_F}{\partial V})$$

14. Neutron stars consist of a Fermi gas of neutrons. The density in a typical neutron star is $5 \times 10^{16} \, kg/m^3$. Calculate the Fermi energy and the Fermi speed of the neutrons.

15. If the temperature is above zero, some of the electrons in the Fermi gas in a metal will be excited to higher states, and the lowest states will not be fully occupied. On the basis of quantum statistical mechanics, it can be shown that the probability for occupation of a state of energy E is

$$f(E) = \frac{1}{e^{(E - E_F)/kT} + 1}$$

where E_F is the Fermi energy for the given temperature T.

(a) Show that $f(E) = \frac{1}{2}$ for $E = E_F$.

(b) Show that if $T \to 0$, the function $f(E)$ approaches a step function: $f(E) = 1$ for $E < E_F$, $f(E) = 0$ for $E > E_F$.

(c) Plot $f(E)$ as a function of E for $E_F = 2.0 \, eV$ and $kT = 0.3 \, eV$.

16. At ordinary temperatures, the specific heat of a Fermi gas of electrons is much smaller than the specific heat of a classical gas. At what temperature would the specific heat of the Fermi gas be equal to 10% of the specific heat of a classical gas? Assume $E_F = 5 \, eV$.

17. According to classical kinetic theory, the relaxation time for the free electrons in a metal should be approximately $\tau = 1/(n_A \pi R^2 \bar{v})$, where n_A is the number of atoms per unit volume, πR^2 the (geometrical) cross section of an atom, and $\bar{v}$ the mean speed of the electrons. Show that if the mean speed of the electrons is calculated from classical statistical mechanics, this expression for the relaxation time would make the conductivity proportional to $T^{-1/2}$. (This classical temperature dependence disagrees with the observed temperature dependence—the conductivity is observed to be proportional to T^{-1}.)

18. According to quantum theory, the scattering of the free electrons by the atoms of the crystal lattice is mainly due to the thermal oscillations of the atoms about their equilibrium positions, and the corresponding relaxation time is approximately $\tau = 1/(n_A \pi R^2 v_F)$, where n_A is the number of atoms per unit volume, R is the root-mean-square amplitude of the thermal oscillations, and v_F is the Fermi velocity [see Eq. (19)]. Show that this expression for the relaxation time implies that the electric conductivity is proportional to T^{-1}. (*Hint:* Use the classical result for the amplitude of the thermal oscillations of the atoms.)

19. The temperature coefficient of resistivity is defined as

$$\alpha = \frac{1}{\rho} \frac{\partial \rho}{\partial T}$$

where $\rho = 1/\sigma$ is the resistivity. For a metal with $\sigma \propto 1/T$ and $\rho \propto T$, evaluate the temperature coefficient of resistivity at 273 K and at 373 K.

20. According to a simple model, the energy of a free electron in a crystal is the following function of the wave number:

$$E = A - B \cos ka$$

where A and B are constants, and a is the distance between adjacent atoms. What value of the effective mass can you deduce from this formula?

21. For a crystal lattice with a spacing $a = 2.0\,\text{Å}$, what is the range of electron wavelengths in the first Brillouin zone? In the second Brillouin zone?

22. Equation (26) is a good approximation provided k is near the bottom of a Brillouin zone, for instance, near the bottom of the first Brillouin zone where $k \ll \pi/a$. This inequality is always satisfied for the free electrons of the lowest kinetic energies, but it is liable to fail for the free electrons of the larger kinetic energies, near the Fermi energy. Is this inequality satisfied for the electrons in silver of an energy equal to the Fermi energy? Assume that these electrons have a speed $\sqrt{2E_F/m_e}$ [compare Eq. (19)], and that the lattice spacing in silver is roughly 3 Å.

23. If an arsenic atom is substituted for one of the silicon atoms in a silicon crystal, the atom tends to release its outermost electron and become an ion, As^+. The ionization energy is quite low because the electron moves in a large orbit that passes through the surrounding silicon material, which acts as a dielectric and therefore brings about a reduction of the Coulomb field of the As^+ ion. Furthermore, the effective mass of such an electron moving in silicon is less than m_e.

Using the Bohr model, calculate the orbital radius and the ionization energy for an electron in the smallest possible orbit around a (pointlike) As^+ ion immersed in silicon. Assume that the Coulomb force has a magnitude $e^2/(4\pi\varepsilon_0\kappa)$, where $\kappa = 11.7$ is the dielectric constant of silicon, and assume that the effective mass of the electron is $m^* = 0.25m_e$. Express the orbital radius in Å and the ionization energy in eV.

24. The transition temperature of a superconductor is inversely proportional to some power of the mass of the ions of the lattice,

$$T_c \propto \frac{1}{M^\alpha}$$

For mercury, $\alpha = 0.50$. Calculate the transition temperature for samples of ^{201}Hg, ^{202}Hg, and ^{204}Hg, given that the transition temperature for ordinary mercury, of an average atomic mass 200.59 u, is 4.153 K.

25. What is the magnitude of the current that will destroy the superconductivity of a long wire of mercury of radius 1.0 mm at (nearly) 0 K? Use the data presented in Fig. 8.32.

26. A very long solenoid has a diameter of 6.0 cm. It is wound with superconducting wire with 1000 turns per meter. If a current of 160 A flows through the wire, what is the magnetic flux through the cross section of the solenoid? How many flux quanta does this amount to?

27. A very long solenoid made of superconducting material has a diameter of 4.0 cm. In consequence of flux quantization, the magnetic field and the current density (current per unit length) in the solenoid will be quantized. What is the magnitude of the quantum of magnetic field? What is the magnitude of the quantum of current density?

28. For each of the superconductors listed in Table 8.4, calculate the energy gap in the spectrum of excited electron states.

29. Consider a narrow region of width Δy within which there is a uniform electric field in the x-direction. A charged particle, originally moving in the y-direction, passes through this region. Calculate the deflection of the particle by means of classical mechanics, using the impulse approximation. Calculate the deflection by means of wave mechanics, using the frequency difference $\Delta\omega = q\,\Delta V/\hbar$ between two portions of the wave at different potential, and verify that your results from classical and from wave mechanics agree. (*Hint:* Examine a given wave front at the times t and $t + \Delta t$.)

30. Measure the lattice spacing of the vortex lines in Fig. 8.36 and from this and from the (average) magnetic field strength, deduce the magnitude of the flux quantum.

Nuclear Structure

From Rutherford's experiments on the bombardment of atoms with alpha particles and from other similar scattering experiments we know that the nucleus of the atom is very small—about 10,000 times smaller than the atom. Yet it contains almost all the mass of the atom. This means that the density of the nuclear material is enormous—about $2 \times 10^{17} \, \text{kg/m}^3$. According to the investigations of Moseley (see Section 4.7), a nucleus of atomic number Z contains Z positive charges. This means that the average charge density in the nuclear material is also enormous—about $10^{25} \, \text{C/m}^3$.

As we will see, the nucleus is made of protons and neutrons very closely packed together. The repulsive Coulomb force between the protons would burst the nucleus apart if there were no extra, attractive force holding it together. This extra force is called the **nuclear force,** or the **"strong" force.** For two adjacent protons in a nucleus, this force is about 100 times stronger than the repulsive Coulomb force. This strong force governs the dynamics of the nuclear protons and neutrons, just as the Coulomb force governs the dynamics of the atomic electrons. Due to the greater strength of the strong force, the excitation energies of the nuclear states are much greater than the excitation energies of the atomic states. The energy differences between atomic states amount to one or several eV, whereas the energy differences between nuclear states amount to one or several MeV. Transitions between atomic states lead to

the emission of visible light or of X rays, whereas transitions between nuclear states lead to the emission of gamma rays.

Unfortunately, the strong force cannot be described by any
simple formula—such as the formula for Coulomb's law or
Newton's law of gravitation—and its behavior as a function of
distance is only imperfectly known. Consequently, nuclear physicists cannot calculate the stationary states of nuclei from first
principles, in the way atomic physicists calculate the states of atoms. Instead, nuclear physicists often rely on theoretical **models**
of the nucleus, such as the liquid-drop model or the shell model.
Such models are caricatures of the real world. They are sketchy
theoretical pictures that incorporate some part of reality and explain some aspect of nuclear structure, but do not give a comprehensive and rigorous account of all aspects.

9.1 **Isotopes**

When the masses of the atoms in a chemically pure sample of some
element are measured with a mass spectrometer (see Section 1.3),
it is found that such a chemically pure sample is a mixture of atoms
of different masses. Atoms that are chemically identical but differ
in mass are called **isotopes.** For example, neon has twelve isotopes
designated ^{16}Ne, ^{17}Ne, ^{18}Ne, ^{19}Ne, ^{20}Ne, ^{21}Ne, ^{22}Ne, ^{23}Ne, ^{24}Ne,
^{25}Ne, ^{26}Ne, and ^{27}Ne, whose masses range from 16.03 to 27.01 u
(see Table 9.1). The superscript placed before the chemical symbol
is called the **mass number;** it equals the mass in atomic units,

TABLE 9.1   The Isotopes of Neon

Isotope	Mass	Number of Protons	Number of Neutrons	Radioactivity
^{16}Ne	16.02575 u	10	6	p
^{17}Ne	17.01769	10	7	β^+
^{18}Ne	18.00571	10	8	β^+, γ
^{19}Ne	19.00188	10	9	β^+
^{20}Ne	19.99244	10	10	stable
^{21}Ne	20.99384	10	11	stable
^{22}Ne	21.99138	10	12	stable
^{23}Ne	22.99447	10	13	β^-, γ
^{24}Ne	23.99361	10	14	β^-, γ
^{25}Ne	24.99768	10	15	β^-
^{26}Ne	26.00047	10	16	
^{27}Ne	27.00725	10	17	

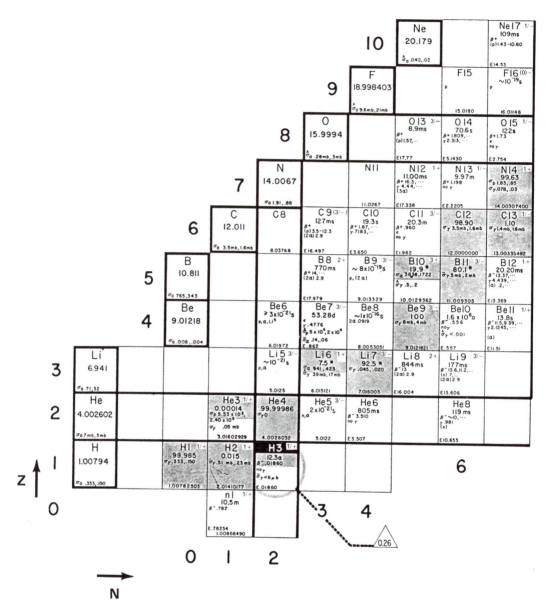

Figure 9-1

rounded off to the nearest integer (more precisely, this superscript is the sum of the number of protons and the number of neutrons in the nucleus; see below.) Naturally occurring samples of neon consist of a mixture of the isotopes ^{20}Ne (90.92%), ^{21}Ne (0.257%), and ^{22}Ne (8.82%). The other isotopes of neon do not occur naturally; they are very unstable and they can only be produced artifi-

Ne18 1.67s β+3.42,··· γ1.042,··· E4.45	**Ne19** 1/+ 17.22s β+2.24,···,ε γ1.357,··· E3.238	**Ne20** 90.51 σᵧ.037,.02 19.992436	**Ne21** 3/+ 0.27 σₐ<1.5 A σᵧ.7,.3 20.993843	**Ne22** 9.22 σᵧ.048,.02 21.991383	**Ne23** 5/+ 37.2s β⁻4.38,3.94,··· γ.440,··· E4.376	**Ne24** 3.38m β⁻1.98,··· γ.472,30,··· E2.47	**Ne25** (1/)+ 0.61s β⁻7.3,··· γ.0895,··· E 7.2	**Ne26** 26.0005	**Ne27** 27.0072	**Ne28**
F17 5/+ 64.7s β+1.74 no γ E2.761	**F18** 1+ 109.8m β+.635 ε E1.655	**F19** 1/+ 100 σᵧ 9.6mb, 21mb 18.9984032	**F20** 2+ 11.0s β⁻5.40,··· γ1.633,··· E7.029	**F21** 5/+ 4.33s β⁻5.34,··· γ.3507,··· E5.69	**F22** 4+ 4.23s β⁻5.5,··· γ1.275, 2.083, 2.166,··· E10.85	**F23** (5/)+ 2.2s β⁻8.3,··· γ 1.701, 2.129,··· 1.822,··· E8.5	**F24** 24.0093	**F25** 25.0138	**F26**	**F27**
O16 99.762 σᵧ.19mb,.3mb 15.99491461	**O17** 5/+ 0.038 σₐ.24,.11 A σᵧ.4mb 16.999131	**O18** 0.200 σᵧ.16mb,.85mb 17.999160	**O19** 5/+ 26.9s β⁻3.25,4.6,··· γ.197,1.357,··· .110,··· E4.819	**O20** 13.5s β⁻2.75 γ1.057 E3.82	**O21** 3.4s β⁻ γ.280–4.584 E8.17	**O22** 22.0101	**O23** 23.0193	**O24**	**18**	
N15 1/- 0.37 σᵧ.02mb 15.00010895	**N16** 2- 7.13s β⁻4.3,10.42,··· γ6.129,7.115,··· (α)1.28,2.02 E10.419	**N17** 1/+ 4.17s β⁻7.81,5.62,··· γ.871,2.184 (n)1.171,.383,··· E8.68	**N18** 1- 0.63s β⁻9.4 γ1.982,1.652, .822,2.474 E14.03	**N19** 0.42s β⁻ γ2.47,··· (n) E12.53	**N20** 20.0238	**N21** 21.0289	**N22**	**N23**		
C14 1/+ 5730a β⁻.156 no γ σᵧ<1μb E.15648	**C15** 1/+ 2.45s β⁻4.51,9.82,··· γ5.298,··· E9.772	**C16** 0.75s β⁻ (n).79,1.7 E8.012	**C17** 17.02257	**C18** 18.0267	**C19** 19.0370	**C20**	**16**			
B13 3/- 17.3ms β⁻13.4,··· γ3.68,··· (n)2.4,3.6,··· E13.436	**B14** 2- 16ms β⁻~14,··· γ6.09,··· E20.64			**B17** 17.0486	**14**					
Be12 24ms β⁻ (n)? E11.71		**Be14** 14.0440	**12**							
Li11 8.7ms β⁻ γ3.368,··· (n) (2n),(3n) E20.7	**10**									

8

Fig. 9.1 Excerpt from a chart of isotopes. The number at the bottom of each box gives the atomic mass in u or, if preceded by the letter E, the energy (in MeV) released in the β decay of some of the radioactive isotopes. The small number in the upper right corner gives the spin of the nucleus, in multiples of $\hbar/2$. The number below the chemical symbol gives the abundance in percent for naturally occurring isotopes (gray) or the half life for artificial radioactive isotopes. Greek letters α, β, and γ indicate radioactive decays; ε indicates electron capture. The numbers following the letters α, β, and γ give the energies (in MeV) of the emitted rays. The numbers following the letter σ give cross sections for diverse reactions initiated by incident neutrons. (From *Chart of the Nuclides*, 13th edition, 1983. Courtesy Knolls Atomic Power Laboratory, Schnectady, NY. Operated by the General Electric Company for the Office of Naval Reactors, the United States Department of Energy.

cially by transmutation of elements, or nuclear "alchemy," in a nuclear reactor or an accelerator. The isotopes ^{20}Ne and ^{22}Ne share the distinction of being the first discovered isotopes of any chemical element. They were identified by mass spectrometry by J. J. Thomson in 1912, and they were soon thereafter separated in laborious diffusion experiments by F. W. Aston.

All chemical elements have several isotopes. Figure 9.1 shows an excerpt from a chart of the isotopes. (Appendix 4 gives a complete list of all the known isotopes.) Hydrogen has three isotopes: ^{1}H (ordinary hydrogen), ^{2}H (deuterium), and ^{3}H (tritium). Helium has five isotopes: ^{3}He, ^{4}He, ^{5}He, ^{6}He, and ^{8}He. Some of these isotopes occur in nature—they are stable or almost stable, having a very long lifetime. Others can only be produced by artificial means—they are unstable, having only a short lifetime.

Since all the atoms of a given chemical element, say, neon, are chemically identical, they must all have the same number of orbital electrons. The mass differences between the isotopes of neon must therefore be due to differences in the structure of the nuclei of these isotopes. The nuclei are made of protons and neutrons (see Table 9.2). All the isotopes of neon have the same number of protons—ten protons—but they have different numbers of neutrons. Thus, ^{17}Ne has seven neutrons, ^{18}Ne has eight, ^{19}Ne has nine, etc. The mass number A is the sum of the number of protons Z and the number of neutrons N in the nucleus,

$$A = Z + N \tag{1}$$

Protons and neutrons are generically called **nucleons;** thus A is the number of nucleons.

It is of some interest to compare the properties of protons and neutrons (summarized in Table 9.2) with those of electrons. The most obvious difference lies in the mass: protons and neutrons are about 1840 times as massive as electrons. Another notable difference lies in the internal structure of the particles: in contrast to electrons, which are pointlike, protons and neutrons are spherical bodies of a small, but finite, extent. The radius of a proton or a neutron is about 1.0×10^{-15} m. The spin of a proton or a neutron is $\frac{1}{2}$, the same as that of an electron. However, the magnetic moment of a proton or a neutron is much smaller than the magnetic moment of an electron; this is as expected from the formula for the magnetic moment of the electron, $\mu_{e,z} = -e\hbar/(2m_e)$, which shows that the magnetic moment is inversely proportional to the mass. By analogy with this formula for the electron, we are tempted to predict that the magnetic moment of the proton is $e\hbar/(2m_p)$; but its actual value is $+2.79e\hbar/(2m_p)$ (the positive sign

TABLE 9.2   The Nucleons

Nucleon	Mass	Spin	Magnetic Moment	Radius
Proton	1.00727647 u	$\frac{1}{2}\hbar$	$2.792846e\hbar/(2m_p)$	1.0×10^{-15} m
Neutron	1.00866501 u	$\frac{1}{2}\hbar$	$-1.913042e\hbar/(2m_p)$	1.0×10^{-15} m

indicates that the moment is parallel to the spin). We are also tempted to predict that the magnetic moment of the neutron is zero, since the charge is zero; but its actual value is $-1.91 e\hbar/(2m_p)$ (the negative sign indicates that the moment is antiparallel to the spin). These **anomalous** values of the magnetic moments are related to the internal structure of protons and neutrons; these particles are charge distributions of finite extent, and their magnetic moments are affected by the amount of rotating charge. In the case of the neutron, the magnetic moment arises from the presence of separate layers of positive and negative charge within the neutron (see the next section for more on the charge distributions within protons and neutrons).

Before the discovery of the neutron by J. Chadwick[*] in 1932, physicists had speculated that nuclei are made of protons and electrons, the only particles known up to that time. The existence of electrons in the nucleus was suggested by beta decay, in which the nucleus ejects an electron (the modern view is that this electron is created during the decay). For instance, according to this picture, the ^{22}Ne nucleus was supposed to be made of 22 protons and 12 electrons, giving the (correct) net electric charge $10e$. However, this picture was found to be incompatible with several items of evidence. For instance, analysis of the relative intensities of the spectral lines in the band spectrum of the ^{14}N^{14}N molecule shows that the spin of the ^{14}N nucleus is $I = 1$;[†] on the basis of the nuclear electron picture, this nucleus would contain 14 protons and 7 electrons, whose net spin would have to be $\frac{1}{2}$, or $\frac{3}{2}$, or $\frac{5}{2}$, or $\ldots \frac{21}{2}$. More evidence comes from the study of magnetic moments of nuclei. If nuclei contained electrons, we would expect their magnetic moments to be of the same order of magnitude as that of an electron. But the observed magnetic moments are typically 10^3 times smaller; they are of the order of magnitude of the magnetic moment of a proton (see Table 9.2). Another instructive argument against the nuclear electron picture emerges from a simple quantum-mechanical estimate of nuclear energies. If an electron were confined in a region of nuclear size ($L \sim 10^{-14}$ m), the energy difference between adjacent energy eigenstates would be of the

---

[*] **James Chadwick,** 1891–1974, English physicist, professor at Liverpool. He began his work with Rutherford, who long suspected the existence of a neutral particle inside the nucleus. For the discovery of the neutron, Chadwick received the Nobel Prize in 1935. During the war he headed the British Mission attached to the Manhattan Project.

[†] By convention, the nuclear spin is designated by the letter **I**, to distinguish it from the electron spin **S**.

order of [see Eq. (6.11)]

$$\Delta E \simeq \frac{\pi^2 \hbar^2}{2m_e L^2} \simeq \frac{\pi^2 \times (1.05 \times 10^{-34}\,\text{J}\cdot\text{s})^2}{2 \times 9.1 \times 10^{-31}\,\text{kg} \times (10^{-14}\,\text{m})^2}$$

$$\simeq 6 \times 10^{-6}\,\text{J} \simeq 4 \times 10^9\,\text{eV} \qquad (2)$$

This is much larger than the observed energy differences between nuclear states, typically 1 to 10 MeV. Note that for a proton, the estimate (2) gives

$$\Delta E \simeq \frac{\pi^2 \hbar^2}{2m_p L^2} \simeq 2 \times 10^6\,\text{eV} \qquad (3)$$

which is of the right order of magnitude.

Chadwick discovered that neutrons are emitted by beryllium when subjected to bombardment by alpha particles, according to the reaction

$$\alpha + {}^9\text{Be} \rightarrow {}^{12}\text{C} + \text{n} \qquad (4)$$

Earlier experimenters had noticed that some kind of neutral "radiation" was emitted in this reaction, but Chadwick deserves the credit for detecting the emitted radiation and for establishing its corpuscular character. He did this by placing a sheet of paraffin (mainly hydrogen) or of some other light atoms behind the beryllium target. Neutrons signalled their impact on the sheet by colliding with the atomic nuclei and ejecting them from the sheet. From a measurement of the energy of the ejected nuclei, Chadwick was able to calculate the mass of the neutron.

This discovery led to the modern picture of the nucleus: the nucleus is made of Z protons and $A - Z$ neutrons. This picture of the nucleus removed all the difficulties listed above. With the development of powerful accelerators, capable of generating beams of energetic electrons or pions, it became possible to explore the internal structure of the nucleus by scattering experiments, analogous to the classic alpha-particle scattering experiments of Rutherford. Such scattering experiments give quite direct evidence for the presence of neutrons in the nucleus.

Neutrons are unstable particles. A free neutron will spontaneously decay in about 15 minutes,* transmuting itself into a proton and creating an electron and an antineutrino,

$$\text{n} \rightarrow \text{p} + \text{e} + \bar{\nu} \qquad (5)$$

---

* This is the mean lifetime for the neutron. The distinction between the mean life and the half life will be discussed in the next chapter.

This reaction is called β decay because it involves the ejection of an electron, or β^- particle. The electron and the antineutrino are ejected with a (shared) energy of roughly 1 MeV. This spontaneous decay raises the question: how can the neutrons be stable within a nucleus? The answer hinges on the Exclusion Principle. Within a nucleus in its ground state, the lowest available energy states are already filled. The extra proton released in the reaction (5) must therefore go into one of the higher energy states. In most nuclei the extra proton does not have enough energy to do this, and hence the Exclusion Principle suppresses the decay of the neutrons, forcing them to remain stable.

If we compare the mass of a nucleus with the sum of the masses of its constituent protons and neutrons, we find that the mass of the nucleus is smaller than the sum of the masses of its constituents. For instance, the helium nucleus has a mass of 4.00151 u,* whereas the sum of the masses of two protons and two neutrons is $2 \times (1.00727\,u) + 2 \times (1.00866\,u) = 4.03186\,u$. Hence the mass of the helium nucleus is smaller by 0.03035 u. This difference is called the **mass defect.** In general,

$$(\text{mass defect}) = Zm_{\text{p}} + Nm_{\text{n}} - m \qquad (6)$$

where m is the mass of the nucleus.

The mass defect arises from the energy released when the constituents bind together to form the nucleus. This energy is called the **binding energy.** According to Einstein's famous energy–mass equivalence [see Eq. (1.149)] any amount of energy ΔE has a mass Δm, with

$$\Delta m = \frac{\Delta E}{c^2} \qquad (7)$$

Thus, the removal of the binding energy from the system of protons and neutrons during the process of nuclear formation implies a corresponding reduction of mass. Because of this direct relation between energy and mass, we can use Eq. (7) to calculate the binding energy B,

$$B = (\text{mass defect})c^2 = (Zm_{\text{p}} + Nm_{\text{n}} - m)c^2 \qquad (8)$$

---

* This number differs from the number listed on the chart of isotopes. The latter number is the mass of a helium atom, which includes the mass of two electrons. To obtain the mass of the ^{4}He nucleus, we must subtract two electron masses from the mass of the atom; this gives $4.0026 - 2 \times 0.00055\,u = 4.00151\,u$.

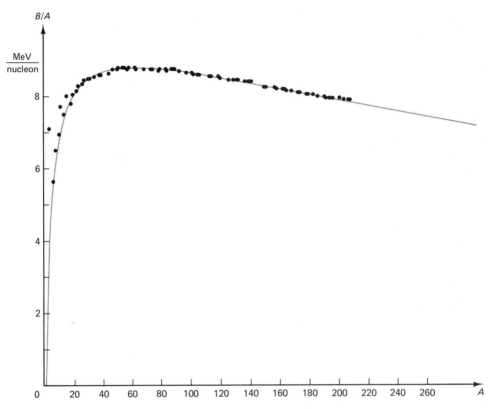

Fig. 9.2 Average binding energy per nucleon vs. mass number, for naturally occurring isotopes. The dots are based on measured binding energies; the smooth curve is based on the liquid-drop model. The curve has a maximum at $A = 56$ for ^{56}Fe.

The binding energy of heavier nuclei is usually greater than that of lighter nuclei. For the purpose of comparing binding energies of different nuclei, it is convenient to consider the average binding energy per nucleon, B/A. Figure 9.2 shows a plot of B/A vs. A for the naturally occurring isotopes. This is called the **curve of binding energy.** For most nuclei, the average binding energy per nucleon is in the vicinity of 8 MeV/nucleon. This is a rather large amount of energy. To put this number in perspective, let us compare it with the rest-mass energy of the nucleons. Each nucleon (neutron or proton) has a rest-mass energy of about $1\,\mathrm{u} \times c^2$; hence A nucleons have a rest-mass energy of about

$$A \times 1\,\mathrm{u} \times c^2 = A \times 1.5 \times 10^{-10}\,\mathrm{J} = A \times 9.4 \times 10^2\,\mathrm{MeV}$$

Thus the ratio of binding energy to rest-mass energy is about $8/(9.4 \times 10^2)$, i.e., the binding energy is almost 1% of the rest-mass energy!

The curve of binding energy has a broad maximum at $A = 56$, at the nucleus of iron. Very heavy nuclei and very light nuclei have appreciably less binding energy than iron. This implies that a heavy nucleus will release energy when we make it split apart, and two light nuclei will release energy when we make them join together. These processes are called **fission** and **fusion,** respectively. We will examine them in the next chapter.

Example 1 Chadwick also investigated neutron production by boron when subjected to bombardment by alpha particles,

$$\alpha + {}^{11}\text{B} \rightarrow {}^{14}\text{N} + \text{n} \tag{9}$$

If the alpha particle has an energy of 5.30 MeV, what is the energy of the neutron? Assume that the ^{14}N nucleus acquires next to no kinetic energy in this reaction.

Solution The energy associated with any mass m is mc^2. Hence the net energy before the reaction is

$$m_\alpha c^2 + K_\alpha + m_B c^2 \tag{10}$$

(where K_α is the kinetic energy of the alpha particle); and the net energy after the reaction is

$$m_N c^2 + m_n c^2 + K_n \tag{11}$$

(where K_n is the kinetic energy of the neutron). Since the energies (10) and (11) must be equal, we obtain

$$K_n = m_\alpha c^2 + K_\alpha + m_B c^2 - m_N c^2 - m_n c^2 \tag{12}$$

The masses 4.00260 u, 11.00931 u, and 14.00307 u for ^4He, ^{11}B, and ^{14}N listed in Fig. 9.1 are atomic masses. To obtain the nuclear masses we must subtract 2 electron masses from the helium mass, 5 electron masses from the boron mass, and 7 electron masses from the nitrogen mass. These electron masses cancel out in Eq. (12). Hence

$$K_n = 4.00260 \, \text{u} \times c^2 + 5.30 \, \text{MeV} + 11.00931 \, \text{u} \times c^2$$

$$- 14.00307 \, \text{u} \times c^2 - 1.00866 \, \text{u} \times c^2$$

$$= 0.00018 \, \text{u} \times c^2 + 5.30 \, \text{MeV}$$

To convert the mass units into energy units, it is most convenient to use the equivalence $1 \, \text{u} = 931.46 \, \text{MeV}/c^2$, which gives

$$K_n = 0.17 \, \text{MeV} + 5.30 \, \text{MeV} = 5.47 \, \text{MeV} \quad \blacksquare \tag{13}$$

The difference between the sum of the initial and the sum of the final rest-mass energies is called the **Q-value** of the reaction,

$$Q = \sum_i m_i c^2 - \sum_i m_i' c^2 \tag{14}$$

where m_i and m_i' are the masses before and after the reaction, respectively. For instance, the Q-value for the reaction (9) is

$$Q = m_\alpha c^2 + m_B c^2 - m_N c^2 - m_n c^2$$

$$= 0.17\,\text{MeV}$$

Reactions with a positive Q-value are called **exoergic;** they release energy. Reactions with a negative Q-value are called **endoergic;** they absorb energy, i.e., these reactions can proceed only if extra energy is supplied, in the form of kinetic energy of an incident projectile. The minimum kinetic energy required to initiate an endoergic reaction is called the **threshold energy.** For a projectile incident on a stationary target, the threshold energy is always larger than $|Q|$, because an appreciable fraction of the kinetic energy of the incident projectile is associated with the motion of the center of mass, and is therefore unavailable for initiating the reaction.

9.2 Nuclear Sizes and Shapes

The earliest determination of the size of a nucleus was due to Rutherford, who noted that the scattering of a particle by a nucleus showed noticeable deviations from what was expected on the basis of Coulomb's law whenever the impact parameter was very small. Rutherford correctly interpreted these deviations as due to contact between the alpha particle and the nucleus, and he deduced a radius of about 3×10^{-15} m for the aluminum nucleus.

Since Rutherford's time, many comprehensive scattering experiments have been performed, which have established that the nuclear radius is proportional to $A^{1/3}$,

$$R = r_0 A^{1/3} \tag{15}$$

with

$$r_0 = 1.2 \times 10^{-15}\,\text{m} \tag{16}$$

or

$$r_0 = 1.2 \text{ fermi} = 1.2 \text{ fm} \tag{17}$$

where the **fermi** is a unit of length, $1 \text{ fm} = 10^{-15} \text{ m}$. Equation (15) implies that the volume of the nucleus is proportional to A,

$$\frac{4\pi}{3} R^3 = \frac{4\pi}{3} r_0^3 A \tag{18}$$

Since the mass of a nucleus is roughly proportional to A, the density of the nuclear material is a constant, the same for all nuclei.

The most comprehensive scattering experiments were carried out by R. Hofstadter* and his associates in the 1950s, using high-energy electrons. These experiments not only determined the nuclear radius, but also explored the charge density throughout the nucleus. Electrons are very suitable for probing the charge density in the nucleus because they do not feel the nuclear force field; they only feel the electric forces exerted by the protons, and they readily penetrate through the nucleus. Figure 9.3 shows the charge density as a function of radial distance for some typical nuclei. From this figure we see that the nuclear surface is not sharply defined—the density gradually tapers toward zero. Thus, the meaning of the nuclear radius given by Eq. (15) is somewhat vague.

Incidentally, electron-scattering experiments have also been used to explore the charge density within individual protons and neutrons. Figure 9.4 shows the charge densities deduced from these experiments. The charge density of the proton decreases roughly exponentially (the charge density is roughly proportional to $e^{-r/a}$, with $a \cong 0.23 \times 10^{-15}$ m). The charge density of the neutron consists of an inner layer of positive charge, surrounded by a layer of negative charge, surrounded by a feeble layer of positive charge; the net charge of the neutron is, of course, zero. Note that the charge distributions of both the proton and the neutron extend somewhat beyond the nominal radius $R = 1.0 \times 10^{-15}$ m listed in Table 9.2.

Several other methods of determination of the nuclear radius give results comparable to Eq. (15). One of these is based on the **isotope shift** of the spectral lines of atoms. The nuclear charge distribution produces a (small) shift of the electronic energy levels of an atom because the electrostatic potential inside the nucleus differs from that of a point charge. Since the nuclei, and the nuclear

---

*   **Robert Hofstadter**, 1915– , American physicist, professor at Stanford. He received the Nobel Prize in 1961 for his pioneering investigations of the structure of nuclei and nucleons by electron scattering.

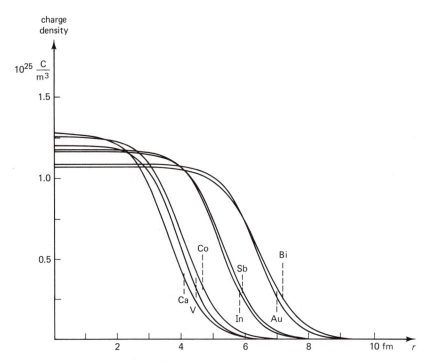

Fig. 9.3 Charge density as a function of radius for several nuclei according to electron scattering experiments by R. Hofstadter. For each nucleus, the radius $R = r_0 A^{1/3}$ is marked by a dashed line. [After R. Hofstadter, *Ann. Rev. Nuc. Sci.* **7,** 231 (1957). Reproduced, with permission, from the *Annual Review of Nuclear and Particle Science*, Volume **7,** © 1957 by Annual Reviews Inc.)

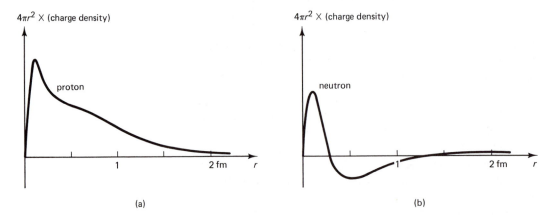

Fig. 9.4 Charge density as a function of radius in the proton (a) and in the neutron (b). Note that, in contrast to Fig. 9.3, the vertical axis represents $4\pi r^2 \times$ (charge density). [After R. M. Littauer, H. F. Schopper, and R. R. Wilson, *Phys. Rev. Lett.* **7,** 144 (1961).]

charge distributions, of different isotopes have different sizes, they will produce shifts of different magnitudes; therefore the spectral lines of the isotopes will display observable, relative shifts. From these spectral shifts, the nuclear sizes can be deduced. A similar method can be used with the isotope shifts of energy levels of muonic atoms, i.e., atoms in which a muon orbits around a nucleus. The muon is a particle with the same charge and spin as an electron, but with a mass 207 times as large (muons are easily produced in the reactions initiated when energetic particles from an accelerator strike a target). The isotope shifts for muonic energy levels in atoms are much greater than those for electronic energy levels—as a consequence of the larger mass, the orbits of muons are 207 times smaller than the orbits of electrons, and this makes the orbits more sensitive to the details of the nuclear charge distribution.

An old and very straightforward method for the determination of the nuclear radius relies on a comparison of the energies of a pair of **mirror nuclei,** in which one nucleus has protons where the other has neutrons, and vice versa. For instance, ^{35}Cl and ^{35}Ar are mirror nuclei; ^{35}Cl has 17 protons and 18 neutrons, whereas ^{35}Ar has 18 protons and 17 neutrons. Since the nuclear force does not make a distinction between protons and neutrons (it is charge independent; see the next section), the nuclear energy is the same in the two mirror nuclei. For instance, the nuclear energy contributed by the 17 protons in ^{35}Cl matches that contributed by the 17 neutrons in ^{35}Ar, and the nuclear energy contributed by the 18 neutrons in ^{35}Cl matches that contributed by the 18 protons in ^{35}Ar. The energy difference between the mirror nuclei must therefore be due to the electrostatic energy of the protons, which depends on the nuclear radius. The measurement of the energy difference between mirror nuclei therefore determines the nuclear radius. In the case of ^{35}Cl and ^{35}Ar, the energy difference can be measured very precisely by the energy released in the radioactive decay reaction that converts ^{35}Ar into ^{35}Cl with the ejection of an antielectron (positron) and a neutrino,

$$^{35}\text{Ar} \rightarrow {}^{35}\text{Cl} + e^+ + \nu \qquad (19)$$

Similar decay reactions take place in the cases of a few other mirror nuclei.

Example 2   (a) Express the difference in electrostatic energy between the ^{35}Ar and the ^{35}Cl nuclei as a function of the nuclear radius.

(b) From measurements of the energy of the positron released in the reaction (19), it is known that this difference in electrostatic energy is 6.2 MeV. Calculate the nuclear radius.

Solution   (a) Each nucleus is a sphere of radius R. From the theory of electricity* we know that the electrostatic potential energy of a sphere with a charge Ze uniformly distributed over its volume is

$$\frac{3}{5}\frac{1}{4\pi\varepsilon_0}\frac{(Ze)^2}{R} \tag{20}$$

For ^{35}Ar, $Z = 18$ and for ^{35}Cl, $Z = 17$. Hence the energy difference is

$$\Delta E = \frac{3}{5}\frac{1}{4\pi\varepsilon_0}\frac{e^2}{R}\left[(18)^2 - (17)^2\right] = \frac{3}{5}\frac{35}{4\pi\varepsilon_0}\frac{e^2}{R} \tag{21}$$

(b) With $\Delta E = 6.2\,\mathrm{MeV} = 9.9 \times 10^{-13}\,\mathrm{J}$, Eq. (21) gives

$$R = \frac{3}{5}\frac{35}{4\pi\varepsilon_0}\frac{e^2}{\Delta E}$$

$$= \frac{3}{5}\frac{35}{4\pi \times 8.85 \times 10^{-12}\,\mathrm{F/m}}\frac{(1.6 \times 10^{-19}\,\mathrm{C})^2}{9.9 \times 10^{-13}\,\mathrm{J}}$$

$$= 4.9 \times 10^{-15}\,\mathrm{m} = 4.9\,\mathrm{fm}$$

This value of the radius is slightly larger than that calculated from Eq. (15). The discrepancy can be attributed to quantum-mechanical corrections that modify our (classical) formula (20) for the electrostatic energy. ∎

Nuclei have also been explored by scattering experiments with neutrons and with protons. The results of these experiments are difficult to interpret because the neutrons and protons interact with the nucleus via the nuclear force, whose dependence on distance is not very well known (see the next section). Furthermore, in contrast to electrons, neither protons nor neutrons are pointlike particles; this complicates their use as probes because the scattering of neutrons and protons depends on both their own internal structure and on the nuclear structure. Determinations of the nuclear radius by neutron and proton scattering give results slightly larger than (15), probably because the nuclear force field extends for some distance beyond the nucleus.

In our discussion we have implicitly assumed that nuclei are spherical. This is true of most nuclei; however, some nuclei are ellipsoidal, with a difference of up to 20% between the major and minor axes. Some of these are oblate ellipsoids (flattened along one axis), some are prolate ellipsoids (elongated along one axis). For example, Fig. 9.5 shows the shape of the isotope ^{176}Lu. The information about nuclear shapes has been extracted from a careful analysis of isotope shifts of spectral lines; as mentioned above, these shifts are sensitive to the nuclear charge distribution.

Fig. 9.5 Shape of the nucleus ^{176}Lu, a prolate ellipsoid.

* See, e.g., H. C. Ohanian, *Physics*, Chapter 26.

9.3 The "Strong" Force

Since the protons in a nucleus are separated by only short distances, the repulsive Coulomb force between them is quite large. To keep the nucleus in equilibrium, this force has to be balanced by an extra, attractive force: the nuclear force or the "strong" force. At its strongest, this force is much stronger than the Coulomb force; for instance, for two protons with a center-to-center distance of 2 fm, the repulsive Coulomb force is 60 N, while the attractive strong force is about 2×10^3 N. However, the strong force is strong only over a limited range; at distances beyond 3 fm, the strong force quickly vanishes.

An important feature of the strong force is its **charge independence:** the force acting between two nucleons is independent of whether they are two protons, two neutrons, or one proton and one neutron.* Another feature is its **spin dependence:** the force between two nucleons of parallel spin is stronger than the force between two nucleons of antiparallel spin.

In contrast to the Coulomb force, with its simple $1/r^2$ dependence on distance, the strong force has a very complicated dependence on distance. In fact, the exact theoretical formula for the dependence on distance is not known, although several (very messy) approximate formulas are available. We will therefore rest content with a qualitative discussion of the dependence on distance of the strong force. The force is usually described by its potential, since this is what enters into the Schrödinger equation. Figure 9.6 is a plot of the strong potential as a function of distance for two nucleons of antiparallel spin and zero orbital angular momentum. The force is strongly attractive in the range 1–2 fm, and it has a repulsive core at distances shorter than 0.8 fm, when the nucleons deeply interpenetrate one another. The approach of the potential toward zero for distances larger than about 2 fm is described by an exponential function,

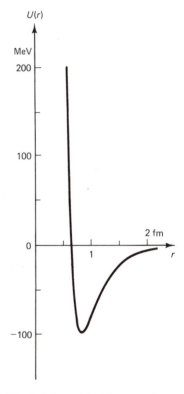

Fig. 9.6 Potential of the strong force for two nucleons of antiparallel spin and zero orbital angular momentum ($^1S_0$ state). [After R. V. Reid, *Ann. Phys.* **50**, 411 (1968).]

$$U(r) \cong -10.46 \text{ MeV} \times \frac{e^{-r/b}}{r/b} \qquad (22)$$

with $b \cong 1.43$ fm. This is called a **Yukawa potential.** In view of this exponential behavior, the potential and the force tend toward zero

---

* According to recent experimental evidence, the charge independence of the nuclear forces is not quite exact. However, the charge-dependent part of the force is very small (less than 1% of the whole) and can usually be neglected.

very quickly whenever the distance becomes large. Hence the nucleons at one end of, say, a uranium nucleus do not directly feel the strong force of the nucleons at the other end, and the nucleons in the nucleus of one atom never feel the strong force of the nucleons in the nucleus of another atom, unless the nuclei come very close in a collision.

Note that the overall shape of the plot for the strong internucleon potential is somewhat similar to that for the interatomic potential shown in Fig. 8.3, although the distance scale in Fig. 9.6 is much smaller and the energy scale much larger. This similarity is no accident; rather, it reflects a similarity in the mechanisms that produce the interatomic and the internucleon forces. The interatomic force is due to the interaction between the constituents of the two atoms. The strong internucleon force is due to the interaction between the constituents of the two nucleons. These subnucleonic constituents are small particles called **quarks.** We will discuss the evidence for quarks in Chapter 11; for now, suffice it to say that each proton or neutron is made of three quarks. As we saw in Section 8.1, the repulsive core of the interatomic force arises via the Exclusion Principle from the overlap between the electron wavefunctions of interpenetrating atoms. Similarly, the repulsive core of the internucleon force arises from the overlap of the quark wavefunctions of the interpenetrating nucleons.

The strong force is a **many-body force,** which means that the force between two nucleons in a nucleus depends on the position of all the other, neighboring nucleons. Consequently, the principle of linear superposition does not hold for the nuclear force: we cannot calculate the net force on a nucleon by taking simply the vector sum of the forces that each of the other nucleons exerts by itself; instead, when calculating the force that each of the other nucleons exerts, we must take into account how this force is modified by the presence of all the other nucleons. This feature of the strong force makes calculations in many-nucleon systems prohibitively difficult, and is largely responsible for the slow progress in the theoretical calculation of nuclear properties.

The simplest system in which we can study the action of the strong force is the **deuteron,** the nucleus of the deuterium atom (^{2}H or ^{2}D),* consisting of one proton and one neutron bound together. Experimentally, the binding energy of the deuteron is found to be

---

* Deuterium was first isolated by **Harold Urey,** 1893– , American chemist, professor at Columbia and at Chicago; for this, he received the Nobel Prize in 1934.

2.22 MeV and the net angular momentum is $J = 1$. This angular momentum is entirely due to the parallel spins of the proton and the neutron. There is no orbital angular momentum, i.e., the system is in an s state.* The time-independent Schrödinger equation for such a system of zero orbital angular momentum reduces to a simple one-dimensional equation involving the radial variable r,

$$\frac{-\hbar^2}{2\mu}\frac{d^2u(r)}{dr^2} + U(r)u(r) = Eu(r) \qquad (23)$$

where μ is the reduced mass for the system (see Section 4.3),

$$\mu = \frac{m_p m_n}{m_p + m_n} \qquad (24)$$

For the sake of simplicity, we will assume that the potential is a square well, of infinite height at $r = 0$, and of finite height U_0 at $r = L$ (see Fig. 9.7). This square well is a crude approximation to the actual potential for the strong force (see Fig. 9.6).

We have already studied the solution of the Schrödinger equation for such a square well in Section 6.3. The energy eigenvalues satisfy Eq. (6.49),

$$k \cot kL = -\kappa \qquad (25)$$

where

$$k = \sqrt{\frac{2\mu E}{\hbar^2}} \qquad (26)$$

and

$$\kappa = \sqrt{\frac{2\mu(U_0 - E)}{\hbar^2}} \qquad (27)$$

$$= \sqrt{\frac{2\mu B}{\hbar^2}} \qquad (28)$$

Equation (25) supplies a relation among E, U_0, and L; if two of these quantities are known, we can calculate the other. As mentioned above, the binding energy of the deuteron is known, $B = 2.22$ MeV. Furthermore, from scattering experiments it is known that $U_0 = 33$ MeV, so that $E = U_0 - B = 30.8$ MeV. We therefore find it expedient to use Eq. (25) to calculate the width of the

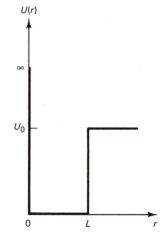

Fig. 9.7 Square-well potential for the deuteron. A constant has been added to the potential so as to make $U(r) = 0$ inside the well, in conformity with Section 6.3.

---

* A more refined analysis shows that the deuteron wavefunction has a small contribution from the p state. We will ignore this refinement here.

square well,

$$
L = \frac{1}{k} \cot^{-1}\left(\frac{-\kappa}{k}\right)
$$

$$
= \frac{\hbar}{\sqrt{2\mu E}} \cot^{-1}\left(-\sqrt{\frac{B}{E}}\right) \tag{29}
$$

$$
= \frac{1.05 \times 10^{-34}\,\text{J} \cdot \text{s}}{\sqrt{2 \times 0.837 \times 10^{-27}\,\text{kg} \times 4.93 \times 10^{-12}\,\text{J}}} \cot^{-1}\left(-\sqrt{\frac{2.22}{30.8}}\right)
$$

$$
= 2.13 \times 10^{-15}\,\text{m} \tag{30}
$$

As we might have expected, this value is in reasonable agreement with the range of the strong potential mentioned above.

Incidentally, a square well with these values of the parameters has only *one* bound state; this is in agreement with the observed absence of excited states in the deuteron.

9.4 The Liquid-Drop Model

The strong force between nucleons is somewhat similar to the intermolecular force between molecules; both forces are attractive over a short range and become strongly repulsive when the molecules or nucleons interpenetrate. Because of this similarity, we expect that an aggregate of a large number of nucleons will behave like an aggregate of a large number of molecules, that is, a solid or a liquid. It turns out that the analogy with a solid is not suitable because the zero-point energy of nucleons confined at definite lattice positions would be prohibitively large; but the analogy with a liquid is quite suitable. The hard repulsive core of the strong force makes the nuclear "fluid" nearly incompressible, while the short-range attraction provides a cohesive force that holds the fluid together; the balance of attraction and repulsion tends to keep nucleons at a definite distance from one another, and gives the fluid a definite, constant density. This provides a qualitative explanation of the constant density of the nuclear material, mentioned in Section 9.2.

The liquid-drop model of the nucleus, originally proposed by Niels Bohr, tries to take advantage of the analogy between a liquid and the nuclear material. This model treats the nucleus as a droplet of nuclear fluid. It ignores the individual nucleons and seeks to explain the properties of nuclei in terms of the gross properties of the nuclear fluid. For example, it explains the spherical shape adopted by most nuclei as follows: any nucleon located on the surface of a droplet of nuclear fluid experiences an inward force pulling it back into the volume and consequently the fluid tends to shrink its exposed surface to the smallest value compatible with its (fixed) volume. In the study of liquids, the force that tends to shrink the amount of exposed surface is called **surface tension;** by analogy, we can speak of a nuclear surface tension. Since a sphere has the least surface area for a given volume, the droplet of liquid will take spherical shape.

The liquid-drop model leads to a simple approximate formula for the binding energy of nuclei, a formula that permits us to understand the general dependence of the binding energy on A and on Z (as shown, e.g., in Fig. 9.2). This formula contains several terms, which we will discuss one by one. The most important term in the binding energy is associated with the bonds that each nucleon forms with its nearest neighbors. The binding energy is proportional to the number of bonds; since every nucleon has roughly as many neighbors as every other nucleon, the total number of bonds is simply proportional to the number of nucleons. Hence the binding energy must contain a term

$$a_1 A \tag{31}$$

where a_1 is a (positive) constant to be adjusted by comparison with the experimental data. However, (31) fails to take into account that nucleons on the nuclear surface have fewer neighbors and therefore form fewer bonds than nucleons inside the nuclear volume. Thus, (31) must be corrected by subtracting a term proportional to the number of nucleons on the nuclear surface; this number is directly proportional to the surface area $4\pi R^2$. Since $R^2 \propto A^{2/3}$ [see Eq. (15)], we can write the correction to the binding energy as

$$-a_2 A^{2/3} \tag{32}$$

where a_2 is another (positive) constant. The minus sign in (32) indicates that this term *reduces* the binding energy.

Next, we must take into account the electrostatic potential energy of the protons in the nucleus. If there are Z protons, the total positive charge within the nucleus is $Q = Ze$. We will assume that this charge is uniformly distributed over the volume of the nucleus; therefore, we can make use of the well-known formula for the electrostatic potential energy of a uniformly charged sphere,

$$\frac{1}{4\pi\varepsilon_0} \frac{3}{5} \frac{Q^2}{R} = \frac{1}{4\pi\varepsilon_0} \frac{3}{5} \frac{Z^2 e^2}{A^{1/3} \times 1.2 \times 10^{-15} \text{ m}}$$

$$= \frac{Z^2}{A^{1/3}} \times 1.14 \times 10^{-13} \text{ J} \tag{33}$$

We can convert this to MeV units ($1\,\text{MeV} = 1.60 \times 10^{-13}\,\text{J}$) and write the contribution of the electrostatic potential energy to the binding energy as

$$-\frac{Z^2}{A^{1/3}} \times 0.710\,\text{MeV} \tag{34}$$

where we have inserted a negative sign because the electrostatic repulsion *reduces* the binding energy. Since our value of the nuclear radius is not completely trustworthy, we will play it safe and rewrite (34) as

$$-a_3 \frac{Z^2}{A^{1/3}} \tag{35}$$

where a_3 is another adjustable constant.

Finally, we have to include a quantum-mechanical correction. Protons and neutrons are fermions, and they obey the Exclusion Principle. The protons and the neutrons have distinct sets of energy states; the energy levels of the proton states and of the neutron states are nearly the same, except for minor energy differences introduced by the Coulomb forces among the protons. In a nucleus with Z protons and $A - Z$ neutrons, the protons will occupy the lowest Z proton states and the neutrons will occupy the lowest $A - Z$ neutron states. If a nucleus of a given mass number A has equal numbers of protons and neutrons ($Z = A/2$, $N = A/2$) then both the proton states and the neutron states will be filled up to the same energy level. If the nucleus has an unequal number of protons and neutrons (say, $Z = A/2 + 3$, $N = A/2 - 3$) then the proton states must be filled up to a higher energy level; this increases the energy of the nucleus. Thus, we see that a nucleus "asymmetric" in protons and neutrons tends to have a higher energy than a nucleus "symmetric" in protons and neutrons. Since a neutron excess will yield the same effect as a proton excess, the energy increase must be an even function of the neutron-proton difference, i.e., it must be a power series in even powers of $(N - Z)$, or $(A - 2Z)$. We will retain only the quadratic term in this power series, so that the energy increase is proportional to $(A - 2Z)^2$. This gives a satisfactory description of the energy increase at a fixed value of A. However, before we include the term $(A - 2Z)^2$ in our binding energy, we must make an extra modification for the dependence on A: the energy increase arising from a given proton or neutron excess is small if A is large. This is so because for large A the relevant energy levels have large quantum numbers and therefore, only small energy intervals. A careful quantum-mechanical analysis shows that the magnitude of the energy increase is inversely proportional to A. Thus we must multiply $(A - 2Z)^2$ by $1/A$, which leads to the following contribution to the binding energy of a nucleus:

$$-\frac{a_4}{4}\frac{(A - 2Z)^2}{A} \tag{36}$$

Here we have inserted a negative sign because an increased energy implies a decrease of binding energy. The term (36) is usually called the **asymmetry energy.**

The total binding energy of the nucleus is the sum of (31), (32), (35), and (36):*

$$B = a_1 A - a_2 A^{2/3} - a_3 \frac{Z^2}{A^{1/3}} - a_4 \frac{(\frac{1}{2}A - Z)^2}{A} \tag{37}$$

The values of the constants a_1, a_2, a_3, and a_4 must be determined from experimental data; if we adjust these constants so as to achieve a best fit between Eq. (37) and the available experimental data on binding energies, we find

$$a_1 = 15.753 \,\text{MeV} \tag{38}$$

$$a_2 = 17.804 \,\text{MeV} \tag{39}$$

$$a_3 = 0.7103 \,\text{MeV} \tag{40}$$

$$a_4 = 94.77 \,\text{MeV} \tag{41}$$

With this, Eq. (37) becomes

$$B = \left[15.753A - 17.804A^{2/3} - 0.7103\frac{Z^2}{A^{1/3}} - 94.77\frac{(\frac{1}{2}A - Z)^2}{A} \right] \text{MeV} \tag{42}$$

This expression is called Weizsäcker's[†] **semiempirical formula** for the binding energy. Note that the general features of this formula are based on theoretical arguments, but the constants are taken from empirical data.

Since nuclear binding energies are commonly measured through the nuclear mass defect, it is useful to transcribe Eq. (42) into mass units. The nuclear binding energy B implies a mass defect $\Delta m =$

---

* One more term usually included in the binding-energy formula is the **pairing energy** which accounts for the empirical observation that nuclei with even numbers of protons and of neutrons are more tightly bound than nuclei with odd numbers of protons or of neutrons. The pairing-energy term is $\frac{1}{2}[1 + (-1)^A](-1)^Z \times 0.010A^{-1/2}$ MeV. We omit this term here because it is quite small and has no simple theoretical justification.

† **Carl Friedrich von Weizsäcker,** 1912– , German theoretical physicist, professor at Hamburg.

B/c^2. Hence the mass of a nucleus with Z protons and $A - Z$ neutrons is $Zm_p + (A - Z)m_n - B/c^2$. Since tables of masses usually list the atomic mass of an isotope (rather than the nuclear mass), it is convenient to add the mass of Z electrons to this expression. The combination $Zm_p + Zm_e$ equals Zm_H. Hence the semiempirical formula for the mass of an isotope is

$$M = Zm_H + (A - Z)m_n - \frac{B}{c^2} \qquad (43)$$

With the values $m_H = 1.007825$ u, $m_n = 1.008665$ u, and $1 \text{ MeV}/c^2 = (1/931.502)$ u, this becomes

$$M = 1.008665A - 0.000840Z$$
$$- \left[0.016911A - 0.019113A^{2/3} - 0.0007625\frac{Z^2}{A^{1/3}} - 0.10174\frac{(\frac{1}{2}A - Z)^2}{A} \right] \qquad (44)$$

The smooth curve in Fig. 9.2 is a plot of the binding energy per nucleon, that is, B/A, as calculated from the semiempirical formula (42). The A and Z values used to evaluate (42) have been taken from a table of stable (or most stable) isotopes. The dots in Fig. 9.2 indicate the actual observed values of the binding energy. Except for nuclei with low values of A—for which the liquid-drop model is not expected to be a good approximation since the number of particles is too small to make up a reasonably uniform liquid—the agreement between Eq. (42) and observation is remarkably good.

The broad maximum in the curve of binding energy near $A = 56$ arises from the combined effects of the surface energy [second term in Eq. (42)] and the electrostatic energy [third term in Eq. (42)]. In light nuclei (small A and Z), a fairly large fraction of the nucleons is on the nuclear surface, and the surface energy is a large fraction of the binding energy; since this surface energy enters into Eq. (42) with a negative sign, the binding energy is small. In heavy nuclei (large A and Z), the electrostatic energy is a large fraction of the binding energy; since this electrostatic energy again enters into Eq. (42) with a negative sign, it reduces the binding energy. Nuclei of intermediate values of A and Z have relatively small surface energies and also small electrostatic energies; they therefore attain the largest values of the binding energy per nucleon.

Note that the electrostatic energy is proportional to Z^2. Thus, if we were to increase Z while holding A fixed, the binding energy would ultimately become negative, i.e., the nucleus would become unbound and burst apart under the disruptive influence of the

Coulomb forces. A nucleus of large Z can only remain bound if it has an even larger value of A so that the first term in Eq. (42) compensates for the third term; this means that the attractive nuclear forces hold the repulsive Coulomb forces in check. Such a large value of A requires the presence of a large number of neutrons in the nucleus; for example, ^{238}U has 146 neutrons and only 92 protons. The neutrons are needed to dilute the concentration of protons in the liquid drop and thereby reduce the effect of the Coulomb repulsion. The situation for ^{238}U is typical; all heavy nuclei need a large excess of neutrons to maintain their stability. For light nuclei, the Coulomb effects are less important and they do not need a large excess of neutrons for their stability.

For a quantitative discussion of the abundance of protons and neutrons, consider those nuclei that have some given mass number A, but different atomic numbers Z (such nuclei are called **isobars**). If one of these nuclei has too many neutrons, it is energetically advantageous to convert a neutron into a proton by the β-decay reaction,

$$n \rightarrow p + e + \bar{\nu} \tag{45}$$

As mentioned in Section 9.1, within a nucleus this reaction is usually inhibited by the Exclusion Principle; but if the number of neutrons is sufficiently larger than the number of protons, then the reaction (45) will be promoted by the reduction of the asymmetry energy. On the other hand, if the nucleus has too many protons, then it is energetically advantageous to convert a proton into a neutron by the following reaction, involving the ejection of an antielectron and neutrino,

$$p \rightarrow n + e^{+} + \nu \tag{46}$$

This reaction is called β^{+} decay. Since the sum of the masses m_n and m_e is larger than the mass m_p, this reaction is impossible for a free proton. However, within a nucleus the reaction is promoted by the reduction of Coulomb energy resulting from the disappearance of a proton.

From this we conclude that nuclei with an excessive number of neutrons tend to undergo β^{-} decay, and nuclei with an excessive number of protons tend to undergo β^{+} decay. Inspection of Fig. 9.1 shows that, indeed, the nuclei below the stable region are β^{-} emitters and those above are β^{+} emitters. For instance, ^{14}C is a β^{-} emitter, and ^{11}C is a β^{+} emitter. The only stable nuclei are those that have insufficient energy for either emission process. For any given value of A the most stable nucleus is the one that has the least energy, or the least mass. We can find this nucleus from our mass formula by setting the derivative $\partial M / \partial Z$ equal to zero:

$$0 = \frac{\partial M}{\partial Z} = -0.000840 + 0.0015250 \frac{Z}{A^{1/3}} - 0.20348 \frac{(\frac{1}{2}A - Z)}{A} \quad (47)$$

The constant term -0.000840 is negligible compared to the other two and can be omitted. Solving for Z, we obtain

$$Z = \frac{A}{2 + 0.01499 A^{2/3}} \quad (48)$$

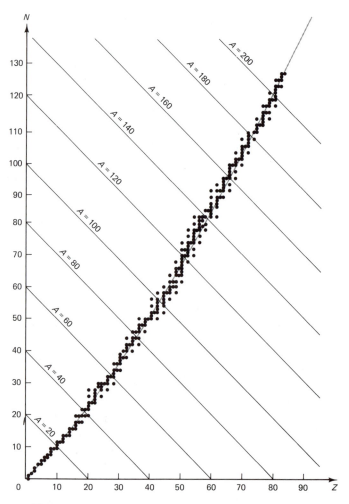

Fig. 9.8 Plot of Z vs. N showing the isobar of least mass. The dots indicate the values of Z and N for all the stable nuclei.

For small values of A, $Z \simeq A/2$; but for large values of A, Z is appreciably smaller than $A/2$, and N is appreciably larger than $A/2$. Figure 9.8 shows a plot of the values of Z and N that satisfy this equation. The dots indicate the values of Z and N for all the stable nuclei. Obviously, the semiempirical curve fits the observed values extremely well. We see that the liquid-drop model gives a good account of the overabundance of neutrons in heavy nuclei.

9.5 **The Shell Model**

Although the liquid-drop model is very successful in explaining the general trends of the binding energy as a function of Z and N, it fails to tell us the detailed properties of nuclei; for this we need a quantum theory of nuclear structure. The importance of quantum effects in the nucleus is evident from the observed quantization of angular momentum: the spin of every nucleus is some multiple of $\hbar/2$, as expected for a system made of protons and neutrons. It is also evident from the observed quantization of energy in the excited states of nuclei. For instance, Fig. 9.9 shows the energy levels of a few of the excited states of ^{111}Cd. Such states are analogous to the excited states of an atom—if the nucleus is one of the upper states it will make a transition to a lower state with the emission of a high-energy photon, or γ ray.

We can also recognize the importance of quantum effects by a simple comparison of the binding energies of nuclei. The observed binding energies are exceptionally large—larger than predicted by the liquid-drop model—if Z or N or both have the values

$$2,\ 8,\ 14,\ 20,\ 28,\ 50,\ 82,\ \text{or } 126 \tag{49}$$

These are called the **magic numbers.** Nuclei with these numbers of Z or N are exceptionally stable; they are roughly analogous to the noble gases. We know that the noble gases are characterized by filled shells of electrons, and we therefore expect that nuclei with magic numbers are characterized by filled shells of protons or neutrons.

The shell model of the nucleus attempts to calculate the energy levels of the protons and neutrons in a nucleus under the assumption that the net nuclear force experienced by each nucleon can be approximated by an average central force. Thus, the shell model attempts to imitate the calculations of the atomic energy

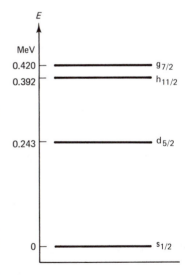

Fig. 9.9 Energy levels of the lowest excited states of ^{111}Cd.

levels of many-electron atoms we described in Section 7.4. But the calculations of the nuclear energy levels suffer from several peculiar complications. For instance, an electron in an atom moves through (mostly) empty space, and therefore has a well-defined orbital state; whereas a nucleon in a nucleus has to shoulder its way through a densely packed crowd of other nucleons, with which it suffers continual collisions. Thus it would seem that the nucleon will be continually scattered, and it cannot have a well-defined orbital state. However, here the Exclusion Principle comes to our rescue: all the quantum states of low energy are filled with other nucleons, and hence the colliding nucleon cannot change its state—it cannot be scattered. Only near the nuclear surface, where other nucleons are less abundant, can the nucleon be scattered. This suggests that the average central potential should resemble a square well, giving next to no force within the nuclear volume and a strong restraining force near the nuclear surface.

Some other complications of the shell model prove less amenable to theoretical resolution. In an atom, the Coulomb attraction generated by the nucleus provides a dominant force that holds all the electrons in an average central configuration, so that the electronic contribution to the potential is also central. In a nucleus, there is no dominant central body. A single orbiting nucleon can therefore distort the average configuration of the other nucleons, and modify the central potential. Furthermore, in an atom the individual contribution of each electron to the potential is known (Coulomb potential), and therefore the net potential of a configuration of electrons can be calculated. In a nucleus, the exact form of the individual contribution of each nucleon is not known. Thus, the shell model must rely on an educated guess for the net potential for which it offers no further theoretical justification.

The early shell-model calculations attempted to fit the nuclear energy levels with a square well of depth about 40 MeV, and a radius comparable to the nuclear radius. The resulting energy levels bore no resemblance to the observed levels. Calculations with potential wells of other shapes fared no better. This problem was finally resolved in 1949 by J. H. D. Jensen and by M. Mayer* who proposed that each nucleon is subject to a very strong spin-orbit interaction, which couples the spin of the nucleon to its own

---

* **J. Hans D. Jensen,** 1907–1973, German physicist, professor at Heidelberg, **Maria Goeppert-Mayer,** 1906–1972, American physicist, professor at Chicago, and **Eugene P. Wigner,** 1902– , American physicist, professor at Princeton, shared the Nobel Prize in 1963 for their work on the theory of nuclear structure.

orbital angular momentum, and which decreases the energy if the spin and the orbital angular momentum are parallel, and increases the energy if they are antiparallel. Thus, in contrast to the electronic spin-orbit interaction, the nuclear spin-orbit interaction relies on the *j-j* coupling scheme rather than the *L-S* coupling scheme (see Section 7.2), and, furthermore, it has the opposite sign. The nuclear spin-orbit interaction does not arise from the magnetic forces between spin and orbital angular momentum, but rather from the spin dependence of the strong force. However, the shell model does not supply any derivation of the spin-orbit interaction from a theory of nuclear forces; it treats the strength of the interaction as an adjustable parameter, chosen to fit the available data.

Figure 9.10 shows the sequential order of the calculated energy levels for a nucleon in a square well, with spin-orbit interaction included. The letters s, p, d, ... indicate the orbital angular momentum of each shell in the usual way. The numbers preceding these letters give the principal quantum number that characterizes the radial wavefunction, as in the case of hydrogen.* The subscripts on the letters give the total angular momentum of a nucleon, $j = l + \frac{1}{2}$ or $l - \frac{1}{2}$. Because of the coupling between the spin and the orbital angular momentum, l is not a good quantum number; but it is nevertheless useful for counting the number of available states.

Since protons and neutrons are distinct particles, they occupy distinct sets of states. Thus, to make the assignment of filled states in a nucleus, we need to draw *two* copies of Fig. 9.10, one for neutrons and one for protons.† In the ground state of a nucleus with N neutrons and Z protons, the lowest N states in the neutron diagram and the lowest Z states in the proton diagram will be occupied. The column of numbers on the right edge of Fig. 9.10 gives the number of available states (including spin states) for each energy level. Inspection of the diagram shows that there are exceptionally large gaps between the energy levels when the cumulative number of neutrons or of protons is 2, 8, 14, 20, 28, 50, 82, or 126. Thus, the shell model yields the observed magic numbers.

The shell model also yields the observed values of the nuclear spin, provided that the nucleons are placed in the shells according

---

* The sequence of principal quantum numbers in the left half of Fig. 9.10 can be remembered by the following simple mnemonic: 1, 1; 2, 1, 1, 2; 1, 2, 3, 1, 2, 3; 1, 2, 3, 4, 1, 2, 3, 4; etc. For each given principal quantum number, the l values follow the increasing order s, p, d, ...

† The energy-level diagram for protons actually differs slightly from that for neutrons because of the extra electric interaction. But we will ignore this correction.

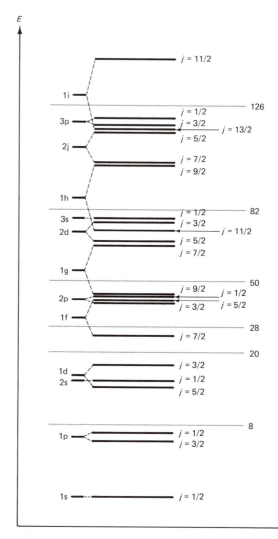

Fig. 9.10 Sequential order of the energy levels of a nucleon in a square well. The left side of the figure gives the energy levels without spin-orbit interaction; the right side gives the levels with spin-orbit interaction. The levels indicate the energy of each shell when all the shells below it are filled and all shells above are empty. The numbers on the far right give the cumulative number of nucleons required to fill the shells below the gray lines. [After P. F. A. Klinkenberg, *Rev. Mod. Phys.* **24,** 63 (1952).]

to the following rule: whenever an even number of neutrons or of protons is in a quantum state of the same angular momentum l and j, their combined angular momentum will be zero, i.e., their net angular momentum will be zero. For instance, if two neutrons are in such a quantum state, then their angular momenta will have opposite directions (opposite values of m_j), and their net angular momentum is zero. This rule implies that for any nucleus with even Z and even N, the spin is necessarily zero; e.g., ^{16}O has zero spin. The rule also implies that for any nucleus with odd Z or odd N, the spin is entirely due to this last, odd nucleon in the highest filled state; e.g., ^{15}O has spin $\frac{1}{2}$. From the values of l and j of the odd nucleon we can calculate the g factor, and the magnetic moment. However, the calculation of the g factor requires a slightly modified version of the Landé formula [Eq. 7.34] because the magnetic moment of the proton (or the neutron) is anomalous, i.e., it differs somewhat from $e\hbar/(2m_p c)$ (see Table 9.2).

The magnetic interaction of the nuclear magnetic moment with the angular momentum of the electrons leads to the **hyperfine structure** of the spectral lines of atoms. The formation of hyperfine-structure multiplets is analogous to the formation of fine-structure multiplets (see Section 7.1), with the nuclear spin **I** playing the role of the electron spin **S**.

Example 3   The analysis of the hyperfine structure of the spectral lines in the light emitted by atoms of ^{14}N shows that the ground state of this atom is split into three closely spaced levels. Deduce the spin of the nucleus of ^{14}N.

Solution   The angular momentum of the electrons in the atom is **J** and the angular momentum of the nucleus is **I**. The total angular momentum is the sum $\mathbf{F} = \mathbf{J} + \mathbf{I}$. According to the usual rules for the addition of angular momenta, the possible values of the quantum number F are

$$J + I, J + I - 1, \ldots, J - I \quad \text{if } I \leqslant J$$

or

$$I + J, I + J - 1, \ldots, I - J \quad \text{if } J < I$$

Thus there are $2I + 1$ possible states of different total angular momenta in the former case and $2J + 1$ possible states in the latter case. The interaction of the nuclear magnetic moment with the angular momentum of the electrons gives these states different energies and causes the observed splitting of the ground state. Since the spectroscopic analysis tells us that the ground state splits into three states, we conclude that either $2I + 1 = 3$ or else $2J + 1 = 3$, i.e., either $I = 1$ or else $J = 1$. But we know that the ground state of the electrons in the nitrogen atom has $J = \frac{3}{2}$ (see Table 7.2). Thus the only acceptable alternative is $I = 1$.   ∎

9.6 Nuclear Magnetic Resonance

Some direct measurements of magnetic moments of nuclei have been performed by means of Stern–Gerlach experiments. However, the most precise measurements of magnetic moments have been performed by the technique of **nuclear magnetic resonance,** which relies on the determination of the resonant frequency for transitions between the states of different spin orientations of a nucleus in a magnetic field. If a nucleus of magnetic moment μ is placed in a uniform magnetic field B_0 in the z-direction, then its energy is*

$$U = -\mu_z B_0 \tag{50}$$

The magnetic moment μ_z is quantized according to the usual rule [compare Eq. (7.33)]

$$\mu_z = gm_I \frac{e\hbar}{2m_p} \qquad m_I = I, I - 1, \ldots, 1 - I, -I \tag{51}$$

where g is the g-factor for the nucleus. Hence the energy levels of the nucleus in the magnetic field are

$$U = -gm_I \frac{e\hbar}{2m_p} B_0 \qquad m_I = I, I - 1, \ldots, 1 - I, -I \tag{52}$$

The energy difference between adjacent energy levels is $\Delta U = -ge\hbar B_0/(2m_p)$ and the corresponding transition frequency is

$$v = \frac{\Delta U}{h} = g\frac{e}{4\pi m_p} B_0 \tag{53}$$

In a magnetic field of 0.1 T, this frequency is of the order of 10 MHz, which lies within the range of radio frequencies. Thus, the downward transitions from an upper to a lower level involve the emission of radio waves, and the upward transitions from a lower to an upper level involve the absorption of radio waves. However, transitions can be triggered not only by the oscillating magnetic field of a radio wave, but equally well by any oscillating magnetic field of the correct frequency.

Nuclear magnetic resonance experiments rely on an oscillating magnetic field arranged at right angles to the steady magnetic field B_0. This oscillating transverse magnetic field is produced by a

---

* Note that we are assuming that the nucleus obeys the "normal" Zeeman effect. This is a good approximation because the coupling between the nuclear magnetic moment and angular momentum of the electrons is fairly weak; even a modest magnetic field (say, 0.01 T) suffices to break this coupling.

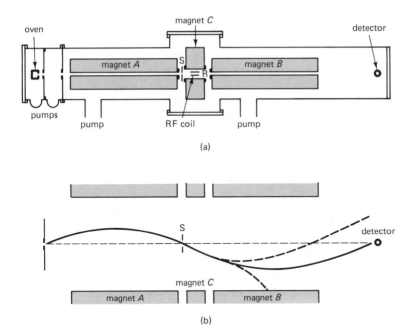

Fig. 9.11 (a) Molecular-beam apparatus used by I. I. Rabi for nuclear magnetic resonance experiments. The molecular beam originates in the oven at the left and passes through the magnets A, C, and B. (b) Path of a molecule in the gaps between the poles of the magnets (vertical distances have been exaggerated for clarity). The solid line shows the path of a molecule whose spin remains unchanged. The dashed lines show the paths of molecules whose spins changes orientation while pasing through the magnetic C. [After I. I. Rabi, *Phys. Rev.* **55**, 526 (1939).]

coil surrounding the sample containing the nuclear spins. The coil is driven by a radio-frequency oscillator. When the frequency of this magnetic field coincides with the natural frequency (53) for transitions from one energy level to the next, the spins resonantly absorb energy from the magnetic field or release energy by stimulated emission, and they make transitions that change their orientation. This resonance condition can be detected by one of several experimental methods: molecular-beam resonance, absorption resonance, or induction resonance.

The **molecular-beam resonance** method was developed by I. I. Rabi* and his collaborators; it is closely related to the Stern-Gerlach experiment. The experimental arrangement, shown in Fig. 9.11, uses two identical long magnets A and B, which produce

---

* **Isidor Isaac Rabi,** 1898– , American experimental physicist, professor at Columbia. He received the Nobel Prize in 1944 for his work on nuclear magnetic resonance with molecular beams.

inhomogeneous magnetic fields, as in the Stern–Gerlach experiment. The second magnet is upside down, so that the magnetic field gradient in this magnet is opposite to that in the first magnet. A short third magnet C is placed between the two long magnets. This magnet provides a uniform vertical magnetic field B_0. A coil or loop of wire placed between the poles of this third magnet provides an oscillating horizontal magnetic field. The beam of molecules travels along the gap between the poles of the magnets, from left to right in Fig. 9.11. If the frequency of the oscillating horizontal field is *not* in resonance with the natural frequency (53), the molecules will pass through the magnet C without any change of orientation of their spins. Such molecules will suffer exactly opposite deflections within the two magnets A and B (see Fig. 9.11b); they will therefore emerge without any net deflection, and strike the detector D. However, if the frequency of the oscillating horizontal magnetic field is in resonance with the natural frequency (53), the orientations of the spins will change while the molecules pass through the magnet C, and the deflections produced by the magnets A and B will fail to cancel. Such molecules will miss the detector. Thus, the resonance condition is signaled by a sharp decrease in the beam intensity reaching the detector. This decrease shows up clearly in a plot of beam intensity as a function of B_0, at a fixed frequency of the oscillating transverse magnetic field. For example, Fig. 9.12 shows such a plot of beam intensity obtained by Rabi and collaborators with a beam of HD molecules. The sharp resonance seen in this plot is due to the protons in the hydrogen atoms. With a fixed frequency of 4.000 MHz, this resonance occurs at 0.0945 T. According to Eq. (53) this implies that the proton has a *g*-factor of

$$g = \frac{4\pi m_p}{eB_0}\nu = \frac{4\pi \times 1.67 \times 10^{-27}\,\text{kg}}{1.60 \times 10^{-19}\,\text{C} \times 0.0945\,\text{T}} \times 4.000 \times 10^6\,\text{Hz}$$

$$= 5.55 \tag{54}$$

and consequently a magnetic moment of

$$\mu_z = g \times \frac{1}{2} \times \frac{e\hbar}{2m_p} = 2.78\,\frac{e\hbar}{2m_p} \tag{55}$$

The more precise value of the magnetic moment listed in Table 9.2 was obtained with protons in a bulk sample of water, rather than protons in a molecular beam. For bulk samples of liquid or solid materials, we must use one of the other two methods for observing nuclear magnetic resonance. In the **absorption resonance** method

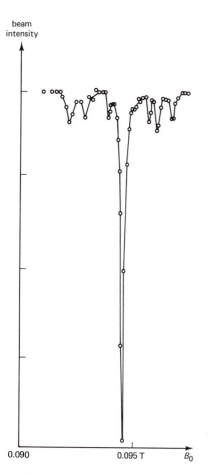

beam
intensity

0.090   0.095 T B_0

Fig. 9.12 Beam intensity of HD molecules as a function of the strength of the steady magnetic field B_0. The frequency of the oscillating transverse magnetic field is held fixed at 4.000 MHz. [From J. M. B. Kellog, I. I. Rabi, N. F. Ramsey and J. R. Zacharias, *Phys. Rev.* **56**, 728 (1939).]

developed by E. M. Purcell,* the sample is placed in a radio-frequency cavity immersed in a steady external magnetic field. The radio waves in the cavity provide the oscillating transverse magnetic field. The resonance condition is signaled by an enhanced attenuation of the radio waves passing through the sample.

In the **induction resonance** method developed by F. Bloch, the sample is immersed in a steady magnetic field and surrounded by a coil which provides the oscillating transverse magnetic field. The resonance condition is signaled by the induced emf that the changing spins in the sample generate in an extra pickup coil placed nearby (see Fig. 9.13).

In nuclear resonance absorption or induction experiments, only a small fraction of the total number of nuclear spins contributes to the observed signal. The reason is that the oscillating transverse magnetic field triggers not only upward transitions (stimulated absorption), but also downward transitions (stimulated emission). In a sample in thermal equilibrium at ordinary temperatures, the number of spins in the lower levels is nearly the same as that in the upper levels [the energy differences given by Eq. (52) are small compared to the thermal energy kT, so that the Boltzmann factor $e^{-\Delta U/kT}$ is nearly 1]. Hence the rates for upward and for downward transitions are almost equal, and the contributions to the observed signal from upward and from downward transitions almost cancel. The residual observed signal is due to the slight excess population in the lower levels, an excess amounting to no more than a few parts in a million. This leads to a slight excess of upward vs. downward transitions.

Figure 9.14 shows a plot of the signal obtained in an induction resonance experiment with hydrogen in a sample of ferric-nitrate solution. In this experiment, the frequency was held fixed, and the magnetic field strength B_0 was swept across the resonance. At a frequency of 30 MHz, the resonance peak occurs at 0.705 T. The width of the resonance peak is about 0.00005 T. This width arises from relaxation processes that disturb the response of the spins to the magnetic field and dissipate the energy absorbed by the spins from the field. There are two such processes: the **spin-lattice relaxation** and the **spin-spin relaxation.** With each of these we can associate a characteristic time that indicates how long the process takes to dissipate a significant amount of energy; these two relaxation

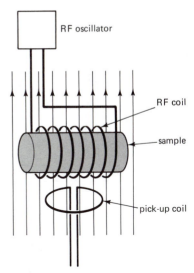

Fig. 9.13 Arrangement for an NMR experiment. The sample under investigation is placed in a uniform steady magnetic field (vertical) and subjected to an oscillating magnetic field (horizontal) produced by an RF coil. A pickup coil registers an induced emf when the nuclear spins suddenly change direction.

---

* **Edward Mills Purcell,** 1912–   , American physicist, professor at Harvard, and **Felix Bloch,** 1905–   , Swiss and later American physicist, professor at Stanford, shared the Nobel Prize in 1952 for their work on nuclear magnetic resonance methods.

induced emf

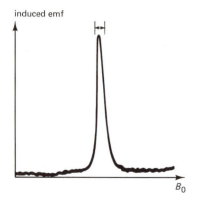

Fig. 9.14 Induced emf generated by a sample of ferric nitrate as a funtion of the strength of the steady magnetic field B_0. The frequency of the oscillating transverse magnetic field is held fixed at 30 MHz. The resonance peak occurs at 0.705 T; the width of this peak is about 0.00005 T. [After N. Bloembergen, E. M. Purcell, and R. V. Pound, *Phys. Rev.* **73**, 679 (1948).]

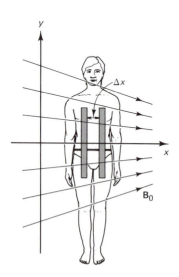

Fig. 9.15 Two slablike volume elements in a magnetic field that increases in proportion to x.

times are called T_1 and T_2. The spin-lattice relaxation arises from the disturbances that the random thermal motions of the atoms exert on the spins. On the average, these disturbances tend to produce thermal equilibrium between the spins and the atoms, removing any energy from the spins in excess of the thermal energy. The characteristic time scale T_1 depends on the environment in which the spin is immersed, i.e., it depends on the kinds of atoms that surround the spin and on their arrangement. In liquids T_1 is typically a few seconds, whereas in crystalline solids it may be as long as a few hours. The spin-spin relaxation arises from the magnetic interactions of neighboring nuclear spins. In liquids, the characteristic time scale T_2 for spin-spin relaxation is typically a few seconds; in solids, it is often as short as a few microseconds.

Nuclear magnetic resonance has found a valuable practical application in medicine, where it is being used in a new technique to produce images of the interior of the human body. This technique hinges on the dependence of the resonant frequency on the strength of the magnetic field [Eq. (53)]. If a human body is immersed in a magnetic field B_0 that is stronger on one side of the body than on the other (a magnetic field with a spatial gradient), then the nuclei in one side of the body will resonate at a higher frequency than those in the other side. Thus, there is a correlation between the positions of the nuclei and their resonant frequencies, and this permits us to map the positions of the nuclei from the observed values of the resonant frequencies. The simple example illustrated in Fig. 9.15 will help to make this clear. The figure shows two slablike volume elements within the body immersed in a magnetic field which increases linearly from one side to the other, $B_0 = \text{const.} \times x$. A very short pulse of oscillating magnetic field is applied by means of the radio-frequency coil. Such a short pulse contains a fairly broad spread of radio frequencies, and is therefore capable of exciting the spins in both volume elements, even though their resonant frequencies are not the same. The signal received in the pick-up coil consists of a superposition of two oscillating emfs of slightly different frequencies. The measured value of the frequency difference $\Delta\nu$ translates into a magnetic field difference ΔB_0, which in turn translates into a position difference Δx between the slabs. Of course, this measurement is sensitive only to the position along the direction of the gradient in the magnetic field. To construct a two-dimensional image of the body, we must change the angular orientation of the magnetic field B_0 or the angular orientation of the body and repeat the measurement several times with different angular orientations. From the data accumulated in such measurements, an image of the density distribution of spins within the body can be synthesized by a computer.

Human tissues have a high abundance of hydrogen, and it is therefore convenient to rely on hydrogen nuclei for the construction of NMR images. However, an image based merely on the density of hydrogen would not reveal sufficient anatomical detail— such an image would lack contrast since the density of hydrogen is fairly uniform throughout the body. To enhance the contrast, it is desirable to make the image sensitive to local differences in the values of the relaxation times T_1 and T_2. The values of the relaxation times depend on the chemical and physical properties of the environment surrounding the spin, and they are therefore different in different tissues. We can achieve the desired sensitivity to the relaxation time by measuring the response of the spins to a series of carefully timed pulses of the oscillating magnetic field, rather than just a single pulse as described above. For instance, if we apply a sequence of pulses with a time interval Δt, then the spins with $T_1 > \Delta t$ will be inhibited from responding to the later pulses because the first pulse raises these spins to the upper level, and they do not have enough time to relax to the lower level between one pulse and the next. Thus, by judicious choice of Δt we can bring out differences between the values of T_1 in different tissues, and thereby increase the contrasts in the image.

Figure 9.16 shows images prepared with the NMR technique. The resolution achieved in this image is comparable to that of the

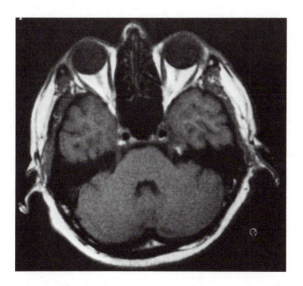
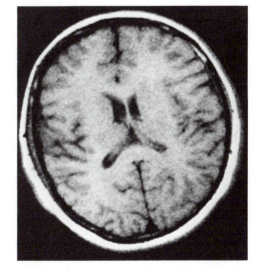

Fig. 9.16 NMR images of cross sections through of human head (Courtesy P. Bottomley, General Electric Research and Development Center.)

best images prepared by X-ray scans. NMR imaging has two important advantages over X rays. Magnetic fields cause no damage to the body, whereas X rays cause ionization damage. Moreover, X-ray images are mostly sensitive to the electron density, whereas NMR images are sensitive to a variety of chemical and physical conditions in the tissues. Thus, X-ray images reveal only gross changes in the sizes and shapes of organs, whereas NMR images also reveal subtle physiological changes.

Chapter 9:   SUMMARY

Magnetic moment of proton:

$$2.79 \frac{e\hbar}{2m_p}$$

Magnetic moment of neutron:

$$-1.91 \frac{e\hbar}{2m_p}$$

Mass defect and binding energy:

$$Zm_p + Nm_n - m = B/c^2$$

Q-value of reaction:

$$Q = \sum_i m_i c^2 - \sum_i m_i' c^2$$

Nuclear radius:

$$R = 1.2\,\text{fm} \times A^{1/3}$$

Yukawa potential:

$$U(r) \propto -\frac{e^{-r/b}}{r}$$

Semiempirical mass formula:

$$M = 1.008665A - 0.000840Z$$

$$-\left[0.016911A - 0.019113A^{2/3} - 0.0007625 \frac{Z^2}{A^{1/3}} - 0.10174 \frac{(\frac{1}{2}A - Z)^2}{A} \right]$$

Magic numbers:

$$2, 8, 14, 20, 28, 50, 82, 126$$

Nuclear magnetic resonance frequency:

$$v = g \frac{e}{4\pi m_p} B_0$$

Chapter 9: PROBLEMS

1. The masses of the atoms ^{16}N, ^{16}O, and ^{16}F are 16.00610 u, 15.99491 u, and 16.01146 u, respectively. Calculate the mass defects for the nuclei of each of these three atoms.

2. According to the alpha particle model of the nucleus, some nuclei can be regarded as "molecules" consisting of several alpha particles bound together. For instance, on the basis of this model, ^{12}C consists of three alpha particles arranged on the corners of a triangle, and ^{16}O consists of four alpha particles arranged on the corners of a tetrahedron. Show that the binding energies of ^{12}C and ^{16}O calculated from this model agree with the actual binding energies determined from the mass defects, provided that the bond between each pair of alpha particles is assigned a binding energy of about 2.42 MeV. (Remember to take into account the binding energy of each alpha particle by itself!)

3. Suppose that you have a radioactive source that yields a stream of particles of unknown mass and unknown, but fixed, energy. Deduce both the mass and the speed of the unknown particles from the following data (which Chadwick obtained in his experiments): when the particles are incident on a hydrogen target, they eject protons with a recoil speed of up to 3.3×10^7 m/s; when the particles are incident on a nitrogen target, they eject nitrogen nuclei with a recoil speed of up to 0.47×10^7 m/s. [*Hint:* Use conservation of energy and of momentum to show that the maximum recoil speed that an (initially stationary) target particle acquires in a head-on elastic collision with an incident particle of speed v_1 is

$$v_2' = v_1 \times \frac{2m_1}{m_1 + m_2}$$

where m_1 and m_2 are the masses of the incident and of the target particle, respectively; then use the hydrogen and the nitrogen data to set up two equations that can be solved for m_1 and v_1 in terms of the known masses of hydrogen and nitrogen.]

4. The bombardment of a beryllium target by alpha particles yields neutrons according to the reaction

$$\alpha + {}^9\text{Be} \rightarrow {}^{12}\text{C} + \text{n}$$

Assume that the alpha particles have an energy of 5.3 MeV and ignore the kinetic energy of the carbon nucleus. Calculate the energy of the neutron; express your answer in MeV.

5. Consider the reactions

$$^2\text{H} + {}^{12}\text{C} \rightarrow {}^{13}\text{N} + \text{n}$$

$$^2\text{H} + {}^{12}\text{C} \rightarrow {}^{13}\text{C} + \text{p}$$

which occur when a carbon target is bombarded by deuterons. If the kinetic energy of the deuteron is 9.0 MeV, what will be the net kinetic energy of the reaction products in each case?

6. Suppose that a projectile of mass m_1 strikes a stationary target nucleus of mass m_2. Taking into account the motion of the center of mass, show that for an endoergic nuclear reaction of a given Q-value (with $|Q| \ll m_1 c^2$, $|Q| \ll m_2 c^2$), the threshold kinetic energy of the projectile is $-Q(m_1 + m_2)/m_2$. Evaluate the threshold kinetic energy for the incident proton in the reaction p + ^{7}Li → ^{7}Be + n. (*Hint:* At threshold the kinetic energy of the reaction products is zero in the center of mass frame.)

7. Calculate the net kinetic energy of the decay products released in the β decay of ^{14}C,

$$^{14}\text{C} \rightarrow {}^{14}\text{N} + \text{e} + \bar{\nu}$$

The antineutrino ($\bar{\nu}$) has zero mass.

8. Using the value of r_0 given in Eq. (17), calculate the density of nuclear material.

9. What radius does Eq. (15) give for the proton? Compare with the actual radius, 1.0 fm.

10. What initial kinetic energy is required so that the classical turning point of an alpha particle aimed at an ^{197}Au nucleus coincides with the nuclear radius?

11. The radius of a proton and of a neutron is about 1.0 fm. What fraction of the volume of a large nucleus is empty? (*Hint:* Compare the mass density of the proton with the mass density of nuclear matter.)

12. The differences between the electrostatic energies of the mirror nuclei

$$^{11}\text{C} - {}^{11}\text{B}, \quad {}^{15}\text{O} - {}^{15}\text{N}, \quad \text{and} \quad {}^{21}\text{Na} - {}^{21}\text{Ne}$$

are 2.79 MeV, 3.48 MeV, and 4.30 MeV, respectively. From these electrostatic energies, calculate the nuclear radius for each of these pairs of mirror nuclei. Are these radii proportional to $A^{1/3}$? What is the constant of proportionality?

13. Suppose that a (hypothetical) heavy particle of mass 10,000 m_e, charge $-e$ is in orbit around the center of a ^{207}Pb nucleus. Suppose that, like the electron, this particle only feels the Coulomb force, not the strong nuclear force; it therefore can pass through nucleons and orbit *inside* the nucleus, as though it were orbiting in vacuum in an electric field. Calculate the radius and the energy of the smallest circular orbit according to a modified Bohr model, taking into account the uniform charge distribution within the volume of the nucleus. Calculate the radius and the energy of the largest Bohr orbit that fits within the nucleus. Calculate the radius and the energy of the smallest Bohr orbit that lies entirely outside of the nucleus.

14. Treating the protons in a nucleus as a Fermi gas of free particles, find the Fermi kinetic energy and the average kinetic energy of the protons in a nucleus of atomic number Z and mass number A. Evaluate these energies (in MeV) for ^{56}Fe and for ^{238}U.

15. The radius of the proton is approximately 1.0 fm. Use Fig. 9.6 to estimate the strong proton–proton force for two protons in contact. Compare with the Coulomb force acting between two protons separated by the same distance. Compare with the gravitational force.

16. According to Fig. 9.6, what is the maximum attractive value of the strong force between two nucleons? Express your answer in newtons.

17. Equation (22) gives the potential for two nucleons separated by more than 2 fm.
(a) According to this potential, what is the force at a distance of 3 fm?
(b) At what distance is the force about ten times smaller?

18. What would you predict for the nuclear magnetic moment of the deuteron? Compare with the experimental value, $0.8574e\hbar/2m_p$.

19. Prove that the square well with the parameters given in the discussion of the deuteron in Section 9.3 has only one bound state.

20. Calculate the masses of the isotopes of neon from the semiempirical mass formula and compare the results with the values listed in Table 9.1.

21. Theoretical speculations suggest that some nuclei of unusually large mass numbers and atomic numbers might be stable. According to the semiempirical mass formula, what would be the binding energy for a nucleus of $Z = 126$ and $N = 182$?

22. If the constant a_3 has the value listed in Eq. (40), what must be the value of the constant r_0 that characterizes the nuclear radius?

23. Suppose that a nucleus of ^{232}Th splits into two equal pieces. Use the semiempirical formula for the binding energy to calculate the energy released in this reaction.

24. Suppose that two nuclei with Z, A fuse to make a single nucleus of $2Z, 2A$. According to the semiempirical formula for the binding energy, what is the energy released in this reaction? Evaluate for the fusion of two ^{12}C nuclei.

25. The smooth curve in Fig. 9.2 was plotted according to the semiempirical formula for the binding energy, with Z as given by Eq. (48). For what value of A and of Z does this curve attain its maximum? (*Hint:* It is easiest to work this out numerically, by preparing a table of values of B/A for A near 56 and Z near 26.)

26. What are the energies and the wavelengths of the gamma rays emitted in the transitions indicated in Fig. 9.9?

27. Since nuclei with magic numbers are exceptionally stable, their average binding energies per nucleon (B/A) are larger than those of nearby nuclei without magic numbers. Verify this by comparing the values of B/A for ^{40}Ca (doubly magic) vs. ^{40}K (not magic); and ^{208}Pb (doubly magic) vs. ^{210}Bi (not magic). The masses of these isotopes are, respectively, 39.96259 u, 39.964 u, 207.97666 u, and 209.98412 u.

28. The isotope ^{16}O is doubly magic—it has $Z = 8$ and $N = 8$. Calculate its mass according to the semiempirical formula and compare with the value listed in the chart of isotopes (Fig. 9.1). Comment on the discrepancy.

29. According to the shell model, what is the spin of each of the isotopes ^{6}Be, ^{7}Be, ^{8}Be, ^{9}Be, ^{10}Be, ^{11}Be, and ^{12}Be?

30. The interaction of the nuclear magnetic moment with the electronic magnetic moment (orbital plus spin) splits each fine-structure level of the atom into a multiplet of hyperfine-structure levels.

(a) In the case of hydrogen, the hyperfine multiplets are doublets, i.e., each of the energy levels shown in Fig. 7.1 is actually a very closely spaced pair of levels. Explain why.

(b) The famous 21-cm line in the radio spectrum of hydrogen is due to a transition between the upper and the lower members of the doublet corresponding to the 1s state. What is the energy separation between the members of this doublet in eV?

31. Some nuclei, such as ^{180}Hf, have excited states that correspond to a more or less rigid rotation of the entire nucleus about an axis. In ^{180}Hf, the observed energies of these excited states are 0.093, 0.3093, 0.6417, and 1.0853 MeV.

(a) Show that these energies roughly fit the same formula as for the energies of a rotating molecule:

$$E = \frac{J(J + 1)\hbar^2}{2I}$$

where the constant I is the moment of inertia and $J = 2, 4, 6, 8$ (by an argument based on the Exclusion Principle, it can be shown that the odd values of the quantum number J are forbidden for the ^{180}Hf nucleus). What value of the moment of inertia I can you deduce from the data?

(b) Assuming rigid rotation, calculate the moment of inertia of the ^{180}Hf nucleus from its mass and radius. Compare with the result obtained in part (a). Is rigid rotation a good approximation?

32. From the data given in Fig. 9.17, calculate the g-factor for the ^{7}Li nucleus. Given that the spin of this nucleus is $\frac{3}{2}$, find the magnetic moment.

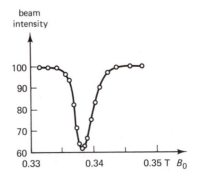

Fig. 9.17 Beam intensity of LiCl molecules as a function of the strength of the steady magnetic field B_0; the frequency of the oscillating transverse magnetic field is held fixed at 5.585 MHz. The beam intensity is minimum at 0.3384 T. This resonance is due to ^{7}Li nuclei in the molecules. [From I. I. Rabi, *Phys. Rev.* **55**, 526 (1939).]

33. For a sample of protons in a magnetic field of $B_0 = 0.01$ T in thermal equilibrium at room temperature, what is the fractional excess population of spins parallel to the magnetic field over spins antiparallel to the magnetic field?

Nuclear Transmutations

In a chemical reaction, the rearrangement of atoms in molecules results in the formation of new molecules. Similarly, in a nuclear reaction, the rearrangement of the protons and neutrons in nuclei results in the formation of new nuclei. Such a formation of new nuclei is a transmutation of elements. For instance, the reaction $\alpha + {}^7\text{Be} \rightarrow {}^{12}\text{C} + \text{n}$, which was mentioned in Section 9.1, transmutes beryllium into carbon. Thus, nuclear physicists have fulfilled the dream of transmutation of elements of ancient alchemists. Almost all of the unstable isotopes listed in Appendix 4 were produced artificially by nuclear transmutations. Many of these isotopes have important industrial and medical applications. Furthermore, all of the heavy elements beyond plutonium were produced artificially by transmutations; these elements do not occur naturally.

The energy released or absorbed in a nuclear reaction is much larger than in a chemical reaction. Typically, the energy change during a rearrangement of atoms in a molecule is of the order of 1 eV, whereas the energy change during a rearrangement of the protons and neutrons in a nucleus is of the order of 1 MeV. This means that many nuclear reactions can be initiated only by bombarding a nucleus with a very energetic projectile, a circumstance which led nuclear physicists to develop accelerators that produce intense beams of such projectiles. The larger energy released in nuclear reactions also led to industrial and military applications: nuclear reactors and bombs.

10.1 Radioactive Decay

Most isotopes are unstable—they decay by spontaneous nuclear reactions and transmute themselves into other, more stable isotopes. The unstable isotopes are **radioactive,** that is, their spontaneous nuclear reactions are accompanied by the emission of α rays, β rays, or γ rays. The α rays are energetic alpha particles (^{4}He nuclei), the β rays are energetic electrons or antielectrons, and the γ rays are energetic photons. Radioactivity was discovered in 1896 by H. Becquerel,[*] who noticed that uranium salts emitted rays which could penetrate sheets of opaque material and make an imprint on a photographic plate. The distinction between α and β rays was first recognized by Rutherford, and the presence of γ rays was first detected by P. Villars.[†] The following are some examples of radioactive decays with emission of α rays and β rays (the numbers on the right give the energies of the rays):

$$^{238}\text{U} \rightarrow {}^{234}\text{Th} + \alpha \qquad (4.20\,\text{MeV}) \qquad (1)$$

$$^{226}\text{Ra} \rightarrow {}^{222}\text{Rn} + \alpha \qquad (4.78\,\text{MeV}, 4.60\,\text{MeV}) \qquad (2)$$

$$^{22}\text{Na} \rightarrow {}^{22}\text{Ne} + \beta^+ \qquad (0.545\,\text{MeV}, 1.82\,\text{MeV}) \qquad (3)$$

$$^{60}\text{Co} \rightarrow {}^{60}\text{Ni} + \beta^- \qquad (0.318\,\text{MeV}) \qquad (4)$$

$$^{90}\text{Sr} \rightarrow {}^{90}\text{Y} + \beta^- \qquad (0.546\,\text{MeV}) \qquad (5)$$

The emission of β rays is accompanied by the emission of neutrinos; but since the latter are usually not detected, they are often omitted when writing reactions. All of these reactions involve transmutations of elements: uranium into thorium, radium into ruthenium, sodium into neon, etc. Transmutation of elements in radioactive decay was discovered in 1902 by Rutherford and F. Soddy.

Instead of emitting an antielectron, some nuclei absorb an electron, which they capture from one of the electronic shells of the atom. An example of such a reaction is

$$^{22}\text{Na} + \beta^- \rightarrow {}^{22}\text{Ne} \qquad (6)$$

Such a reaction is called **electron capture** (EC). Note that the reactions (3) and (6) lead to the same transmutation of the nucleus,

---

[*] **Antoine Henri Becquerel,** 1852–1908, French physicist, professor at the École Polytechnique. For his discovery of radioactivity, he received the Nobel Prize in 1903.

[†] **Paul Villars,** 1860–1934, French chemist at the École Normale Supérieure.

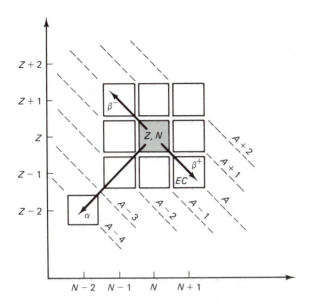

Fig. 10.1 Nuclear transmutations produced by α decay, β decay, and EC. Each isotope is represented by a box positioned in the Z–N plane, as in the chart of isotopes (Fig. 9.1). A nuclear transmutation changes Z and N, and can be represented by a displacement in the Z–N plane.

but the radioactive emissions are different. The disappearance of an electron from one of the shells of the atom (usually the K shell) leaves a hole, which is immediately filled by the transition of an electron from an adjacent shell, with the emission of a characteristic X ray (usually K_α). Thus, EC is always accompanied by X rays. Figure 10.1 summarizes the nuclear transmutations resulting from α decay, β decay, and EC.

In many cases of α decay, β decay, and EC the nucleus emerges from the reaction with an excess of internal energy, i.e., the nucleus emerges in an excited state. In nuclear physics, an excited state that lasts for a measurable interval of time is usually called an **isomeric state.** From the excited state, the nucleus will then make a transition to a lower state or to the ground state, emitting a γ ray. For example, Figure 10.2 shows the energy-level diagram for the ^{60}Ni nuclei formed in the β^- decay of ^{60}Co. This nucleus makes two transitions in succession (in "cascade"), emitting two γ rays. The emission of γ rays in transitions between excited states of a nucleus is analogous to the emission of light or X rays in transitions between excited states of an atom.

Alternatively, a nucleus in an excited state can make a transition to a lower state by transferring its excitation energy to an

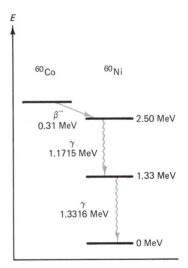

Fig. 10.2 Energy-level diagram for the ^{60}Ni nucleus formed in the decay ^{60}Co $\rightarrow$ ^{60}Ni $+ \beta^-$.

atomic electron. Such an energy transfer is called **internal conversion.** The electron that suddenly acquires the nuclear energy will be ejected from the atom at high speed, i.e., it will be ejected as a β ray.

In some cases of decay of an unstable isotope, the decay product is itself an unstable isotope. Thus, the first decay is succeeded by another, and another, etc. This results in a sequential series of decays. For example, the decay of ^{238}U initiates such a series:

$$
\begin{aligned}
^{238}\text{U} &\to {}^{234}\text{Th} + \alpha \\
&\quad \hookrightarrow {}^{234}\text{Pa} + \beta^- \\
&\qquad \hookrightarrow {}^{234}\text{U} + \beta^- \\
&\qquad\quad \hookrightarrow {}^{230}\text{Th} + \alpha \\
&\qquad\qquad \hookrightarrow {}^{226}\text{Ra} + \alpha \\
&\qquad\qquad\quad \hookrightarrow \text{etc.}
\end{aligned} \tag{7}
$$

The decays continue step by step until a stable isotope is reached. In the case of the ^{238}U series, the final stable decay product is lead. Many of the α and β decays are accompanied by the emission of γ rays because the intermediate decay products are usually in excited states; for the sake of simplicity these γ rays have not been indicated in Eq. (7).

If we initially have some given amount of some radioisotope material, then this amount will gradually decrease in time as more and more of the nuclei decay. Measurements show that the quantitative law that describes this decay process is very simple: if a certain fraction of the initial amount of radioactive material decays in a certain time interval, then the same fraction of the remainder decays in the next (equal) time interval, and the same fraction of the new remainder decays in the next (equal) time interval, etc. For example, suppose we initially have 1 g of radioactive strontium. This decays by β decay,

$$
^{90}\text{Sr} \to {}^{90}\text{Y} + \beta^- \tag{8}
$$

In this reaction, strontium is called the **parent** material and yttrium is the **daughter** material. Measurements show that it takes 29 years for one half of the initial amount of parent to decay. It then follows from the law of radioactive decay that during the next 29 years, one half of the remainder will decay, etc. Hence the amounts of parent left at time $t = 0, 29, 58, 87$ years, etc. will be, respectively, $1, \frac{1}{2}, \frac{1}{4}, \frac{1}{8}$ g, etc. Thus, the amounts left after the lapse of equal time intervals form a geometric progression. If $n(t)$ represents the number of strontium nuclei at time t and n_0 the number at time zero, then

$$n(t) = n_0 \left(\frac{1}{2}\right)^{t/(29 \text{ yr})} \tag{9}$$

The time required for one half of the parent material to decay is called the **half-life,** or $t_{1/2}$. In terms of the half-life we can write the law of radioactive decay as

$$n(t) = n_0 \left(\frac{1}{2}\right)^{t/t_{1/2}} \tag{10}$$

The formula (10) is not only valid at the times $t = 0, t_{1/2}, 2t_{1/2}, 3t_{1/2}$, etc., but, by continuity, it is also valid at all intermediate times. Figure 10.3 is a plot of n vs. time.

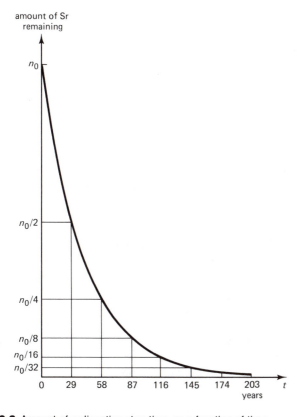

Fig. 10.3 Amount of radioactive strontium as a function of time.

By means of the identity $2 = e^{\ln 2} = e^{0.693\cdots}$, we can put Eq. (10) into the convenient form

$$n(t) = n_0 e^{-(\ln 2)t/t_{1/2}} \tag{11}$$

or

$$n(t) = n_0 e^{-t/\tau} \tag{12}$$

where

$$\tau = \frac{t_{1/2}}{\ln 2} \tag{13}$$

is called the **mean life.** Equation (12) is the usual mathematical form for the **law of radioactive decay.**

Example 1   Prove that the quantity τ defined by Eq. (13) is the mean lifetime of the decaying nuclei.

Solution   According to Eq. (12), the number of nuclei that decay in a time t is

$$-dn = \frac{1}{\tau} n_0 e^{-t/\tau} dt \tag{14}$$

Hence the mean lifetime is

$$\bar{t} = \frac{1}{n_0} \int_0^\infty t(-dn) \tag{15}$$

$$= \frac{1}{n_0} \int_0^\infty \frac{t}{\tau} n_0 e^{-t/\tau} dt = \frac{1}{\tau}\left[-\tau t e^{-t/\tau} - \tau^2 e^{-t/\tau} \right]_0^\infty = \tau \quad \blacksquare \tag{16}$$

Some radioisotopes have extremely long half-lives (e.g., 4.5×10^9 years for ^{238}U), whereas others have extremely short half-lives (e.g., $0.017\,s$ for ^{31}Na). Table 10.1 gives the half-lives of some important radioisotopes; note that the γ rays listed in this table are not emitted by the parent nucleus, but by the daughter.

In practice, we are often interested not so much in the amount of parent material, but in the **decay rate** of the material since this is what determines the rate of emission of α, β, or γ rays. The decay rate is simply the rate of change of n, i.e., it is the derivative of n with respect to t. From Eq. (12) we obtain

$$\frac{dn}{dt} = -\frac{1}{\tau} n_0 e^{-t/\tau} \tag{17}$$

TABLE 10.1 Some Radioisotopes and Their Half-Lives

Radioisotope	Radioactivity	Half-Life
^{14}C	β^-,	5730 yr
^{22}Na	β^+, γ	2.6 yr
^{40}K	β^-, γ	1.3×10^9 yr
^{60}Co	β^-, γ	5.24 yr
^{90}Sr	β^-, γ	28.8 yr
^{131}I	β^-, γ	8.05 days
^{226}Ra	α, γ	1620 yr
^{238}U	α, γ	4.5×10^9 yr

or

$$\frac{dn}{dt} = -\frac{1}{\tau} n \tag{18}$$

Hence the decay rate at any time is directly proportional to the amount of radioactive material left at that time (which is already fairly obvious from the original statement of the law of radioactive decay). This means that a plot of the radioactivity—the rate of emission of α, β, or γ rays—of a radioisotope looks similar to Fig. 10.3. The radioactivity decreases steeply with time in the same way as the amount of parent material decreases; whenever the time increases by one half-life, the radioactivity is reduced to one half its initial value.

The decay rate of a radioisotope is usually expressed in **curie** (Ci),*

$$1 \text{ curie} = 1 \text{ Ci} \equiv 3.7 \times 10^{10} \frac{\text{disintegrations}}{\text{s}} \tag{19}$$

---

* Named after **Marie Sklodowska Curie,** 1867–1934, and **Pierre Curie** (her husband), 1859–1906, French physicists and chemists, professors at the Sorbonne. For their discovery of the radioactive elements radium and polonium, they shared the 1903 Nobel Prize with Becquerel. On the death of Pierre, in an accident, Marie succeeded him at the Sorbonne, and continued her research on radioactivity. In 1911 she received a second Nobel Prize, in Chemistry, for further work on radium. Her daughter, Irène Joliot-Curie, 1897–1956, continued the family tradition of research on radioactivity and shared the 1935 Nobel Prize with Frédéric Joliot-Curie (her husband), 1900–1958, for the artificial production of radioactive substances by bombardment with alpha particles.

In the recent version of the metric system, the curie has been replaced by the **becquerel,**

$$1 \text{ becquerel} = 1 \text{ Bq} = 1 \text{ disintegration/s}$$

For example, 1 g of ^{131}I contains 4.6×10^{21} atoms and therefore has a decay rate

$$-\frac{dn}{dt} = \frac{1}{\tau}n = \frac{\ln 2}{t_{1/2}}n$$

$$= \frac{\ln 2}{8.05 \text{ days}} \times 4.6 \times 10^{21}$$

$$= 4.6 \times 10^{15} \text{ disintegrations/s} \qquad (20)$$

Expressing this in curies, we find

$$-\frac{dn}{dt} = 4.6 \times 10^{15} \frac{\text{disintegrations}}{\text{s}}$$

$$\times \frac{1 \text{ Ci}}{3.7 \times 10^{10} \dfrac{\text{disintegrations}}{\text{s}}} = 1.2 \times 10^{5} \text{ Ci} \qquad (21)$$

This is an extremely large disintegration rate, i.e., 1 g of ^{131}I is extremely radioactive. For instance, in medical applications (e.g., thyroid scan) the amount of ^{131}I introduced in the human body is typically only 10^{-9} g, giving a decay rate of about 10^{-4} Ci.

Example 2   The isotope ^{14}C is used for the radioactive dating of organic materials.* Samples of fresh carbon from trees, in equilibrium with the CO_2 of the atmosphere, have an abundance of 98.89% ^{12}C, 1.11% ^{13}C, and 1.3×10^{-10}% ^{14}C (the ^{14}C in the atmosphere is continually replenished by reactions initiated by high-energy cosmic rays incident on the Earth from outer space). After the tree dies, the abundance of ^{12}C and ^{13}C in the wood remains constant, but the abundance of ^{14}C decreases because of radioactive decay. Suppose that a piece of wood from an Egyptian tomb of the Old Kingdom contains 1 g of carbon whose measured activity is 3.9×10^{-12} Ci. How old is the wood?

Solution   The number of nuclei in 1 g of carbon is $1 \text{ g} \times (1 \text{ mole}/12 \text{ g}) \times 6.02 \times 10^{23}$ nuclei/mole $= 5.0 \times 10^{22}$ nuclei. Thus, 1 g of fresh carbon would contain $n_0 = 5.0 \times 10^{22} \times 1.3 \times 10^{-12} = 6.5 \times 10^{10}$ nuclei of ^{14}C with an activity

$$-\frac{dn}{dt}\bigg|_{t=0} = \frac{\ln 2}{t_{1/2}}n\bigg|_{t=0} = \frac{\ln 2}{5730 \text{ yr}} \times 6.5 \times 10^{10}$$

$$= 0.25/\text{s} = 6.8 \times 10^{-12} \text{ Ci} \qquad (22)$$

---

* The carbon-14 dating method was developed by **Willard Frank Libby,** 1908– , American chemist, professor at Chicago and the University of California, who received the 1960 Nobel Prize in chemistry for it.

The measured activity of the sample is smaller than this by a factor of $3.9 \times 10^{-12}/6.8 \times 10^{-12}\,\text{Ci} = 0.57$. Since the activity is proportional to the amount of radioactive material, $n(t)/n_0 = 0.57$. With this, the law of radioactive decay leads to

$$0.57 = e^{-(\ln 2)t/t_{1/2}} \tag{23}$$

or

$$t = -\frac{t_{1/2}}{\ln 2}\ln 0.57 = -\frac{5730\,\text{yr}}{\ln 2}\ln 0.57 = 4600\,\text{yr} \tag{24}$$

This means the wood is about as old as the Great Pyramid. ■

Finally, a few general remarks about the theoretical basis of the law of radioactive decay. This law is actually a probabilistic law rooted in the probabilistic features of the quantum mechanics of the nucleus. *On the average*, one half of the nuclei initially present in a sample of radioactive material will have decayed after one half-life; however, sometimes a few more nuclei will decay, sometimes a few less. These uncertainties are quite insignificant if the number of nuclei in the sample is large, but they are crucial if the number is small. In the extreme case of a single nucleus, it is impossible to predict exactly when this nucleus will decay; it is only possible to predict that the chance for the occurrence of decay increases with time in a definite, exponential manner—within a time interval equal to $t_{1/2}$, the chance for the occurrence of decay is 0.5. It is as though the decay of each nucleus were controlled by the toss of a coin: if the coin shows heads, the nucleus decays in the time interval $t_{1/2}$; if the coin shows tails, the nucleus does not decay in this time interval. In the latter case, the coin must be tossed again to decide whether the nucleus decays in the next time interval, between $t = t_{1/2}$ and $t = 2t_{1/2}$, etc.

This probabilistic aspect of the law of radioactive decay shows up experimentally in deviations between the simple mathematical law (12) for n and actual measurements. On the average the measurements agree with the value n predicted by Eq. (12), but they display fluctuations—typically the number of decays deviates from the predicted number by the square root of that number. For example, if 100 decays are expected to occur in a certain time interval, then the actual number will typically be between 90 and 110. With a careful statistical study of these deviations, we can confirm that these are random fluctuations, exactly as we would expect in a process governed by probabilities. In the next section we will examine the quantum-mechanical theory of α decay, and we will see why radioactive decay is governed by probabilities.

10.2 **Alpha Decay**

The escape of an alpha particle from a nucleus involves barrier penetration. The potential barrier experienced by the alpha particle during its escape is due to the forces acting between the alpha particle and the daughter nucleus. Figure 10.4 shows the potential energy as a function of distance from the center of the nucleus. For large r, the potential energy is simply the repulsive Coulomb energy of the alpha particle (charge $2e$) in the field of the daughter nucleus (charge Ze):

$$U(r) = \frac{2Ze^2}{4\pi\varepsilon_0 r} \qquad r > R \qquad (25)$$

At the nuclear surface ($r = R$), the potential energy drops sharply due to the action of the nuclear force. Inside the nucleus, the potential is approximately constant,

$$U(r) = -U_0 \qquad r < R \qquad (26)$$

The value of U_0 is not critical for the calculation of the barrier-penetration probability; we will assume that $U_0 \simeq 10\,\text{MeV}$. The horizontal line in Fig. 10.4 indicates the energy of a typical alpha particle. Experimentally it is found that in all cases of α decay, the energy of the alpha particle is substantially below the maximum height of the barrier (at $r = R$).

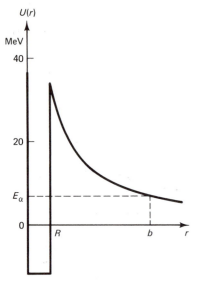

Fig. 10.4 The potential energy of an alpha particle interacting with the daughter nucleus. The Coulomb barrier shown here corresponds to ^{232}Th.

Example 3 The nucleus ^{238}U emits alpha particles of energy $4.20\,\text{MeV}$ [see Eq. (1)]. Compare this with the maximum height of the Coulomb barrier.

Solution With $r = R = r_0 A^{1/3}$, Eq. (25) becomes

$$U(R) = \frac{2Ze^2}{4\pi\varepsilon_0 r_0 A^{1/3}} \qquad (27)$$

In this equation, Z and A are the atomic number and the mass number of the *daughter* nucleus (^{234}Th with $Z = 90$, $A = 234$). Hence

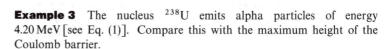

$$U(R) = \frac{2 \times 90 \times (1.6 \times 10^{-19}\,\text{C})^2}{4\pi\varepsilon_0 \times 1.2 \times 10^{-15}\,\text{m} \times (234)^{1/3}} \qquad (28)$$

$$= 5.6 \times 10^{-12}\,\text{J} = 35\,\text{MeV}$$

This is much larger than the energy of the emitted alpha particle. Thus, we see that the alpha particle tunnels through the Coulomb barrier quite near its base, where the barrier is thick. The very long half-life of 4.5×10^9 years for ^{238}U is a consequence of the low probability for such a tunneling process. ∎

For a quantitative discussion, we must make use of the results derived in Section 6.4. The probability for tunneling through the nuclear barrier is approximately given by the Gamow factor, Eq. (6.87),

$$P = e^{-\frac{4\pi}{\hbar} \frac{Ze^2}{4\pi\varepsilon_0} \frac{1}{v_\alpha}} \tag{29}$$

where v_α is the final speed of the ejected alpha particle. Since this probability is quite small, the alpha particle is very unlikely to escape at the first attempt. Instead, the alpha particle will bounce back and forth within the nucleus, making repeated impacts on the barrier at the nuclear surface. Each impact counts as an escape attempt. The time between these attempts is $2R/v_{in}$, where v_{in} is the speed of the alpha particle inside the nucleus. The typical value of this time is $2R/v_{in} \simeq 10^{-21}$ s, based on a typical kinetic energy $K_{in} \simeq 10 - 20$ MeV and a typical radius $R \simeq 10^{-14}$ m for a heavy nucleus. Hence the rate of escape for alpha particles is

$$\begin{pmatrix} \text{number of escapes} \\ \text{per second} \end{pmatrix} = \begin{pmatrix} \text{rate of escape} \\ \text{attempts} \end{pmatrix} \times \begin{pmatrix} \text{probability} \\ \text{for escape} \end{pmatrix} \tag{30}$$

$$= \frac{v_{in}}{2R} \times P \tag{31}$$

The number of escapes per second equals the inverse of the mean life. Consequently,

$$\tau = \frac{2R}{v_{in}} \times \frac{1}{P} = \frac{2R}{v_{in}} e^{\frac{4\pi}{\hbar} \frac{Ze^2}{4\pi\varepsilon_0} \frac{1}{v_\alpha}} \tag{32}$$

To compare this with the experimental data, it is convenient to take logarithms,

$$\ln \tau = \ln\left(\frac{2R}{v_{in}}\right) + \frac{4\pi}{\hbar} \frac{Ze^2}{4\pi\varepsilon_0} \frac{1}{v_\alpha} \tag{33}$$

The important term in this equation is the second term on the right side; this term gives by far most of the dependence of the mean life on the alpha-particle energy, whereas the first term is merely an additive constant that has nearly the same value for all the heavy nuclei.

Before we look at the experimental data, we ought to make a correction in Eq. (31). In our calculation we have implicitly assumed that the alpha particle exists inside the parent nucleus even *before* the decay occurs. The rationale behind this assumption is that the alpha particle (or ^{4}He nucleus) is an extremely stable configuration for two protons and two neutrons; thus, it is quite likely

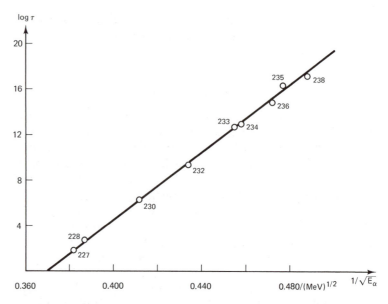

Fig. 10.5 Log τ vs. the reciprocal of the square root of the energy for alpha particles emitted by some isotopes of uranium (τ is expressed in seconds). The straight line is a least-squares fit to the data points.

that some of the protons and neutrons in a nucleus will spontaneously adopt the alpha-particle configuration. Of course, any such alpha particle is ephemeral—it will fall apart after a few trips back and forth within the nucleus. But for our purposes this is irrelevant. All we need to know is the average number of alpha particles to be found in a nucleus at a given instant of time. The available data suggest that this number is between 0.1 and 1, and hence we ought to multiply the right side of Eq. (31) by some such factor. This adds another (roughly) constant term to the right side of Eq. (33), but does not affect the energy dependence.

Figure 10.5 is a plot of the observed values of $\ln \tau$ vs. the reciprocal of the square root of the observed kinetic energy $[1/\sqrt{E_\alpha} = 1/(\sqrt{\frac{1}{2}m_\alpha}\,v_\alpha)]$ for alpha particles emitted by some nuclei of the same charge and roughly the same radius. According to Eq. (33), the plot should be a straight line, and indeed this theoretical prediction is in agreement with the experiment points. Note that the mean lives in this figure span a range from less than 1 minute to more than 10^9 years! This extreme range of times makes the agreement quite remarkable. The agreement can be improved by a more careful evaluation of the integral in Eq. (6.77). Such a more careful evaluation reveals that P includes a (slight) dependence on the nuclear radius.

10.3 Beta Decay

The simplest β-decay reaction is the decay of the neutron, yielding a proton, an electron, and an antineutrino:

$$n \to p + \beta^- + \bar{\nu} \tag{34}$$

For free neutrons, this decay proceeds with a half life of 10.6 min. However, as we mentioned in Section 9.1, for neutrons confined in a stable nucleus, the decay is inhibited. More generally, for neutrons in an unstable nucleus, the decay may be either faster or slower than for a free neutron. For instance, the β decay of the isotope ^{31}Na has a half life of only 0.017 s. The short half-life is in part due to the presence of several neutrons, each of which can decay; but it is mostly due to the availability of an excess of nuclear energy, which promotes the decay.

Other β reactions, closely related to Eq. (34), are the following:*

$$p \to n + \beta^+ + \nu \tag{35}$$

$$\beta^- + p \to n + \nu \tag{36}$$

$$\bar{\nu} + p \to n + \beta^+ \tag{37}$$

The first of these is β^+ decay. This reaction is energetically impossible for a free proton; however, it is possible for a proton in an unstable nucleus that can supply extra energy for the reaction.

The second reaction brings about electron capture (EC) in nuclei. As a general rule, any nucleus susceptible to β^+ decay is also susceptible to EC. Whether the nucleus undergoes the former reaction or the latter is a matter of chance. For instance, in a sample of ^{107}Cd nuclei, 0.31% will decay by β^+, and 99.69% by EC. In some nuclei, the decay energy is sufficient for EC, but not for β^+ decay; such nuclei decay exclusively by EC.

Example 4 Express the Q-value for the β^+-decay reaction and the EC- decay reaction of an arbitrary isotope in terms of the initial and final masses.

Solution We can write the β^+ decay as

$$^A[Z] \to {}^A[Z-1] + \beta^+ + \nu \tag{38}$$

---

* Note that the symbol β^+ for the antielectron (or positron) could equally well be written $\bar{e}$, according to the standard convention that an overbar indicates an antiparticle. Usually, the symbols e and $\bar{e}$ are preferred in the study of nuclear reactions, whereas the symbols β^- and β^+ are preferred in the study of radioactivity. Of course, these two subjects overlap.

where the symbol $^A[Z]$ stands for the isotope of atomic number Z and mass number A. As in the preceding chapter, we will express the Q-value in terms of the mass of the atom. If M_Z stands for the mass of the atom, then the nuclear mass is $M_Z - Zm_e$. The neutrino mass is zero (see below). Hence, the Q value is

$$Q = (M_Z - Zm_e)c^2 - (M_{Z-1} - (Z-1)m_e)c^2 - m_e c^2$$
$$= M_Z c^2 - M_{Z-1}c^2 - 2m_e c^2 \tag{39}$$

From this we see that the β^+ decay is energetically possible if the initial mass of the atom exceeds the final mass by at least two electron masses.

Likewise, we can write the EC reaction as

$$\beta^- + {}^A[Z] \to {}^A[Z-1] + \nu \tag{40}$$

for which the Q-value is

$$Q = m_e c^2 + (M_Z - Zm_e)c^2 - (M_{Z-1} - (Z-1)m_e)c^2$$
$$= M_Z c^2 - M_{Z-1}c^2 \tag{41}$$

We see that EC is energetically possible if the initial mass of the atom exceeds the final mass. This means that in some nuclei EC is energetically viable, but not β^+ emission. ■

The interaction that brings about the reactions (34)–(37) is a new kind of interaction called the **"weak" interaction.** It is one of the four fundamental interactions of matter: the electromagnetic, strong, weak, and gravitational interactions. Whereas the electromagnetic and the strong interactions play an important role in the structure of matter by providing the forces that bind together the particles within the atom or the nucleus, the weak interaction does not participate in this. It only makes its appearance in the β reactions (34)–(37) and in some similar reactions involving other elementary particles (see Chapter 11). It is called "weak" because it is much weaker than the electromagnetic or the strong interaction. However, in the realm of atomic and nuclear physics it is much stronger than the gravitational interaction, which is the weakest of all.

A remarkable feature of β decay is that the electrons or antielectrons ejected from the nuclei of a given isotope sometimes emerge with one energy, and sometimes with another. For instance, Fig. 10.6 shows the energy spectrum of electrons ejected in the decay of a sample of ^{210}Bi. The electron energies span the range from 0 to 1.16 MeV, with a broad maximum around 0.15 MeV. This variability of the energy of β rays is in stark contrast to the constancy of energy of α rays—the energy spectrum of α rays ejected in the decay of a heavy nucleus displays unique, sharply defined energies. We can summarize this difference between the α and the β

number of β rays
per MeV

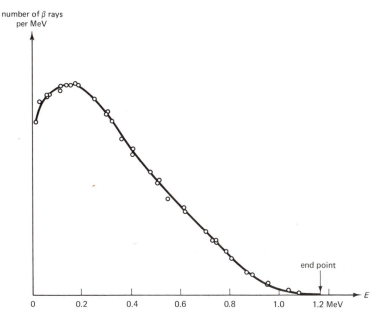

Fig. 10.6 Energy spectrum of β rays ejected by the radioactive decay of a sample of ^{210}Bi. [After G. J. Neary, *Proc. Roy. Soc.* (London) **A175**, 71 (1940).]

rays thus: the energy distribution of the α rays is a line spectrum, whereas that of the β rays is a continuous spectrum.

In the 1920s, the variability of the energy of β rays posed a puzzle for theoretical physicists, and led some of them to speculate that perhaps energy is not conserved. Whenever a β ray is emitted with less than the maximum available energy, what happens to the energy deficit? The answer is that the energy deficit represents the energy carried away by an extra particle ejected by the nucleus: the neutrino. This explanation of the apparent failure of conservation of energy was proposed by W. Pauli in 1931. At the time there was no experimental evidence for the existence of any extra particle in β decay, and Pauli's proposal was an act of faith, based on nothing but a firm belief in the conservation laws.

Direct experimental evidence for the existence of neutrinos was not obtained until 1953 when F. Reines and C. L. Cowan* exposed targets containing a large amount of hydrogen to the very intense flux of antineutrinos produced by a nuclear reactor, and observed

---

* **Frederick Reines,** 1918– , American physicist, professor at the University of California, Irvine. **Clyde L. Cowan,** 1919– , American physicist, professor at The Catholic University.

the reaction (37), which converts a proton into a neutron plus an antielectron. They detected these neutrons and antielectrons in coincidence.

Neutrinos (and antineutrinos) are particles of zero mass or nearly zero mass, zero electric charge, and spin $\frac{1}{2}$. Neutrinos are very hard to detect because their interactions with other particles are extremely weak—they only interact via the "weak" force. The mass of the neutrino can be determined by a careful examination of the upper limit of the energy spectrum of the β rays (Fig. 10.6). The shape of the spectral curve near E_{max} depends on the neutrino mass. The available experimental data establish that the mass of a neutrino is less than 6×10^{-8} u, or less than 6×10^{-4} MeV/c^2. This leaves open the possibility that the mass may be very small, but not quite zero.*

Theoretical calculations by Enrico Fermi[†] showed that the neutrino hypothesis led to the observed spectral curve for the β-ray energies. The electron and the neutrino ejected in a β decay share the available energy at random. This makes it unlikely that the electron will receive all the energy, or that the neutrino will receive all of the energy; hence instances of decays with the extreme values $E = E_{max}$ or $E = 0$ of the electron energy are rare, in qualitative agreement with the spectral curve shown in Fig. 10.6.

10.4 Low-Energy Nuclear Reactions; the Compound Nucleus

The experimental investigation of nuclear reactions began in 1919 when Rutherford discovered that the bombardment of nitrogen with alpha particles led to the formation of oxygen and of protons,

$$^4\text{He} + {}^{14}\text{N} \rightarrow {}^{17}\text{O} + {}^1\text{H} \tag{42}$$

---

* Some new, unconfirmed measurements appear to indicate that the neutrino mass is different from zero.

[†] **Enrico Fermi,** 1901–1954, Italian and later American physicist, professor at Rome and at Chicago. Fermi was a brilliant experimenter, and also a gifted theoretician. Among his most important contributions were the experimental and theoretical investigations of β decay, his production of artificial isotopes by neutron bombardment, which earned him the Nobel Prize in 1938, and his work on the fission of uranium. Fermi was one of the leaders of the Manhattan Project and his experimental demonstration of chain reactions with the Chicago reactor was the crucial step in the development of the bomb.

In these experiments, and in his earlier experiments, Rutherford used a naturally radioactive material as source of his beam of α particles. This experimental technique was drastically improved in 1932 when J. D. Cockcroft and E. T. S. Walton* used an electrostatic accelerator (see Fig. 10.7) to generate a beam of energetic protons, with which they bombarded a target of lithium and observed the first nuclear reaction initiated by artificially accelerated projectiles. The accelerator built by Cockcroft and Walton was similar to that built by R. Van de Graaff† a year earlier. This machine accelerates the projectiles by means of a strong electrostatic field, produced by a large amount of electrostatic charge accumulated on a spherical capacitor at high voltage. A different machine, called a cyclotron, was developed at about the same time by E. O. Lawrence.‡ The cyclotron holds the projectiles in a spiral orbit in a magnetic field and accelerates them by means of repeated pushes from an oscillating electric field (see Figs. 10.8 and 1.2).

As nuclear physicists built bigger and better accelerators, the bombardment of targets with artificially produced beams of projectiles became the standard technique for the experimental investigation of nuclear reactions. The energy of the projectiles used to initiate nuclear reactions is usually in the range from a few MeV to 140 MeV.§ This range of energies characterizes the realm of low-energy nuclear physics, in contrast to a range from 140 MeV to several hundred MeV for high-energy nuclear physics, and a range of several hundred MeV to many thousand MeV for elementary-particle physics. Figure 10.9 shows a typical experimental arrangement. The beam of energetic protons, alpha particles, or deuterons travels from the accelerator to the target chamber via an evacuated pipe. The target consist of a very thin foil of whatever material

Fig. 10.7 The accelerator of Cockroft and Walton, being operated by Walton. (Courtesy University of Cambridge, Cavendish Laboratory.)

---

* **Sir John Douglas Cockcroft,** 1897–1967, and **Ernest Thomas Sinton Walton,** 1903– , British physicists. For their pioneering work on nuclear reactions with artificially accelerated particles, they shared the Nobel Prize in 1951. Cockcroft was the first director of the Atomic Energy Research Establishment at Harwell.

† **Robert Jemison Van de Graaff,** 1901–1967, American physicist, professor at the Massachusetts Institute of Technology. Van de Graaff later modified his generator so that it could be used for the production of very hard X rays for medical applications.

‡ **Ernest Orlando Lawrence,** 1901–1958, American experimental physicist, professor at Berkeley, and director of the Radiation Laboratory (now Lawrence Berkeley Laboratory). He was awarded the Nobel Prize in 1939 for his invention and development of the cyclotron. During the war he organized the Los Alamos Laboratory and worked on the separation of ^{235}U and ^{239}Pu isotopes required for the bomb.

§ 140 MeV is the threshold for pion production.

(a)   (b)

Fig. 10.8 (a) Lawrence's 37-inch cyclotron; (b) the Dees of the 37-inch
cyclotron. (Courtesy Lawrence Berkeley Laboratory.)

is under investigation. Low-energy nuclear reactions are usually
accompanied by the ejection of protons, neutrons, alpha particles,
deuterons, and γ rays from the target nuclei. These particles are
identified and counted by detectors surrounding the target: gas-
filled detectors, such as proportional counters or Geiger counters
that detect the ions produced by the passage of the particle through
a gas; scintillation detectors, such as plastics or sodium iodide that,
when struck by a particle, give off a small flash of light which is
registered by a photomultiplier tube; or semiconductor detectors,
such as germanium doped with lithium, that detect electron–hole
pairs produced by the passage of the particle through a semiconductor junction.

Experimental physicists like to express their results of determinations of nuclear reaction rates in terms of the **cross section** σ
which is defined as the ratio of reaction rate per nucleus to the
incident flux of projectiles,

$$\sigma = \frac{\text{(reaction rate per nucleus)}}{\text{(incident flux)}} \tag{43}$$

The reaction rate is the number of reactions per second; the unit
for the numerator in Eq. (43) is s^{-1}. The incident flux is the
number of projectiles incident per m^2 per second; the units for the
denominator in Eq. (43) are $m^{-2}s^{-1}$. Therefore, the cross section
has the units m^2, i.e., it has the units of an area. The cross section

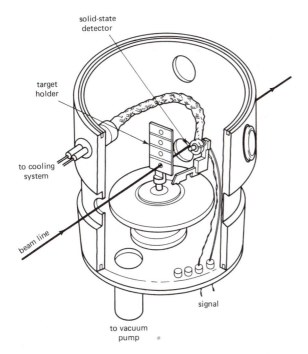

Fig. 10.9 Scattering chamber used for the bombardment of a target by a beam of protons. (Courtesy P. Stoler, Rensselaer Polytechnic Institute.)

is a measure of the probability of occurrence of the reaction. If the projectiles were classical particles traveling along straight lines and if every projectile approaching a nucleus with impact parameter less than the nuclear radius were to trigger the reaction, then the cross section for that reaction would be equal to the geometrical cross-sectional area of the nucleus. Since the projectile is a quantum-mechanical particle whose wavefunction can overlap the nucleus and interact even when the classical impact parameter is more than the nuclear radius, and since the probability to trigger the reaction can be less than 100%, the cross section can be either more than or less than the geometrical cross-sectional area. Nuclear cross sections are often measured in **barn,**

$$1 \text{ barn} = 10^{-28} \text{ m}^2$$

Typical nuclear cross sections have magnitudes ranging from a fraction of a barn to several barns.

Nuclear reactions are much more complicated than chemical reactions because all the nucleons in the nucleus participate, whereas in a chemical reaction only the outermost electrons of the atom participate. However, as Bohr pointed out in 1936, the general participation of all nucleons in the reaction permits us to contrive a very simple model for the theoretical description of nuclear reactions. According to this model, the incident projectile is absorbed by the nucleus, and quickly shares its energy with all the nucleons. This merger of the projectile and the original nucleus forms a **compound nucleus,** whose mass number and charge is the sum of the mass numbers and charges of the projectile and the original nucleus. For instance, the bombardment of ^{63}Cu with protons leads to the formation of the compound nucleus ^{64}Zn*,

$$p + {}^{63}\text{Cu} \rightarrow {}^{64}\text{Zn*} \tag{44}$$

The asterisk on the chemical symbol indicates that the nucleus is in an excited state. The same compound nucleus can also be formed by other means. For instance, the bombardment of ^{60}Ni with alpha particles yields

$$^{4}\text{He} + {}^{60}\text{Ni} \rightarrow {}^{64}\text{Zn*} \tag{45}$$

If the energy of the alpha particle is suitably adjusted, then the energy of the compound nuclei formed in the reaction (44) and (45) will be the same.

Because of the energy brought in by the projectile, the compound nucleus is in a highly excited, unstable state. After a short while it decays with the ejection of one or several nucleons. The decay occurs so quickly—within about 10^{-16} s—that the compound nucleus can never be observed directly. Thus the compound nucleus is only an ephemeral intermediate stage of the nuclear reaction. The decay of the compound nucleus can occur via one of several alternative routes, or "channels." For instance, the compound nucleus ^{64}Zn* has been observed to decay via the alternative channels

$$^{64}\text{Zn*} \rightarrow {}^{63}\text{Zn} + \text{n} \tag{46}$$

$$\rightarrow {}^{62}\text{Zn} + \text{n} + \text{n} \tag{47}$$

$$\rightarrow {}^{62}\text{Cu} + \text{p} + \text{n} \tag{48}$$

The choice among these channels is governed by probabilities, which depend on the energy. However, the decay probabilities do not depend on the manner in which the compound nucleus was formed. Whether the energy was brought into the compound nucleus by a proton [via reaction (44)], or by an alpha particle [via reaction (45)] makes no difference—the compound nucleus retains no memory of how it was formed.

Thus, according to the compound-nucleus model, a nuclear reaction is a two-step process. In the first step the projectile merges with the target nucleus forming a compound nucleus, and in the second step the compound nucleus decays. The two steps are probabilistically independent events, in the sense of probability theory. Hence the net probability for the reaction is the product of the probability for the formation of the compound nucleus and the probability for the decay.

We can test this prediction of the compound-nucleus model against experiment by comparing the measured cross sections for the reactions

$$p + {}^{63}Cu \rightarrow \text{final channels} \tag{49}$$

with the measured cross sections for the reactions

$$\alpha + {}^{60}Ni \rightarrow \text{final channels} \tag{50}$$

where "final channels" stands for the three alternatives listed in Eqs. (46)–(48). If proton bombardment results in a high cross section for, say, p + n ejection relative to n + n ejection, then alpha bombardment ought to do likewise. Figure 10.10 shows the measured cross sections as a function of energy. The data confirm the prediction of the model: at any given energy of the compound nucleus, the cross sections for final channels obtained via proton bombardment are, roughly, in the same ratio as the cross sections for these final channels obtained via alpha bombardment.

In all examples above the compound nucleus was formed by bombardment of a target with charged particles (protons or alpha particles). This leads to the formation of a compound nucleus of rather high excitation energy, since the bombarding particle must be launched with a substantial amount of initial kinetic energy so that it can overcome (or penetrate) the Coulomb barrier. It is also possible to form a compound nucleus by bombardment with neutrons.* This has the advantage that neutrons of very low energy may be employed, since there is no Coulomb barrier. It turns out that neutrons of very low energy are more easily absorbed by a nucleus than neutrons of high energy. (If the neutron moves at low speed, it spends more time inside the nucleus and is more likely to engage in a reaction. It is therefore not surprising that the cross section for absorption of a low-energy neutron by a nucleus is inversely proportional to the speed; this is known as

---

* Intense beams of neutrons are available from nuclear reactors, in which fission reactions (see below) release an abundant flux of neutrons. Alternatively, neutrons can be generated by exposing targets of light nuclei to bombardment by energetic protons or deuterons from an accelerator; this permits the generation of neutrons by reactions such as $p + {}^{7}Li \rightarrow {}^{7}Be + n$ or $D + {}^{9}Be \rightarrow {}^{10}Be + n$.

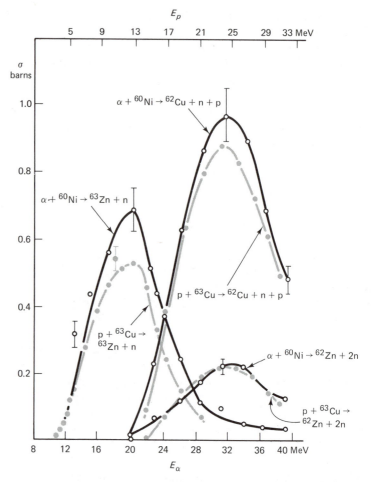

Fig. 10.10 Cross sections for reactions initiated by p + ^{63}Cu and by α + ^{60}Ni. The upper scale gives the proton energy and the lower scale the alpha energy. [From S. N. Ghoshal, *Phys. Rev.* **80**, 939 (1950).]

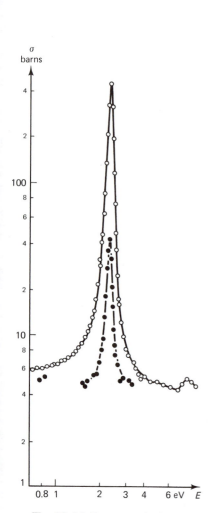

Fig. 10.11 Resonance in the cross sections for n + ^{123}Te → ^{124}Te + γ (dots) and n + ^{123}Te → ^{123}Te + n (circles). (After D. J. Hughes and J. A. Harvey, *Neutron Cross Sections*.)

the $1/v$ law.) Hence neutron bombardment gives us access to the low-lying energy states of the compound nucleus. These low-lying states display resonance phenomena: the cross section for the formation of the compound nucleus becomes extremely large whenever the neutron energy coincides exactly with the energy required to put the nucleus into one of its excited states. Consequently, the cross section for any nuclear reaction that has this excited compound nucleus as an intermediate stage will be large at this energy. For example, Fig. 10.11 shows the cross sections for the reactions

$$n + {}^{123}\text{Te} \rightarrow {}^{124}\text{Te} + \gamma \tag{51}$$

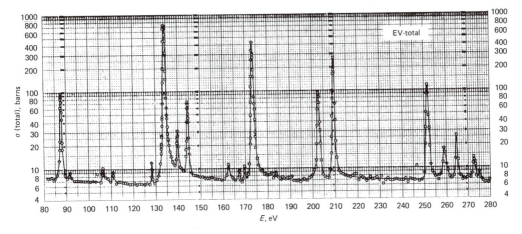

Fig. 10.12 Resonances seen in the total cross section for neutrons incident on ^{47}Ag. (From Brookhaven National Laboratory, *Neutron Cross Sections*.)

and

$$n + {}^{123}\text{Te} \rightarrow {}^{123}\text{Te} + n \tag{52}$$

The first of these reactions is **radiative capture;** the second is **elastic scattering.** Although the latter reaction does not lead to a nuclear transmutation, it does involve the formation of a compound nucleus (^{124}Te*) and the subsequent decay of this compound nucleus by ejection of a neutron. Experimentally, the occurrence of the reaction can be distinguished from the nonoccurrence because the ejected neutron has a direction of motion different from that of the incident neutron. Both the radiative-capture channel and the elastic-scattering channel display sharp resonances at a neutron energy of 2.33 eV, which corresponds to an excited state in ^{123}Te.

Sharp resonances are a ubiquitous feature of nuclear reactions at very low energies (see Fig. 10.12). These resonances tend not to show up at higher energies—several MeV—because there the compound nucleus has available a wide variety of possible decay channels, and the elastic-scattering channel is relatively insignificant. We can understand why the availability of abundant decay channels tends to suppress the resonance, if we think of an analogy between the excited state of the compound nucleus and a classical oscillator. An abundance of decay channels that can remove energy from the excited state is analogous to a large frictional resistance in the oscillator, which prevents the amplitude from building up and makes the resonance broad and indistinct. Furthermore, at the higher energies, the intervals between resonances tend to be small; therefore, the resonances overlap, and they become difficult or impossible to observe.

Although the compound-nucleus model successfully accounts for a wide variety of low-energy nuclear reactions, it fails to account for some inelastic scattering reactions. The disagreeement between the predictions of the model and experiment is especially conspicuous in reactions involving charged projectiles, such as protons and alpha particles. On the basis of the compound model, we would expect that when a fairly energetic charged projectile, say an alpha particle of 30 or 40 MeV, penetrates the nucleus, it would quickly share its energy with the nucleons, and then find itself trapped within the surrounding Coulomb barrier (see Section 10.2). In the compound nucleus formed in this way, the most likely process of deexcitation would be the subsequent ejection of one or several neutrons. Since tunneling through the Coulomb barrier is a slow process, the ejection of an alpha particle would be a rare event, i.e., the cross section for inelastic scattering of alpha particles would be small. However, this prediction of the compound model disagrees with experiment: the observed value of the cross section for inelastic scattering of alpha particles is fairly large (of the order of 10% of the total reaction cross section), and the angular distribution of scattered alpha particles exhibits a sequence of sharp maxima and minima as a function of scattering angle (see Fig. 10.13). This strong dependence on angle also disagrees with the

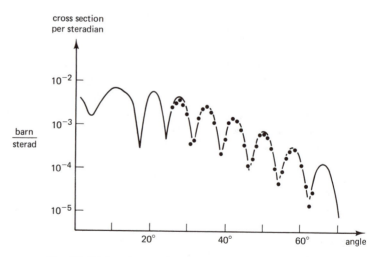

Fig. 10.13 Angular distribution of 43-MeV alpha particles inelastically scattered by nuclei of ^{154}Sm. [After D. L. Hendrie et al., *Phys. Letters* **26B**, 127 (1968).]

compound model, which predicts a fairly weak dependence on angle, with symmetry about 90° (naively, we would expect that in the reference frame of the compound nucleus the probability for ejection of a particle is more or less independent of angle).

The explanation of these properties of inelastic scattering involves a **direct-reaction** mechanism: the incident alpha particle interacts with only a small portion of the nucleus, suffers some loss of energy and some deflection, and emerges from the nucleus without further delay. The alpha particles that engage in such direct reactions are those that pass through the rim of the nucleus in a glancing collision—alpha particles that attempt to pass through the middle interact with many nucleons and usually initiate the formation of a compound nucleus. The maxima and minima in the angular distribution of alpha particles (see Fig. 10.13) result from an interference effect: the emerging alpha-particle wave receives contributions from opposite ends of the rim of the nucleus, and these contributions can interfere constructively or destructively depending on the angle. Thus, the pattern of maxima and minima in the angular distribution of alpha particles scattered by a nucleus is analogous to the interference pattern of light waves scattered by a pair of thin slits. Similar patterns of maxima and minima also have been observed in the angular distributions of protons and of neutrons scattered by nuclei. This indicates that the inelastic scattering of these particles also proceeds via the direct-reaction mechanism described above.

10.5 Fission

Although in most nuclei the binding energy per nucleon is in the vicinity of 8 MeV, in heavy nuclei (say, $A > 200$) the binding energy per nucleon is somewhat smaller (see Fig. 9.2). The decrease of the binding energy is due to the increasing importance of the electrostatic repulsion. Therefore, it is energetically favorable for a heavy nucleus to split up into two fragments, forming two lighter nuclei. For example, consider the spontaneous **fission** of a ^{238}U nucleus into two equal fragments,

$$^{238}U \rightarrow {}^{119}Pd + {}^{119}Pd \tag{53}$$

This is a **symmetric** fission reaction: a nucleus of mass number A and charge number Z splits into two nuclei, each of mass number

$A/2$ and charge number $Z/2$. According to the semiempirical formula [Eq. (9.42)] the binding energy of the original nucleus is

$$B_{A,Z} = \left(15.753A - 17.804A^{2/3} - \frac{0.7103Z^2}{A^{1/3}} \right) \text{MeV} \qquad (54)$$

and the binding energy of *each* final nucleus is

$$B_{A/2,Z/2} = \left[15.753\frac{A}{2} - 17.804\left(\frac{A}{2}\right)^{2/3} - \frac{0.7103(Z/2)^2}{(A/2)^{1/3}} \right] \text{MeV}$$

$$(55)$$

In these expressions we have neglected the small contributions made by the asymmetry energy. The energy released in the fission reaction is then approximately

$$2B_{A/2,Z/2} - B_{A,Z} = \left(-4.6A^{2/3} + \frac{0.26Z^2}{A^{1/3}} \right) \text{MeV} \qquad (56)$$

The meaning of the two terms on the right side of Eq. (56) is as follows: The first term represents the work done by the nuclear "surface tension" force; this term is negative because this force opposes the fission, i.e., it opposes the increase in surface area that necessarily accompanies the splitting of a single drop of fluid into two drops. The second term represents the work done by the Coulomb force; this term is positive because the electrostatic repulsion favors fission by pushing the two fragments apart. If this second term dominates over the first term, then the fission reaction releases energy and is viable. The necessary criterion for fission is therefore

$$-4.6A^{2/3} + 0.26\frac{Z^2}{A^{1/3}} > 0 \qquad (57)$$

i.e.,

$$\frac{Z^2}{A} > 18 \qquad (58)$$

In the case of ^{238}U, we have $Z = 92$, $A = 238$, and

$$\frac{Z^2}{A} = \frac{(92)^2}{238} = 35.6 \qquad (59)$$

so that the criterion (58) is amply satisfied.

Note that the criterion (58) is necessary for fission, but not sufficient. Although the uranium nucleus has enough energy for fission, it is metastable and remains in its initial configuration for a very long time. When fission finally does occur, it begins with a slight elongation of the nucleus (see Fig. 10.14b); the elongation then increases, developing two bulges joined by a waist (Fig. 10.14c, d); finally, the waist pinches off and the two fragments separate (Fig. 10.14e, f). Obviously, the nuclear surface-tension forces resist the elongation of the nucleus and it is only when the nucleus splits (Fig. 10.14e,f) that these nuclear forces irrevocably lose out to the Coulomb forces. Figure 10.15 shows a plot of the potential energy of the drop of nuclear fluid as a function of its deformation; zero deformation corresponds to Fig. 10.14a; extreme deformation corresponds to Fig. 10.14f in which the drop has split into two well-separated drops. Although the energy of the underformed drop is higher than the energy of the separated drops, the intermediate stages of deformation have the highest energy; thus, there is a potential barrier which opposes fission. The nucleus must tunnel through this barrier. Therefore, the quantum-mechanical theory of fission is quite similar to the theory of α decay (in fact, α decay can be regarded as an extreme case of fission: the nucleus splits into a small ^{4}He nucleus and a large daughter nucleus). The probability for the occurrence of a spontaneous fission can be calculated from the shape of the potential-energy curve plotted in Fig. 10.15.

In the case of ^{238}U, spontaneous fission is a rare occurrence; only one nucleus in 2×10^6 will decay by spontaneous fission, the others will decay by α emission. Furthermore, when spontaneous fission does occur, it is not always symmetric [as in Eq. (53)]; more often than not the fission is **asymmetric,** the nucleus splitting into two unequal fragments (see Fig. 10.16). However, the mass difference between the fragments is usually not all that large (typically 30%) and the energy released is not all that far from the value calculated for symmetric fission.

According to Eq. (56), the energy released in the symmetric fission of ^{238}U is

$$\left[-4.6(238)^{2/3} + \frac{0.26(92)^2}{(238)^{1/3}} \right] \text{MeV} = (-180 + 360)\,\text{MeV} \quad (60)$$

$$= 180\,\text{MeV} \quad (61)$$

This energy will emerge in the form of kinetic energy of the fission fragments, kinetic energy which is impressed on the fragments by the repulsive Coulomb force that pushes them apart during the fission process. In the fission of a nucleus of ^{238}U, the Coulomb force does 360 MeV of work on the fragments [see the second term in brackets in Eq. (60)]; simultaneously, the strong force does 180 MeV of work against the fragments [see the first term in brackets in Eq. (60)]; on balance this leaves the fragments with a net gain

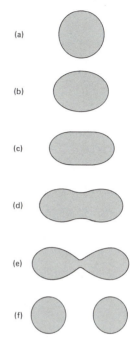

(a)

(b)

(c)

(d)

(e)

(f)

Fig. 10.14 Fission of a drop of nuclear fluid. [After D. L. Hill and J. A. Wheeler, *Phys. Rev.* **89**, 1102 (1953).]

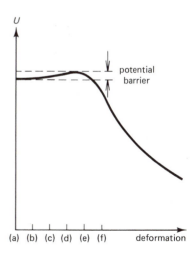

Fig. 10.15 Potential energy as a function of deformation (the letters give a rough correspondence with Fig. 10.14.)

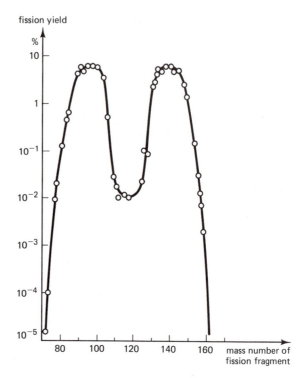

fission yield

Fig. 10.16 Distribution of mass numbers of fragments produced by the fission of the nucleus ^{236}U formed when ^{235}U absorbs a thermal neutron. The most probable fission products are one nucleus of mass number ~ 96 and one nucleus of mass number ~ 140. The distribution produced in the fission of ^{238}U is similar. [From J. M. Siegel et al., *Rev. Mod. Phys.* **18**, 513 (1946).]

of 180 MeV in kinetic energy. Thus, the source of the energy released in fission in a nuclear reactor or bomb is actually the electrostatic interaction, not the strong interaction.

The kinetic energy of each fragment will be about 90 MeV. This is a relatively large amount of energy as can be seen by comparing it with the rest-mass energy of the fragments. The rest-mass energy mc^2 of a ^{119}Pd nucleus is about 1.1×10^5 MeV; hence the ratio of kinetic energy to rest-mass energy is

$$\frac{K}{mc^2} = \frac{90\text{ MeV}}{1.1 \times 10^5\text{ MeV}} = 0.080 \times 10^{-2} \qquad (62)$$

The kinetic energy is about 0.08% of the rest-mass energy! The speed corresponding to such a large kinetic energy is also very large and is best expressed in terms of the speed of light:

$$K = \frac{1}{2} mv^2 = \frac{1}{2} mc^2 \left(\frac{v^2}{c^2} \right) \tag{63}$$

i.e.,

$$\frac{v}{c} = \sqrt{2K/mc^2} = \sqrt{2 \times 0.08 \times 10^{-2}} = 0.040 \tag{64}$$

This means that the speed of each fragment is 4% of the speed of light.

The total energy released as a consequence of the fission of a uranium nucleus is actually somewhat larger than 180 MeV. The fission fragments are very unstable—they contain both an unacceptable excess of internal energy (excitation energy) and also an unacceptable excess of neutrons. To eliminate these excesses, the fission fragments undergo a series of radioactive decays involving the ejection of neutrons, β rays, γ rays, and neutrinos. Some of these particles are ejected almost instantaneously, while others suffer a slight delay. These secondary emissions release roughly an extra 20 MeV and bring the total energy yield per fission to about 200 MeV; thus, the total energy yield is about 0.090% of the initial rest-mass energy. Table 10.2 shows how the total energy is distributed, on the average, among fission fragments and other ejecta. Note that the average kinetic energy of the fission fragments is somewhat lower than was calculated above [see Eq. (61)]; this is so because asymmetric fission—which predominates—yields somewhat lower kinetic energies than symmetric fission.

Example 5   What is the amount of energy released in the complete fission of 1 kg of uranium (a lump about as large as an egg)?

Solution   The energy released is 0.090% of the rest-mass energy of 1 kg, i.e.,

$$0.090 \times 10^{-2} mc^2 = 0.090 \times 10^{-2} \times 1 \, kg \times (3 \times 10^8 \, m/s)^2 = 8.1 \times 10^{13} \, J \tag{65}$$

TABLE 10.2   Distribution of energy in fission

Fission fragments (kinetic energy)	165 ± 5 Mev
Neutrons (kinetic energy)	5 ± 0.5
γ rays, instantaneous	7 ± 1
β particles, delayed	7 ± 1
γ rays, delayed	6 ± 1
Neutrinos, delayed	10
Total energy per fission	200 ± 6 Mev

This is equivalent to the energy released in the explosion of about 20,000 tons of TNT.* ∎

10.6 Chain Reactions

Although spontaneous fission in ^{238}U is rare, this nucleus is susceptible to induced fission when subjected to bombardment by neutrons. The collision of a neutron with a nucleus and its absorption initiates violent vibrations of the nucleus, which are likely to split it apart. This results in **neutron-induced fission reaction,** such as

$$n + {}^{238}U \rightarrow {}^{145}Ba + {}^{94}Kr \qquad (66)$$

Incidentally: the reaction (66) is of some historical interest, since it led to the discovery of fission in 1938 by O. Hahn and F. Strassmann,[†] who detected barium in a sample of uranium irradiated by neutrons.

Both of the fission fragments released in the reaction (66) are very neutron rich, and they almost immediately eject two or three of their excess neutrons. The net reaction in neutron-induced fission of ^{238}U can therefore be summarized as

$$n + {}^{238}U \rightarrow \text{fission fragments} + 2n \text{ or } 3n \qquad (67)$$

Practical applications of fission rely on the neutrons released in this reaction to induce further fission reactions. If we trigger a first fission in a sample of uranium, the neutrons released in this first fission will collide with other uranium nuclei and induce their fission, and the neutrons released there will induce further fissions, etc. (see Fig. 10.17). The result is a self-sustaining **chain reaction.** Provided no neutrons, or only few neutrons, are lost from this chain, the result is an avalanche of neutrons and of fissions. In such an avalanche the number of neutrons and the number of fissions

---

* 1 ton of TNT releases 10^9 cal $= 4.2 \times 10^9$ J.

[†] **Otto Hahn,** 1879–1968, German chemist, director of the Kaiser-Wilhelm Institute. At first, Hahn and his pupil **Fritz Strassmann,** 1902– , were reluctant to believe that the uranium nucleus could have split apart. But these doubts were quickly removed by the work of **Lise Meitner,** 1878–1968, and **Otto Frisch,** 1904–1979. For his discovery of nuclear fission, Hahn was awarded the Nobel Prize in Chemistry in 1944.

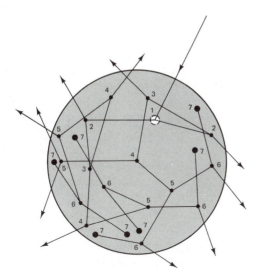

Fig. 10.17 Chain reaction initiated by a single fission at the point 1.

in successive steps of the chain grow geometrically. For example, if on the average two neutrons are released per fission reaction and each of these neutrons induces a further fission reaction, then the number of fissions in successive steps of the chain will be 2, 4, 8, 16, 64, ... This geometric growth of the reaction rate leads to an explosive release of energy.

The cross section for neutron-induced fission reactions depends on the neutron energy. In the case of ^{238}U, the minimum neutron energy required for initiating the fission reaction is 1.2 MeV. Neutrons of this energy or a larger energy are called **fast neutrons.** The neutrons released in fission of ^{238}U are initially fast (see Table 10.1), but they are likely to suffer several consecutive inelastic collisions with nuclei before they are finally absorbed in one of these collisions. In such inelastic collisions, the neutrons lose most of their kinetic energy, and when they are finally absorbed, their residual energy is insufficient to trigger a fission. Thus, the inelastic collisions effectively remove neutrons from the fission chain, and thereby inhibit the chain reaction. The most likely final fate of a low-energy neutron is absorption by a uranium nucleus according to the reaction

$$n + {}^{238}U \rightarrow {}^{239}U \tag{68}$$

Hence ^{238}U tends to soak up neutrons without undergoing fission, and it will not sustain a chain reaction.

In the case of ^{235}U, there is no minimum neutron energy required for fission. This nucleus is rather unstable, and it will fission even if the energy of the incident neutron is very low, because the binding energy that becomes available when the neutron is captured by the nucleus is by itself sufficient to initiate the fission reaction. In fact, the cross section for induced fission is largest for incident neutrons of the lowest energy ($1/v$ law), because the slowest neutrons spend the longest time passing through the nucleus, which enhances the probability that an absorption reaction will occur. Thus, ^{235}U will sustain a chain reaction—it is a **fissile** material.

The abundance of ^{235}U in naturally occurring uranium ores is quite low—such ores consists of about 99.3% ^{238}U and only 0.7% ^{235}U. Since the mass difference between these isotopes is small, their separation is difficult. It can be accomplished by converting them into uranium hexafluoride, a gaseous compound, in which the two isotopes can be segregated by diffusion through membranes or by centrifugation.

Besides ^{235}U, there exist two other isotopes suitable for chain reactions. Both resemble ^{235}U in that the spontaneous decay rate is low (so they can be held in storage without serious loss) and in that the fissions can be induced by neutrons of low energy: one is ^{233}U and the other is an isotope of plutonium, ^{239}Pu. The latter is very fissile, but it is not found in ores on the Earth; it can only be obtained by artifical means, by nuclear transmutation.

The geometric growth of the reaction rate in a chain reaction is described mathematically by the **multiplication factor,** i.e., the factor by which the number of neutrons increases in successive steps along the fission chain. If no neutrons are lost from the fission chain, then the multiplication factor is simply equal to the average number of neutrons released per fission; but if some neutrons are lost, then the multiplication factor will be smaller. The mass of fissile material is said to be in a **critical** condition if the multiplication factor is unity; the chain reaction then merely proceeds at a constant rate—as in a nuclear reactor. The mass is in a **supercritical** condition if the multiplication factor exceeds unity; the chain reaction then proceeds at a geometrically increasing rate, leading to an explosion—as in a nuclear bomb.

Neutrons can be lost from the fission chain by two mechanisms: they can be absorbed by impurities in the fissile material and they can escape from the volume of the fissile material. Even the purest "weapons-grade" uranium contains an admixture of several percent ^{238}U, which absorbs neutrons by the transmutation reaction (68). However, the dominant loss mechanism is usually the escape of the neutrons across the surface of the fissile material. The rate of loss of neutrons depends on the shape and size of the material. The most efficient shape is a sphere, since in this case the surface is as distant as possible from the bulk of the material. Obviously, neutrons are less likely to escape from a large sphere than from a small sphere—in the former a neutron released at some average point within the bulk of the material has a longer distance to travel to the surface and is therefore more likely to be absorbed by a nucleus before it reaches the surface. To attain the critical condition, the sphere of fissile material must have a minimum size. For ^{235}U of normal density, the sphere of minimum size has a diameter of 18 cm; the corresponding minimum mass, or the **critical mass,** is 53 kg. For ^{239}Pu, the critical mass is about one third as large.

The critical mass is significantly diminished if the fissile material is surrounded by a neutron "reflector" that inhibits the loss of neutrons. The reflector must be made of some substance whose nuclei strongly scatter neutrons, but do not absorb them. In a nuclear bomb a thick shell of beryllium metal can serve as a neutron reflector; neutrons that escape from the sphere of fissionable material collide with the beryllium nuclei and bounce back into the fissile material. The critical mass is further diminished if the fissile material is compressed to higher than normal densities. For example, compression of the material to twice its normal density diminishes the critical mass by a factor of 4. In a denser material, a neutron encounters more nuclei per unit length along its path, and therefore is more likely to be absorbed before it can escape.

A simple fission bomb, or **"atomic" bomb,** consists of two pieces of ^{235}U such that separately their masses are less than the critical mass, but jointly their masses add up to more than the critical mass. To detonate such a bomb, the two pieces of ^{235}U, initially at a safe distance from one another, are suddenly brought closely together. The assembly of the two subcritical masses into a single supercritical mass must be done very quickly; if it is done too slowly, a partial explosion (predetonation) will push the masses apart prematurely, before the chain reaction can develop fully. In the first such bomb, the device for the assembly of the two pieces of uranium consisted of a gun which propelled one piece of uranium toward the other at high speed (see Fig. 10.18); the propellant was an ordinary chemical explosive. As Fig. 10.18 shows, the uranium is surrounded by a thick layer of some material that serves as neutron reflector. This massive layer of material not only diminishes the amount of uranium needed to attain criticality, but also increases the efficiency of the explosion—the inertia of the massive layer delays the expansion of the exploding uranium and thereby permits the chain reaction to proceed for a longer time and with

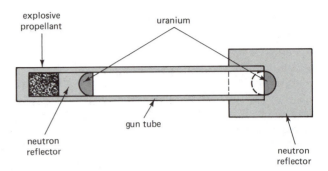

Fig. 10.18 Schematic diagram of a gun device used in ^{235}U bomb.

a larger net release of energy. This secondary action of the reflector is in principle no different from the action of the "tamper" used when blasting with ordinary chemical explosives.

Once a supercritical mass of ^{235}U has been assembled, any stray neutron (from spontaneous fission or from cosmic rays) can initiate the chain reaction. However, it is more efficient to supply the uranium with a fairly large burst of neutrons from a separate source at the instant the subcritical masses are joined—if the chain reaction starts with a large number of neutrons, it will take less time to reach explosive proportions. The source of neutrons is called an **initiator;** it consists of small amounts of two substances, such as radioactive polonium and lithium, the first of which is an α-ray emitter, and the second of which reacts with α rays to produce neutrons. If two such substances are suddenly mixed together, a burst of neutrons will be generated.

A more sophisticated fission bomb consists of a barely sub-critical mass of ^{239}Pu; if this mass is suddenly compressed to a higher than normal density it will become supercritical. The sudden compression is achieved by the preliminary explosion of a chemical high-explosive. If this explosive has been carefully layered in several segments, or explosive "lenses," arranged in a shell around a sphere of ^{239}Pu (see Fig. 10.19), then its detonation will crush the sphere of plutonium into itself. This implosion of the plutonium suddenly brings its density to the supercritical value and triggers the chain reaction. The implosion technique is used with ^{239}Pu because this isotope has a strong tendency to predetonate, and the gun technique would not be able to assemble the critical mass quickly enough.

In a nuclear reactor, the fission chain reaction proceeds under controlled conditions, at a constant rate rather than at an explosively increasing rate. The fissile material in the reactor is in a

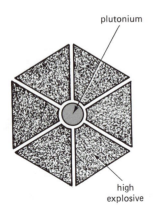

Fig. 10.19 Schematic diagram of an implosion device used in ^{239}Pu bomb.

critical rather than a supercritical condition. Most reactors operate with "enriched" uranium, consisting of a few percent ^{235}U mixed with 90-odd percent of ^{238}U. Such a uranium mixture cannot by itself sustain a chain reaction—the ^{238}U absorbs too many of the neutrons. However, if the uranium is surrounded by a substance capable of slowing down fast neutrons, then a chain reaction becomes viable. Slow neutrons are much more efficient at sustaining a chain reaction in ^{235}U than fast neutrons, because the cross section for induced fission by slow neutrons increases with decreasing speed (the $1/v$ law, already mentioned above).

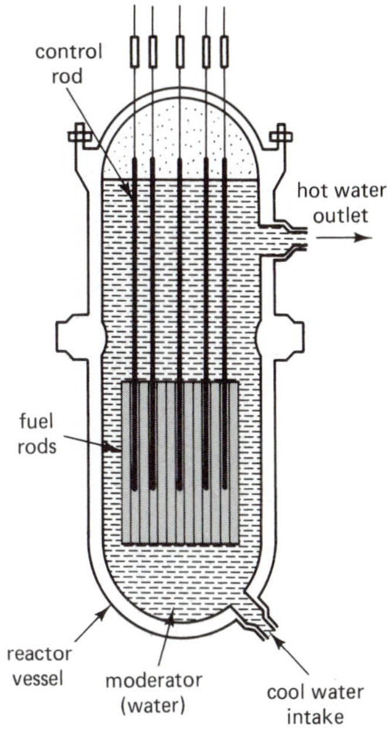

The substance used for slowing down the neutrons is called the **moderator.** Inside the reactor the uranium is usually placed in long fuel rods and these are immersed in the bulk of the moderator (see Fig. 10.20). Fast neutrons released by fissions diffuse from the fuel rods into the moderator; there they lose their kinetic energy by collisions with the moderator's nuclei; and then they diffuse back into one or another of the fuel rods and induce further fissions. A good moderator must absorb the kinetic energy of the neutrons, but not absorb the neutrons themselves. The three best moderators are ordinary water (H_2O), heavy water (D_2O), and graphite (pure carbon). These moderators are very efficient at slowing down neutrons because they contain a high proportion of nuclei of low mass, i.e., nuclei of a mass equal to, or not much larger than, the mass of the neutron.* As is known from classical mechanics, a head-on collision between two particles of equal mass transfers all of the kinetic energy to the initially stationary particle, whereas a collision between particles of unequal mass transfers only a fraction of the kinetic energy. In a few collisions with low-mass moderator nuclei, the neutrons lose almost all their kinetic energy. They only retain a kinetic energy equal to that of the random thermal motion of the moderator nuclei, i.e., a kinetic energy of $\frac{3}{2}kT$, where T is the temperature of the moderator. At typical reactor temperatures, this kinetic energy is of the order of 0.1 eV. Neutrons of such low energy are called **thermal neutrons;** they are extremely efficient at inducing fissions.

Heavy water and graphite are such good moderators that in their presence even natural uranium, with its small percentage of ^{235}U, can sustain a chain reaction. A graphite moderator and natural uranium fuel were used in the first nuclear reactor built in 1942 under the direction of Enrico Fermi at the University of

Fig. 10.20 Schematic diagram of a nuclear reactor.

Labels on figure: control rod, hot water outlet, fuel rods, reactor vessel, moderator (water), cool water intake

* A high-density gas of H_2 or D_2 would also be a good moderator, but it would be an impractical material.

Chicago. Several similar reactors were built at Hanford, Washington, soon thereafter, as part of the Manhattan Project. These reactors were used as **converters,** i.e., they were used to convert ^{238}U into ^{239}Pu by the following sequence of reactions: Within the rods of natural uranium, some of the neutrons from the fission of ^{235}U are absorbed by ^{238}U transmuting it into ^{239}U [see Eq. (68)]:

$$n + {}^{238}U \rightarrow {}^{239}U \tag{69}$$

The isotope ^{239}U then spontaneously undergoes two successive decays, which transmute it into ^{239}Pu:

$$^{239}U \rightarrow {}^{239}Np + \beta^- \rightarrow {}^{239}Pu + \beta^- \tag{70}$$

This is how the few kilograms of ^{239}Pu for the first atomic plutonium bombs were obtained.

Today reactors are used extensively to provide intense beams of neutrons for research; to produce radioisotopes for scientific, industrial, and medical applications; and as sources of heat for the commercial generation of electric power. The latter reactors are called **power reactors.** The main advantage of nuclear power plants over conventional powers plants is that the uranium fuel is somewhat cheaper than an equivalent amount of coal or oil. But this advantage is nearly balanced by higher construction and maintenance costs.

Power reactors now in commercial operation obtain their energy from the fission of the isotope ^{235}U. In view of the limited supply of this isotope, it would be desirable to develop reactors that consume some other fissile material. The obvious choice for an alternative reactor fuel is the very fissile isotope ^{239}Pu. Although this isotope does not occur naturally, it can be readily manufactured by transmutation of the abundant isotope ^{238}U [see Eqs. (69) and (70)]. Since the fuel of all power reactors is a mixture of ^{238}U and ^{235}U, such a manufacture of ^{239}Pu is an automatic side effect of the operation of these reactors; the ^{239}Pu gradually accumulates in the fuel rods, and can subsequently be extracted by chemical reprocessing of the spent fuel. A reactor using ^{239}Pu as fuel not only makes good use of a material that would otherwise have to be regarded as dangerous radioactive waste, but—if the core of the reactor is surrounded by a mantle of ^{238}U—it can also manufacture its own ^{239}Pu fuel. The number of neutrons released in the fission of ^{239}Pu is sufficiently large so that in an efficiently designed reactor, slightly more than one of the neutrons released per fission reaction can be diverted from the fission chain to the transmutation of ^{238}U. This means that the reactor produces more ^{239}Pu (from ^{238}U) than it consumes (from its original supply). A reactor that produces more fissile material than it consumes is called a **breeder.** Once we start the fuel cycle of breeder reactors with an initial load of ^{239}Pu, we need only supply the abundant and cheap ^{238}U to keep the fuel cycle going. Since breeders extract energy indirectly from ^{238}U, they would be able to generate power for as long as our supply of this abundant isotope lasts. Furthermore, breeding is also possible with ^{232}Th, which is even more abundant than ^{238}U, and could satisfy our energy requirements for several thousand years. Although a few experimental and prototype breeder reactors have been built, most of them in Europe, they have not yet reached a commercial stage.

10.7 Fusion

In light nuclei (say, $A < 20$), the binding energy per nucleon is small (see Fig. 9.2) because a large fraction of the nucleons are at or near the nuclear surface, where they cannot form as many bonds as nucleons deep within the nuclear volume. Light nuclei release energy when they are brought together and are made to fuse into a heavier nucleus. The source of the energy released during fusion is the strong interaction which does work on the merging nuclei, while the electrostatic interaction does work against them and therefore absorbs some energy.

If two nuclei are to fuse, they must be smashed together at high speed; otherwise, their Coulomb repulsion would push them apart before the strong attraction has a chance to act. To achieve fusion reactions on a large scale, we need a gas at extremely high temperature (a plasma) where the violent random thermal motions result in frequent high-speed collisions. Such fusion reactions in a plasma are called **thermonuclear reactions.** The temperature needed to trigger fusion must be comparable to the temperature at the center of the Sun—15×10^6 K or more. However, even at such temperatures, the initial kinetic energy of colliding nucleons is not sufficient to permit them to pass over the top of the Coulomb barrier. Instead, fusion involves a barrier-penetration process. The potential barrier opposing fusion has the same shape as the potential barrier opposing α decay (see Fig. 10.4); only the height of the barrier is different since it is proportional to the product of the charges of the nuclei. The probability for tunneling through the barrier is independent of whether the particle approaches the barrier from the inside (as in α decay) or from the outside (as in fusion). Hence the probability for fusion of two colliding nuclei is given approximately by the usual Gamow factor, with the charge of the alpha particle ($2e$) replaced by the charge of the approaching nucleus ($Z'e$):

$$P \simeq e^{-\frac{2\pi}{\hbar} \frac{ZZ'e^2}{4\pi\varepsilon_0} \frac{1}{v}} \tag{71}$$

The typical thermal speed of a nucleus of mass m in a gas at temperature T is

$$v = v_{\text{rms}} = \sqrt{\frac{3kT}{m}} \tag{72}$$

Hence

$$P \simeq e^{-\frac{2\pi}{\hbar} \frac{ZZ'e^2}{4\pi\varepsilon_0} \sqrt{\frac{m}{3kT}}} \tag{73}$$

This formula shows that the probability for tunneling is much smaller for nuclei of high or medium Z than for nuclei of low Z; and the probability is much smaller at low temperatures than at high temperatures. Thus, thermonuclear reactions are most likely for nuclei of very low Z, such as hydrogen, in a plasma at extremely high temperature.

Life on Earth depends on the thermonuclear reactions that occur at the center of the Sun, releasing the energy that ultimately emerges from the surface of the Sun in the form of heat and light. The Sun is a giant thermonuclear reactor that fuses hydrogen nuclei into helium nuclei. This thermonuclear "burning" of hydrogen proceeds through a sequence of three steps, called the **proton–proton chain:**

$$^{1}\text{H} + {}^{1}\text{H} \rightarrow {}^{2}\text{H} + \bar{\text{e}} + \nu \qquad (1.19\,\text{MeV}) \tag{74}$$

$$^{1}\text{H} + {}^{2}\text{H} \rightarrow {}^{3}\text{He} + \gamma \qquad (5.49\,\text{MeV}) \tag{75}$$

$$^{3}\text{He} + {}^{3}\text{He} \rightarrow {}^{4}\text{He} + {}^{1}\text{H} + {}^{1}\text{H} \qquad (12.85\,\text{MeV}) \tag{76}$$

The net effect of this sequence of steps is the transmutation of hydrogen into helium, with the release of the amounts of energy listed in Eqs. (74)–(76). The first step is the fusion of two protons, resulting in the formation of deuterium and the simultaneous ejection of an antielectron and a neutrino. The antielectron almost immediately collides with one of the many electrons in the plasma, and annihilates with that electron, emitting two γ rays [the energy listed in Eq. (74) includes the energy of these γ rays]. The next step is the fusion of hydrogen and deuterium, with the formation of ^{3}He. The third step is the fusion of two ^{3}He nuclei, with the formation of ^{4}He (ordinary helium) and the simultaneous ejection of two energetic protons. Since the final step requires two ^{3}He nuclei, each of the preceding steps must occur twice before the final step can occur once. Thus, the proton–proton chain consumes four protons to make one ^{4}He nucleus. The energy released per proton consumed is about 6.6 MeV. Since the rest-mass energy of a proton is 938 MeV, the energy released corresponds to $(6.6/938) \times 100\% = 0.70\%$ of the initial rest mass. This means that fusion releases more energy than fission per unit mass of "fuel" consumed.

Note that the first step of the proton–proton chain releases a neutrino. Thus, the center of the Sun is not only a source of heat, but also a source of a copious flow of neutrinos. Since the interaction of neutrinos with matter is very weak, the matter in the Sun (and in the Earth) is nearly transparent to neutrinos—they stream outward from the center of the Sun without hindrance. At the Earth, the flux of these neutrinos is about 10^{15} per square meter per second.

Alternatively, the thermonuclear "burning" of hydrogen can proceed through a sequence of six steps, called the **carbon cycle:**

$$^1\text{H} + {}^{12}\text{C} \rightarrow {}^{13}\text{N} + \gamma \qquad (1.95\,\text{MeV}) \qquad (77)$$

$$^{13}\text{N} \rightarrow {}^{13}\text{C} + \bar{\text{e}} + \nu \qquad (2.22\,\text{MeV}) \qquad (78)$$

$$^1\text{H} + {}^{13}\text{C} \rightarrow {}^{14}\text{N} + \gamma \qquad (7.54\,\text{MeV}) \qquad (79)$$

$$^1\text{H} + {}^{14}\text{N} \rightarrow {}^{15}\text{O} + \gamma \qquad (7.35\,\text{MeV}) \qquad (80)$$

$$^{15}\text{O} \rightarrow {}^{15}\text{N} + \bar{\text{e}} + \nu \qquad (2.71\,\text{MeV}) \qquad (81)$$

$$^1\text{H} + {}^{15}\text{N} \rightarrow {}^{12}\text{C} + {}^4\text{He} \qquad (4.96\,\text{MeV}) \qquad (82)$$

Note that the last step regenerates the carbon destroyed in the first step. Thus the carbon goes through a cycle; it merely acts as a catalyst whose average amount remains constant. The energy released per proton consumed is about the same as for the proton chain. The theory of the carbon cycle and the proton–proton chain first emerged in 1939 from calculations by H. Bethe,* who made a careful theoretical investigation of all the possible fusion reactions that might occur in the Sun. Bethe's original version of the proton–proton chain did not include the reaction (76), but instead several other alternative reactions, which are now believed to be less important than (76).

---

* **Hans Albrecht Bethe,** 1906– , German and later American theoretical physicist, professor at Cornell. During the war he worked on the Manhattan Project as director of the Theoretical Physics Division at Los Alamos. He received the Nobel Prize in 1967 for his investigations of nuclear reactions in stars.

In the Sun, the predominant fusion process is the proton–proton chain; but in stars hotter than the Sun, the predominant process is the carbon cycle. The reason for this shift of predominance is that at higher temperatures (and higher speeds), the protons find it easier to overcome the strong Coulomb barrier of ^{12}C, leading to an enhanced rate for the reaction (77). Of course, higher temperatures also lead to an enhanced rate for the reaction (74), but this latter reaction suffers from the disadvantage of requiring a simultaneous fusion and β decay, which reduces the reaction probability below the value given by the Gamow factor.

On the Earth, artificial fusion reactions have so far been achieved on a large scale only in "hydrogen" bombs, or thermonuclear bombs. In these devices, deuterium and tritium are made to fuse into helium. By starting the reaction with deuterium and tritium, we bypass the rather slow hydrogen reaction (74), and we can achieve fusion at pressures much less than those at the center of the Sun. The high temperatures required to trigger the fusion reactions are generated by exploding an ordinary atomic bomb next to a container filled with deuterium and tritium. In most thermonuclear bombs, the deuterium and tritium are surrounded by a blanket of natural uranium; the abundant flux of neutrons released by the fusion reactions then produces fissions in this uranium blanket, enhancing the explosive yield. This kind of hydrogen bomb is a fission–fusion–fission device; typically, one-half of the total energy yield is due to fusion, one-half is due to fission. The total energy yield of a hydrogen bomb is of the order of one or several megatons—roughly a thousand times more than an atomic bomb.

For the peaceful use of fusion energy it is not economically viable to start the reaction with tritium, since this is an extremely rare artificial isotope, costing about $1 million per pound. The most practical scheme is to start the reaction with deuterium and a small amount of lithium and beryllium. Through a sequence of steps, known as the DT cycle, these nuclei are made to fuse into helium and tritium, with the release of about 9 MeV per deuterium nucleus consumed. The temperature required to trigger this fusion reaction is 1.1×10^8 K—almost ten times hotter than the center of the Sun. Such temperatures have been attained in plasmas, but the big obstacle to the practical exploitation of controlled nuclear fusion lies in the confinement of such a plasma for a time long enough to permit the fusion reaction to proceed. Two schemes are under investigation: the **magnetic-confinement** scheme attempts to hold the plasma suspended in space by means of a clever configuration of magnetic fields. The **inertial-confinement** scheme attempts to vaporize small pellets of fuel by means of intense laser beams and to trigger fusion very suddenly, so that the pellet reacts before it has time to disperse. Both of these schemes are at an experimental stage. The experiments have demonstrated that fusion reactions can be initiated, but they have not yet managed to extract useful amounts of energy.

Fusion reactors have two important advantages over fission reactors as sources of mechanical and electric power: the fuel for fusion reactors (deuterium and lithium) is much more abundant on Earth than the fuel for fission reactors (uranium or thorium), and fusion reactors do not produce the large amount of radioactive residues that are the baneful by-product of the operation of fission reactors.

Chapter 10:   SUMMARY

Law of radioactive decay:

$$n(t) = n_0 e^{-t/\tau}$$

Mean life:

$$\tau = \frac{t_{1/2}}{\ln 2}$$

curie:

$$1 \, \text{Ci} = 3.7 \times 10^{10} \text{ disintegrations/s}$$

becquerel:

$$1 \, \text{Bq} = 1 \text{ disintegration/s}$$

Mean life for α decay:

$$\ln \tau = \ln \left(\frac{2R}{v_{\text{in}}} \right) + \frac{4\pi}{\hbar} \frac{Ze^2}{4\pi\varepsilon_0} \frac{1}{v_\alpha}$$

Cross section:

$$\sigma = \frac{\text{(reaction rate per nucleus)}}{\text{(incident flux)}}$$

barn:

$$1 \text{ barn} = 10^{-28} \, \text{m}^2$$

Energy released per ^{235}U nucleus in fission:

$$200 \, \text{MeV}$$

Energy released per proton in fusion:

$$6.6\,\text{MeV}$$

Chapter 10:   PROBLEMS

1. The complete series of decays that begins with the decay of ^{238}U involves the sequential emission of alpha and beta rays, as follows: α, β^-, β^-, α, α, α, α, β^-, α, β^-, α, β^-, β^-, α. List the series of reactions that release these alpha and beta rays; give the chemical symbol and the mass number of each isotope in these reactions, as in Eq. (7).

2. A radioactive decay has a mean life τ. What is the rms deviation of the lifetime from the mean? What is the most probable lifetime?

3. From the half-life of radium listed in Table 10.1, calculate the activity of 1 g of pure radium. The curie was originally defined as the activity of 1 g of pure radium; is this consistent with your result?

4. The isotope ^{60}Co is commonly used as a source of γ rays in high-level industrial irradiation cells. How many kg of ^{60}Co are needed to achieve an activity of 10^6 curie? After what time interval will the activity of such a cell have decreased to 10% of its initial value?

5. A beam of thermal neutrons, of energy 0.055 eV, emerges from a nuclear reactor. What fraction of these neutrons will decay within 30 m from the reactor?

6. The age of a uranium-rich rock can be determined by measuring how much of the uranium has decayed. The most suitable isotope for such measurements is ^{238}U, which has a half-life of 4.47×10^9 years. The stable end product of the series of radioactive decays initiated by uranium [see Eq. (7)] is ^{208}Pb. Suppose that a sample of the very oldest rocks found on Earth contains 1.000 g of ^{238}U and 0.900 g of ^{208}Pb, and no other isotopes of lead whatsoever.
(a) Calculate the age of this rock.
(b) How do you know that all the ^{208}Pb is due to the decay of ^{238}U, i.e., how do you know that the rock initially contained no lead at all?

7. Most of the chemical elements were created by nuclear reactions in the interior of stars that lived and died long before the Sun was born. The relative abundance of radioactive elements can tell us the time of element formation. For instance, theoretical calculations indicate that the isotopes ^{238}U and ^{232}Th must originally have been produced with a relative abundance of $1:1.9$. At present, the relative abundance of these isotopes is $1:3.9$. Given that the half-life of ^{238}U is 4.47×10^9 years and that of ^{232}Th is 1.40×10^{10} years, calculate how long ago these elements were formed.

8. What is the slope of the straight line drawn through the experimental points in Fig. 10.5? Does this value of the slope agree with the value calculated from the theoretical formula (33)?

9. By comparing the data in Fig. 10.5 with the theoretical formula (33) deduce an effective value of v_{in} for alpha particles within the nuclei listed in this figure. Use your value of v_{in} to predict the mean lives of ^{234}U and ^{238}U, which emit alpha particles of 4.78 MeV and 4.20 MeV, respectively.

10. Use the uncertainty principle to estimate the speed v_{in} of an alpha particle confined in a nucleus measuring about 10^{-14} m across.

11. Both ^{243}Am and ^{244}Cm decay by alpha decay. The former nucleus emits an alpha particle of 5.27 MeV and has a half-life of 7.37×10^3 years. The latter nucleus emits an alpha particle of 5.80 MeV. What is the theoretical prediction for the half-life of this nucleus?

12. Alpha particles emitted by ^{226}Ra have a kinetic energy of 4.7845 MeV. What is the recoil kinetic energy of the daughter atom? What is the net energy released in the alpha-decay reaction? The mass of the daughter atom ^{222}Rn is 222.018 u.

13. What is the maximum energy available for the neutrino in the β^+ decay of ^{11}C? In the EC decay of ^{11}C? Use the masses listed in the chart of isotopes in Fig. 9.1; express your answer in MeV.

14. The measured energies of alternative beta decays can be used to determine the mass difference between the daughter nuclei. For instance, the nucleus of ^{64}Cu can decay into ^{64}Zn (by β decay) or into ^{64}Ni (by β^+ decay). The measured Q-value for the former reaction is 0.57 MeV, and that for the latter reaction is 0.66 MeV. From these data, deduce the mass difference between ^{64}Zn and ^{64}Ni; express your answer in u.

15. The total capture cross section for thermal neutrons in ^{27}Al is $\sigma = 0.233$ barn. How far can a beam of such neutrons penetrate a slab of aluminum (of density 2.70 g/cm^3) before one-half of it is absorbed? (*Hint:* Show that the number of neutrons absorbed in a thickness dx is $dN = -nN\sigma dx$, where n is the number of aluminum nuclei per unit volume.)

16. For incident alpha particles of energy of 20 MeV, the cross section for the reaction

$$\alpha + {}^{60}\text{Ni} \rightarrow {}^{63}\text{Zn} + n$$

is 0.68 barn. Suppose that a beam of alpha particles is aimed at a sheet of nickel. The alpha particles arrive at the rate of 2.0×10^{13}; the sheet of nickel has a density of 8.9 g/cm^3 and is 0.050 mm thick. What is the rate of formation of ^{63}Zn?

17. Suppose that an alpha particle of energy 14 MeV incident on a ^{60}Ni nucleus forms the compound nucleus ^{64}Zn* via the reaction (45).
   (a) What is the excitation energy of ^{64}Zn* relative to the ground state ^{64}Zn?
   (b) The same compound nucleus can also be formed via the reaction (44) when a proton is incident on ^{63}Cu. If you want to attain the same excitation energy in ^{64}Zn*, what must be the energy of the incident proton?
   For the sake of simplicity, ignore the recoil energy of the ^{64}Zn* nucleus (it is less than 1 MeV). The masses of ^{60}Ni, ^{63}Cu, and ^{64}Zn are 59.9308 u, 62.9296 u, and 63.9291 u, respectively.

18. Consider the reaction

$$\alpha + {}^{10}\text{B} \rightarrow {}^{14}\text{N*} \rightarrow {}^{13}\text{N} + \text{n}$$

Suppose that the energy of the incident alpha particle is 1.83 MeV. What is the excitation energy of the ${}^{14}\text{N*}$ nucleus (relative to the ground state ${}^{14}\text{N}$)? What is the energy of the neutron ejected in this reaction? Take into account the recoil energies of the nuclei, but assume that the ${}^{14}\text{N*}$ nucleus dissipates its initial recoil energy before it ejects the neutron. Values of the masses are given in the chart of isotopes, Fig. 9.1.

19. The energy of one of the excited states of ${}^{14}\text{N}$ is 8.62 MeV (relative to the ground state).

(a) If a proton incident on ${}^{13}\text{C}$ is to form this excited state via the reaction $\text{p} + {}^{13}\text{C} \rightarrow {}^{14}\text{N*}$, what must be the energy of the proton? Use the masses given in the chart of isotopes, and remember to take into account the recoil energy of the ${}^{14}\text{N*}$ nucleus.

(b) If this ${}^{14}\text{N*}$ nucleus decays to the ground state by emission of a single gamma ray, what will be the energy of this gamma ray? Assume that the ${}^{14}\text{N*}$ nucleus is at rest when it emits the gamma ray, and take into account the recoil energy the nucleus acquires during the emission.

20. Check that the angular separation between the minima in Fig. 10.12 is *roughly* in accord with the usual condition $d \sin \theta = n\lambda$ for the minima in a diffraction pattern, with d equal to the diameter of the nucleus.

21. Combine Eqs. (58) and (9.48) to find the mass number of the smallest nucleus for which symmetric fission is energetically viable. What nucleus is this?

22. The first atomic bomb exploded at Alamogordo on July 17, 1945, had an explosive yield equivalent to 18 kilotons of TNT. How many kg of plutonium actually underwent fission in this explosion? (The energy released per kg of Pu is about the same as that released per kg of U; see Example 5.)

23. The effects of the blast wave and of the thermal radiation of a nuclear bomb obey scaling laws that permit us to predict the effects of an explosion of any arbitrary yield from the known effects of a "reference" explosion of some other yield. The overpressure of the blast wave obeys a cube-root scaling law: if a reference explosion of W_1 kilotons generates a given overpressure at a distance D_1, then an arbitrary explosion of W kilotons generates the same overpressure at a distance D such that

$$\frac{D}{D_1} = \left(\frac{W}{W_1}\right)^{1/3}$$

The thermal radiation obeys a square-root scaling law: if a reference explosion of W_1 kilotons generates a given thermal exposure (energy per unit area) at a distance D_1, then an arbitrary explosion of W kilotons generates the same thermal exposure at a distance D given by

$$\frac{D}{D_1} = \left(\frac{W}{W_1}\right)^{1/2}$$

At Hiroshima, buildings were wrecked by the blast wave at a distance of 1.0 mi from the center of the explosion; at what distance would a 1-megaton explosion produce the same effect? At Hiroshima, wooden telephone poles were ignited by the thermal pulse at a distance of 1.25 mi; at what distance would a 1-megaton explosion produce the same effect?

24. The Sun contains about 1.5×10^{30} kg of hydrogen and it radiates heat and light at the rate of 3.9×10^{26} W. At this rate, how long will it take the Sun to burn all of its hydrogen?

25. The proton–proton chain can also proceed via the alternative set of reactions:

$$^{1}\text{H} + {}^{1}\text{H} \rightarrow {}^{2}\text{H} + \bar{e} + \nu \qquad (1.19 \text{ MeV})$$

$$^{1}\text{H} + {}^{2}\text{H} \rightarrow {}^{3}\text{He} + \gamma \qquad (5.49 \text{ MeV})$$

$$^{3}\text{He} + {}^{4}\text{He} \rightarrow {}^{7}\text{Be} + \gamma$$

$$^{7}\text{Be} + e \rightarrow {}^{7}\text{Li} + \gamma + \nu \qquad (0.05 \text{ MeV})$$

$$^{7}\text{Li} + {}^{1}\text{H} \rightarrow {}^{4}\text{He} + {}^{4}\text{He}$$

(This is the version of the proton–proton chain originally proposed by Bethe.) Use the masses given in Appendix 4 or in the chart of isotopes calculate the energy released in the third and fifth reactions for which no energy has been listed above. How does the net energy released compare with that of the set of reactions (74)–(76)?

26. (a) The temperature at the center of the Sun is 15×10^{6} K. According to Eq. (73), what is the probability for fusion of two colliding protons at this temperature? What is the probability for fusion of a proton colliding with a ^{12}C nucleus?
(b) At what temperature would the probability for the latter fusion equal the probability for the former fusion at 15×10^{6} K?

27. For the sake of simplicity, assume that all the energy released by the Sun is due to the proton-proton chain [Eqs. (74)–(76)]. Given that the Sun radiates heat and light at the rate of 3.9×10^{26} W, calculate the flux of neutrinos (or number of neutrinos per unit area and unit time) incident on the Earth.

Elementary Particles

All the particles we have become acquainted with in the preceding chapters—electrons, protons, neutrons, photons, and neutrinos—are abundant in our natural environment. Electrons, protons, and neutrons are the constituents of the atoms in our own bodies and in everything that surrounds us. Photons and neutrinos are emitted copiously by the Sun; furthermore, high-energy photons and neutrinos are emitted by natural radioactive substances. All these particles, with the exception of free neutrons, are stable: if left to themselves, they live forever. In the 1930s, physicists began to discover some new, extraordinary particles that are created by the conversion of energy into mass in high-energy collisions between the everyday particles. Most of these new particles are extremely unstable; their lifetimes are very short, ranging from 10^{-6} to 10^{-23} s. A few are stable when left to themselves; but they are made of anti-matter, which annihilates when it comes into contact with the ordinary matter of our immediate environment. Hence all these new particles lead an ephemeral existence—they can only be observed if they are intercepted soon after they are created in a high-energy reaction.

The first of these new particles were discovered in experimental investigations of cosmic rays. But they were subsequently produced artificially by the bombardment of targets with beams of high-energy protons or electrons from an accelerator. During the last thirty years, the construction of more and more powerful accelerators has led to the discovery of more and more new particles. At last count, there were well over 300 such particles. They are often called **elementary particles,** but this is not intended to mean that they are truly elementary; it merely means that they are more elementary than nuclei. Most of these particles are more massive and more complex than protons and neutrons; hence they cannot play the role of elementary, fundamental building blocks of matter.

Up to the 1960s, high-energy experimentation held the lead over theory. Nobody knew what to make of the chaotic diversity of hundreds of particles discovered by the experimental physicists. But in the last two decades, theoretical physicists have come to recognize that all the particles—with the exception of electrons, photons, and a few particles closely related to these two—are made of small building blocks, called quarks. Protons, neutrons, and all the other heavy particles are systems of quarks bound tightly together. This new picture of the internal structure of particles has given us a clear understanding of the chaotic diversity of particles, and is making it possible to calculate many of the properties of these particles. Furthermore, theorists have developed a new unified picture of interactions, treating different kinds of force fields as closely related aspects of one single underlying field, in much the same way as electric and magnetic fields are closely related aspects of an underlying electromagnetic field. This unified theory of fields has established an intimate connection between the weak and electromagnetic interactions, and has given us a deep insight into these interactions. Attempts at a unified theory that also encompasses the strong interactions are in progress. Theory has now taken the lead over experimentation, and the predictions of theorists provide guidance and challenge to the experimenters. We are on the threshold of achieving an understanding of particle dynamics and structure comparable to our understanding of atomic structure and nuclear structure.

11.1 The Discoveries of the Antielectron, Muon, and Pion

The energies required to initiate a reaction between particles are of the order of one or several GeV. In lack of better, the early experimenters in high-energy physics relied on cosmic rays to initiate reactions. **Cosmic rays** are very energetic particles—mostly protons—crisscrossing interstellar space. The provenance of these

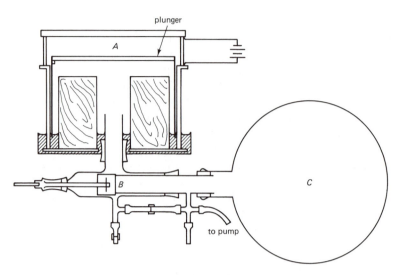

Fig. 11.1 Schematic diagram of Wilsons's cloud chamber. The expansion chamber *A* contains saturated water vapor. Its bottom is closed off by a plunger, which jumps downward when the pressure at its underside is reduced suddenly by opening the valve *B* of the evacuated flask *C*. The sudden adiabatic expansion of the chamber cools the water vapor, leaving it in a supersaturated state. Any ions that happen to be present in the chamber then trigger the formation of water droplets. [After C. T. R. Wilson, *Proc. Roy. Soc. (London)* **A87**, 277 (1912).]

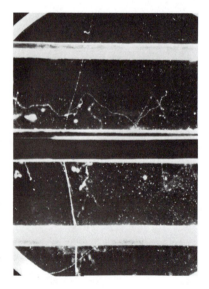

Fig. 11.2 Cosmic-ray tracks seen in a cloud chamber. [From T. H. Johnson, J. G. Barry, and R. P. Shutt, *Phys. Rev.* **57**, 1047 (1940).]

particles is not reliably known. During flares, ordinary stars, such as the Sun, emit cosmic rays; but supernova explosions are the most likely source of the bulk of the cosmic rays. Typically, the energy of cosmic-ray particles is a few GeV, although some extreme cases of energies as high as 10^{11} GeV have been observed. When these primary cosmic rays collide with nuclei in the upper layers of the atmosphere, they generate showers of secondary cosmic rays. These are gradually absorbed as they travel through the atmosphere. But even at sea level there remains an appreciable flux of cosmic rays; for instance, a human body intercepts a few dozen cosmic rays per second. At higher altitudes the flux of cosmic rays is greater.

Cosmic rays were discovered in 1912 by V. F. Hess,* who carried some ionization chambers to high altitude in a balloon. The tracks of cosmic rays can be made visible in a cloud chamber (Fig. 11.1),

---

* **Viktor Franz Hess,** 1883–1964, Austrian and later American physicist, professor at Innsbruck and at Fordham University. For his discovery of cosmic rays he was awarded the Nobel Prize in 1936.

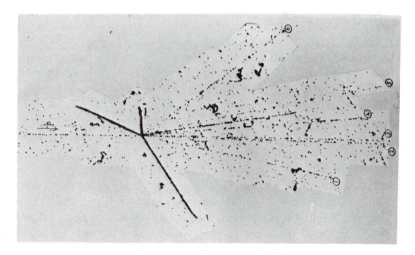

Fig. 11.3 A "star" in a photographic emulsion. The faint track coming in from the left was made by the incident primary cosmic ray. [From R. H. Brown et al., *Phil. Mag.* **40,** 862 (1940).]

invented by C. T. R. Wilson and developed into a practical instrument by P. M. S. Blackett.[†] In such a chamber, the passage of a charged particle ionizes molecules of gas along its track; these ions act as condensation nuclei, triggering the formation of small water droplets, which mark the track of the particle. Figure 11.2 shows an example of a cosmic-ray track in a cloud chamber. The tracks can also be made visible in a photographic emulsion. A charged particle passing through such an emulsion triggers the formation of silver grains, and therefore yields a trail of black spots in the developed emulsion, which can be seen with a microscope. Figure 11.3 shows the impact of a primary cosmic ray on a nucleus in a photographic emulsion. The violent collision created a large number of new particles by conversion of energy into mass.

The first new particles discovered in cosmic rays were the antielectron, the muon, and the pion. In the late 1920s, P. A. M. Dirac formulated a new wave equation for electrons, which incorporated relativity into quantum mechanics. As we already mentioned in Chapter 7, Dirac's equation explains the spin of the electron as an angular momentum arising from a circulating flow of energy and

---

[†] **Patrick Maynard Stuart Blackett,** 1897–1974, English physicist, professor at Birbeck College, Manchester, and Imperial College. He was awarded the Nobel Prize in 1948 for his development of the cloud chamber and its extensive application in research in nuclear physics and cosmic rays.

momentum within the electron wave. His equation also predicts the relation (7.9) between the magnetic moment and the spin of the electron. However, these successes of Dirac's relativistic theory were marred by one defect: the relativistic equation had solutions of negative energy. Any electron of positive energy would then presumably be unstable; the electron could make an immediate transition to one of these states of negative energy, emitting a γ-ray photon. To avoid this difficulty, Dirac proposed that all of the negative energy states are already full, even in what we normally call a vacuum; consequently, the Exclusion Principle prevents the disappearance of ordinary electrons into negative-energy states. The electrons filling the negative-energy states are called the "Dirac sea." These electrons are not directly observable, since the Exclusion Principle prevents them from reacting to external forces. However, if one of the negative-energy electrons is absent, leaving a hole in the sea of filled states, then this hole can react to external forces. The hole behaves like a particle of positive charge, in much the same way as a hole in the valence band of a semiconductor.

In 1932, C. D. Anderson* discovered the tracks of electronlike particles of positive charge while observing cosmic-ray particles in a cloud chamber. Anderson had placed his cloud chamber in a strong magnetic field, so that he could discriminate between particles of negative and of positive charge by their directions of deflection, and he found that some of the electronlike tracks deflected in the direction corresponding to positive charge (see Fig. 11.4). The electronlike particles came to be called **antielectrons,** or **positrons.** Subsequent measurements established that they have the same mass, same spin, and same magnetic moment as electrons—they are identical with electrons in every respect, except for their positive electric charge.

Although the impact of cosmic rays on the nuclei in the Earth's atmosphere creates a fair number of antielectrons, they do not stay around very long. As soon as an antielectron encounters an electron, both annihilate, releasing two γ rays:

$$e + \bar{e} \rightarrow 2\gamma \tag{1}$$

(The reaction proceeds with the emission of two γ rays rather than one because momentum conservation would be impossible with one γ ray.) This annihilation reaction can be regarded as a transi-

---

* **Carl David Anderson,** 1905–   , American physicist, professor at the California Institute of Technology. He received the Nobel Prize in 1936 for his discovery of the positron. In this same year he discovered the muon.

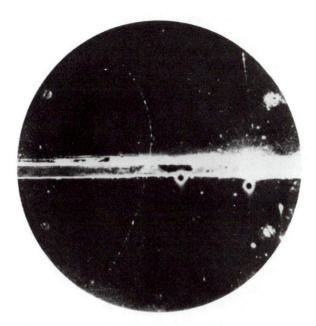

Fig. 11.4 The first photograph of the track of an antielectron in a cloud chamber. [From C. D. Anderson, *Phys. Rev.* **43**, 491 (1933).]

tion of a positive-energy electron into an available, empty negative-energy hole in the Dirac sea.* The reaction usually occurs with negligible initial kinetic energy (the antielectron usually is slowed down by preliminary collisions before it annihilates). Thus, the energy carried away by the γ rays is simply the rest-mass energy $2m_e c^2$ of two electrons; each γ ray has an energy $m_e c^2$, or 0.511 MeV.

Two other new particles were discovered in cosmic rays: the **muon** (or μ meson[†]) and the **pion** (or π meson). The existence of mesons, whose name was originally intended to indicate that the value of their mass is intermediate between that of an electron and

---

* Modern quantum theory has rejected the intuitive picture of the filled Dirac sea in favor of a more formal treatment of negative-energy states, because it was discovered that *all* particles have antiparticles. Since bosons do not obey the Exclusion Principle, antibosons cannot be regarded as holes in a filled sea. The formal approach relies on the truism that only the *changes* of energy, momentum, or charge associated with changes in the particle states are observable; hence we can regard the creation of a negative-energy particle as the destruction of a positive-energy antiparticle, and the destruction of a negative-energy particle as the creation of a positive-energy antiparticle. This formal equivalence permits us to eliminate all reference to negative-energy states from the mathematical description of reactions.

[†] In the strict modern meaning of the word *meson*, the muon is not a meson; but the name lingers.

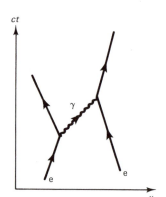

Fig. 11.5 Feynman diagram showing emission of a photon by the electron on the left and absorption by the electron on the right.

a proton, was first proposed on theoretical grounds by H. Yukawa* in 1935.

By then, the study of the quantum theory of fields had led to the conclusion that the Coulomb force between charged particles is due to a mechanism of exchange of photons. For example, consider two electrons separated by some distance. According to the quantum theory of fields, each electron sends out a succession of photons, which are captured by the other electron. This process of emission, propagation, and absorption of photons can be represented in a spacetime diagram, shown in Fig. 11.5; such a diagram is called a **Feynman diagram.** The mutual Coulomb force between the two electrons is simply due to the recoils suffered by the electrons during continual emissions and absorptions.[†] The photons that participate in this exchange mechanism may be regarded as the carriers of the Coulomb force—they "carry" the force from one electron to the other, over the intervening distance. These photons are called **virtual photons** because they are never observed directly. They violate the law of conservation of energy; for instance, if an electron at rest emits a photon, the energy after the emission exceeds the energy before by the photon energy plus the recoil energy of the electron. But this violation of energy conservation is hidden from view by the uncertainty principle—the virtual photon only lasts a short time Δt, so that the energy excess is less than the energy uncertainty, $\Delta E \simeq \hbar/\Delta t$. The relativistic theory of electromagnetic interactions based on this picture of photon exchange is called **quantum electrodynamics,** or **QED.** It was fully developed by R. P. Feynman, J. Schwinger, and S. Tomonaga,[‡] and other theoreticians in the 1940s, and stands as the most accurate theory in all of physics—for instance, the prediction of quantum electrodynamics for the magnetic moment of the electron has been tested, and confirmed, to within ten significant figures.

---

* **Hideki Yukawa,** 1907–1981, Japanese physicist, professor at Kyoto. For his work on the theory of nuclear forces, he received the Nobel Prize in 1949.

[†] At first sight, it would seem that this recoil mechanism can never give an *attractive* electric force, such as the force between an electron and a proton. But we must remember that a photon is not described by a localized orbit, but by a wave extended over a wide region. The particle on the left in Fig. 11.4 can emit a photon wave of momentum directed toward the left, and yet this wave can be absorbed by the particle on the *right.* This gives an attractive force.

[‡] **Richard P. Feynman,** 1918– , American physicist, professor at the California Institute of Technology, **Julian Schwinger,** 1918– , American physicist, professor at Harvard, and **Sin-itiro Tomonaga,** 1901–1979, Japanese physicist, director of the Research Institute for Fundamental Physics at Kyoto, shared the Nobel Prize in 1965 for their contributions to quantum electrodynamics.

Yukawa proposed that the strong nuclear force arises from a similar exchange mechanism, i.e., protons and neutrons emit and absorb quanta of the strong force field, and these quanta carry the force from one to the other. According to the quantum theory of fields, the short range of the strong force is to be blamed on the mass of the carrier—it can be established that the range of the force and the mass of the carrier are inversely proportional. We can understand this inverse proportionality by the following simple argument. Suppose the carrier particle has a mass m. Then the emission of such a particle leads to an energy excess of at least mc^2. If this energy excess is to be (barely) hidden by the energy uncertainty, we need $\Delta E \simeq mc^2$ and hence

$$\Delta t \simeq \frac{\hbar}{\Delta E} \simeq \frac{\hbar}{mc^2} \qquad (2)$$

Since the speed of the carrier particle cannot exceed the speed of light, it can travel at most a distance $c\,\Delta t$ in a time interval Δt. According to Eq. (2), this distance is

$$c\,\Delta t \simeq \frac{\hbar}{mc} \qquad (3)$$

This distance is to be identified with the range of the strong force. We know from Chapter 9 that the range of this force is $\sim 2 \times 10^{-15}$ m; thus

$$2 \times 10^{-15}\,\text{m} \simeq \frac{\hbar}{mc} \qquad (4)$$

and

$$m \simeq \frac{\hbar}{c}\frac{1}{2 \times 10^{-15}\,\text{m}} \simeq 2 \times 10^{-28}\,\text{kg} \sim 0.1\,\text{u} \qquad (5)$$

This means that the mass of the carrier particles for the strong force is about 200 electron masses.

Yukawa's proposal appeared to receive experimental support when in 1937 C. D. Anderson and his collaborators discovered the muon in cosmic rays. This particle has a mass of 0.113 u, just about the value estimated above. But in 1946 experiments established that the muon interacts only weakly with protons and neutrons, much too weakly to act as carrier for the strong force. Attempting to clear up this puzzle, several theoreticians proposed that the carrier must be a different kind of meson, and this proposal was confirmed within a year with the discovery of pions by

Fig. 11.6 Tracks of a pion, a muon, and an electron in a photographic emulsion. The track of the pion ends at point *A*, where the pion decayed with the creation of a muon. The track of the muon ends at *B*, where it decayed with the creation of an electron. This photograph is a mosaic assembled from many micrographs focused at different depths in the emulsion. [From R. H. Brown et al., *Nature* **163**, 47 (1949).]

C. F. Powell and G. P. S. Occhialini.* These pions were discovered by the tracks they made in a photographic emulsion exposed to cosmic rays at high altitude. Figure 11.6 shows the track of a pion in an emulsion. The pion decayed into a muon, and this decayed into an electron, according to the sequence of reactions

$$\pi^- \to \mu + \bar{\nu} \tag{6}$$
$$\to e + \nu + \bar{\nu} \tag{7}$$

The neutrinos are neutral; hence they leave no tracks in the emulsion. The first pions that were discovered had positive or negative charge, and a mass of

$$m_{\pi^\pm} = 0.150 \, \text{u} \tag{8}$$

Later, a neutral pion (π^0) was discovered in accelerator experiments; its mass is somewhat smaller,

$$m_{\pi^0} = 0.145 \, \text{u} \tag{9}$$

In high-energy physics it is customary to express particle masses in MeV/c^2. In these units, the pion masses are

$$m_{\pi^\pm} = 140 \, \text{MeV}/c^2 \tag{10}$$

and

$$m_{\pi^0} = 135 \, \text{MeV}/c^2 \tag{11}$$

---

* **Cecil Frank Powell,** 1903–1969, English physicist, professor at Bristol. He was awarded the Nobel Prize in 1950 for his work on detecting particle tracks in photographic emulsions and for his discovery of the pion. **Guiseppe P. S. Occhialini,** 1907– , Italian physicist, professor at Milan.

Most of the large accelerators of today are descendants of the cyclotron built by E. O. Lawrence in 1931. With his first cyclotron, measuring only a few inches across, he accelerated ions to 80 keV. In the following years, Lawrence and his collaborators at the Radiation Laboratory at Berkeley (now the Lawrence Berkeley Laboratory) built a series of progressively larger cyclotrons (see Fig. 11.7). With these they achieved energies of up to 27 MeV with proton projectiles.

The early cyclotrons relied on the constant orbital frequency of a charged particle moving in a circular orbit in a magnetic field. We know from Section 1.1 that a particle of charge e and momentum p moves in an orbit of radius

$$r = \frac{p}{eB} \tag{12}$$

with an orbital frequency, or cyclotron frequency,

$$v = \frac{eB}{2\pi m} \tag{13}$$

As the energy of the particle increases, the orbital radius increases, but the frequency remains constant. Thus, the particle returns to the same sector of the cyclotron at constant time intervals, and can

Fig. 11.7 The 60-inch cyclotron at Berkeley, built in 1939. (Courtesy Lawrence Berkeley Laboratory.)

therefore be accelerated by an alternating electric field of the same constant frequency as that of the orbital motion. Unfortunately, for particles with speeds larger than $\sim 20\%$ of the speed of light, the approximation of a constant orbital frequency breaks down. For such a particle, a relativistic calculation shows that Eq. (12) for the orbital radius remains valid, but Eq. (13) for the orbital frequency must be replaced by (see Example 2.4)

$$\nu = \frac{eB}{2\pi m} \sqrt{1 - \frac{v^2}{c^2}} \tag{14}$$

The frequency decreases as the speed increases. Hence the acceleration of relativistic particles requires either an adjustment of the frequency of the electric field, to match the gradual decrease of frequency of the orbital motion; or else an adjustment of the strength of the magnetic field, so as to keep the product $B\sqrt{1 - v^2/c^2}$ in Eq. (14) constant. Machines relying on the first tactic are called **synchrocylotrons,** and machines relying on the second are called **synchrotrons.** In both kinds of machines, the particles are accelerated in bunches. For each bunch the machine must go through a cycle of adjustment of the electric fields or the magnetic fields, and then reset itself for the next bunch.

All the high-energy proton accelerators are synchrotrons. They use an increasing magnetic field not only to keep the orbital frequency constant, but also to keep the orbital radius (nearly) constant. Hence, they only need to produce a magnetic field within an evacuated circular beam pipe that serves as "racetrack" for the accelerated particles.

Fig. 11.8 The Bevatron, built in 1954. This is one of the oldest accelerators still in use. (Courtesy Lawrence Berkeley Laboratory.)

TABLE 11.1   Some Accelerators

Accelerator	Start of Operation	Particles	Energy
Brookhaven Alternating Gradient Synchrotron (AGS), New York	1961	Protons	33 GeV
Stanford Linear Accelerator (SLAC), California	1961	Electrons	22
Cornell Electron Synchrotron, New York	1967	Electrons	12
Serpukhov Proton Synchrotron, USSR	1967	Protons	76
Fermilab Main Ring, Illinois	1972	Protons	500
Deutsches Elektronen Synchrotron (DESY), Germany	1974	Electrons	22
CERN Super Proton Synchrotron (SPS), Switzerland–France	1976	Protons	500
Fermilab Tevatron, Illinois	1984	Protons	1,000
Japanese National Laboratory (KEK)	1986	Electrons	30
CERN Large Electron-Positron Storage Ring (LEP)	1988?	Electrons	60
Serpukhov UNK, USSR	1990?	Protons	3,000
Superconducting Super Collider (SSC), U.S.A	1994?	Protons	20,000

The first large synchrotron was the Bevatron at Berkeley, which produced protons of an energy of up to 6.4 GeV (see Fig. 11.8). By bombarding a target with this proton beam, E. G. Segré and O. Chamberlain* succeeded in creating antiprotons in 1955.

Table 11.1 is a list of some accelerators now in operation or under development. The two largest accelerators in operation are

---

* **Emilio Gino Segré,** 1905–   , Italian and later American physicist, professor at Berkeley. **Owen Chamberlain,** 1920–   , American physicist, professor at Berkeley. For their discovery of the antiproton, they shared the Nobel Prize in 1959.

Fig. 11.9 Main ring and experimental areas (lower right) at Fermilab. (Courtesy Fermilab.)

the proton synchrotrons at the Fermi National Accelerator Laboratory (Fermilab, near Chicago) and at the Centre Européen de la Recherche Nucléaire (CERN, on the Swiss–French border near Geneva). Figure 11.9 shows an overall view of Fermilab; the accelerator is buried underground in a circular tunnel with a radius of 1 km. The CERN accelerator is slightly larger, with a radius of about 1.2 km. The Fermilab accelerator produces beams of protons with an energy of 1000 GeV, or 1 TeV, and a speed of 99.99995% of the speed of light. Figure 11.10 shows the tunnel at Fermilab, with the magnets that hold the protons in their circular orbit; the beam pipe is inside the magnets, hidden from view. At regular intervals, the beam pipe is interrupted by cavities in which oscillating electric fields impel the protons to higher energy. The upper row of magnets consists of conventional electromagnets; the lower row consists of superconducting magnets capable of producing a stronger magnetic field. The protons are first accelerated in an orbit inside the upper magnets, and they are then steered into a new orbit inside the lower magnets where they attain their final energy. Incidentally: before the protons are allowed to enter the giant circular accelerator they must pass through several smaller preliminary accelerators. At Fermilab there are three preliminary accelerators: the protons generated by a proton gun are first given

Fig. 11.10 Magnets surrounding the beam pipe in the underground tunnel at Fermilab. The magnets of the upper row are electromagnets; those of the lower row are superconducting magnets. (Courtesy Fermilab.)

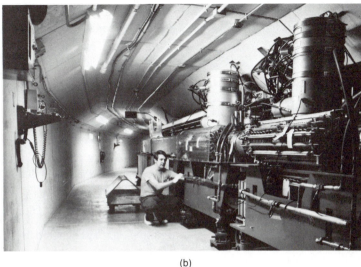

(a)

(b)

Fig. 11.11 (a) Linear accelerator at Fermilab. (b) RF generators (top) and cavities (center) in the booster accelerator at Fermilab. The oscillating electric fields in these cavities kick the protons when they pass through, and impel them to higher energies. (Courtesy Fermilab.)

Fig. 11.12 The accelerator gallery and the experimental areas (lower edge) at the Stanford Linear Accelerator. (Courtesy Stanford Linear Accelerator Center.)

an energy of about 1 MeV by an electrostatic generator; then their energy is raised to 200 MeV by a linear accelerator; and then raised further to 8 GeV by a "small" circular booster accelerator (see Fig. 11.11). Only after the protons have passed through these preliminary stages do they enter the main ring.

The next largest accelerating machine is the Stanford Linear Accelerator (SLAC), 2 miles long. Figure 11.12 shows the gallery sitting above the tunnel in which the machine is buried. This accelerator produces a beam of electrons with an energy of 22 GeV and a speed that is within 8 cm/s of the speed of light. For electrons of such a speed, a circular accelerator would be impractical because the centripetal acceleration would cause the electrons to lose a prohibitive amount of energy by electromagnetic radiation (synchrotron radiation).

Once the projectiles have been given their maximum energy, they are guided out of the accelerator by steering magnets and directed against a target consisting of a block of metal or a tankful of liquid. The reactions that occur in the collisions between the projectiles and the protons and neutrons of the target create an abundance of new particles by conversion of energy into mass. However, not all of the kinetic energy of the incident projectiles can participate in these reactions: only the center-of-mass energy is available for reactions and can be used up in reactions. According to Newtonian physics, for a proton of kinetic energy K colliding with a stationary proton, the available kinetic energy relative to the center of mass is only $K/2$; with the inclusion of the rest-mass energies of both protons, the net available energy becomes $2m_\text{p}c^2 + K/2$. According to relativistic physics—which governs the behavior of high-energy protons—the net available energy is even less; it is only

$$\sqrt{2m_\text{p}c^2(2m_\text{p}c^2 + K)}$$

For instance, a proton with a kinetic energy $K = 1\,\text{TeV} = 10^3\,\text{GeV}$ gives a net available energy of only

$$\sim\sqrt{2\,\text{GeV}(2\,\text{GeV} + 10^3\,\text{GeV})} \simeq 45\,\text{GeV}$$

By far the largest part of the energy of motion of the projectile is tied up in the motion of the center of mass and cannot participate in reactions. Thus, collisions between a high-energy proton and a stationary proton are quite inefficient.

The efficiency improves drastically if two high-energy protons of equal energies and opposite momenta are made to collide head on. The net available energy is then simply the sum of the energies of the two protons. At the CERN Super Proton Synchrotron (SPS), a beam of protons and a beam of antiprotons are made to travel in opposite directions around the accelerator in adjacent orbits that intersect at two locations, where head-on collisions occur (see Fig. 11.13). The energy of each beam is 270 GeV, giving a net center-of-mass energy of 540 GeV.

Example 1   If we want to attain a center-of-mass energy of 540 GeV with a high-energy proton incident on a stationary proton, what incident kinetic energy do we need?

Solution   According to the relativistic formula, the center-of-mass energy is

$$\sqrt{2m_\text{p}c^2(2m_\text{p}c^2 + K)}, \quad \text{or approximately} \quad \sqrt{2m_\text{p}c^2 K}$$

Hence

$$\sqrt{2m_\text{p}c^2 K} \simeq 540\,\text{GeV} \tag{15}$$

and

$$K \simeq \frac{(540\,\text{GeV})^2}{2\,\text{GeV}} \simeq 1.5 \times 10^5\,\text{GeV} \quad \blacksquare \tag{16}$$

Experiments at accelerator laboratories make use of a variety of detectors to detect the new particles created in high-energy collisions. These detectors fall into two broad groups: counters, such as scintillation counters and Cherenkov counters, that register the

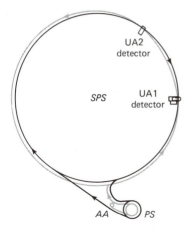

Fig. 11.13 Intersecting orbits of protons (black) and antiprotons (gray) in the SPS accelerator at CERN. The protons move around the accelerator in a clockwise direction, and the antiprotons in a counterclockwise direction. The antiprotons are produced by the impact of 26-GeV protons from a proton synchrotron on a metal target. These antiprotons are accumulated for several days in an accumulator ring (*AA*), and then dumped into the main ring, in which they are accelerated to their final energy of 270 GeV. They collide headon with protons of the same energy at the intersection points UA1 and UA2. These intersection points are surrounded by large arrays of particle detectors (see Fig. 11.20).

passage of each charged particle; and track-recording devices, such as bubble chambers, spark chambers, streamer chambers, and proportional chambers, that provide pictures of the paths of the particles.

Scintillation counters consist of a special plastic or liquid material that gives off a brief and faint flash of light when struck by a charged particle; photomultiplier tubes pick up this flash of light and convert it into a pulse of current to be counted by an electronic circuit. **Cherenkov counters** consist of a volume of dielectric (usually a tank filled with gas at high pressure) within which the speed of light is less than 3×10^8 m/s. When a high-energy charged particle enters this dielectric, its speed will exceed the speed of light. Under these conditions the particle will emit an electromagnetic shock wave, analogous to the sonic boom emitted by a supersonic aircraft. This Cherenkov radiation can be picked up by a photomultiplier tube.

The **bubble chamber,** originally invented by D. A. Glaser,* is a tank filled with a liquid such as liquid hydrogen, helium, or freon at a temperature slightly below its boiling point. The liquid is suddenly decompressed adiabatically by means of an expansion bellows or piston attached to the tank. This lowers the boiling point, placing the liquid in a superheated condition. The liquid is then unstable; it is about to start boiling, but will not start for a few fractions of a second unless some disturbance supplies energy for the formation of the first few bubbles. A high-energy charged particle passing through the chamber ionizes molecules of liquid along

---

* **Donald Arthur Glaser,** 1926– , American physicist, professor at the University of Michigan and at Berkeley. His invention of the bubble chamber gained him the Nobel Prize in 1960.

its track, kicking electrons out of them. These electrons quickly deposit their energy in the liquid and trigger the formation of bubbles. Thus, a fine trail of bubbles marks the track of the charged particle. High-speed cameras take photographs of these trails of bubbles. Then, within a few hundredths of a second, the chamber is recompressed; this quenches the bubbles and readies the chamber for the next cycle of operation.

Figure 11.14 shows the large bubble chamber at CERN. The bubble chamber is surrounded by a massive electromagnet whose magnetic field curves the orbits of the particles and permits the determination of their momenta from the observed radii of curvature [see Eq. (12)]. Figure 11.15 shows the interior of a bubble chamber during construction. The portholes are for cameras and for flashlamps. Photographs of the tracks of particles passing through the bubble chamber are taken simultaneously with several cameras at different angles so as to obtain a stereoscopic view of the tracks. Bubble chambers are expensive to operate and maintain, but they attain a higher spatial resolution than any other track-recording device. Careful measurements of the photographs deter-

Fig. 11.14 The Big European Bubble Chamber (BEBC) at CERN. The chamber is hidden from view inside the large magnet. (Photo CERN.)

Fig. 11.15 The interior of the GARGAMELLE bubble chamber during construction. (Photo CERN.)

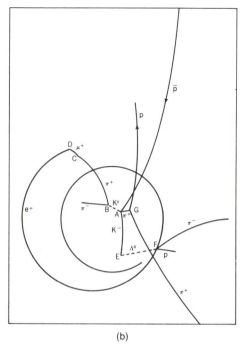

(a)

(b)

Fig. 11.16 (a) Photograph of tracks in a bubble chamber showing a sequence of reactions initiated by the impact of an antiproton in a bubble chamber. (Photo CERN.) (b) In this tracing, the dashed lines indicate the trajectories of neutral particles, not visible in the photograph.

mine track positions to within 0.05 mm, and the curvature of the track in the magnetic field determines the momentum of the particle to within about 0.1%.

Figure 11.16 is an exceptionally interesting photograph of tracks of particles made visible with a bubble chamber at CERN. It shows a sequence of events involving the creation of several particles in the collision of an antiproton and a proton, and the subsequent decays and collisions of these particles. The antiproton ($\bar{p}$) enters the field of view from above; its track is slightly curved to the left by the magnetic field. This antiproton was produced outside of the bubble chamber by the impact of a beam of protons on a target of metal. At the point A indicated in Fig. 11.16b, the antiproton collided with a proton at rest in the liquid of the bubble chamber. The antiproton and the proton were destroyed in this collision, which created two kaons (K^0, K^-) and two pions (π^+, π^0). By a remarkable coincidence, all of these particles, except for the π^0, caused further events within the field of view of the bubble chamber.

The K^0 particle is electrically neutral and hence leaves no visible track in the bubble chamber; nevertheless, it is possible to reconstruct the trajectory of the K^0 particle because, after a short while, it spontaneously decayed (at B) into two pions that did leave tracks. One of these pions then decayed into an antimuon and a neutrino (at C), and the antimuon in turn decayed into an antielectron and two neutrinos (at D). Meanwhile the K^- created in the original antiproton-proton collision suffered a collision with another proton at rest in the liquid of the bubble chamber (at E). This collision led to the creation of a lambda particle (Λ^0) and a pion (π^0). The lambda particle is neutral and leaves no visible track; but we can see that at F it decayed into a pion (π^-) and a proton. Furthermore, one of the pions created in the original antiproton-proton collision suffered an elastic collision with another proton at rest (at G), which caused this proton to recoil. The complete sequence of reactions illustrated by the tracks in Fig. 11.16 can be summarized as follows:

$$\bar{p} + p \rightarrow K^0 + K^- + \pi^+ + \pi^0 \qquad \text{(A)}$$
$$\hookrightarrow + p \rightarrow \pi^+ + p \quad \text{(G)}$$
$$\hookrightarrow + p \rightarrow \Lambda^0 + \pi^0 \quad \text{(E)}$$
$$\hookrightarrow \pi^- + p \quad \text{(F)}$$
$$\hookrightarrow \pi^+ + \pi^- \qquad \text{(B)}$$
$$\hookrightarrow \mu^+ + \nu \qquad \text{(C)}$$
$$\hookrightarrow \bar{e} + \nu + \bar{\nu} \quad \text{(D)} \qquad (17)$$

Spark chambers yield somewhat cruder pictures of particle tracks than bubble chambers, but they are much simpler. A spark chamber consists of many parallel screens or thin plates of metal, each separated from the next by a gap of about a centimeter. The space between the plates is filled with a gas, usually neon. Alternate plates are connected to the positive and the negative terminals of a high-voltage supply, which produces an electric field of about 10^6 V/m between the plates. A high-energy charged particle passing through the chamber ionizes the gas along its track. The electrons released by this ionization are accelerated in the strong electric field, strike gas molecules, and release more electrons, which release even more electrons, etc. This develops into an electron avalanche and produces an electric discharge between the plates, with a visible spark. Thus, a succession of sparks marks the passage of the particle through the chamber (see Fig. 11.17). Cameras record these sparks photographically; usually mirrors are placed around the

(a)

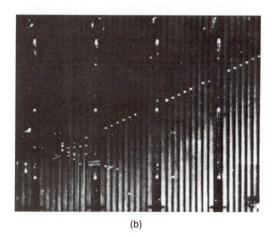

(b)

Fig. 11.17 (a) A spark chamber; (Courtesy Brookhaven National Laboratory.) (b) tracks of particles made visible by sparks in a spark
chamber. (Photo CERN.)

spark chamber so that a single photograph can simultaneously record several views of the tracks of sparks seen from several angles.

Although spark chambers do not have the high spatial resolution of bubble chambers, they have a much higher time resolution. The high voltage is usually applied to the chamber for only a brief interval of time, about 10^{-6} s; the chamber is therefore sensitive only during this brief interval. This permits the chamber to capture the track of one individual particle, provided it is triggered at a suitable moment. The trigger is provided by auxiliary scintillation counters placed around the chamber; these make preliminary identifications of the arriving particles and trigger the spark chamber whenever there is an interesting event.

Streamer chambers are similar to spark chambers, but they use only one pair of widely separated plates instead of the many closely spaced pairs used in the spark chamber. A very brief pulse of high voltage, lasting only about 10^{-8} s, is applied to these plates. Under these conditions the electrons released by the ionization of the gas in the chamber do not have enough time to develop a full-sized spark from one plate to the other; instead they only give rise to small, faint protosparks, or streamers, which delineate the track of the particle. Streamer chambers are capable of good spatial resolution; when photographed from a direction perpendicular to the plates the streamers determine the positions of points on the track of the particle to within 0.1 mm.

Proportional chambers use a grid of thin parallel wires at positive potential sandwiched between two plates or screens at negative potential (see Fig. 11.18). Electrons released by the ionization of the gas in the chamber drift to the nearest positive wire, and give rise to an electric discharge in the strong electric field in the immediate vicinity of the wire. This electric discharge registers as a current pulse on the wire (the magnitude of the current pulse is proportional to the amount of ionization, hence the name *proportional* for these chambers). Each wire is connected to its own electric circuit, and therefore each wire constitutes an independent detector, which signals the location of the discharge. In multiwire proportional chambers, good spatial resolution is attained by filling the entire chamber with a large number of wires in a dense array (typically one wire per millimeter). In drift proportional chambers, the spacings between the wires are much larger, but excellent spatial resolution is attained by measuring the time delay between arrival of the high-energy particle in the chamber and detection of the current pulses on the wires (the instant of arrival of the particle in the chamber is determined by an auxiliary scintillation counter placed next to the chamber). Since the electrons released by ion-

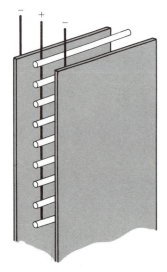

Fig. 11.18 Arrangement of positive and negative electrodes in a proportional chamber.

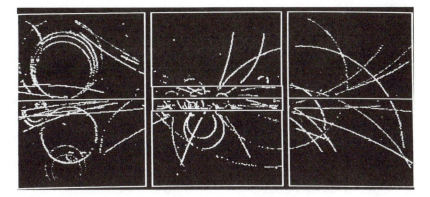

Fig. 11.19 Images of tracks of particles reconstructed by a computer from data collected by a multiwire proportional chamber. (Photo CERN.)

ization drift toward the positive wires at a known speed, the measurement of the time delays permits computation of the precise distances that these electrons had to travel from their points of origin to the wires. Drift proportional chambers can determine track positions with an accuracy of up to 0.05 mm.

Proportional chambers do not provide photographs of the tracks of particles. Instead, they record the tracks as a sequence of electric signals from their detector wires. These signals are fed into a computer which constructs an image of the tracks (see Fig. 11.19). Because they have good spatial resolution, and because they deliver data in a form that can be directly manipulated by computers, proportional chambers have become very popular in high-energy physics, largely supplanting spark chambers. Figure 11.20

Fig. 11.20 The large UA1 particle detector at CERN. The tracks shown in Fig. 11.19 were obtained with this detector. (Photo CERN.)

shows an example of a large particle detector consisting of arrays of scintillation counters combined with drift chambers containing a total of about 100,000 wires. The detector also contains a magnet, so that the momentum of charged particles can be determined by their deflection in the magnetic field. The output of the detector is directly fed into a computer, which analyzes the track data, displays the tracks, and immediately identifies the types of particles passing through the detector.

11.3 Leptons, Baryons, and Mesons

The known particles fall into three main groups: the leptons, the baryons, and the mesons. The last two groups are collectively designated as **hadrons,** or **strongly interacting particles,** because they interact via the strong force. Besides these three main groups, there is one other; the intermediate bosons. This group consists of particles, such as the photon, that serve as carriers of the interactions by means of the exchange mechanism described in Section 11.1.

There are six different **leptons.** Table 11.2 lists these leptons, their masses, spins, electric charges, lifetimes, and principal decay modes. The circumstances surrounding the discovery of the muon have already been discussed in Section 11.1 The tauon was dis-

TABLE 11.2   The Leptons

Particle	Mass	Spin	Electric Charge	Mean Life	Principal Decay Modes Mode	Fraction (%)
e	$0.511003 \, \text{MeV}/c^2$	1/2	-1	stable		
μ	105.659	1/2	-1	$2.19703 \times 10^{-6} \, \text{s}$	$e^- \nu \bar{\nu}$	100
τ	1784	1/2	-1	$3.4 \times 10^{-13} \, \text{s}$	$\mu^- \nu \bar{\nu}$	17.6
					$e^- \nu \bar{\nu}$	17.4
					hadrons, neutrals	51.6
ν_e	0^a	1/2	0	stable		
ν_μ	0	1/2	0	stable		
ν_τ	0	1/2	0	stable		

[a] According to some recent, tentative evidence, the mass of the neutrino may be different from zero, about $20 \, \text{eV}/c^2$.

covered in 1975 by M. Perl* and his associates in electron–anti-electron collision experiments at the Stanford Linear Accelerator. The muon and the tauon are essentially heavy versions of the electron; both are unstable and decay into electrons. Because of its large mass, the tauon is also capable of decaying into hadrons; this is its preferred decay mode.

The neutrinos (v_e, v_μ, and v_τ) are particles of zero mass and zero electric charge. The electron neutrino v_e is the one we became acquainted with in earlier chapters; it is emitted in β decay. The muon and tauon neutrinos v_μ and v_τ are emitted in the decays of the muon and the tauon, respectively. For instance, one of the neutrinos emitted in the reaction (7) is a muon neutrino and the other an electron antineutrino,

$$\mu \rightarrow e + v_\mu + \bar{v}_e$$

Both the v_e neutrino and the v_μ neutrino have been detected directly in reactions that absorb these neutrinos. The v_τ neutrino has not been detected directly, but its existence has been inferred from energy and momentum conservation in the decay of the tauon.

Besides the leptons of Table 11.2 there also exist six antileptons: the antielectron, the antimuon, the antitauon and the three anti-neutrinos. These antiparticles have the opposite electric charge, but they have exactly the same mass and spin as the corresponding particles.

The **baryons** are the most numerous group of particles. In Table 11.3 each entry includes several baryons that are listed together because of close similarities. The first column of this table gives the baryon name with a number in parentheses; this number is the average mass (in MeV/c^2) of the several baryons. For instance, the entry N(939) represents the proton and the neutron; the average mass of this pair is 939 MeV/c^2. The second column gives the spin and the intrinsic parity (to be discussed in Section 11.5). The third column gives electric charges of the several baryons that share a common name. The fourth column gives the isospin (to be discussed in Section 11.5). The next column gives the lifetime or the energy uncertainty (to be discussed below). And the last column gives the principal decay modes.

For every baryon in Table 11.3 there exists an antibaryon. As in the case of leptons, these antiparticles have the opposite charge but the same mass and spin as the corresponding particles.

---

* **Martin L. Perl,** 1948– , American physicist, professor at Stanford.

TABLE 11.3 The Baryons[a]

Particle (and mass)	Spin[(Parity)]	Electric Charge	Isospin	Mean Life or ΔE	Partial Decay Modes	
					Mode	Fraction (%)
$S = 0$, $T = 1/2$ NUCLEON RESONANCES (N)						
N(939)	$1/2^+$	1, 0	1/2	p: $> 10^{32}$ yr n: 898 s	$pe\bar{\nu}$	100
N(1440)	$1/2^+$	1, 0	1/2	200 MeV	$N\pi$ $N\pi\pi$	50–70 30–50
N(1520)	$3/2^-$	1, 0	1/2	125	$N\pi$ $N\pi\pi$	50–60 40–50
N(1535)	$1/2^-$	1, 0	1/2	150	$N\pi$ $N\eta$	35–50 45–55
N(1650)	$1/2^-$	1, 0	1/2	150	$N\pi$ ΛK $N\pi\pi$	55–65 ~8 20–35
N(1675)	$5/2^-$	1, 0	1/2	155	$N\pi$ $N\pi\pi$	35–40 60–65
N(1680)	$5/2^+$	1, 0	1/2	125	$N\pi$ $N\pi\pi$	55–65 35–45
N(1700)	$3/2^-$	1, 0	1/2	100	$N\pi$ $N\pi\pi$	5–15 80–90
N(1710)	$1/2^+$	1, 0	1/2	110	$N\pi$ $N\eta$ ΛK ΣK $N\pi\pi$	10–20 ~25 ~15 2–10 <50
N(1720)	$3/2^+$	1, 0	1/2	200	$N\pi$ $N\pi\pi$	10–20 <75
N(2190)	$7/2^-$	1, 0	1/2	350	$N\pi$	~14
N(2220)	$9/2^+$	1, 0	1/2	400	$N\pi$	~18
N(2250)	$9/2^-$	1, 0	1/2	300	$N\pi$	~10
N(2600)	$11/2^-$	1, 0	1/2	400	$N\pi$	~5
$S = 0$, $T = 3/2$ DELTA RESONANCES (Δ)						
Δ(1232)	$3/2^+$	2, 1, 0, −1	3/2	115	$N\pi$	99.4
Δ(1620)	$1/2^-$	2, 1, 0, −1	3/2	140	$N\pi$ $N\pi\pi$	25–35 ~70
Δ(1700)	$3/2^-$	2, 1, 0, −1	3/2	250	$N\pi$ $N\pi\pi$	10–20 80–90
Δ(1900)	$1/2^-$	2, 1, 0, −1	3/2	150	$N\pi$ ΣK	5–15 ~10
Δ(1905)	$5/2^+$	2, 1, 0, −1	3/2	300	$N\pi$ $N\pi\pi$	5–15 <75
Δ(1910)	$1/2^+$	2, 1, 0, −1	3/2	220	$N\pi$ ΣK $N\pi\pi$	15–25 2–20 <75
Δ(1920)	$3/2^+$	2, 1, 0, −1	3/2	250	$N\pi$	14–20
Δ(1930)	$5/2^-$	2, 1, 0, −1	3/2	250	$N\pi$	4–14

TABLE 11.3 The Baryons (continued)

Particle (and mass)	Spin$^{\text{(Parity)}}$	Electric Charge	Isospin	Mean Life or ΔE	Partial Decay Modes Mode	Partial Decay Modes Fraction (%)
$\Delta(1950)$	$7/2^+$	$2,1,0,-1$	$3/2$	240 MeV	$N\pi$ $N\pi\pi$	35–45 <40
$\Delta(2420)$	$11/2^+$	$2,1,0,-1$	$3/2$	300	$N\pi$	5–15
				$S=-1, T=0$ LAMBDA RESONANCES (Λ)		
$\Lambda(1116)$	$1/2^+$	0	0	2.63×10^{-10} s	$p\pi^-$ $n\pi^0$	64.2 35.8
$\Lambda(1405)$	$1/2^-$	0	0	40 MeV	$\Sigma\pi$	100
$\Lambda(1520)$	$3/2^-$	0	0	15.6	$N\bar{K}$ $\Sigma\pi$ $\Lambda\pi\pi$	45 42 10
$\Lambda(1600)$	$1/2^+$	0	0	150	$N\bar{K}$ $\Sigma\pi$	15–30 10–60
$\Lambda(1670)$	$1/2^-$	0	0	35	$N\bar{K}$ $\Sigma\pi$ $\Lambda\eta$	15–25 20–60 15–35
$\Lambda(1690)$	$3/2^-$	0	0	60	$N\bar{K}$ $\Sigma\pi$ $\Lambda\pi\pi$ $\Sigma\pi\pi$	20–30 20–40 ~25 ~20
$\Lambda(1800)$	$1/2^-$	0	0	300	$N\bar{K}$	25–40
$\Lambda(1800)$	$1/2^+$	0	0	150	$N\bar{K}$ $\Sigma\pi$ $N\bar{K}^*(892)$	20–50 10–40 30–60
$\Lambda(1820)$	$5/2^+$	0	0	80	$N\bar{K}$ $\Sigma\pi$ $\Sigma(1385)\pi$	55–65 8–14 5–10
$\Lambda(1830)$	$5/2^-$	0	0	95	$N\bar{K}$ $\Sigma\pi$ $\Sigma(1385)\pi$	3–10 35–75 >15
$\Lambda(1890)$	$3/2^+$	0	0	100	$N\bar{K}$ $\Sigma\pi$	20–35 3–10
$\Lambda(2100)$	$7/2^-$	0	0	200	$N\bar{K}$ $N\bar{K}^*(892)$	25–35 10–20
$\Lambda(2110)$	$5/2^+$	0	0	200	$N\bar{K}$ $\Sigma\pi$ $N\bar{K}^*(892)$	5–25 10–40 10–60
$\Lambda(2350)$	$9/2^+$	0	0	150	$N\bar{K}$ $\Sigma\pi$	~12 ~10
				$S=-1, T=1$ SIGMA RESONANCES (Σ)		
$\Sigma(1193)$	$1/2^+$	$1,0,-1$	1	$\Sigma^+: 0.800\times10^{-10}$ s $\Sigma^0: 6\times10^{-20}$ s $\Sigma^-: 1.48\times10^{-10}$ s	$p\pi^0$ $n\pi^+$ $\Lambda\gamma$ $n\pi^-$	51.6 48.4 100 100
$\Sigma(1385)$	$3/2^+$	$1,0,-1$	1	35 MeV	$\Lambda\pi$ $\Sigma\pi$	88 12

TABLE 11.3   The Baryons (continued)

Particle (and mass)	Spin(Parity)	Electric Charge	Isospin	Mean Life or ΔE	Partial Decay Modes Mode	Partial Decay Modes Fraction (%)
$\Sigma(1660)$	$1/2^+$	$1, 0, -1$	1	100 MeV	$N\bar{K}$	10–30
$\Sigma(1670)$	$3/2^-$	$1, 0, -1$	1	60	$N\bar{K}$	7–13
					$\Lambda\pi$	5–15
					$\Sigma\pi$	30–60
$\Sigma(1750)$	$1/2^-$	$1, 0, -1$	1	90	$N\bar{K}$	10–40
					$\Sigma\eta$	15–55
$\Sigma(1775)$	$5/2^-$	$1, 0, -1$	1	120	$N\bar{K}$	37–43
					$\Lambda\pi$	14–20
					$\Sigma(1385)\pi$	8–12
					$\Lambda(1520)\pi$	17–23
$\Sigma(1915)$	$5/2^+$	$1, 0, -1$	1	120	$N\bar{K}$	5–15
$\Sigma(1940)$	$3/2^-$	$1, 0, -1$	1	220	$N\bar{K}$	<20
$\Sigma(2030)$	$7/2^+$	$1, 0, -1$	1	180	$N\bar{K}$	17–23
					$\Lambda\pi$	17–23
					$\Sigma(1385)\pi$	5–15
					$\Lambda(1520)\pi$	10–20
					$\Delta(1232)\bar{K}$	10–20
$\Sigma(2250)$	?	$1, 0, -1$	1	100	$N\bar{K}$	<10

$S = -2, T = 1/2$ CASCADE RESONANCES (Ξ)

Particle (and mass)	Spin(Parity)	Electric Charge	Isospin	Mean Life or ΔE	Mode	Fraction (%)
$\Xi(1318)$	$1/2^+$	$0, -1$	1/2	Ξ^0: 2.9×10^{-10} s	$\Lambda\pi^0$	100
				Ξ^-: 1.64×10^{-10} s	$\Lambda\pi^-$	100
$\Xi(1530)$	$3/2^+$	$0, -1$	1/2	9.1 MeV	$\Xi\pi$	100
$\Xi(1820)$	$3/2$	$0, -1$	1/2	26	$\Lambda\bar{K}$	~45
					$\Xi(1530)\pi$	~45
$\Xi(2030)$	?	$0, -1$	1/2	20	$\Lambda\bar{K}$	~20
					$\Sigma\bar{K}$	~80

OTHER BARYONS

Particle (and mass)	Spin(Parity)	Electric Charge	Isospin	Mean Life or ΔE	Mode	Fraction (%)
$\Omega^-(1672)$	$3/2^+$	-1	0	0.82×10^{-10} s	ΛK^-	67.8
					$\Xi^0\pi^-$	23.6
$\Lambda c^+(2282)$	$1/2^+$	1	0	2.3×10^{-13} s	Λ anything	33

[a] Based on *Review of Particle Properties* by the Particle Data Group, April 1986.

Finally, the **mesons** are another numerous group of particles. They are listed in Table 11.4. Note that all mesons have integer spin—they are bosons; whereas all baryons have half-integer spin—they are fermions. This is the distinctive difference between mesons and baryons. For every meson there exists an antimeson. These antiparticles have already been included in Table 11.4. For instance, the antiparticle for the π^+ is the π^-, and vice versa. The antiparticle for the π^0 is the π^0 itself; this means that when two π^0 meet, they can annihilate each other.

TABLE 11.4   The Mesons[a]

Particle (and mass)	Spin[(Parity)]	Electric Charge	Isospin	Mean life or ΔE	Partial Decay Modes Mode	Fraction (%)
			$S = 0$	NONSTRANGE MESONS		
$\pi(138)$	0^-	$1, 0, -1$	1	$\pi^\pm$: 2.603×10^{-8} s π^0: 0.87×10^{-16} s	$\mu^\pm \nu$ $\gamma\gamma$	100 98.8
$\eta(549)$	0^-	0	0	6×10^{-19} s	neutral charged	70.9 29.1
$\rho(770)$	1^-	$1, 0, -1$	1	154 MeV	$\pi\pi$	≈ 100
$\omega(783)$	1^-	0	0	9.9	$\pi^+\pi^-\pi^0$	89.6
$\eta'(958)$	0^-	0	0	0.24	$\eta\pi\pi$ $\rho^0\gamma$	65.3 30.0
$S(975)$	0^+	0	0	33	$\pi\pi$ $K\bar{K}$	78 22
$\delta(980)$	0^+	$1, 0, -1$	1	54	$\eta\pi$ $K\bar{K}$	seen seen
$\phi(1020)$	1^-	0	0	4.2	K^+K^- $K_L K_S$ $\pi^+\pi^-\pi^0$	49.5 34.3 14.8
$h_1(1190)$	1^+	0	0	320	$\rho\pi$	seen
$B(1235)$	1^+	$1, 0, -1$	1	150	$\omega\pi$	dominant
$f(1270)$	2^+	0	0	176	$\pi\pi$	84.3
$A(1270)$	1^+	$1, 0, -1$	1	315	$\rho\pi$	dominant
$D(1285)$	1^+	0	0	26	$K\bar{K}\pi$ $\eta\pi\pi$	11 49
$\varepsilon(1300)$	0^+	0	0	150–400	$\pi\pi$ $K\bar{K}$	~ 90 ~ 10
$A_2(1320)$	2^+	$1, 0, -1$	1	110	$\rho\pi$ $\eta\pi$ $\omega\pi\pi$	70.1 14.5 10.6
$E(1420)$	1^+	0	0	56	$K\bar{K}\pi$	seen
$\iota(1440)$	0^-	0	0	76	$K\bar{K}\pi$	seen
$f'(1525)$	2^+	0	0	70	$K\bar{K}$	dominant
$\rho(1600)$	1^-	$1, 0, -1$	1	260	4π $\pi\pi$	60 23
$\omega(1670)$	3^-	0	0	166	3π	seen
$A(1680)$	2^-	$1, 0, -1$	1	250	$f\pi$ $\rho\pi$	53 34
$\phi(1680)$	1^-	0	0	130	$K^*\bar{K} + \bar{K}^*K$	dominant
$g(1690)$	3^-	$1, 0, -1$	1	200	2π 4π	23.8 70.9
$\theta(1690)$	2^+	0	0	134	$\eta\eta$	seen
$\phi(1850)$	3^-	0	0	96	$K\bar{K}$	seen
$h(2030)$	4^+	0	0	200	$\pi\pi$	17
$\eta_c(2980)$	0^-	0	0	11	$\eta'\pi^+\pi^-$	4.1
$J/\psi(3100)$	1^-	0	0	0.063	hadrons + radiative	85

TABLE 11.4   The Mesons (continued)

Particle (and mass)	Spin$^{(Parity)}$	Electric Charge	Isospin	Mean life or ΔE	Partial Decay Modes Mode	Fraction (%)
$\chi(3415)$	0^+	0	0		$2(\pi^+\pi^-)$	3.8
					$\pi^+\pi^- K^+ K^-$	2.9
$\chi(3510)$	1^+	0	0	< 1.0 MeV	$\gamma J/\psi(3100)$	25.8
$\chi(3555)$	2^+	0	0	2.9	$\gamma J/\psi(3100)$	14.8
$\psi(3685)$	1^-	0	0	0.215	hadrons + radiative	98.1
$\psi(3770)$	1^-	0		25	$D\bar{D}$	dominant
$\psi(4030)$	1^-	0		52	hadrons	dominant
$\psi(4160)$	1^-	0		78	hadrons	dominant
$\psi(4415)$	1^-	0		43	hadrons	dominant
$\Upsilon(9460)$	1^-	0		0.044	$\mu^+\mu^-$	2.9
					e^+e^-	2.8
					$\tau^+\tau^-$	3.2
$\chi_b(9860)$		0			$\gamma\Upsilon(9460)$	seen
$\chi_b(9895)$		0			$\gamma\Upsilon(9460)$	35
$\chi_b(9915)$		0			$\gamma\Upsilon(9460)$	22
$\Upsilon(10023)$	1^-	0		0.029	$\Upsilon(9460)\pi\pi$	18.7
$\chi_b(10255)$		0			$\gamma\Upsilon(9460)$	seen
					$\gamma\Upsilon(10023)$	seen
$\chi_b(10270)$		0			$\gamma\Upsilon(9460)$	seen
					$\gamma\Upsilon(10023)$	seen
$\Upsilon(10355)$	1^-	0		0.012	$\gamma\chi_b(10255)$	15.6
				24	$\gamma\chi_b(10270)$	12.7
$\Upsilon(10575)$	1^-	0				
$\Upsilon(10860)$	1^-	0		110		
$\Upsilon(11020)$	1^-	0		71		

$S = \pm 1$ STRANGE MESONS

Particle (and mass)	Spin$^{(Parity)}$	Electric Charge	Isospin	Mean life or ΔE	Mode	Fraction (%)
$K(496)$	0^-	1, 0	1/2	$K^\pm: 1.237 \times 10^{-8}$ s	$\mu^\pm \nu$	63.5
					$\pi^\pm \pi^0$	21.2
$\bar{K}(496)$	0^-	0, -1	1/2	$K_L^0: 5.18 \times 10^{-8}$ s	$\pi^0\pi^0\pi^0$	21.5
					$\pi^+\pi^-\pi^0$	12.4
					$\pi^\pm \mu^\mp \nu$	27.1
					$\pi^\pm e^\mp \nu$	38.7
				$K_S^0: 0.892 \times 10^{-10}$ s	$\pi^+\pi^-$	68.6
					$\pi^0\pi^0$	31.4
$K^*(892)$	1^-	1, 0	1/2	51 MeV	$K\pi$	≈ 100
$\bar{K}^*(892)$	1^-	0, -1	1/2			
$Q(1280)$	1^+	1, 0	1/2	90	$K\rho$	42
$\bar{Q}(1280)$	1^+	0, -1	1/2		$\kappa(1350)\pi$	28
					$K^*(892)\pi$	16
					$K\omega$	11
$\kappa(1350)$	0^+	1, 0	1/2	~ 250	$K\pi$	seen
$\bar{\kappa}(1350)$	0^+	0, -1	1/2			
$Q(1400)$	1^+	1, 0	1/2	184	$K^*(892)\pi$	94
$\bar{Q}(1400)$	1^+	0, -1	1/2			

TABLE 11.4   The Mesons (continued)

Particle (and mass)	Spin$^{(Parity)}$	Electric Charge	Isospin	Mean life or ΔE	Partial Decay Modes Mode	Partial Decay Modes Fraction (%)
K*(1430)	2$^+$	1, 0	1/2	100 MeV	Kπ	44.8
K̄*(1430)	2$^+$	0, −1	1/2		K*(892)π	23.6
					K*(892)ππ	13.5
L(1770)	2$^-$	1, 0	1/2	~200	K*(1430)π	dominant
L̄(1770)	2$^-$	0, −1	1/2			
K*(1780)	3$^-$	1, 0	1/2	150	Kππ	large
K̄*(1780)	3$^-$	0, −1	1/2		K π	17

S = 0 CHARMED NONSTRANGE MESONS

Particle (and mass)	Spin$^{(Parity)}$	Electric Charge	Isospin	Mean life or ΔE	Partial Decay Modes Mode	Partial Decay Modes Fraction (%)
D(1867)	0$^-$	1, 0	1/2	D$^\pm$: 9.2 × 10^{-13} s	e$^\pm$ anything	19
					K$^\pm$ anything	16
					K^0 or K̄^{0} anything	48
D̄(1867)	0$^-$	0, −1	1/2	D^0: 4.4 × 10^{-13} s	K$^-$ anything	44
					K^0 or K̄^{0} anything	33
D*(2010)	1$^-$	1, 0	1/2	D$^\pm$: <2 MeV	D$^0\pi^+$	49
					D$^+\pi^0$	34
D̄*(2010)	1$^-$	0, −1	1/2	D^0: <5	D$^0\pi^0$	51
					D$^0\gamma$	48

S = ±1 CHARMED STRANGE MESONS

Particle (and mass)	Spin$^{(Parity)}$	Electric Charge	Isospin	Mean life or ΔE	Partial Decay Modes Mode	Partial Decay Modes Fraction (%)
F$^+$(1971)	0$^-$	1	0	2.8 × 10^{-13} s	$\phi\pi^\pm$	seen
F$^-$(1971)	0$^-$	−1	0			

S = 0 BOTTOM NONSTRANGE MESONS

Particle (and mass)	Spin$^{(Parity)}$	Electric Charge	Isospin	Mean life or ΔE	Partial Decay Modes Mode	Partial Decay Modes Fraction (%)
B(5272)	0$^-$	1, 0	1/2	B$^\pm$: ?	D*$^\pm\pi^\pm\pi^\mp$	2.7
B̄(5272)	0$^-$	0, −1	1/2	B^0: 14 × 10^{-13} s	D̄$^0\pi^+\pi^-$	7

[a] Based on *Review of Particle Properties* by the Particle Data Group, April 1986.

Almost all of the known particles are unstable; they decay, spontaneously falling apart into several other particles. The only absolutely stable particles are the electron, the proton, the photon, and the neutrinos. Figure 11.16 shows some examples of decays: the decay of a kaon and that of a lambda particle. Each of these particles lives for about 10^{-10} s from the instant of production to the instant of decay. By the standards of high-energy physics a time interval of 10^{-10} s is a rather long time interval. If a particle lives that long, physicists can do experiments on it and with it; for example, kaons live long enough to be assembled in a beam and shot at a target. In the terminology of high-energy physics, any particle with a lifetime of 10^{-10} or even 10^{-14} s is regarded as **stable.** Most of the particles listed in Tables 11.3 and 11.4 have

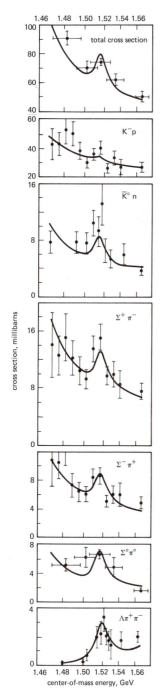

cross section, millibarns

1.46   1.48   1.50 1.52 1.54 1.56
center-of-mass energy, GeV

Fig. 11.21 Cross sections for diverse reactions initiated by a K⁻-p collision, as a function of the center-of-mass energy. [After M. B. Watson et al., *Phys. Rev.* **131**, 2248 (1963)]

lifetimes of only about 10^{-23} s. Such particles are regarded as unstable.

If a particle lasts only 10^{-23} s, it cannot be detected directly. Since the maximum speed is the speed of light, in a time interval of 10^{-23} s, the particle can travel at most a distance

$$10^{-23}\,\text{s} \times 3 \times 10^8\,\text{m/s} \simeq 10^{-15}\,\text{m} \tag{18}$$

This distance is much smaller than the diameter of an atom—it is roughly the same as the diameter of a proton.* Obviously, such a particle does not travel far enough to leave a visible track in a bubble chamber or any other detector. The short-lived particles cannot be detected directly, but their existence can be deduced from circumstantial evidence. Consider the case of the short-lived lambda particle $\Lambda(1520)$, found in collisions between kaons and protons. If a beam of negative kaons impacts on the protons in a bubble chamber, a variety of reactions will occur. Here are some of these:

$$K^- + p \rightarrow K^- + p \tag{19}$$

$$K^- + p \rightarrow \bar{K}^0 + n \tag{20}$$

$$K^- + p \rightarrow \Sigma^+ + \pi^- \tag{21}$$

$$K^- + p \rightarrow \Sigma^- + \pi^+ \tag{22}$$

$$K^- + p \rightarrow \Sigma^0 + \pi^0 \tag{23}$$

$$K^- + p \rightarrow \Lambda + \pi^+ + \pi^- \tag{24}$$

The first of these reactions is elastic scattering. The other reactions involve the creation of new particles by the conversion of some kinetic energy into mass (the Σ particles in these reactions are those designated by $\Sigma(1193)$ in Table 11.3]. The reaction cross section for each of the reactions (19)–(24) is some function of energy. Figure 11.21 shows how the cross sections vary as function of the center-of-mass energy. The key feature in these plots is the local maximum in the cross sections at a center-of-mass energy of 1520 MeV; at this energy *all* the cross sections have a peak. This coincidence leads us to suspect that all these maxima are due to the same cause. In each instance the colliding particles merge to create a new particle:

$$K^- + p \rightarrow \Lambda(1520) \tag{25}$$

---

* The relativistic time-dilation effect (see Chapter 2) will stretch out the lifetime of a fast-moving particle and permit it to travel farther. However, for the energies attainable by our accelerators, the time-dilation effect is usually at most a factor of 10 or 100; this does not help very much.

and this new particle subsequently decays into one or another of the pairs of particles that emerge from the collision:

$$K^- + p \to \Lambda(1520) \to \ K^- + p \tag{26}$$

$$\to \ \bar{K}^0 + n \tag{27}$$

$$\to \ \Sigma^+ + \pi^- \tag{28}$$

$$\to \ \Sigma^- + \pi^+ \tag{29}$$

$$\to \ \Sigma^0 + \pi^0 \tag{30}$$

$$\to \ \Lambda + \pi^+ + \pi^- \tag{31}$$

The production of a lambda will occur only if the energy of the kaon relative to the center of mass is just right—according to Eq. (25), the energy of the kaon plus the mass–energy of the proton must add up to the mass–energy of the lambda. When the energy is right, the cross section for the production of the lambda will have a maximum and consequently the cross sections for all the reactions (26)–(31) will also have maxima since they all involve the production of a lambda as a first step. Obviously, the lambda particle in the reactions (26)–(31) plays a role analogous to that of the compound nucleus in nuclear reactions.

Note that the plots of Fig. 11.21 show that the cross section is not only large when the center-of-mass energy is exactly 1520 MeV, but is also quite large at energies somewhat below and somewhat above 1520 MeV. This indicates that there is an uncertainty in the energy for the production of a lambda. The magnitude of this uncertainty is about ± 16 MeV. The uncertainty in the production energy implies an uncertainty in the mass–energy of the lambda; the mass is uncertain by ± 16 MeV/c^2,

$$M_{\Lambda(1520)} = 1520 \,\text{MeV}/c^2 \pm 16 \,\text{MeV}/c^2$$

This uncertainty is *not* an experimental error—the equipment used by high-energy physicists is sometimes a bit crude, but it is not *that* crude. This is a quantum-mechanical uncertainty arising from the energy–time uncertainty relation—the lambda has a very short lifetime, hence a large uncertainty of energy. In fact, we can estimate the lifetime from the uncertainty in energy:

$$\Delta t \simeq \frac{\hbar}{\Delta E} \simeq \frac{1.05 \times 10^{-34}\,\text{J} \cdot \text{s}}{16\,\text{MeV}} \simeq 4 \times 10^{-23}\,\text{s}$$

Short-lived particles such as the lambda are often called **resonances**. Thus one speaks of the $\Lambda(1520)$ as a K^-p resonance. But with just as much justice one could say that the $\Lambda(1520)$ is a $\bar{K}^0 n$ resonance, or a $\Sigma^+ \pi^-$ resonance, etc. Physicists prefer to call such an object a resonance rather than a particle because it is nothing but a (hypothetical) intermediate state in a reaction chain [see Eqs. (26)–(31)].

11.4 **Fundamental Interactions**

In earlier chapters we have become acquainted with the strong interaction, which acts among the protons and neutrons in a nucleus, and the weak interaction, which is responsible for β decay and other decay reactions. These are two of the four fundamental interactions; the other two are the familiar electromagnetic interaction and the gravitational interaction. Table 11.5 lists these four fundamental interactions, their strengths, and their ranges. For the purposes of this table, the strengths of the interactions are reckoned for neutrons or protons in close contact, and all strengths are expressed relative to that of the strong interaction, which is arbitrarily assigned a strength of 1 (in general, the strengths of the interactions depend on energy; the values quoted in the table are appropriate for particles of low energy).

According to the quantum theory of fields, all the interactions rely on the mechanism of exchange of quanta. Each force field has its own quanta—analogous to photons—which act as carriers of the interaction. Thus, all the forces are transmitted from one particle to another by successive processes of emission, propagation, and absorption of such carriers.

The carriers of the gravitational interaction are the **gravitons,** the quanta of the gravitational field. Figure 11.22 shows an example of a Feynman diagram with graviton exchange. Since the gravitational interaction is very weak—it is the weakest of all the interactions—detection of individual gravitons is far beyond the sensitivity of our instruments; we can only detect the cooperative effects of the exchange of a large number of gravitons. However, we can deduce the properties of the gravitons from the known

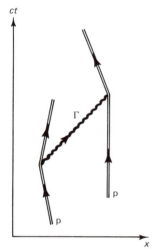

Fig. 11.22 Exchange of a virtual graviton between two protons, leading to the gravitational attraction between the protons.

TABLE 11.5 The Four Fundamental Interactions

Interaction	Strength	Range	Carrier
Gravitational	10^{-38}	∞	Graviton
Weak	10^{-6}	10^{-18} m	$W^{\pm}$ and Z° bosons
Electromagnetic	10^{-2}	∞	Photon
Strong	1	10^{-15} m	Gluons

properties of the gravitational force. Gravitons have zero mass and spin 2.

The carriers of the weak interaction are the charged **W$^+$** and **W$^-$ bosons,** and the neutral **Z^0 boson.** The masses of the W$^+$ and the W$^-$ particles are 81,800 MeV/c^2; the mass of the Z^0 particle is 92,600 MeV/c^2. These enormous masses account for the extremely short range of the weak force [see Eq. (3)]. The spins of the W$^\pm$ and Z^0 are 1. Figure 11.23 shows some examples of Feynman diagrams involving the exchange of W$^\pm$ and Z^0 bosons in diverse weak processes.

As already discussed in Section 11.1, the carriers of the electromagnetic interaction are the photons, of mass zero and spin 1.

And the carriers of the strong interaction between protons and neutrons are the pions and some other mesons. However, as we will see in Section 11.6, protons, neutrons, and pions are composite bodies made out of quarks. Thus, at a more fundamental level, we should consider the strong force between quarks, rather than that between neutrons or protons. The carriers of this more fundamental strong force are the **gluons,** of mass zero and spin 1. They have not been detected directly, since they are permanently hidden from view within the hadrons.

All the carriers of interactions have integer spins. Because of their role as intermediaries of forces, they are called **intermediate bosons.** Table 11.6 lists the intermediate bosons.

The W$^\pm$ and Z^0 intermediate bosons were discovered only recently, in experiments with colliding proton and antiproton beams at CERN in 1984. They are the heaviest particles ever detected—their masses are more than 80 times the mass of a proton. The discovery of the W$^\pm$ and Z^0 bosons was a spectacular confirmation of the unified theory of electromagnetic and weak interactions, or **electroweak theory,** formulated in 1968 by S. Weinberg, A. Salam, and S. Glashow.* This theory achieved the unification of the electromagnetic and weak interactions, treating them as aspects of one single underlying interaction. This unification may be regarded as analogous to that achieved by Maxwell and Einstein in classical electromagnetism: Maxwell formulated his field equations and thereby established the intimate relationship between the electric

---

* **Steven Weinberg,** 1933– , American physicist, professor at Harvard and at the University of Texas, **Abdus Salam,** 1926– , Pakistani physicist, director of the International Center for Theoretical Physics at Trieste, and **Sheldon Lee Glashow,** 1932– , American physicist, professor at Harvard, shared the Nobel Prize in 1979 for their electroweak theory.

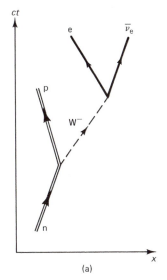

(a)

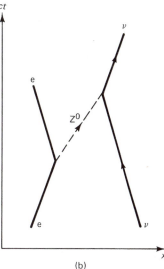

(b)

Fig. 11.23 (a) Emission of a virtual W$^-$ boson by a neutron. In this process the neutron transmutes into a proton, and the W$^-$ boson subsequently creates an electron and an antineutrino. The net result is the reaction n → p + e + $\bar{\nu}_e$, i.e., the β decay of the neutron. (b) Exchange of a Z^0 boson between an electron and a neutrino, leading to electron-neutrino scattering.

TABLE 11.6   The Intermediate Bosons

Boson	Mass	Spin	Electric Charge	Mean Life or ΔE	Principal Decay Modes	
Graviton	0	2	0	stable	—	
W^+, W^-	81,800 MeV/c^2	1	1, −1	<7 GeV	$e\nu$	seen
Z^0	92,600	1	0	<5 GeV	e^+e^-	seen
					$\mu^+\mu^-$	seen
Photon	0	1	0	stable	—	
Gluon	0	1	0	(does not exist as free particle)		

and magnetic fields; and Einstein discovered the Lorentz-transformation properties of the fields and thereby showed that the electric and magnetic fields are components of a single six-component field (the "field tensor"), components which transform into one another under Lorentz transformations.

Analogously, the electroweak unification hinges on treating the photon and the $W^\pm$ and Z^0 bosons as different components of a single underlying multicomponent quantum field. The photon and the $W^\pm$ and Z^0 bosons are related by rotations in a fictitious, abstract space; these rotations transform the different field components into one another. The field equations are symmetric with respect to these rotations, and consequently we would expect that the photon and the $W^\pm$ and Z^0 bosons have the same mass. But in the electroweak theory, the $W^\pm$ and Z^0 bosons acquire a large mass by a cunning contrivance: the scheme of **spontaneously broken symmetry.** The essential feature of this scheme is that the *solutions* of the field equations are contrived so as to have less symmetry than the equations themselves. As a simple example of a classical system in which there is such a spontaneous breakdown of symmetry, consider a pencil standing vertically on its tip on a table. The equation of motion and the initial configuration of the system are rotationally symmetric, i.e., both are invariant under rotations around the vertical axis. However, the equilibrium is unstable, and when the pencil topples over and settles into a stable horizontal configuration flat on the table, it breaks the rotational symmetry by selecting a definite direction of fall. Note that this spontaneous breakdown of symmetry results from the energy difference between the initial symmetric configuration and the final asymmetric configuration—the symmetric configuration has high energy, and the asymmetric configuration has low energy. The electroweak theory takes advantage of a similar energy difference to trigger a breakdown of symmetry—the field energy is contrived in such a way that the initial symmetric configuration with equal (zero) masses for the photon and the $W^\pm$ and Z^0 bosons is unstable, whereas the final asymmetric configuration with unequal masses is stable. The fields spontaneously settle into the latter configuration, with large masses for the $W^\pm$ and Z^0 bosons. Thus, the electroweak theory tells us that the photon and the $W^\pm$ and Z^0 bosons are closely related, but their relationship is hidden from view. The relationship can be brought into the open by restoring the symmetry. This can be done by giving these particles very high energies, in excess of 100 GeV; at such energies, the particles would attain their symmetric, unstable equilibrium configuration. The symmetry between photons and $W^\pm$ and Z^0 bosons would then become explicit, and all these particles would behave in the same way. The high energies required for this restoration of explicit symmetry are not available in laboratories on the Earth, but they were available during the early stages (Big Bang) of the evolution of the universe— at that time the symmetry was unbroken.

The large masses of the $W^\pm$ and Z^0 bosons not only explain the short range of the weak interaction, but also its weakness: the electromagnetic and the weak interactions intrinsically have the same strength, but the latter appear weaker because the interacting particles find it much more difficult to emit a very massive $W^\pm$ or Z^0 boson than a massless photon. The electroweak theory predicted the existence of the intermediate bosons, and it predicted their masses: $83,000 \text{ MeV}/c^2$ and $93,000 \text{ MeV}/c^2$, in excellent agreement with the experimental results (see Table 11.6).

The $W^\pm$ and Z^0 bosons discovered at CERN were produced by the reaction

$$p + \bar{p} \rightarrow W^\pm \text{ or } Z^0 + \text{(other particles)}$$

In order to attain the center of mass energy required for this reaction, C. Rubbia* and his associates modified the CERN accelerator, converting it into a proton–antiproton collider, in which protons and antiprotons of energy 270 GeV each travel around the accelerator ring in opposite directions and engage in head-on collisions. The production of a sufficient number of antiprotons posed a challenging technical obstacle in this experiment. These antiprotons are manufactured in preliminary collisions of protons in a smaller auxiliary accelerator. The antiprotons are then stored in a circular

---

* **Carlo Rubbia,** 1934–   , Italian physicist at CERN and professor at Harvard, and **Simon van der Meer,** 1925–   , Dutch physicist at CERN, shared the Nobel Prize in 1984 for the development of the proton–antiproton collider and the discovery of the $W^\pm$ and Z^0 particles.

orbit in an accumulator ring. Upon production, the antiprotons have appreciable random momentum components, and they tend to leak out of the ring. This problem was solved by a clever correction system invented by S. van der Meer. Sensors along the ring detect the deviations in energy and momentum of the antiprotons at a point of their circular orbit, and immediately send signals to the diametrically opposite point of the orbit where an adjustable electric field applies corrective impulses to these antiprotons when they arrive; since the antiprotons move at a speed close to the speed of light, this correction system requires split-hair timing. In the experiments, bunches of about 10^{11} antiprotons—the greatest concentrations of antimatter ever manufactured on Earth—were accumulated over intervals of several days, and then suddenly dumped into the main accelerator ring, into an orbit of opposite direction to that of the protons in this ring. The $W^{\pm}$ and Z^0 bosons created in collisions were identified by their decay products: $\bar{e} + \nu$ or $e + \bar{\nu}$ for the charged bosons, and $e + \bar{e}$ for the neutral boson.

Of late, theoreticians have made attempts at a **grand unified theory,** or **GUT,** of the electromagnetic, weak, and strong interactions. In this unified theory, the photon, the $W^{\pm}$ and Z^0 bosons, and the carriers of the strong interaction are all treated as components of one single underlying multicomponent field. One remarkable prediction of the grand unified theory is that protons are unstable; they decay into leptons with a lifetime of some 10^{30} or 10^{33} years (there are different versions of GUTs, with somewhat different predictions for the value of the lifetime of the protons). Several experiments that attempt to detect the decay of the proton are in progress; preliminary results indicate that the lifetime is longer than 10^{32} years. Another remarkable consequence of grand unified theories is that they provide us with an explanation of charge conservation. It turns out that if the electromagnetic interaction is an aspect of the strong interaction (and vice versa), then the charges of all the hadrons must be exactly integer multiples of the charges of the leptons. Thus, these theories finally give us an explanation of why the charge of the proton is exactly equal in magnitude to the charge of the electron.

11.5 Conserved Quantities

The reactions among particles obey several conservation laws. Some of these conservation laws are already familiar from classical physics, and they are based on general theoretical principles. Some other conservation laws are new, purely empirical rules, without any obvious theoretical justification. Such empirical conservation laws were proposed by particle physicists who sought to account for the puzzling *absence* of diverse hypothetical reactions. In general, we expect that any reaction not forbidden by conservation laws will occur, although perhaps at a low rate.* This expectation is based on the tunneling phenomenon of quantum mechanics, which ensures that the transition from any given initial state to any given final state is bound to occur sooner or later, unless this transition is incompatible with a conservation law. Thus, the absence of certain kinds of reactions, otherwise not forbidden by any known conservation law, suggests the intervention of a new conservation law involving some new conserved quantity. However, when we formulate a new conservation law on such purely empirical grounds, we cannot claim to have achieved an explanation of why the reactions in question do not occur—we have merely achieved a concise codification of the observed facts. Although such a codification has no explanatory value, it has predictive value, since it empowers us to make predictions as to which reactions are possible. Furthermore, the empirical conservation laws provide us with valuable guidance in the construction of theories of interactions.

Some of the conservation laws are absolute—they are obeyed by all reactions under all circumstances. Others are approximate—they are obeyed by some reactions, but not by all. It might appear that an approximate conservation law—obeyed sometimes, and violated at other times—would be quite useless. But, as we will see, the violations of the approximate conservation laws do not happen at random. These violations are directly correlated with the kind of interaction involved in the reaction, and they follow an orderly, predictable pattern.

The absolutely conserved quantities are energy, momentum, angular momentum, electric charge, lepton number, and baryon number. The approximately conserved quantities are isospin, strangeness, parity, C parity, and G parity. The first four of these quantities—energy, momentum, angular momentum, and electric charge—are familiar; their conservation is a consequence of the basic laws of classical or quantum mechanics and electromagnetism. The other quantities are less familiar; we will discuss them one by one.

---

* This brings to mind the words of T. H. White in *The Once and Future King:* "Everything not forbidden is compulsory."

Lepton Number. There are three separate conservation laws for lepton number, corresponding to the three separate varieties of leptons: e, μ, and τ. The conservation law for electron–lepton number simply states that the net number of electron-type leptons remains constant in any reaction. Antileptons are reckoned as making a negative contribution to the net lepton number. In essence, this means that whenever a lepton is created or destroyed in a reaction, a corresponding antilepton must be created or destroyed. For example, consider the decay of the neutron,

$$n \rightarrow p + e + \bar{\nu}_e \tag{32}$$

The net electron–lepton number before the reaction is zero, and the net electron–lepton number after the reaction is $1 + (-1)$, still zero. Thus, the reaction conserves electron–lepton number.

The conservation laws for muon–lepton number and for tauon–lepton number are formulated similarly. Since the conservation laws for lepton numbers are absolute, any reaction that violates these laws is absolutely forbidden. For instance, the hypothetical reaction

$$\mu \rightarrow e + \bar{\nu}_e \tag{33}$$

is absolutely forbidden—it violates the conservation law for muon–lepton number. This reaction has never been observed. The conservation of lepton numbers has no obvious theoretical foundation, but is supported by abundant empirical evidence.

Baryon Number. The conservation law for baryon number states that the net number of baryons remains constant. Antibaryons are reckoned as making a negative contribution to the net baryon number. (Note that in contrast to the three varieties of lepton number, there is only one variety of baryon number, and only one conservation law.) Like the conservation of lepton number, the conservation of baryon number is an empirical law, without theoretical foundation.

The observed stability of the proton can be regarded as a corollary of the conservation laws for energy and for baryon number. Since the proton is the lightest of all baryons, energy conservation requires that the hypothetical decay products be nonbaryons, and the decay would violate baryon number conservation. As remarked in the preceding section, GUT theories suggest that the proton is in fact unstable. But since the experiments that have sought to detect proton decay have not met with any success, all the available empirical evidence is (so far) consistent with baryon number conservation. If and when any proton decays are detected, the law of baryon number conservation will have to be demoted from an absolute conservation law to an approximate law.

Isospin. A complete mathematical definition of the remaining conserved quantities—isospin, strangeness, and parity—falls beyond the limits imposed on this book; we will merely present a brief descriptive outline of each. The concept of **isospin** is rooted in the charge independence of the strong interaction. As we know from Chapter 9, the strong interaction does not distinguish between protons and neutrons: it treats protons and neutrons identically, as nucleons. Formally, we can think of the proton and the neutron as two states of the nucleon, and we can label these two states with a new quantum number, the isospin, which is mathematically analogous to ordinary spin, but has nothing whatsoever to do with angular momentum. The isospin vector is designated by the letter **T.** This is not a vector in ordinary three-dimensional space, but rather a vector in a fictitious, abstract three-dimensional space. The magnitude and the third component (T_z, or T_3) of the isospin vector obey the usual quantization conditions familiar from our study of angular momentum. The isospin quantum number for the nucleon is $T = \frac{1}{2}$, and the magnitude of the isospin vector is

$$\sqrt{T(T + 1)} = \sqrt{\tfrac{1}{2}(\tfrac{1}{2} + 1)} = \sqrt{\tfrac{3}{4}}$$

The proton and the neutron correspond to different orientations of this isospin vector: the proton corresponds to isospin "up" ($T_3 = +\frac{1}{2}$), and the neutron to isospin "down" ($T_3 = -\frac{1}{2}$). The symmetry of protons and neutrons with regard to the strong force can then be thought of as a rotational symmetry in isospin space— if we rotate a nucleon from spin "up" to spin "down," we do not change the interaction energy of the nucleon. This isospin symmetry of the strong interactions was first formulated by Heisenberg, who recognized that mirror nuclei and some other isobaric nuclei were related by isospin rotations.

The proton and the neutron are said to form an isospin doublet. Other baryons and mesons can likewise be grouped into **isospin multiplets.** For instance, Fig. 11.24 shows the nucleon doublet, the Λ singlet, the Σ triplet, and the Ξ doublet in an energy-level diagram. Note that the energy intervals between the multiplets are much larger than those within the multiplets. Furthermore, within each multiplet, the energy increases with decreasing electric charge. Note that in Tables 11.3 and 11.4, the particles have been grouped into isospin multiplets.

At first sight, this isospin formalism looks like nothing but a fancy mathematical scheme for labeling particles. (We might with

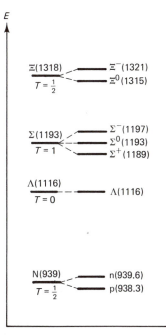

Fig. 11.24 Isospin multiplets: the nucleon doublet, the Λ singlet, the Σ triplet, and the Ξ doublet.

equal justification, or lack thereof, label the members of a soccer team as a multiplet with $T = 5$, and with T_3 ranging from -5 to 5.) But the mathematical complication pays off when we examine the reactions that proceed via the strong interaction: it turns out that all such reactions conserve isospin, i.e., they conserve T and T_3. This implies that certain reactions are forbidden to proceed via the strong interaction (although they may proceed via some other interaction). For instance, if deuterons are bombarded with deuterons, the reaction

$$^2\text{H} + {}^2\text{H} \rightarrow {}^4\text{He} + \pi^0 \tag{34}$$

is forbidden. In this reaction, the deutron and the alpha particle have isospin $T = 0$, whereas the pion has isospin $T = 1$; thus, the isospin is not conserved. Experimentally, this reaction has never been observed.

The conservation law for isospin is approximate because, in contrast to the strong interaction, the electromagnetic and weak interactions violate this conservation law. Thus, any reaction creating (or destroying) photons and any reaction creating (or destroying) leptons violates isospin conservation, and it is therefore meaningless to assign isospin quantum numbers to photons and leptons. For electromagnetic reactions, the violation of isospin conservation is partial: T is not conserved, but T_3 is conserved. For weak reactions the violation is total: T and T_3 are not conserved. Incidentally: the small mass differences between the members of isospin multiplets (see Fig. 11.24) are due to the electromagnetic interaction. This interaction distinguishes between, say, the proton and the neutron by their charge and gives them different masses. Since this destroys the isospin symmetry between the proton and the neutron, the isospin symmetry is said to be **broken** by the electromagnetic interaction.

Strangeness and Hypercharge. The conservation law for strangeness was introduced to account for an apparent inconsistency in the behavior of some particles, such as the Λ and Σ baryons, and the K mesons. These came to be called **strange** particles because they are produced at a fast rate in high-energy collisions, but they decay at a slow rate, i.e., they have exceptionally long lifetimes. This suggested that these particles are endowed with a new conserved quantity—**strangeness.** The strong and electromagnetic interactions obey the conservation law for strangeness, but the weak interaction does not. The production reaction of strange particles always involves the creation of *two* such particles of opposite strangeness ("associated production"); it therefore obeys strangeness conservation and can proceed via the strong interaction. The decay reaction involves the destruction of a single strange particle; it therefore violates strangeness conservation and must proceed via the weak interaction. Thus, the decay reaction proceeds slowly.

The strangeness can be expressed in terms of the electric charge, the baryon number, and the third component of isospin:

$$S = 2(Q - T_3) - B$$

where the electric charge is measured in units of e. Since Q and B are absolutely conserved, we can regard the conservation of strangeness as equivalent to the conservation of the third component of isospin. Thus, the conservation of strangeness is not a new, independent conservation law, but a consequence of other conservation laws.

Tables 11.3 and 11.4 include assignments of strangeness numbers for all the baryons and mesons; note that in this table the particles are listed in groups according to their strangeness. For example, the proton and the neutron have $S = 0$, the Λ baryons have $S = -1$, the pions have $S = 0$, the K mesons have $S = \pm 1$, etc. Thus, in the production reaction

$$\pi^- + p \rightarrow K^0 + \Lambda \tag{35}$$

the net strangeness is 0 before the reaction, and 0 after the reaction. In the decay reaction

$$\Lambda \rightarrow p + \pi^- \tag{36}$$

the strangeness is -1 before the reaction, and 0 after. This violation of strangeness conservation prevents the reaction from proceeding via the strong interaction, and gives the Λ its relatively long lifetime, about 10^{-10} s.

In modern terminology, the strangeness number S has been largely superseded by the **hypercharge** number, which equals the strangeness plus the baryon number, $Y = S + B$. Conservation of hypercharge is equivalent to conservation of strangeness.

Parity. The concept of parity arises when we examine the behavior of a wavefunction under the operation of inversion of coordinates, i.e., the replacement of x by $-x$, y by $-y$, and z by $-z$. If the function remains unchanged under this operation, it is said to have even parity, or $P = +1$; if it changes sign, it has odd parity, or $P = -1$. The wavefunctions for spherically symmetric potentials, such as the Coulomb potential, have one parity or another. For instance, the wavefunctions for the electron in a hydrogen atom have parity $P = (-1)^l$. This parity of the wave-function of a particle may be regarded as the orbital parity; besides this, some particles also have an **intrinsic parity.** The dichotomy of orbital and intrinsic parity is analogous to that of orbital angular momentum and intrinsic angular momentum. The proton has intrinsic parity $+1$, the pion -1, the kaon -1, the lambda -1, etc. Parity is a multiplicative, rather than an additive, quantum number: the net parity of a system of particles is the product of all the orbital and intrinsic parities. The strong and electromagnetic interactions conserve parity, but the weak interaction does not. Since the reactions that create leptons are weak, the leptons do not have intrinsic parities. Thus, the electron in the hydrogen atom has a well-defined orbital parity, but no intrinsic parity.

The first indications of a violation of parity conservation by the weak interaction were seen in the puzzling decay of the K^+, which was found to decay sometimes into states of positive parity, sometimes into states of negative parity. This puzzle led T. D. Lee and C. N. Yang* to make a critical examination of the validity of parity conservation. They concluded that there was unequivocal evidence for parity conservation in the strong and electromagnetic interactions, but no evidence whatsoever in the weak interactions. They proposed a direct experimental test of parity conservation in the β decay of a nucleus. Consider a nucleus whose spin is aligned with the z-axis; for such a nucleus, the parity operation is equivalent to a reversal of the z-axis. If the nucleus decays, the initial and the final parities of the nucleus are well defined, i.e., the parities are either positive or negative. Hence parity conservation would require that the wavefunction of the emitted electron also has a well-defined parity, i.e., the wavefunction has to be either symmetric or antisymmetric under reversal of the z-axis. In either case the probability distribution for the emitted electron would be symmetric, and therefore a sample of nuclei with aligned spins would emit equal numbers of electrons parallel to the nuclear spin direction and antiparallel to this direction. The experiment was first performed by C. S. Wu[†] et al., using a sample of ^{60}Co nuclei cooled to 0.01 K and placed in a strong magnetic field, so as to achieve alignment of the nuclear spins. The β rays emitted from

---

* **Tsung Dao Lee,** 1926– , **Chen Ning Yang,** 1922– , American physicists, professors at Columbia University and at Stony Brook, respectively. For their theoretical work on parity conservation, they received the Nobel Prize in 1957.

† **Chien Shiung Wu,** 1913– , Chinese and later American physicist, professor at Columbia.

this sample were found to emerge preferentially in the direction opposite to that of the nuclear spin. This asymmetry signaled the overthrow of parity.

The nonconservation of parity implies that nature makes an absolute distinction between right-handed electrons and left-handed electrons, that is to say, electrons with spin parallel to their momentum and electrons with spin antiparallel to their momentum. In the decay of ^{60}Co, the spin of the residual nucleus is one unit smaller than that of the original nucleus; hence the emitted electron and the neutrino must carry away this spin difference. Since the electron emerges in a direction opposite to that of the nuclear spin (see Fig. 11.25), the electron has its spin antiparallel to its momentum, i.e., this electron is left-handed. As a general rule, the leptons emitted in decays governed by the weak interaction are preferentially left-handed, and the antileptons, right-handed (for neutrinos and antineutrinos this preference is total; the emitted neutrinos are *always* left-handed, and the antineutrinos are *always* right-handed).

C parity and G Parity. These two conserved quantities are multiplicative quantities, like ordinary (spatial) party. Both are applicable only to mesons.* C parity refers to the behavior under the transformation of particles into antiparticles, and G parity refers to the behavior under this same transformation compounded with a rotation by 180° in isospin space. The strong interaction conserves C and G. The electromagnetic interaction conserves C, but not G. The weak interaction conserves neither C nor G.

When the failure of parity conservation was discovered, it was at first thought that the product CP would be conserved by the weak interaction, even though the individual factors C and P are not. The decay of the K_L^0 meson provides a very sensitive test of the conservation of CP: it can be shown that if CP is conserved, this meson can decay into three pions, but not into two. An experimental search for the two-pion decay was undertaken by J. W. Cronin and V. L. Fitch,[†] and they found that about 0.3% of the K_L^0 mesons decay into two pions. This established that the weak interaction does not conserve CP, although the violation is rather rare.

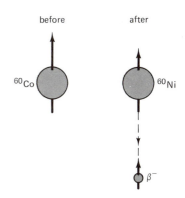

before   after

^{60}Co ^{60}Ni

β^-

Fig. 11.25 Emission of the electron in the β decay of ^{60}Co. The heavy arrows indicate the directions of the spins. The electron is emitted preferentially in a direction opposite to that of the nuclear spin.

---

* If special restrictive conditions are met, then C and G parity are also applicable to suitable systems of several baryons. But they are undefined for individual baryons.

† **James Watson Cronin,** 1931– , and **Val L. Fitch,** 1923– , American physicists, professors at Princeton. They received the Nobel Prize in 1980 for their discovery of the violation of CP symmetry in the decay of the K^0 mesons.

TABLE 11.7   Interactions and Conservation Laws

Interaction	Approximately Conserved Quantity					
	Isospin	Hypercharge	P Parity	C Parity	G Parity	CP Parity
Strong	yes	yes	yes	yes	yes	yes
Electromagnetic	no	yes	yes	yes	no	yes
Weak	no	no	no	no	no	no[a]

[a] The weak interaction violates CP conservation only in the decay of the K_L^0 meson, and in some other rare instances.

The violation of CP conservation carries an interesting consequence for time-reversal invariance of reactions. In essence, time-reversal invariance means that reactions are equally capable of proceeding forward and backward. A general theorem—the CPT theorem—of quantum field theory asserts that any violation of CP conservation must be compensated by a violation of time-reversal invariance. Thus the experimental evidence for the violation of CP conservation in the decay of the K_L^0 meson implies that nature is not time symmetric. We are, of course, accustomed to time asymmetry in the behavior of macroscopic systems, in which the second law of thermodynamics holds sway. But time asymmetry in the behavior of microscopic systems came as an unhappy surprise, and the consequences are not yet fully understood.

Table 11.7 lists the approximate conservation laws obeyed by the strong, electromagnetic, and weak interactions (the gravitational interaction has been left out of this table, since it plays no role in the reactions of interest to particle physicists). This table can help us to determine what interaction drives a given reaction. The reaction will proceed via the strongest interaction compatible with the conservation laws.

Example 2   Consider the decay of the N(1535) resonance, the decay of the π^0 meson, and the decay of the Σ^+ baryon, according to the reactions

$$N(1535)^+ \rightarrow p + \eta \tag{37}$$

$$\pi^0 \rightarrow \gamma + \gamma \tag{38}$$

$$\Sigma^+ \rightarrow p + \pi^0 \tag{39}$$

What interaction is involved in each of these reactions?

Solution   The reaction (37) involves strongly interacting particles only, and it conserves isospin and strangeness [the N(1535) resonance has isospin $T = \frac{1}{2}$, and the η meson has $T = 0$]. Thus, this reaction is consistent with the strong interaction. It will therefore proceed via the strong interaction, regardless of the other two interactions, because the former dominates over the latter two.

Since the reaction (38) involves photons, we can immediately conclude that it proceeds via the electromagnetic interaction—we need not even bother with an examination of the conservation laws.

The reaction (39) violates the strangeness conservation law. Hence it is consistent only with the weak interaction.

The strong interaction gives the highest reaction rates and the shortest lifetimes, whereas the weak interaction gives the lowest reaction rates and the longest lifetimes. Thus, the N(1535) resonance ought to be short lived, and the Σ^+ baryon (relatively) long lived. The observed lifetimes for the decays (37), (38), and (39) are about 10^{-23} s, 0.83×10^{-16} s, and 0.80×10^{-10} s, respectively (these values for lifetimes are fairly typical for decays proceeding via the strong, electromagnetic, and weak interactions). ■

Table 11.7 also tells us something about the symmetries of the strong, electromagnetic, and weak interactions. There is a profound connection between conservation laws obeyed by a physical system and symmetries. This connection is contained in **Noether's theorem:** *to every symmetry there corresponds a conservation law.* In this context, a symmetry of the system is any mathematical operation that leaves the potential energy unchanged. For example, consider a system of particles interacting with each other, and suppose that the system has translational symmetry, i.e., suppose that the potential energy is independent of the position of the system in relation to its surroundings. The derivatives of the potential energy with respect to the coordinates x, y, z describing the overall position of the system are then zero, which implies the external force is zero (e.g., $F_x = -\partial U/\partial x = 0$). Hence the net momentum of the system is constant. This is one instance of Noether's theorem. Likewise, the conservation of angular momentum emerges from rotational symmetry, and the conservation of energy emerges from time-translation symmetry.

The approximate conservation laws listed in Table 11.7 are also associated with symmetry operations: parity is associated with reflection symmetry in ordinary space, isospin with rotation symmetry in isospin space, and C and G parity with different kinds of particle–antiparticle exchange symmetries. Thus, Table 11.7 may be regarded as a chart of the symmetries of the interactions. These symmetries place severe restrictions on these interactions, and thereby provide us with guidance for the construction of theories of interactions. Much of the effort in high-energy theoretical physics is directed toward the search for symmetries and their exploitation.

In the preceding section we saw that particles can be grouped into isospin multiplets whose members are related by isospin symmetry. This classification of particles into isospin families helps to bring some order into the confusing diversity of the more than 300 known particles. We would like to broaden the classification scheme so as to group several isospin multiplets into supermultiplets. But the search for such broader families is hampered by the large mass differences between isospin multiplets, which tends to hide the family relationships. Several classification schemes were proposed in the 1960s, the most successful of which turned out to be the "**Eightfold Way**" of M. Gell-Mann and Y. Ne'eman.* This scheme groups the isospin multiplets into **supermultiplets** whose members are related by a mathematical symmetry called SU(3), which combines isospin symmetry and hypercharge symmetry. The particles in a given supermultiplet differ in isospin and in hypercharge, but all their other quantum numbers—spin, baryon number, parity— are identical.

Figure 11.26 shows the members of the spin-$\frac{1}{2}$ baryon supermultiplet, or the **baryon octet**, on an energy-level diagram. If the SU(3) symmetry were perfect, then the masses of all the members of the supermultiplet would be identical, as indicated by the energy level shown on the far left of Fig. 11.26. The energy differences between the members of this supermultiplet are blamed on a breaking of the SU(3) symmetry. The symmetry is broken because the strong interaction includes a portion ("semistrong interaction") that distinguishes between particles of different hypercharge. We can also display the supermultiplet on a plot of hypercharge vs. isospin, as in Fig. 11.27. The hexagonal pattern of this plot is a characteristic feature of the SU(3) symmetry.

Figures 11.28 and 11.29 display two other supermultiplets of the Eightfold Way: the **baryon decuplet** and the **meson octet**. The Eightfold Way accommodates all known baryons and mesons into supermultiplets of the kinds shown in the figures. However, some of the supermultiplets are incomplete, i.e., some positions in the supermultiplet are left open. Presumably this means that some supermultiplet members remain yet to be discovered. Hence the

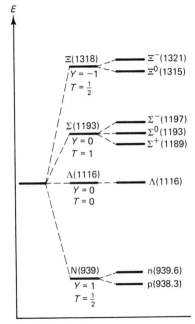

Fig. 11.26 The baryon octet.

E

$\Xi(1318)$ —— $\Xi^-(1321)$
$Y = -1$ —— $\Xi^0(1315)$
$T = \frac{1}{2}$

$\Sigma(1193)$ —— $\Sigma^-(1197)$
$Y = 0$ —— $\Sigma^0(1193)$
$T = 1$ —— $\Sigma^+(1189)$

$\Lambda(1116)$ —— $\Lambda(1116)$
$Y = 0$
$T = 0$

$N(939)$ —— $n(939.6)$
$Y = 1$ —— $p(938.3)$
$T = \frac{1}{2}$

* **Murray Gell-Mann,** 1929– , American theoretical physicist, professor at the California Institute of Technology; for his work on the classification of particles, he received the Nobel Price in 1969. **Yuval Ne'eman,** 1925– , Israeli physicist, professor at Tel-Aviv.

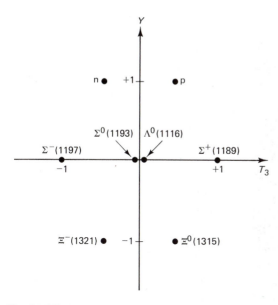

Fig. 11.27 Hypercharge vs. isospin for the baryon octet.

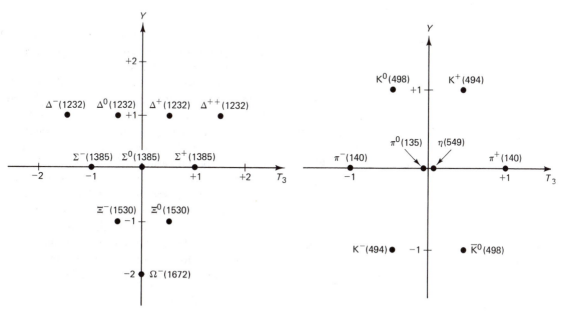

Fig. 11.28 Hypercharge vs. isospin for the baryon decuplet.

Fig. 11.29 Hypercharge vs. isospin for the meson octet.

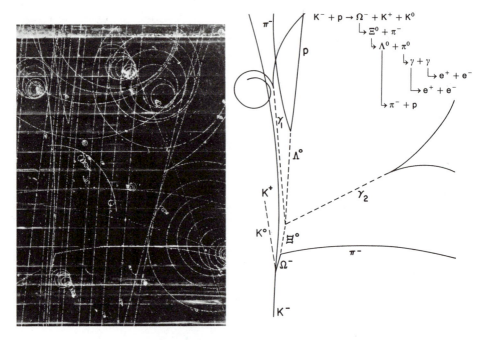

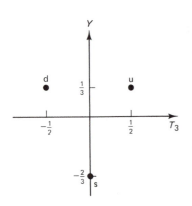

Fig. 11.30 Bubble-chamber picture showing track of Ω^- particle. (Courtesy Brookhaven National Laboratory.)

Fig. 11.31 Hypercharge vs. isospin for the quark triplet.

Eightfold Way predicts the existence of new particles. One remarkably successful instance of such a prediction was the case of the omega particle. When theoreticians first identified the baryon decuplet, the position in the lowest corner of Fig. 11.28 was still open—the omega had not yet been discovered. The Eightfold Way also predicted the mass of this particle, about $1675 \, \text{MeV}/c^2$. Experimenters saw this as a challenge and immediately began to search for the missing particle. Within a few months, experimenters at Brookhaven National Laboratory found it. Figure 11.30 is a bubble-chamber photograph of a track made by the first omega ever detected. Its mass agreed precisely with the predicted value.

The smallest nontrivial supermultiplet of the Eightfold Way is a triplet consisting of one isospin doublet and one isospin singlet (see Fig. 11.31).* None of the elementary particles in our list of over 300 particles appears to belong to this triplet. From a study of the SU(3) symmetry it can be shown that all the larger super-

---

* The only supermultiplet smaller than the triplet is a singlet consisting of just one particle. This is a trivial multiplet, of no interest in the present context.

TABLE 11.8   The Quark Triplet

Quark	Mass	Spin	Electric Charge	Baryon Number	Hypercharge
u	5 MeV/c^2	1/2	2/3	1/3	1/3
d	9	1/2	$-1/3$	1/3	1/3
s	180	1/2	$-1/3$	1/3	$-2/3$

multiplets can be regarded as combinations of such triplets. This led M. Gell-Mann and G. Zweig to suggest that all the known particles are composite systems made of two or three truly elementary particles belonging to a fundamental triplet. The members of the fundamental triplet are called **quarks,** a word borrowed by Gell-Mann from an obscure line in a book by J. Joyce. The three quarks are designated by the symbols u, d, and s, meaning **up, down,** and **strange.** All the quarks have spin $\frac{1}{2}$ and they have fractional electric charges: the u, d, and s quarks have charges of $2e/3$, $-e/3$, and $-e/3$, respectively (see Table 11.8). Each quark, like any other particle, has an antiparticle. These antiparticles belong to a separate supermultiplet, the antitriplet with the antiquarks $\bar{u}$, $\bar{d}$, and $\bar{s}$ (see Fig. 11.32).

We can construct all the known particles out of quarks, by joining quarks together in diverse ways. For example, a proton is made of two u quarks and one d quark (see Fig. 11.33). A neutron is made of two d quarks and one u quark (see Fig. 11.34). A positive pion is made of one u quark and one d antiquark (see Fig. 11.35), etc. When constructing a particle out of quarks, we must also specify the quantum state of the quarks. For the supermultiplets of lowest energy and spin, the quarks will be in the ground state; for the multiplets of higher energy and spin, the quarks will be in excited states. For instance, for the spin-$\frac{1}{2}$ baryon octet, the quarks are in the ground state; whereas for the spin-$\frac{3}{2}$ baryon octet, the quarks are in an excited state. The extra energy and extra orbital angular momentum of the quarks account for

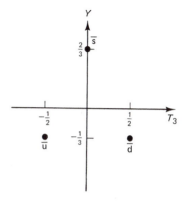

Fig. 11.32 Hypercharge vs. isospin for the antiquark triplet.

Fig. 11.33 Quark structure of the proton. The heavy arrows indicate the spins.

Fig. 11.34 Quark structure of the neutron.

Fig. 11.35 Quark structure of the positive pion.

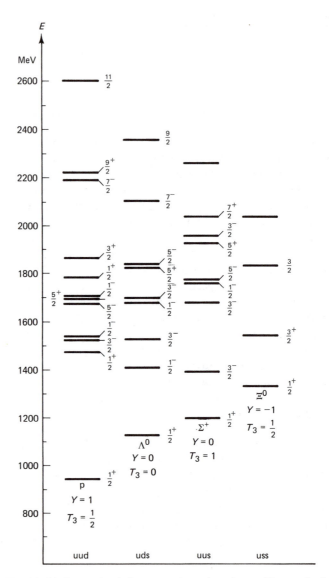

Fig. 11.36 Energy-level diagrams of quark systems. The quarks in each system are listed below the diagram.

the difference between these two octets. Figure 11.36 shows the energy-level diagrams for some of the different quark systems involved in the baryon octet. Note that each of these diagrams is a genuine energy-level diagram for a given system of quarks, in contrast to the purely formal diagram of Fig. 11.26.

The family relationships among the particles belonging to a supermultiplet of the Eightfold Way emerge from this construction as a direct consequence of the quark content of the particles. Furthermore, from the quark structure of the particles we can calculate the mass differences between adjacent isospin multiplets, and we can calculate the magnetic moments of the particles, and also some of their reaction cross sections. The good agreement of such calculations with the available experimental data lends support to the quark model.

Unfortunately, there is no direct experimental evidence for the existence of quarks—all searches for free quarks have been unsuccessful. Experimenters have searched for quarks in the debris of high-energy collisions of particles from accelerators and in the debris of the even higher-energy collisions initiated by cosmic rays. They have searched for quarks in samples of materials collected on the Earth and also on the Moon. In all such searches, the distinctive feature that would have permitted the unambiguous identification of a quark is its fractional charge, of magnitude $e/3$ or $2e/3$. But only in a few exceptional instances did the searches find a possible indication of a fractional charge, and in these exceptional instances the evidence was not sufficiently firm for a positive identification of a quark.

This seems to suggest that quarks are permanently confined inside the ordinary particles, so that there is no way to break them out of, say, a proton. According to recent theoretical schemes, the quarks are held in place by a strong force (the "color" force; see the next section), which prevents their escape because it does *not* decrease with distance. In the absence of free, individual quarks, we must use other means to prove the existence of quarks. If we cannot use brute-force collision experiments to fragment protons into their individual building blocks, we can use somewhat more subtle collision experiments to obtain evidence that elementary building blocks do indeed exist inside protons: we can use high-energy electrons to probe the internal structure of the proton, just as Rutherford used alpha particles to probe the internal structure of the atom. The probing of a target with such high-energy projectiles is called **deep inelastic scattering,** because the violence of the collisions usually leads to the creation of new particles, with a corresponding energy loss of the projectile. Deep inelastic scattering of electron projectiles on proton and neutron targets was studied intensively at the Stanford Linear Accelerator around 1970. The maximum available electron momentum was $22\,\text{GeV}/c$, giving a de Broglie wavelength of

$$\lambda = \frac{h}{p} = \frac{6.6 \times 10^{-34}\,\text{J} \cdot \text{s}}{22 \times 10^{9} \times 1.6 \times 10^{-19}\,\text{J}/(3.0 \times 10^{8}\,\text{m/s})}$$

$$= 5.7 \times 10^{-17}\,\text{m}$$

As mentioned in Section 5.1 in connection with the electron microscope, this wavelength determines the smallest detail that an electron beam can resolve. Thus, electrons of $22\,\text{GeV}/c$ can resolve details down to a length scale of a few times $10^{-17}\,\text{m}$.* Since the radius of the proton is about $10^{-15}\,\text{m}$, such electrons are adequate to "feel" the internal structure of the proton.

The Stanford experiments showed that in collisions of fairly low energy, in which the electrons were only probing the average charge distribution, the proton behaved like a ball of charge, with an exponentially decreasing charge density as a function of radius (compare Fig. 9.4). But in collisions of the highest energy, in which the electrons were resolving the details of the protonic structure, the proton behaved like a composite system made of pointlike charged constituents. The deflections suffered by the electrons in these experiments indicated the presence of small, hard kernels in the interior of the proton, just as the deflections of alpha particles in Rutherford's experiments indicated the presence of the nucleus in the interior of the atom. Careful examination of the dependence of the scattering on the energy and deflection angle of the projectiles revealed that these constituents have spin $\frac{1}{2}$. And comparison of the scattering data for proton and neutron targets showed that the cross sections are consistent with the fractional charges that the quark model assigns to the constituents of the proton and neutron. However, the experiments also revealed that only about half of the net momentum of the proton is stored in the quarks; the other half is stored in electrically neutral particles within the proton. These neutral particles are the gluons, or carriers of the color force. They make a substantial contribution to the matter content of a proton or any other hadron.

In contrast to protons and neutrons, electrons and the other leptons seem to be elementary particles without any internal structure. Experiments with colliding electron–antielectron beams at Stanford, and subsequent experiments with even higher-energy colliding beams at the DESY accelerator, established that the electrons as well as the muons and tauons produced in these collisions behave like quantum-mechanical point charges. These experiments probed the leptons down to a length scale of $10^{-18}\,\text{m}$, but found no structure, except for the electromagnetic "structure"

---

* Actually, the electron does not directly probe the proton. It merely serves as source of a virtual photon (see Fig. 11.5), and the latter penetrates the proton and probes its interior. Hence the relevant momentum is not that of the electron, but that of the photon, which equals the momentum *change* of the electron; but we will ignore this complication.

that a point charge is supposed to possess according to quantum electrodynamics.

11.7 **Color and Charm**

The simple quark model with three quarks (and three antiquarks) described in the preceding section suffers from some deficiencies, which became apparent as soon as theoretical physicists began to study the precise arrangement of quarks within the known particles and the forces that hold these quarks together. The quarks have spin $\frac{1}{2}$ and hence should obey the Exclusion Principle. This leads to an immediate difficulty in the case of particles, such as the omega, made of three identical quarks. The omega is made of three s quarks, with parallel spins. These quarks are presumably in the ground state, since the spin-$\frac{3}{2}$ decuplet to which the omega belongs is the decuplet of lowest energy. But the occupation of the same orbital and spin state by three identical particles conflicts with the Exclusion Principle! We can get out of this difficulty by a simple trick: we postulate that these three quarks are not identical, i.e., we postulate that each of the three quarks is a different variety of quark. These varieties differ by a new label which has been designated by the name **"color."** Of course, this color has nothing to do with real color—it is merely a somewhat unimaginative name for a new property of matter. The different quark colors are "red," "green," and "blue." Thus, there is a red u quark, a green u quark, and a blue u quark, etc.

Color is a very elusive property of matter; it usually remains hidden inside the normal particles. All the normal particles are colorless—they consist of several quarks with an equal mixture of all three colors. Nevertheless, color plays a very important role in the theory of the force that holds the quarks together. This force is called the **color force,** because the source of this force is color, just as the source of gravity is mass, and the source of the electric force is electric charge. The theory of the color force is called **quantum chromodynamics,** or **QCD.** This theory is not yet as well understood as quantum electrodynamics, but many of its broad features have already become clear. The color force relies on the usual exchange mechanism involving quanta of the color field, which carry the force from one quark to the other. These carriers are the **gluons,** particles of zero mass and spin 1.

From a fundamental point of view, the strong force between hadrons is nothing but a side effect of the color force, in much the same way as the van der Waals force between molecules is a side effect of the electromagnetic force. The similarity between the mechanisms underlying this strong force and the van der Waals force accounts for the similarity between the variations of these forces with distance, as already mentioned in Section 9.3.

The color force has a very remarkable property. All the fundamental forces listed in Table 11.5 decrease with distance, but the color force remains constant or even increases as the distance between the quarks increases. This persistence of the force brings about the confinement of quarks. If one of the quarks in, say, a proton is somewhat separated from its companions by the violent impact of a collision, the color attraction pulls it back into its original position as soon as the collision ends. However, the color force allows the quarks to move rather freely within certain limits. This situation has been described by the picturesque phrases "infrared slavery" and "ultraviolet freedom"; in this context "infrared" refers to long distances and "ultraviolet" refers to short distances. Crudely, we can think of the color force as acting like a rubber string tying the quark to its companions. The quark can move freely within the limits set by the length of the string, but if the quark moves farther than this length and stretches the string, strong elastic restraining forces come into play.

In an extremely violent collision, a quark may become separated from its companions by such a large distance that the potential energy stored in the color interaction equals the energy required to create a quark-antiquark pair. If such a creation occurs, a meson will emerge from the scene of the collision, but not an isolated quark. In terms of the crude analogy to rubber strings, we can think of the creation as involving the breakage of the string tying the quark to its companions; the new quark and antiquark are created at the site of the break, each attached to one of the two new ends of string. Evidence for the creation of particles by breakage of strings has been observed in high-energy electron-antielectron collisions at the DESY accelerator. When the electron and the antielectron meet, they annihilate, and their energy becomes available for the creation of a quark-antiquark pair joined by a string. However, the quark system created in this way is likely to have a large energy excess, with large quark velocities. Since the momentum of the colliding electron-antielectron pair is zero, the momentum of the quark-antiquark pair must also be zero, i.e., the velocities of the quarks must be collinear, but in opposite directions. The string joining them breaks not only once, but several times, giving birth to a bunch of particles, all of them with more or less

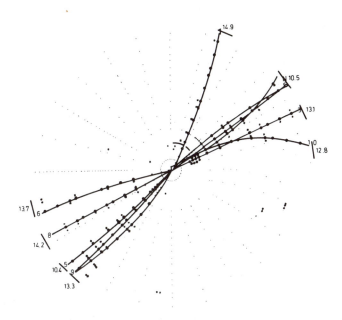

Fig. 11.37 A two-jet event observed in an electron-antielectron collision. Before their collision and annihilation, the electron and the antielectron were moving in opposite directions along a line perpendicular
to the plane of the page. (Courtesy DESY.)

collinear motion along the line of the velocities of the first quark-
antiquark pair. Thus, two oppositely directed "jets" of particles
spurt out from the scene of the collision. Figure 11.37 shows a two-
jet event of this kind observed in an electron–antielectron collision.
Sometimes the collisions have been observed to produce three jets.
The third jet is believed to be due to a gluon radiated by one of
the quarks, by a mechanism analogous to the radiation of a photon
by an electric charge. This gluon quickly converts its energy into
a bunch of hadrons, which are observed as the third jet.

A further modification of the quark model arose from the
electroweak theory described in Section 11.4. When this theory is
combined with the quark model, the rates of the weak decays of
hadrons can be calculated from the weak interactions of the quarks.
Unfortunately, this led to a serious discrepancy in the case of the
decay of the K_L^0, for which such calculations yielded an excessively
large rate of decay into muons, in conflict with the experimental
data. To avoid this conflict, theoreticians found it necessary to
postulate that besides the three quarks u, d, s, there exists a fourth
quark. This fourth quark has been called the **charm** quark, or c.
It differs from the other quarks by a new (approximately) conserved
quantity called charm.

TABLE 11.9 The Six Quarks

Quark	Mass*	Electric Charge	Strangeness	Charm	Bottomness	Topness
u	5 MeV/c^2	2/3	0	0	0	0
d	9	−1/3	0	0	0	0
s	180	−1/3	1	0	0	0
c	1,500	2/3	0	1	0	0
b	4,800	−1/3	0	0	1	0
t	~40,000	2/3	0	0	0	1

* The quark mass depends on the distance (or energy) scale at which it is measured. The values given here are appropriate for short distances.

The hypothesis of the c quark soon received firm experimental support. In 1974 a team of experimenters under the direction of S. Ting at the Brookhaven accelerator and an independent team under the direction of J. B. Richter* at the Stanford accelerator discovered several new particles while investigating high-energy collisions: the J/ψ meson and some other related ψ mesons (see Table 11.4). These mesons consist of a charm quark and a charm antiquark bound together. The heavier of these mesons were found to decay into two D mesons, each consisting of one charm quark or antiquark and one up or down quark.

But the proliferation of quarks did not stop with four quarks. In 1977 a team of experimenters at Fermilab discovered the Υ mesons. These are by far the most massive hadrons known (see Table 11.4). It seems that each of these contains another new kind of very massive quark and antiquark bound together. This fifth kind of quark has been labeled **bottom,** or b; it is endowed with a conserved quantity called bottomness. And there is some recent, tentative experimental evidence for a sixth quark, labeled **top,** or *t*, which is endowed with yet another conserved quantity called topness. Theoretical speculations regarding such a sixth quark had emerged earlier in a comparison of quarks and leptons. As pointed out in Section 11.6, the leptons are believed to be elementary particles, whereas the hadrons are composite particles made out of elementary quarks. This suggests an analogy between the elementary leptons and the equally elementary quarks: since there are six of the former, there ought to be six of the latter (not counting color).

Table 11.9 lists the six varieties of quarks: up, down, strange, charm, bottom, and top. These are sometimes called the varieties

---

* **Samuel Chao Chung Ting,** 1936– , and **Burton Richter,** 1931– , American physicists, professors at the Massachusetts Institute of Technology and at Stanford, respectively, were awarded the Nobel Prize in 1976 for their discovery of the J/Ψ.

of "flavors" of quarks. Each flavor comes in three varieties of color: red, green, and blue. Furthermore, for each flavor of quark, there exists an antiquark, which comes in three varieties of anticolor: antired, antigreen, antiblue. If we count all these quarks, we find there are 36 distinct quarks and antiquarks altogether. But the proliferation of quarks will apparently not stop here—theoreticians have proposed schemes with several dozen flavors of quarks, all with their color varieties and their antiquarks. This excessive proliferation of quarks has led to speculations that perhaps quarks are not, after all, the ultimate building blocks of matter. Some new theoretical schemes try to assemble quarks out of smaller units, variously called prequark, preons, or rishons. Will we ever reach the end of our quest for the ultimate constitutents of matter?

Chapter 11:   SUMMARY

Range of force produced by carrier:

$$\frac{\hbar}{mc}$$

Particles:

$$\left.\begin{array}{l} \text{Leptons} \\ \text{Baryons} \\ \text{Mesons} \\ \text{Intermediate bosons} \end{array}\right\} \text{hadrons}$$

Interactions:

	Strengths:	
Gravitational	10^{-38}
Weak	10^{-6}
Electromagnetic	10^{-2}
Strong	1

Conserved quantities:

Absolute:		Approximate:	
	Energy		Isospin
	Momentum		Strangeness or
	Angular momentum		hypercharge
	Electric charge		P parity
	Baryon number(?)		C parity
	Lepton number		G parity
			CP parity

Quarks:

$$u, d, s, c, b, t$$

Chapter 11: PROBLEMS

1. An electron and an antielectron of unequal momenta collide and annihilate each other. Show that one single emitted photon can not carry away the momentum of both electrons.

2. The $27\frac{1}{2}$-in. cyclotron illustrated in Fig. 11.7 accelerated deuterons to an energy of 5 MeV. What magnetic field did this cyclotron require to keep the deuterons of this energy in an orbit of diameter $27\frac{1}{2}$ in.?

3. In the Fermilab Tevatron, protons of momentum 1000 GeV/c, or 1 TeV/c, move in an orbit of radius 1 km. What is the magnetic field required to hold the protons in this orbit? What is the frequency of the motion?

4. Suppose that an electron of 22 GeV emerging from the Stanford Linear Accelerator is made to race against a light signal along a straight track 100 m long. How far ahead of the electron will the light signal be at the end of this race?

5. Lambda particles [$\Lambda(1116)$] and kaons are to be produced by the reaction

$$p + p \rightarrow p + \Lambda + K^+$$

According to the relativistic formula for the available energy in the center of mass, what is the minimum kinetic energy required for a proton incident on a stationary proton?

6. A particle of kinetic energy K collides with a stationary particle of the same mass. Show that, according to Newtonian nonrelativistic physics, the net kinetic energy of the two particles relative to the reference frame moving with the center of mass is $K/2$.

7. What is the total number of known baryons listed in Table 11.3? What is the total number of known mesons listed in Table 11.4? What is the total number of known particles, including baryons and antibaryons, mesons and antimesons, leptons and antileptons?

8. Calculate the range of the weak interaction from the mass of the $W^\pm$ boson listed in Table 11.6.

9. Suppose that the mean lifetime of the proton is 10^{33} years. How many proton decays are there per year in a mass of 10^3 tons of water?

10. Which of the following reactions are forbidden by an absolute conservation law?

$$\pi^+ + p \rightarrow \Lambda + K^0$$
$$K^- + p \rightarrow K^- + p + \pi^0$$
$$\pi^- + \eta \rightarrow \pi^- + \pi^0 + \pi^0$$
$$K^- + \eta \rightarrow \Sigma^- + \pi^0$$
$$e + \nu \rightarrow \pi^- + \pi^0$$

11. Is hypercharge conserved in the following reactions?

$$\pi^+ + \eta \rightarrow \Lambda + K^+$$

$$\Lambda \rightarrow p + \pi^-$$

$$K^0 \rightarrow \pi^+ + \pi^-$$

12. The Ξ^- particle decays according to the reaction

$$\Xi^- \rightarrow \Lambda + \pi^-$$

and also according to the reaction

$$\Xi^- \rightarrow \Lambda + \mu + \bar{\nu}_\mu$$

Which interaction is involved in each of these reactions?

13. The lambda particle decays according to several alternative reactions, among which are the following:

$$\Lambda \rightarrow p + \pi^-$$

$$\Lambda \rightarrow p + e + \bar{\nu}_e$$

$$\Lambda \rightarrow p + \pi^- + \gamma$$

Which interaction is involved in each case?

14. Consider the sequence of reactions shown in Fig. 11.30. Identify the interaction involved in each step of this sequence.

15. What is the parity of each of the wavefunctions in Eqs. (6.133)–(6.137)?

16. The masses of the members of the baryon decuplet obey an **equal-spacing rule:** the mass differences between all the adjacent rows are equal. This equal-spacing rule was used to predict the mass of the Ω^- before its discovery. Using the masses of the other members of the decuplet (see Fig. 11.28), what mass would you have predicted for the Ω^-?

17. If we ignore the small mass differences within the isospin multiplets, then there are only four distinct masses in the baryon octet: the mass of the nucleon doublet (m_N), Σ triplet (m_Σ), Λ singlet (m_Λ), and Ξ doublet (m_Ξ). According to the Eightfold Way, these masses obey the **Gell-Mann–Okubo mass formula**

$$m_a = m_0 + m_1 Y + m_2 \left[T(T+1) - \tfrac{1}{4}Y^2 \right]$$

where a stands for N, Σ, Λ, or Ξ; T is the corresponding isospin quantum number; Y the hypercharge; and m_0, m_1, and m_2 are constants. Show that this mass formula implies

$$(m_N + m_\Xi) = \tfrac{1}{2}(3m_\Lambda + m_\Sigma)$$

and compare this theoretical result with the measured masses.

18. (a) By inspection of Fig. 11.28, verify that the isospin and the hypercharge of the members of the decuplet are related by $T = \frac{1}{2}Y + 1$.

(b) Substitute this relation into the Gell-Mann–Okubo mass formula given in Problem 17 and above that this leads to a mass formula of the form $m_a + m_0' + m_1' Y$, which coincides with the equal-spacing rule stated in Problem 16.

19. Each of the particles n, p, Σ^-, Σ^0, Λ, Σ^+, Ξ^- and Ξ^0 of the baryon octet consists of three quarks. List the three quarks inside each of these particles.

20. Each of the particles K^0, K^+, π^-, π^+, K^-, and $\bar{K}^0$ of the meson octet consists of a quark u, d, or s and an antiquark. List the quark and the antiquark inside of each of these particles.

21. The particle π^0 consists of a quark and an antiquark. Which quark-antiquark combinations are possible?

22. According to the quark model, the magnetic moment of the proton is $\frac{3}{2}$ times the magnetic moment of the u quark, and the magnetic moment of the neutron is -1 times the magnetic moment of the u quark. Is this consistent with the measured values of the magnetic moments of the proton and the neutron?

APPENDIX *1*   **Best Values of Fundamental Constants**

Compiled by E. R. Cohen and B. N. Taylor under the auspices of the CODATA Task Group on Fundamental Constants. This set has been officially adopted by CODATA and is taken from J. Phys. Chem. Ref. Data, Vol. 2, No. 4, p. 663 (1973) and CODATA Bulletin No. 11 (December 1973).

Quantity	Symbol	Numerical Value *	Uncert. (ppm)	SI †	← Units →	cgs ‡
Speed of light in vacuum	c	299792458(1.2)	0.004	$m \cdot s^{-1}$		$10^2\ cm \cdot s^{-1}$
Permeability of vacuum	μ_0	4π =12.5663706144		$10^{-7}\ H \cdot m^{-1}$ $10^{-7}\ H \cdot m^{-1}$		
Permittivity of vacuum, $1/\mu_0 c^2$	ϵ_0	8.854187818(71)	0.008	$10^{-12}\ F \cdot m^{-1}$		
Fine-structure constant, $[\mu_0 c^2/4\pi](e^2 \hbar c)$	α α^{-1}	7.2973506(60) 137.03604(11)	0.82 0.82	10^{-3}		10^{-3}
Elementary charge	e	1.6021892(46) 4.803242(14)	2.9 2.9	$10^{-19}\ C$		$10^{-20}\ emu$ $10^{-10}\ esu$
Planck constant	h $\hbar = h/2\pi$	6.626176(36) 1.0545887(57)	5.4 5.4	$10^{-34}\ J \cdot s$ $10^{-34}\ J \cdot s$		$10^{-27}\ erg \cdot s$ $10^{-27}\ erg \cdot s$
Avogadro constant	N_A	6.022045(31)	5.1	$10^{23}\ mol^{-1}$		$10^{23}\ mol^{-1}$
Atomic mass unit, $10^{-3} kg \cdot mol^{-1} N_A^{-1}$	u	1.6605655(86)	5.1	$10^{-27}\ kg$		$10^{-24}\ g$
Electron rest mass	m_e	9.109534(47) 5.4858026(21)	5.1 0.38	$10^{-31}\ kg$ $10^{-4}\ u$		$10^{-28}\ g$ $10^{-4}\ u$
Proton rest mass	m_p	1.6726485(86) 1.007276470(11)	5.1 0.011	$10^{-27}\ kg$ u		$10^{-24}\ g$ u
Ratio of proton mass to electron mass	m_p/m_e	1836.15152(70)	0.38			
Neutron rest mass	m_n	1.6749543(86) 1.008665012(37)	5.1 0.037	$10^{-27}\ kg$ u		$10^{-24}\ g$ u
Electron charge to mass ratio	e/m_e	1.7588047(49) 5.272764(15)	2.8 2.8	$10^{11}\ C \cdot kg^{-1}$		$10^7\ emu \cdot g^{-1}$ $10^{17}\ esu \cdot g^{-1}$
Magnetic flux quantum, $[c]^{-1}(hc/2e)$	Φ_0 h/e	2.0678506(54) 4.135701(11) 1.3795215(36)	2.6 2.6 2.6	$10^{-15}\ Wb$ $10^{-15}\ J \cdot s \cdot C^{-1}$		$10^{-7}\ G \cdot cm^2$ $10^{-7}\ erg \cdot s \cdot emu^{-1}$ $10^{-17}\ erg \cdot s \cdot esu^{-1}$
Josephson frequency-voltage ratio	$2e/h$	4.835939(13)	2.6	$10^{14}\ Hz \cdot V^{-1}$		
Quantum of circulation	$h/2m_e$ h/m_e	3.6369455(60) 7.273891(12)	1.6 1.6	$10^{-4}\ J \cdot s \cdot kg^{-1}$ $10^{-4}\ J \cdot s \cdot kg^{-1}$		$erg \cdot s \cdot g^{-1}$ $erg \cdot s \cdot g^{-1}$
Faraday constant, $N_A e$	F	9.648456(27) 2.8925342(82)	2.8 2.8	$10^4\ C \cdot mol^{-1}$		$10^3\ emu \cdot mol^{-1}$ $10^{14}\ esu \cdot mol^{-1}$
Rydberg constant, $[\mu_0 c^2/4\pi]^2(m_e e^4/4\pi\hbar^3 c)$	R_∞	1.097373177(83)	0.075	$10^7\ m^{-1}$		$10^5\ cm^{-1}$
Bohr radius, $[\mu_0 c^2/4\pi]^{-1}(\hbar^2/m_e e^2)=\alpha/4\pi R_\infty$	a_0	5.2917706(44)	0.82	$10^{-11}\ m$		$10^{-9}\ cm$
Classical electron radius, $[\mu_0 c^2/4\pi](e^2/m_e c^2)=\alpha^3/4\pi R_\infty$	$r_e = \alpha \lambdabar_C$	2.8179380(70)	2.5	$10^{-15}\ m$		$10^{-13}\ cm$
Thomson cross section, $(8/3)\pi r_e^2$	σ_e	0.6652448(33)	4.9	$10^{-28}\ m^2$		$10^{-24}\ cm^2$
Free electron g-factor, or electron magnetic moment in Bohr magnetons	$g_e/2 = \mu_e/\mu_B$	1.0011596567(35)	0.0035			
Free muon g-factor, or muon magnetic moment in units of $[c](e\hbar/2m_\mu c)$	$g_\mu/2$	1.00116616(31)	0.31			
Bohr magneton, $[c](e\hbar/2m_e c)$	μ_B	9.274078(36)	3.9	$10^{-24}\ J \cdot T^{-1}$		$10^{-21}\ erg \cdot G^{-1}$
Electron magnetic moment	μ_e	9.284832(36)	3.9	$10^{-24}\ J \cdot T^{-1}$		$10^{-21}\ erg \cdot G^{-1}$
Gyromagnetic ratio of protons in H_2O	γ'_p $\gamma'_p/2\pi$	2.6751301(75) 4.257602(12)	2.8 2.8	$10^8\ s^{-1} \cdot T^{-1}$ $10^7\ Hz \cdot T^{-1}$		$10^4\ s^{-1} \cdot G^{-1}$ $10^3\ Hz \cdot G^{-1}$
γ'_p corrected for diamagnetism of H_2O	γ_p $\gamma_p/2\pi$	2.6751987(75) 4.257711(12)	2.8 2.8	$10^8\ s^{-1} \cdot T^{-1}$ $10^7\ Hz \cdot T^{-1}$		$10^4\ s^{-1} \cdot G^{-1}$ $10^3\ Hz \cdot G^{-1}$
Magnetic moment of protons in H_2O in Bohr magnetons	μ'_p/μ_B	1.52099322(10)	0.066	10^{-3}		10^{-3}
Proton magnetic moment in Bohr magnetons	μ_p/μ_B	1.521032209(16)	0.011	10^{-3}		10^{-3}
Ratio of electron and proton magnetic moments	μ_e/μ_p	658.2106880(66)	0.010			
Proton magnetic moment	μ_p	1.4106171(55)	3.9	$10^{-26}\ J \cdot T^{-1}$		$10^{-23}\ erg \cdot G^{-1}$
Magnetic moment of protons in H_2O in nuclear magnetons	μ'_p/μ_N	2.7927740(11)	0.38			
μ'_p/μ_N corrected for diamagnetism of H_2O	μ_p/μ_N	2.7928456(11)	0.38			
Nuclear magneton, $[c](e\hbar/2m_p c)$	μ_N	5.050824(20)	3.9	$10^{-27}\ J \cdot T^{-1}$		$10^{-24}\ erg \cdot G^{-1}$
Ratio of muon and proton magnetic moments	μ_μ/μ_p	3.1833402(72)	2.3			
Muon magnetic moment	μ_μ	4.490474(18)	3.9	$10^{-26}\ J \cdot T^{-1}$		$10^{-23}\ erg \cdot G^{-1}$
Ratio of muon mass to electron mass	m_μ/m_e	206.76865(47)	2.3			

Quantity	Symbol	Numerical Value [*]	Uncert. (ppm)	SI [†]	← Units →	cgs [‡]
Muon rest mass	m_μ	1.883566(11)	5.6	10^{-28} kg		10^{-25} g
		0.11342920(26)	2.3	u		u
Compton wavelength of the electron, $h/m_e c = \alpha^2/2R_\infty$	λ_C	2.4263089(40)	1.6	10^{-12} m		10^{-10} cm
	$\lambdabar_C = \lambda_C/2\pi = a a_0$	3.8615905(64)	1.6	10^{-13} m		10^{-11} cm
Compton wavelength of the proton, $h/m_p c$	$\lambda_{C,p}$	1.3214099(22)	1.7	10^{-15} m		10^{-13} cm
	$\lambdabar_{C,p} = \lambda_{C,p}/2\pi$	2.1030892(36)	1.7	10^{-16} m		10^{-14} cm
Compton wavelength of the neutron, $h/m_n c$	$\lambda_{C,n}$	1.3195909(22)	1.7	10^{-15} m		10^{-13} cm
	$\lambdabar_{C,n} = \lambda_{C,n}/2\pi$	2.1001941(35)	1.7	10^{-16} m		10^{-14} cm
Molar volume of ideal gas at s.t.p.	V_m	22.41383(70)	31	10^{-3} m³·mol⁻¹		10^3 cm³·mol⁻¹
Molar gas constant, $V_m p_0/T_0$ ($T_0 \equiv 273.15$ K; $p_0 \equiv 101325$ Pa$\equiv$1atm)	R	8.31441(26)	31	J·mol⁻¹·K⁻¹		10^7 erg·mol⁻¹·K⁻¹
		8.20568(26)	31	10^{-5} m³·atm·mol⁻¹·K⁻¹		10 cm³·atm·mol⁻¹·K⁻¹
Boltzmann constant, R/N_A	k	1.380662(44)	32	10^{-23} J·K⁻¹		10^{-16} erg·K⁻¹
Stefan-Boltzmann constant, $\pi^2 k^4/60\hbar^3 c^2$	σ	5.67032(71)	125	10^{-8} W·m⁻²·K⁻⁴		10^{-5} erg·s⁻¹·cm⁻²·K⁻⁴
First radiation constant, $2\pi hc^2$	c_1	3.741832(20)	5.4	10^{-16} W·m²		10^{-5} erg·cm²·s⁻¹
Second radiation constant, hc/k	c_2	1.438786(45)	31	10^{-2} m·K		cm·K
Gravitational constant[1]	G	6.6720(41)	615	10^{-11} m³·s⁻²·kg⁻¹		10^{-8} cm³·s⁻²·g⁻¹
Ratio, kx-unit to ångström, $\Lambda = \lambda(Å)/\lambda(kxu)$; $\lambda(CuK\alpha_1) \equiv 1.537400$ kxu	Λ	1.0020772(54)	5.3			
Ratio, Å* to ångström, $\Lambda^* = \lambda(Å)/\lambda(Å^*)$; $\lambda(WK\alpha_1) \equiv 0.2090100$ Å*	Λ^*	1.0000205(56)	5.6			

ENERGY CONVERSION FACTORS AND EQUIVALENTS

Quantity	Symbol	Numerical Value [*]	Units	Uncert. (ppm)
1 kilogram (kg·c²)		8.987551786(72)	10^{16} J	0.008
		5.609545(16)	10^{29} MeV	2.9
1 Atomic mass unit (u·c²)		1.4924418(77)	10^{-10} J	5.1
		931.5016(26)	MeV	2.8
1 Electron mass $m_e \cdot c^2$)		8.187241(42)	10^{-14} J	5.1
		0.5110034(14)	MeV	2.8
1 Muon mass ($m_\mu \cdot c^2$)		1.6928648(96)	10^{-11} J	5.6
		105.65948(35)	MeV	3.3
1 Proton mass ($m_p \cdot c^2$)		1.5033015(77)	10^{-10} J	5.1
		938.2796(27)	MeV	2.8
1 Neutron mass ($m_n \cdot c^2$)		1.5053738(78)	10^{-10} J	5.1
		939.5731(27)	MeV	2.8
1 Electron volt		1.6021892(46)	10^{-19} J	2.9
			10^{-12} erg	2.9
	1 eV/h	2.4179696(63)	10^{14} Hz	2.6
	1 eV/hc	8.065479(21)	10^5 m⁻¹	2.6
			10^3 cm⁻¹	2.6
	1 eV/k	1.160450(36)	10^4 K	31
Voltage-wavelength conversion, hc		1.986478(11)	10^{-25} J·m	5.4
		1.2398520(32)	10^{-6} eV·m	2.6
			10^{-4} eV·cm	2.6
Rydberg constant	$R_\infty hc$	2.179907(12)	10^{-18} J	5.4
			10^{-11} erg	5.4
		13.605804(36)	eV	2.6
	$R_\infty c$	3.28984200(25)	10^{15} Hz	0.075
	$R_\infty hc/k$	1.578885(49)	10^5 K	31
Bohr magneton	μ_B	9.274078(36)	10^{-24} J·T⁻¹	3.9
		5.7883785(95)	10^{-5} eV·T⁻¹	1.6
	μ_B/h	1.3996123(39)	10^{10} Hz·T⁻¹	2.8
	μ_B/hc	46.68604(13)	m⁻¹·T⁻¹	2.8
			10^{-2} cm⁻¹·T⁻¹	2.8
	μ_B/k	0.671712(21)	K·T⁻¹	31
Nuclear magneton	μ_N	5.505824(20)	10^{-27} J·T⁻¹	3.9
		3.1524515(53)	10^{-8} eV·T⁻¹	1.7
	μ_N/h	7.622532(22)	10^6 Hz·T⁻¹	2.8
	μ_N/hc	2.5426030(72)	10^{-3} m⁻¹·T⁻¹	2.8
			10^{-4} cm⁻¹·T⁻¹	2.8
	μ_N/k	3.65826(12)	10^{-4} K·T⁻¹	31

[*] Note that the numbers in parentheses are the one standard-deviation uncertainties in the last digits of the quoted value computed on the basis of internal consistency, that the unified atomic mass scale $^{12}C = 12$ has been used throughout, that u=atomic mass unit, C=coulomb, F=farad, G=gauss, H=henry, Hz=hertz=cycle/s, J=joule, K=kelvin (degree Kelvin), Pa=pascal=N·m⁻², T=tesla (10^4 G), V=volt, Wb=weber= T·m², and W=watt. In cases where formulas for constants are given (e.g., R_∞), the relations are written as the product of two factors. The second factor, in parentheses, is the expression to be used when all quantities are expressed in cgs units, with the electron charge in electrostatic units. The first factor, in brackets, is to be included only if all quantities are expressed in SI units. We remind the reader that with the exception of the auxiliary constants which have been taken to be exact, the uncertainties of these constants are correlated, and therefore the general law of error propagation must be used in calculating additional quantities requiring two or more of these constants.

[†] Quantities given in u and atm are for the convenience of the reader; these units are not part of the International System of Units (SI).

[‡] In order to avoid separate columns for "electromagnetic" and "electrostatic" units, both are given under the single heading "cgs Units." When using these units, the elementary charge e in the second column should be understood to be replaced by e_m or e_s, respectively.

[1] According to a recent, more precise measurement, $G=6.6726(5) \times 10^{-11}$ m³·s⁻²·kg⁻¹. The uncertainty is 65 ppm.

Conversion Factors

The units for each quantity are listed alphabetically, but the SI unit is always listed first. All numbers are adapted from "American National Standard; Metric Practice" published by the Institute of Electrical and Electronics Engineers, 1982.

Angle

1 radian $= 57.30° = 3.438 \times 10^3{}' = \frac{1}{2\pi}$ rev $= 2.063 \times 10^5{}''$

1 degree (°) $= 1.745 \times 10^{-2}$ radian $= 60' = 3600'' = \frac{1}{360}$ rev

1 minute of arc (') $= 2.909 \times 10^{-4}$ radian $= \frac{1}{60}° = 4.630 \times 10^{-5}$ rev $= 60''$

1 revolution (rev) $= 2\pi$ radian $= 360° = 2.160 \times 10^4{}' = 1.296 \times 10^6{}''$

1 second of arc ('') $= 4.848 \times 10^{-6}$ radian $= \frac{1}{3600}° = \frac{1}{60}' = 7.716 \times 10^{-7}$ rev

Length

1 meter (m) $= 1 \times 10^{10}$ Å $= 100$ cm $= 1 \times 10^{15}$ fm $= 3.281$ ft $= 39.37$ in. $=$
1×10^{-3} km $= 1.057 \times 10^{-16}$ light-year $= 1 \times 10^6 \, \mu$m $=$
6.214×10^{-4} mi $= 1.094$ yd

1 angstrom (Å) $= 1 \times 10^{-10}$ m $= 1 \times 10^{-8}$ cm $= 1 \times 10^5$ fm $=$
3.281×10^{-10} ft $= 1 \times 10^{-4} \, \mu$m

1 centimeter (cm) $= 0.01$ m $= 1 \times 10^8$ Å $= 1 \times 10^{13}$ fm $= 3.281 \times 10^{-2}$ ft $=$
0.3937 in. $= 1 \times 10^{-5}$ km $= 1.057 \times 10^{-18}$ light-year $= 1 \times 10^4 \, \mu$m

1 fermi (fm) $= 1 \times 10^{-15}$ m $= 1 \times 10^{-13}$ cm $= 1 \times 10^5$ Å

1 foot (ft) $= 0.3048$ m $= 30.48$ cm $= 12$ in. $= 3.048 \times 10^5 \, \mu$m $=$
1.894×10^{-4} mi $= \frac{1}{3}$ yd

1 inch (in.) $= 2.540 \times 10^{-2}$ m $= 2.54$ cm $= \frac{1}{12}$ ft $= 2.54 \times 10^4 \, \mu$m $= \frac{1}{36}$ yd

1 kilometer (km) $= 1 \times 10^3$ m $= 1 \times 10^5$ cm $= 3.281 \times 10^3$ ft $=$
0.6214 mi $= 1.094 \times 10^3$ yd

1 light-year $= 9.461 \times 10^{15}$ m $= 9.461 \times 10^{17}$ cm $= 9.461 \times 10^{12}$ km $=$
5.879×10^{12} mi

1 micron, or micrometer (μm) $= 1 \times 10^{-6}$ m $= 1 \times 10^{4}$ Å $= 1 \times 10^{-4}$ cm $=$ 3.281 $\times 10^{-6}$ ft $= 3.937 \times 10^{-5}$ in.

1 statute mile (mi) $= 1.609 \times 10^{3}$ m $= 1.609 \times 10^{5}$ cm $= 5280$ ft $=$ 1.609 km $= 1760$ yd

1 yard (yd) $= 0.9144$ m $= 91.44$ cm $= 3$ ft $= 36$ in. $= \frac{1}{1760}$ mi

Time

1 second (s) $= 1.157 \times 10^{-5}$ day $= \frac{1}{3600}$ h $= \frac{1}{60}$ min $=$ 1.161 $\times 10^{-5}$ sidereal day $= 3.169 \times 10^{-8}$ yr

1 day $= 8.640 \times 10^{4}$ s $= 24$ h $= 1440$ min $= 1.003$ sidereal days $=$ 2.738 $\times 10^{-3}$ yr

1 hour (h) $= 3600$ s $= \frac{1}{24}$ day $= 60$ min $= 1.141 \times 10^{-4}$ yr

1 minute (min) $= 60$ s $= 6.944 \times 10^{-4}$ day $= \frac{1}{60}$ h $= 1.901 \times 10^{-6}$ year

1 sidereal day $= 8.616 \times 10^{4}$ s $= 0.9973$ day $= 23.93$ h $= 1.436 \times 10^{3}$ min $=$ 2.730 $\times 10^{-3}$ yr

1 year (yr) $= 3.156 \times 10^{7}$ s $= 365.24$ days $= 8.766 \times 10^{3}$ h $=$ 5.259 $\times 10^{5}$ min $= 366.24$ sidereal days

Mass

1 kilogram (kg) $= 6.024 \times 10^{26}$ u $= 1000$ g $= 1 \times 10^{-3}$ t $= 35.27$ oz.-mass $=$ 2.205 lb-mass $= 1.102 \times 10^{-3}$ short ton-mass $= 6.852 \times 10^{-2}$ slug

1 atomic mass unit (u) $= 1.6605 \times 10^{-27}$ kg $= 1.6605 \times 10^{-24}$ g

1 gram (g) $= 1 \times 10^{-3}$ kg $= 6.024 \times 10^{23}$ u $= 1 \times 10^{-6}$ t $=$ 3.527 $\times 10^{-2}$ oz.-mass $= 2.205 \times 10^{-3}$ lb-mass $=$ 1.102 $\times 10^{-6}$ short ton-mass $= 6.852 \times 10^{-5}$ slug

1 metric ton (t) $= 1 \times 10^{3}$ kg $= 1 \times 10^{6}$ g $= 2.205 \times 10^{3}$ lb-mass $=$ 1.102 short ton-mass $= 68.52$ slugs

1 pound-mass (lb-mass)* $= 0.4536$ kg $= 453.6$ g $= 4.536 \times 10^{-4}$ t $=$ 16 oz.-mass $= \frac{1}{2000}$ short ton-mass $= 3.108 \times 10^{-2}$ slug

1 slug $= 14.59$ kg $= 1.459 \times 10^{4}$ g $= 32.17$ lb-mass.

Area

1 square meter (m^2) $= 1 \times 10^{4}$ cm^2 $= 10.76$ ft^2 $= 1.550 \times 10^{3}$ in.^{2}

1 barn $= 1 \times 10^{-28}$ m^2 $= 1 \times 10^{-24}$ cm^2

1 square centimeter (cm^2) $= 1 \times 10^{-4}$ m^2 $= 1.076 \times 10^{-3}$ ft^2 $= 0.1550$ in.^{2}

1 square foot (ft^2) $= 9.290 \times 10^{-2}$ m^2 $= 929.0$ cm^2 $= 144$ in.^{2}

1 square inch (in.^{2}) $= 6.452 \times 10^{-4}$ m^2 $= 6.452$ cm^2 $= \frac{1}{144}$ ft^2

---

* This is the "avoirdupois" pound. The "troy" or "apothecary" pound is 0.3732 kg, or 0.8229 lb avoirdupois.

Volume

1 cubic meter (m^3) $= 1 \times 10^6 \, cm^3 = 35.31 \, ft^3 = 6.102 \times 10^4 \, in.^3 = 1 \times 10^3$ liters
1 cubic centimeter (cm^3) $= 1 \times 10^{-6} \, m^3 = 3.531 \times 10^{-5} \, ft^3 =$
   $6.102 \times 10^{-2} \, in.^3 = 1 \times 10^{-3}$ liter
1 cubic foot (ft^3) $= 2.832 \times 10^{-2} \, m^3 = 2.832 \times 10^4 \, cm^3 = 1728 \, in.^3 = 28.32$ liters
1 cubic inch ($in.^3$) $= 1.639 \times 10^{-5} \, m^3 = 16.39 \, cm^3 = 5.787 \times 10^{-4} \, ft^3$
1 liter (l) $= 1 \times 10^{-3} \, m^3 = 1000 \, cm^3 = 3.531 \times 10^{-2} \, ft^3$

Density

1 kilogram per cubic meter (kg/m^3) $= 1 \times 10^{-3} \, g/cm^3 =$
   $6.243 \times 10^{-2} \, \text{lb-mass}/ft^3 = 3.613 \times 10^{-5} \, \text{lb-mass}/in.^3 = 1.940 \times 10^{-3} \, slug/ft^3$
1 gram per cubic centimeter (g/cm^3) $= 1 \times 10^3 \, kg/m^3 = 62.43 \, \text{lb-mass}/ft^3 =$
   $3.613 \times 10^{-2} \, \text{lb-mass}/in.^3 = 1.940 \, slug/ft^3$
1 lb-mass per cubic foot ($\text{lb-mass}/ft^3$) $= 16.02 \, kg/m^3 =$
   $1.602 \times 10^{-2} \, g/cm^3 = 3.108 \times 10^{-2} \, slug/ft^3$
1 slug per cubic foot ($slug/ft^3$) $= 515.4 \, kg/m^3 = 0.5154 \, g/cm^3 = 32.17 \, \text{lb-mass}/ft^3$

Speed

1 meter per second (m/s) $= 100 \, cm/s = 3.281 \, ft/s = 3.600 \, km/h =$
   $2.237 \, mi/h$
1 centimeter per second (cm/s) $= 0.01 \, m/s = 3.281 \times 10^{-2} \, ft/s$
1 foot per second (ft/s) $= 0.3048 \, m/s = 30.48 \, cm/s = 1.097 \, km/h =$
   $0.6818 \, mi/h$
1 kilometer per hour (km/h) $= 0.2778 \, m/s = 27.78 \, cm/s = 0.9113 \, ft/s =$
   $0.6214 \, mi/h$
1 mile per hour (mi/h) $= 0.4470 \, m/s = 44.70 \, cm/s = 1.467 \, ft/s = 1.609 \, km/h$

Acceleration

1 meter per second squared (m/s^2) $= 100 \, cm/s^2 = 3.281 \, ft/s^2 = 0.1020$ gee
1 centimeter per second squared (cm/s^2) $= 0.01 \, m/s^2 =$
   $3.281 \times 10^{-2} \, ft/s^2 = 1.020 \times 10^{-3}$ gee
1 foot per second squared (ft/s^2) $= 0.3048 \, m/s^2 = 30.48 \, cm/s^2 =$
   3.108×10^{-2} gee
1 gee $= 9.807 \, m/s^2 = 980.7 \, cm/s^2 = 32.17 \, ft/s^2$

Force

1 newton (N) $= 1 \times 10^5$ dynes $= 0.1020 \, kp = 0.2248$ lb
1 dyne $= 1 \times 10^{-5} \, N = 1.020 \times 10^{-6} \, kp = 2.248 \times 10^{-6}$ lb
1 kilopond, or kilogram force (kp) $= 9.807 \, N =$
   9.807×10^5 dynes $= 2.205$ lb
1 pound (lb) $= 4.448 \, N = 4.448 \times 10^5$ dynes $= 0.4536 \, kp$
1 short ton $= 8.896 \times 10^3 \, N = 8.896 \times 10^8$ dynes $= 907.2 \, kp = 2000$ lb

Energy

1 joule (J) $= 9.478 \times 10^{-4}$ Btu $= 0.2388$ cal $= 1 \times 10^7$ ergs $=$
6.242×10^{18} eV $= 0.7376$ ft·lb $= 2.778 \times 10^{-7}$ kW·h

1 British thermal unit (Btu)* $= 1.055 \times 10^3$ J $= 252.0$ cal $=$
1.055×10^{10} ergs $= 778.2$ ft·lb $= 2.931 \times 10^{-4}$ kW·h

1 calorie (cal)† $= 4.187$ J $= 3.968 \times 10^{-3}$ Btu $= 4.187 \times 10^7$ ergs $=$
3.088 ft·lb $= 1 \times 10^{-3}$ kcal $= 1.163 \times 10^{-6}$ kW·h

1 erg $= 1 \times 10^{-7}$ J $= 9.478 \times 10^{-7}$ Btu $= 2.388 \times 10^{-8}$ cal $=$
6.242×10^{11} eV $= 7.376 \times 10^{-8}$ ft·lb $= 2.778 \times 10^{-14}$ kW·h

1 electron-volt (eV) $= 1.602 \times 10^{-19}$ J $= 1.602 \times 10^{-12}$ erg $=$
1.182×10^{-19} ft·lb

1 foot-pound (ft·lb) $= 1.356$ J $= 1.285 \times 10^{-3}$ Btu $= 0.3239$ cal $=$
1.356×10^7 ergs $= 8.464 \times 10^{18}$ eV $= 3.766 \times 10^{-7}$ kW·h

1 kilowatt-hour (kW·h) $= 3.600 \times 10^6$ J $= 3412$ Btu $=$
8.598×10^5 cal $= 3.6 \times 10^{13}$ ergs $= 2.655 \times 10^6$ ft·lb

Power

1 watt (W) $= 0.2388$ cal/s $= 1 \times 10^7$ ergs/s $=$
0.7376 ft·lb/s $= 1.341 \times 10^{-3}$ hp

1 calorie per second (cal/s) $= 4.187$ W $= 4.187 \times 10^7$ erg/s $=$
3.088 ft·lb/s $= 5.615 \times 10^{-3}$ hp

1 erg per second (erg/s) $= 1 \times 10^{-7}$ W $= 2.388 \times 10^{-8}$ cal/s $=$
7.376×10^{-8} ft·lb/s $= 1.341 \times 10^{-10}$ hp

1 foot-pound per second (ft·lb/s) $= 1.356$ W $= 0.3238$ cal/s $=$
1.356×10^7 ergs/s $= 1.818 \times 10^{-3}$ hp

1 horsepower (hp) $= 745.7$ W $= 178.1$ cal/s $= 550$ ft·lb/s

1 kilowatt (kW) $= 1 \times 10^3$ W $= 238.8$ cal/s $= 737.6$ ft·lb/s $= 1.341$ hp

Pressure

1 newton per square meter (N/m²), or **pascal** (Pa) $= 9.869 \times 10^{-6}$ atm $=$
7.501×10^{-4} cmHg $= 10$ dynes/cm² $= 2.089 \times 10^{-2}$ lb/ft² $=$
1.450×10^{-4} lb/in.² $= 7.501 \times 10^{-3}$ torr

1 atmosphere (atm) $= 1.013 \times 10^5$ N/m² $= 76.00$ cmHg $=$
1.013×10^6 dynes/cm² $= 2.116 \times 10^3$ lb/ft² $= 14.70$ lb/in.²

1 centimeter of mercury (cmHg) $= 1.333 \times 10^3$ N/m² $= 1.316 \times 10^{-2}$ atm $=$
1.333×10^4 dynes/cm² $= 27.85$ lb/ft² $= 0.1934$ lb/in.² $= 10$ torr

1 dyne per square centimeter (dyne/cm²) $= 0.1$ N/m² $=$
9.869×10^{-7} atm $= 7.501 \times 10^{-5}$ cmHg $= 2.089 \times 10^{-3}$ lb/ft² $=$
1.450×10^{-5} lb/in.²

---

* This is the "International Table" Btu; there are several other Btus.

† This is the "International Table" calorie, which equals exactly 4.1868 J. There
are several other calories; for instance, the thermochemical calorie, which equals
4.184 J.

1 kilopond per square centimeter (kp/cm^2) $= 9.807 \times 10^4 \, N/m^2 =$
0.9678 atm $= 9.807 \times 10^5$ dynes/cm$^2 = 14.22 \, lb/in.^2$

1 pound per square inch ($lb/in.^2$, or psi) $= 6.895 \times 10^3 \, N/m^2 =$
6.805×10^{-2} atm $= 6.895 \times 10^4$ dynes/cm$^2 = 7.031 \times 10^{-2} \, kp/cm^2$

1 torr, or millimeter of mercury (mmHg) $= 1.333 \times 10^2 \, N/m^2 =$
0.1 cmHg

Electric Charge*

1 coulomb (C) $\Leftrightarrow 2.998 \times 10^9$ statcoulombs, or esu of charge $\Leftrightarrow$
0.1 abcoulomb, or emu of charge

Electric Current

1 ampere (A) $\Leftrightarrow 2.998 \times 10^9$ statamperes, or esu of current $\Leftrightarrow$
0.1 abampere, or emu of current

Electric Potential

1 volt (V) $\Leftrightarrow 3.336 \times 10^{-3}$ statvolt, or esu of potential $\Leftrightarrow$
1×10^8 abvolts, or emu of potential

Electric Field

1 volt per meter (V/m) $\Leftrightarrow 3.336 \times 10^{-5}$ statvolt/cm $\Leftrightarrow 1 \times 10^6$ abvolts/cm

Magnetic Field

1 tesla (T), **or weber per square meter** ($Wb/m)^2 \Leftrightarrow 1 \times 10^4$ gauss

Electric Resistance

1 ohm (Ω) $\Leftrightarrow 1.113 \times 10^{-12}$ statohm, or esu of resistance $\Leftrightarrow$
1×10^9 abohms, or emu of resistance

Electric Resistivity

1 ohm-meter ($\Omega \cdot m$) $\Leftrightarrow 1.113 \times 10^{-10}$ statohm-cm $\Leftrightarrow 1 \times 10^{11}$ abohm-cm

Capacitance

1 farad (F) $\Leftrightarrow 8.988 \times 10^{11}$ statfarads, or esu of capacitance $\Leftrightarrow$
1×10^{-9} abfarad, or emu of capacitance

Inductance

1 henry (H) $\Leftrightarrow 1.113 \times 10^{-12}$ stathenry, or esu of inductance $\Leftrightarrow$
1×10^9 abhenrys, or emu of inductance

---

* The dimensions of the electric quantities in SI units, electrostatic units (esu), and electromagnetic units (emu) are different; hence the relationships among these units must be regarded as correspondences ($\Leftrightarrow$) rather than equalities ($=$).

The Periodic Table of the Chemical Elements*

IA	IIA	IIIB	IVB	VB	VIB	VIIB	VIII			IB	IIB	IIIA	IVA	VA	VIA	VIIA	0
1 **H** 1.0079																	2 **He** 4.00260
3 **Li** 6.94	4 **Be** 9.01218											5 **B** 10.81	6 **C** 12.011	7 **N** 14.0067	8 **O** 15.9994	9 **F** 18.998403	10 **Ne** 20.179
11 **Na** 22.98977	12 **Mg** 24.305											13 **Al** 26.98154	14 **Si** 28.0855	15 **P** 30.97376	16 **S** 32.06	17 **Cl** 35.453	18 **Ar** 39.948
19 **K** 39.0983	20 **Ca** 40.08	21 **Sc** 44.9559	22 **Ti** 47.88	23 **V** 50.9415	24 **Cr** 51.996	25 **Mn** 54.9380	26 **Fe** 55.847	27 **Co** 58.9332	28 **Ni** 58.69	29 **Cu** 63.546	30 **Zn** 65.39	31 **Ga** 69.72	32 **Ge** 72.59	33 **As** 74.9216	34 **Se** 78.96	35 **Br** 79.904	36 **Kr** 83.80
37 **Rb** 85.4678	38 **Sr** 87.62	39 **Y** 88.9059	40 **Zr** 91.22	41 **Nb** 92.9064	42 **Mo** 95.94	43 **Tc** 98.9062	44 **Ru** 101.07	45 **Rh** 102.9055	46 **Pd** 106.42	47 **Ag** 107.8682	48 **Cd** 112.41	49 **In** 114.82	50 **Sn** 118.71	51 **Sb** 121.75	52 **Te** 127.60	53 **I** 126.9045	54 **Xe** 131.29
55 **Cs** 132.9054	56 **Ba** 137.33	57–71 Rare Earths	72 **Hf** 178.49	73 **Ta** 180.9479	74 **W** 183.85	75 **Re** 186.207	76 **Os** 190.2	77 **Ir** 192.22	78 **Pt** 195.08	79 **Au** 196.9665	80 **Hg** 200.59	81 **Tl** 204.383	82 **Pb** 207.2	83 **Bi** 208.9804	84 **Po** (209)	85 **At** (210)	86 **Rn** (222)
87 **Fr** (223)	88 **Ra** 226.0254	89–103 Acti-nides	104 **Rf** (260)	105 **Ha** (260)	106 (263)	107 (263)		109 (266)									

Rare Earths (Lanthanides)*

57 **La** 138.9055	58 **Ce** 140.12	59 **Pr** 140.9077	60 **Nd** 144.24	61 **Pm** (145)	62 **Sm** 150.36	63 **Eu** 151.96	64 **Gd** 157.25	65 **Tb** 158.9254	66 **Dy** 162.50	67 **Ho** 164.9304	68 **Er** 167.26	69 **Tm** 168.9342	70 **Yb** 173.04	71 **Lu** 174.967

Actinides

89 **Ac** (227)	90 **Th** 232.0381	91 **Pa** 231.0359	92 **U** 238.029	93 **Np** 237.0482	94 **Pu** (244)	95 **Am** (243)	96 **Cm** (247)	97 **Bk** (247)	98 **Cf** (251)	99 **Es** (254)	100 **Fm** (257)	101 **Md** (258)	102 **No** (259)	103 **Lr** (260)

* In each box, the upper number is the *atomic number.* The lower number is the *atomic mass,* i.e., the mass (in grams) of one mole or, alternatively, the mass (in u) of one atom. Numbers in parentheses denote the atomic masses of the most stable or best-known isotope of the element; all other numbers represent the average masses of a mixture of several isotopes as found in naturally occurring samples of the element.

The Isotopes

The following table lists all the known isotopes and all the known isomers (excited nuclear states) with a lifetime of one second or longer. The table is reproduced from the *Nuclear Wallet Cards* by J. K. Tuli (Brookhaven National Laboratory, 1985).

The first column of each page lists the atomic number Z, the chemical symbol, and the mass number A for each isotope or isomer (the letter m after the mass number indicates an isomer).

The second column gives the spin and the intrinsic parity (+ or −) of the nucleus.

The third column gives the difference Δ between the atomic mass of the isotope and $A \times (1\,\text{u})$, expressed in MeV. Since 1 MeV is equivalent to $1.073535 \times 10^{-3}\,\text{u}$, the atomic mass is

$$M = A \times (1\,\text{u}) + \Delta \times 1.073535 \times 10^{-3}\,\text{u}$$

For example, for ^{16}O, the value of Δ is -4.737 MeV, so that

$$M = 16 \times (1\,\text{u}) - 4.737 \times 1.073535 \times 10^{-3}\,\text{u} = 15.99491466\,\text{u}$$

The fourth column gives the abundance for naturally occurring isotopes or the half life for artificially produced isotopes. The *italic* number on the right gives the uncertainty in the last decimal, e.g., for ^{16}O the abundance is 99.762% and its uncertainty is $\pm 0.015\%$.

The fifth column gives the decay modes. Most of the symbols in this column are self-explanatory. ε stands for electron capture or β^+ decay, IT for isomeric transition from one excited state to another, SF for spontaneous fission, and combined symbols, such as εp, for an initial decay followed by a delayed, secondary decay.

Top-left table:

Z	El	A	Jπ	Δ (MeV)	T1/2 or Abundance	Decay Mode
0	n	1	1/2+	8.071	10.25 m 20	β-
1	H	1	1/2+	7.289	99.985% 1	
		2	1+	13.136	0.015% 1	
		3	1/2+	14.950	12.33 y 6	β-
		4?	2-	25.840		
2	He	3	1/2+	14.931	0.000138% 3	
		4	0+	2.425	99.999862% 3	
		5	3/2-	11.390	0.60 MeV 2	n, α
		6	0+	17.592	806.7 ms 15	β-
		7	(3/2)-	26.110	160 keV 30	n
		8	0+	31.598	119.0 ms 15	β-, β-n 16%
		9		40.810		n
3	Li	4		25.120		
		5	3/2-	11.680	≈1.5 MeV	p, α
		6	1+	14.086	7.5% 2	
		7	3/2-	14.907	92.5% 2	
		8	2+	20.945	838 ms 6	β-2α
		9	3/2-	24.954	178.3 ms 4	β-, β-n 49.5%, β-n2α
		10		33.830	1.2 MeV 3	n
		11	(1/2-)	40.900	8.7 ms 1	β-, β-n 60.8%, β-t 0.01%
4	Be	6	0+	18.374	92 keV 6	p, α
		7	3/2-	15.769	53.29 d 7	ε
		8	0+	4.942	6.8 eV 17	2α
		9	3/2-	11.348	100%	
		10	0+	12.607	1.6·10⁶ y 2	β-
		11	1/2+	20.174	13.81 s 8	β-, β-α 3.1%
		12	0+	25.077	24.4 ms 30	β-, β-n<1%
		13		34.950s		
		14	0+	41.020s		
5	B	7	(3/2-)	27.870	1.3 MeV 2	p, α
		8	2+	22.920	770 ms 3	ε2α
		9	3/2-	12.416	0.54 keV 21	p2α
		10	3+	12.051	19.9% 2	
		11	3/2-	8.668	80.1% 2	
		12	1+	13.370	20.20 ms 2	β-, β-3α 1.58%
		13	3/2-	16.562	17.36 ms 16	β-, β-n 0.28%
		14	2-	23.664	16.1 ms 12	β-
		15		28.970		β-
		16		37.640s		
		17		44.010s		
6	C	8	0+	35.095	230 keV 50	α, p
		9	(3/2-)	28.913	126.5 ms 9	εp2α
		10	0+	15.702	19.255 s 53	ε
		11	3/2-	10.650	20.385 m 20	ε
		12	0+		98.90% 3	
		13	1/2-	3.125	1.10% 3	
		14	0+	3.020	5730 y 40	β-
		15	1/2+	9.873	2.449 s 5	β-
		16	0+	13.694	0.747 s 8	β-, β-n≥98.8%
		17		21.030		
		18	0+	24.890		

1

Top-right table:

Z	El	A	Jπ	Δ (MeV)	T1/2 or Abundance	Decay Mode
11	Na	21	3/2+	-2.189	22.48 s 3	ε
		22	3+	-5.185	2.602 y 2	ε
		23	3/2+	-9.531	100%	
		24	4+	-8.419	15.020 h 7	β-
		24m	1+	-7.947	20.18 ms 10	IT, β-≈0.003%
		25	5/2+	-9.359	59.6 s 7	β-
		26	3+	-6.906	1.072 s 9	β-
		27	5/2+	-5.650	302 ms 7	β-, β-n 0.08%
		28	1+	-1.140	30.5 ms 4	β-, β-n 0.58%
		29	3/2	2.640	42.9 ms 15	β-, β-n 15.1%
		30	2	8.200	53 ms 3	β-, β-n 33%
		31		11.810	16.9 ms 7	β-, β-n 30%
		32		16.530	13.5 ms 15	β-, β-n 39%
		33		21.450	8.0 ms 6	β-, β-n 77%
		34		26.640	5.5 ms 10	β-, β-n
		35			1.5 ms 5	β-, β-n
12	Mg	20	0+	17.572	0.1 s	ε, εp
		21	(3/2,5/2)	10.914	122 ms 3	ε, εp 29.3%
		22	0+	-0.397	3.857 s 9	ε
		23	3/2+	-5.473	11.317 s 11	ε
		24	0+	-13.933	78.99% 3	
		25	5/2+	-13.192	10.00% 1	
		26	0+	-16.214	11.01% 2	
		27	1/2+	-14.586	9.462 m 11	β-
		28	0+	-15.019	20.90 h 3	β-
		29	3/2+	-10.728	1.38 s 13	β-
		30	0+	-9.100	0.33 s	β-
		31		-3.790s	0.23 s 3	β-, β-n 1.7%
		32	0+	-1.770	120 ms 20	β-, β-n 2.7%
		33		3.930s	90 ms 20	β-, β-n 17%
		34	0+	6.940s	20 ms 10	β-, β-n
		35		13.560s		
13	Al	22	4+	18.040	70 ms	ε, εp
		23		6.767	0.47 s 3	ε, εp
		24	4+	-0.055	2.066 s 10	ε, εα 0.0067%
		24m	1+	0.384	130 ms 4	IT 93%, ε 7%, εα
		25	5/2+	-8.915	7.183 s 12	ε
		26	5+	-12.210	7.2·10⁵ y 3	ε
		26m	0+	-11.982	6.345 s 3	ε
		27	5/2+	-17.197	100%	
		28	3+	-16.851	2.2406 m 5	β-
		29	5/2+	-18.215	6.56 m 6	β-
		30	3+	-15.890	3.60 s 6	β-
		31	(3/2,5/2)+	-15.090	0.644 s 25	β-
		32	1+	-11.180s	35 ms 15	β-
		33		-9.270s		
		34		-4.360s		
		35		-1.440s		
		36		3.910s		
14	Si	24	0+	10.755	0.10 s	ε, εp
		25	5/2+	3.827	220 ms 3	ε, εp
		26	0+	-7.144	2.210 s 21	ε
		27	5/2+	-12.385	4.16 s 2	ε
		28	0+	-21.492	92.23% 1	

3

Bottom-left table:

Z	El	A	Jπ	Δ (MeV)	T1/2 or Abundance	Decay Mode
6	C	19		32.760s		
		20	0+	38.030s		
7	N	10		39.700s		
		11	1/2-	24.910	0.74 MeV 10	p
		12	1+	17.338	11.000 ms 16	ε, ε3α 3.44%
		13	1/2-	5.345	9.965 m 4	ε
		14	1+	2.863	99.634% 9	
		15	1/2-	0.101	0.366% 9	
		16	2-	5.682	7.13 s 2	β-, β-α 0.0012%
		17	1/2-	7.871	4.173 s 4	β-, β-n 95%
		18	1-	13.117	624 ms 12	β-
		19		15.873		
		20		22.100s		
		21		26.050s		
8	O	12	0+	32.060	400 keV 250	p
		13	(3/2-)	23.111	8.90 ms 20	ε
		14	0+	8.007	70.606 s 18	ε
		15	1/2-	2.856	122.24 s 16	ε
		16	0+	-4.737	99.762% 15	
		17	5/2+	-0.809	0.038% 3	
		18	0+	-0.782	0.200% 12	
		19	5/2+	3.332	26.91 s 8	β-
		20	0+	3.796	13.51 s 5	β-
		21		8.130	3.4 s	β-
		22	0+	9.440		
		23		17.460s		
9	F	14		33.610s		
		15	(1/2+)	16.770	1.0 MeV 2	p
		16	(0)-	10.680	40 keV 20	p
		17	5/2+	1.951	64.49 s 16	ε
		18	1+	0.873	109.77 m 5	ε
		19	1/2+	-1.487	100.%	
		20	2+	-0.017	11.00 s 2	β-
		21	5/2+	-0.048	4.32 s 3	β-
		22	(3,4)+	2.830	4.23 s 4	β-
		23	(3/2,5/2)+	3.350	2.23 s 14	β-
		24		8.750s		
		25		12.540s		
10	Ne	16	0+	23.989	≈50 keV	p
		17	1/2-	16.480	109.0 ms 10	ε, εp
		18	0+	5.319	1.672 s 5	ε
		19	1/2+	1.751	17.22 s 2	ε
		20	0+	-7.046	90.51% 9	
		21	3/2+	-5.735	0.27% 2	
		22	0+	-8.027	9.22% 9	
		23	5/2+	-5.155	37.24 s 12	β-
		24	0+	-5.950	3.38 m 2	β-
		25	(1/2,3/2)+	-2.160	602 ms 8	β-
		26	0+	0.440		
		27		6.750s		
11	Na	18		25.320s		
		19		12.929	0.03 s ?	p
		20	2+	6.841	446 ms 3	ε, εα 21%

2

Bottom-right table:

Z	El	A	Jπ	Δ (MeV)	T1/2 or Abundance	Decay Mode
14	Si	29	1/2+	-21.895	4.67% 1	
		30	0+	-24.433	3.10% 1	
		31	3/2+	-22.950	2.62 h 1	β-
		32	0+	-24.081	105 y 13	β-
		33	(3/2+)	-20.570	6.11 s 21	β-
		34	0+	-19.860	2.77 s 20	β-
		35		-14.540s		
		36	0+	-12.760s		
		37		-7.000s		
		38	0+	-4.660s		
15	P	26	(3+)	11.260s	20 ms	ε, εp, ε2p?
		27		-0.750		
		28	3+	-7.161	270.3 ms 5	ε
		29	1/2+	-16.951	4.142 s 15	ε
		30	1+	-20.207	2.498 m 4	ε
		31	1/2+	-24.441	100%	
		32	1+	-24.306	14.26 d 4	β-
		33	1/2+	-26.338	25.34 d 12	β-
		34	1+	-24.558	12.43 s 8	β-
		35	(1/2,3/2)+	-24.940	47.3 s 7	β-
		36		-20.890	5.9 s	β-
		37		-19.100s		
		38		-14.660s		
		39		-12.300s		
		40		-7.620s		
16	S	28	0+	4.130		
		29	5/2+	-3.160	0.187 s 4	ε, εp
		30	0+	-14.063	1.24 s 4	ε
		31	1/2+	-19.045	2.584 s 18	ε
		32	0+	-26.016	95.02% 9	
		33	3/2+	-26.586	0.75% 1	
		34	0+	-29.932	4.21% 8	
		35	3/2+	-28.847	87.51 d 12	β-
		36	0+	-30.664	0.02% 1	
		37	7/2-	-26.897	5.05 m 2	β-
		38	0+	-26.862	2.84 h 1	β-
		39		-23.000s	11.5 s	β-
		40		-22.520		
		41	0+	-17.870s		
		42	0+	-16.420s		
17	Cl	30		4.840s		
		31		-7.070	0.15 s	ε, εp
		32	1+	-13.330	298 ms 2	ε, εp≈0.007%, εα≈0.01%
		33	3/2+	-21.004	2.511 s 3	ε
		34	0+	-24.440	1.5262 s 25	ε
		34m	3+	-24.294	32.23 m 14	ε 53.1%, IT 46.9%
		35	3/2+	-29.014	75.77% 5	
		36	2+	-29.522	3.01·10⁵ y 2	β- 98.1%, ε 1.9%
		37	3/2+	-31.762	24.23% 5	
		38	2-	-29.798	37.24 m 5	β-
		38m	5-	-29.127	715 ms 3	IT
		39	3/2+	-29.804	55.6 m 2	β-
		40	2-	-27.540	1.35 m 2	β-

4

508

Z	El	A	Jπ	Δ (MeV)	T1/2 or Abundance	Decay Mode
17	Cl	41	(1/2,3/2)+	-27.400	34 s 3	β-
		42	2-	-24.420s	6.8 s 3	β-
		43		-23.130	3.3 s 2	β-
		44		-20.010s		
18	Ar	32	0+	-2.180	0.1 s	ε, εp
		33	1/2+	-9.380	173 ms 2	ε, εp 34%
		34	0+	-18.379	845 ms 3	ε
		35	3/2+	-23.049	1.775 s 4	ε
		36	0+	-30.231	0.337% 3	
		37	3/2+	-30.948	35.04 d 1	ε
		38	0+	-34.715	0.063% 1	
		39	7/2-	-33.242	269 y 3	β-
		40	0+	-35.040	99.600% 3	
		41	7/2-	-33.067	1.827 h 7	β-
		42	0+	-34.420	32.9 y 11	β-
		43		-31.980	5.37 m 6	β-
		44	0+	-32.262	11.87 m 5	β-
		45	(7/2-)	-29.720	21.48 s 15	β-
		46	0+	-29.720	8 s 1	β-
19	K	34		-1.480s		
		35	3/2+	-11.168	0.19 s	ε, εp
		36	2+	-17.426	342 ms 2	ε, εp, εα
		37	3/2+	-24.799	1.226 s 7	ε
		38	3+	-28.802	7.636 m 18	ε
		38m	0+	-28.671	924.6 ms 15	ε
		39	3/2+	-33.807	93.2581% 30	
		40	4-	-33.535	1.277·10⁹ y 8 0.0117% 1	β- 89.3%, ε 10.7%
		41	3/2+	-35.560	6.7302% 30	
		42	2-	-35.023	12.360 h 3	β-
		43	3/2+	-36.592	22.3 h 1	β-
		44	2-	-35.810	22.13 m 19	β-
		45	3/2+	-36.611	17.3 m 6	β-
		46	(2-)	-35.420	107 s 10	β-
		47	1/2+	-35.698	17.5 s 3	β-
		48	(2-)	-32.124	6.9 s 2	β-
		49		-30.790	1.3 s	β-, β-n
		50			472 ms 4	β-, β-n 29%
		51			365 ms 5	β-, β-n
		53	(3/2+)		30 ms 5	β-, β-n
20	Ca	36	0+	-6.440	0.1 s	ε, εp
		37	3/2+	-13.160	175 ms 3	ε, εp
		38	0+	-22.060	447 ms 10	ε
		39	3/2+	-27.276	859.6 ms 14	ε
		40	0+	-34.847	96.941% 13	
		41	7/2-	-35.138	1.03·10⁵ y 4	ε
		42	0+	-38.548	0.647% 3	
		43	7/2-	-38.409	0.135% 3	
		44	0+	-41.470	2.086% 5	
		45	7/2-	-40.813	163.8 d 18	β-
		46	0+	-43.138	0.004% 3	
		47	7/2-	-42.343	4.535 d 4	β-
		48	0+	-44.216	≥2·10¹⁶ y 0.187% 3	β-

5

Z	El	A	Jπ	Δ (MeV)	T1/2 or Abundance	Decay Mode
23	V	56		-46.110s		
24	Cr	44	0+	-13.220		
		45	(7/2-)	-19.460	50 ms 6	ε, εp>25%
		46	0+	-29.472	0.26 s 6	ε, εp
		47	3/2-	-34.554	508.0 ms 10	ε
		48	0+	-42.818	21.56 h 3	ε
		49	5/2-	-45.329	42.09 m 15	ε
		50	0+	-50.258	4.345% 9	
		51	7/2-	-51.448	27.704 d 4	ε
		52	0+	-55.415	83.789% 12	
		53	3/2-	-55.283	9.501% 11	
		54	0+	-56.931	2.365% 5	
		55	3/2-	-55.106	3.497 m 3	β-
		56	0+	-55.291	5.94 m 10	β-
		57	3/2-	-52.690s	21 s	β-
		58	0+	-52.050s		
25	Mn	46		-12.470s		
		47		-22.650s		
		48		-29.220s		
		49	(5/2-)	-37.611	0.38 s	ε
		50	0+	-42.626	283.0 ms 4	ε
		50m	5+	-42.399	1.75 m 3	ε
		51	5/2-	-48.240	46.2 m 1	ε
		52	6+	-50.703	5.591 d 3	ε
		52m	2+	-50.325	21.1 m 2	ε 98.32%, IT 1.75%
		53	7/2-	-54.687	3.7·10⁶ y 4	ε
		54	3+	-55.554	312.5 d 5	ε
		55	5/2-	-57.709	100%	
		56	3+	-56.908	2.5785 h 6	β-
		57	5/2-	-57.488	1.45 m	β-
		58	3+	-55.830	65.3 s 7	β-
		58m	(0+)	-55.830	3.0 s 1	β-
		59	3/2-,5/2-	-55.477	4.6 s 1	β-
		60	3+	-52.900	1.79 s 10	β-
		62	(3+)		0.9 s	β-
26	Fe	49	(7/2-)	-24.470	75 ms 10	ε, εp
		50	0+	-34.470		
		51	(5/2-)	-40.218	0.25 s	ε
		52	0+	-48.331	8.275 h 8	ε
		52m	(12+)	-41.491	46 s 2	ε 80%, IT 20%
		53	7/2-	-50.944	8.51 m 2	ε
		53m	19/2-	-47.903	2.58 m 6	IT
		54	0+	-56.251	5.8% 1	
		55	3/2-	-57.477	2.68 y 2	ε
		56	0+	-60.604	91.72% 30	
		57	1/2-	-60.179	2.2% 1	
		58	0+	-62.152	0.28% 1	
		59	3/2-	-60.662	44.496 d 7	β-
		60	0+	-61.407	1.49·10⁶ y 27	β-
		61	3/2-,5/2-	-58.919	5.98 m 6	β-
		62	0+	-58.896	68 s 2	β-
		63	(5/2-)	-55.190	4.9 s	β-
27	Co	51		-27.420s		

7

Z	El	A	Jπ	Δ (MeV)	T1/2 or Abundance	Decay Mode
20	Ca	49	3/2-	-41.291	8.716 m 11	β-
		50	0+	-39.571	13.9 s 6	β-
		51		-35.940	10 s	β-, β-n?
		53	(3/2-,5/2-)		90 ms 15	β-
21	Sc	38		-4.460s		
		39		-14.180s		
		40	4-	-20.527	182.3 ms 7	ε, εp
		41	7/2-	-28.643	596.3 ms 17	ε
		42	0+	-32.124	681.3 ms 7	ε
		42m	(7+)	-31.507	61.6 s 4	ε
		43	7/2-	-36.189	3.891 h 12	ε
		44	2+	-37.815	3.927 h 8	ε
		44m	6+	-37.544	58.6 h 1	IT 98.8%, ε 1.2%
		45	7/2-	-41.070	100%	
		45m	3/2+	-41.058	0.32 s 1	IT
		46	4+	-41.759	83.83 d 2	β-
		46m	1-	-41.616	18.70 s 5	IT
		47	7/2-	-44.331	3.345 d 3	β-
		48	6+	-44.493	43.7 h 1	β-
		49	7/2-	-46.555	57.4 m	β-
		50	5+	-44.538	1.710 m 8	β-
		50m	2+,3+	-44.281	0.35 s 3	IT 98.7%, β- 1.3%
		51	(7/2-)	-43.220	12.4 s 1	β-
		52		-40.040s		
22	Ti	40	0+	-9.064		
		41	3/2+	-15.700	80 ms 2	ε, εp
		42	0+	-25.122	199 ms 6	ε
		43	7/2-	-29.321	513 ms 8	ε
		44	0+	-37.549	54.2 y 21	ε
		45	7/2-	-39.007	3.08 h 1	ε
		46	0+	-44.126	8.0% 1	
		47	5/2-	-44.932	7.3% 1	
		48	0+	-48.487	73.8% 1	
		49	7/2-	-48.558	5.5% 1	
		50	0+	-51.426	5.4% 1	
		51	3/2-	-49.727	5.76 m 1	β-
		52	0+	-49.464	1.7 m 1	β-
		53	(3/2)-	-46.830	32.7 s 9	β-
		54	0+	-45.430s		
23	V	42		-8.220s		
		43		-17.920s		
		44		-23.800s	90 ms 25	ε, εα
		45	7/2-	-31.875	539 ms 18	ε
		46	0+	-37.075	422.33 ms 20	ε
		47	3/2-	-42.005	32.6 m 3	ε
		48	4+	-44.472	15.974 d 3	ε
		49	7/2-	-47.956	330 d 15	ε
		50	6+	-49.220	1.5·10¹⁷ y +3-7 0.250% 2	ε>70%, β-<30%
		51	7/2-	-52.200	99.750% 2	
		52	3+	-51.440	3.75 m 1	β-
		53	7/2-	-51.847	1.61 m 4	β-
		54	3+,4+,5+	-49.889	49.8 s 5	β-
		55	(7/2-)	-49.150	6.54 s 15	β-

6

Z	El	A	Jπ	Δ (MeV)	T1/2 or Abundance	Decay Mode
27	Co	52		-34.300s		
		53	(7/2-)	-42.640	262 ms 25	ε
		53m	(19/2-)	-39.450	0.25 s	ε 98.5%, p≈1.5%
		54	0+	-48.009	193.23 ms 14	ε
		54m	(7+)	-47.811	1.48 m 2	ε
		55	7/2-	-54.026	17.53 h 3	ε
		56	4+	-56.038	78.76 d 12	ε
		57	7/2-	-59.343	270.9 d 6	ε
		58	2+	-59.844	70.916 d 15	ε
		58m	5+	-59.819	9.15 h 10	IT
		59	7/2-	-62.227	100%	
		60	5+	-61.647	5.271 y 1	β-
		60m	2+	-61.588	10.47 m 4	IT 99.75%, β- 0.25%
		61	7/2-	-62.897	1.650 h 5	β-
		62	2+	-61.424	1.50 m 4	β-
		62m	5+	-61.402	13.91 m 5	β-, IT<1%
		63	(7/2-)	-61.839	27.4 s 5	β-
		64	1+	-59.791	0.30 s 3	β-
		65		-59.160		β-
28	Ni	53	(7/2-)	-29.410	45 ms 15	ε, εp
		54	0+	-39.210		
		55	7/2-	-45.330	189 ms 5	ε
		56	0+	-53.904	6.10 d 2	ε
		57	3/2-	-56.077	36.08 h 9	ε
		58	0+	-60.225	68.27% 1	
		59	3/2-	-61.154	7.5·10⁴ y 13	ε
		60	0+	-64.471	26.10% 1	
		61	3/2-	-64.220	1.13% 1	
		62	0+	-66.746	3.59% 1	
		63	1/2-	-65.513	100.1 y 20	β-
		64	0+	-67.098	0.91% 1	
		65	5/2-	-65.125	2.520 h 2	β-
		66	0+	-66.028	54.6 h 4	β-
		67		-63.742	21 s 1	β-
		68	0+	-63.482		β-?
		69		-60.460		
29	Cu	55		-31.630s		
		56		-38.500s		
		57		-47.380s	0.18 s ?	ε
		58	1+	-51.662	3.204 s 7	ε
		59	3/2-	-56.353	81.5 s 5	ε
		60	2+	-58.344	23.2 m 3	ε
		61	3/2-	-61.981	3.408 h 10	ε
		62	1+	-62.797	9.74 m 2	ε
		63	3/2-	-65.579	69.17% 2	
		64	1+	-65.424	12.701 h 2	ε 62.9%, β- 37.1%
		65	3/2-	-67.261	30.83% 2	
		66	1+	-66.256	5.10 m 2	β-
		67	3/2-	-67.303	61.92 h 9	β-
		68	1+	-65.560	31 s 1	β-
		68m	(6-)	-64.838	3.75 m 5	IT 86%, β- 14%
		69	3/2-	-65.741	3.0 m 1	β-
		70	1+	-62.982	4.5 s 1	β-

509

8

Top left

Z El A		Jπ	Δ (MeV)	T1/2 or Abundance	Decay Mode
29 Cu	70m	(5-)	-62.842	46 s 5	β-
	71	(3/2-)	-62.820s	20 s	β-
	72	(1+)		6.6 s	β-
	73			3.9 s	β-
30 Zn	57	(7/2-)	-32.610	40 ms 10	ε, εp
	58	0+	-42.210		ε
	59	3/2-	-47.260	183.7 ms 23	ε, εp
	60	0+	-54.185	2.38 m 5	ε
	61	3/2-	-56.343	89.1 s 2	ε
	62	0+	-61.170	9.26 h 2	ε
	63	3/2-	-62.212	38.1 m 3	ε
	64	0+	-66.002	48.6% 3	
	65	5/2-	-65.910	243.9 d 1	ε
	66	0+	-68.899	27.9% 2	
	67	5/2-	-67.879	4.1% 1	
	68	0+	-70.006	18.8% 4	
	69	1/2-	-68.417	55.6 m 16	β-
	69m	9/2+	-67.978	13.76 h 2	IT 99.97%, β- 0.03%
	70	0+	-69.560	0.6% 1	
	71	(1/2-)	-67.322	2.45 m 10	β-
	71m	(9/2)+	-67.165	3.94 h 5	β-
	72	0+	-68.134	46.5 h 1	β-
	73	(3/2-)	-65.410	23.5 s 10	β-
	74	0+	-65.707	95 s 1	β-
	75		-62.700	10.2 s 3	β-
	76	0+	-62.460	5.7 s 3	β-
	77	(7/2)+	-58.910s	1.4 s 3	β-
	78	0+	-57.960s	1.47 s 15	β-
	79			2.63 s 9	β-, β-n?
	80	0+			
31 Ga	61		-47.540s		
	62	0+	-51.999	116.12 ms 26	ε
	63	3/2-,5/2-	-56.690	32.4 s 5	ε
	64	0+	-58.837	2.630 m 11	ε
	65	3/2-	-62.654	15.2 m 2	ε
	66	0+	-63.724	9.49 h 7	ε
	67	3/2-	-66.878	3.261 d 1	ε
	68	1+	-67.085	68.1 m 3	ε
	69	3/2-	-69.323	60.1% 2	
	70	1+	-68.905	21.15 m 5	β- 99.59%, ε 0.41%
	71	3/2-	-70.142	39.9% 2	
	72	3-	-68.591	14.10 h 1	β-
	73	3/2-	-69.705	4.87 h 3	β-
	74	(4-)	-68.000	8.1 m 1	β-
	74m	1+	-68.000	9.5 s 10	IT, β-?
	75	3/2-	-68.466	2.10 m 3	β-
	76	(3-)	-66.440	32.6 s 6	β-
	77	1/2-,3/2-	-66.410s	13.2 s 2	β-
	78	(3)	-63.560s	5.09 s 5	β-
	79	(3/2-)	-62.760	3.00 s 9	β-, β-n 0.098%
	80		-59.380	1.66 s 9	β-, β-n 0.84%
	81		-57.990	1.23 s 1	β-, β-n 12%

9

Top right

Z El A		Jπ	Δ (MeV)	T1/2 or Abundance	Decay Mode
33 As	87			0.75 s 6	β-, β-n
34 Se	67		-46.860s		
	68	0+	-54.080s	1.6 m	ε
	69		-56.290	27.4 s 2	ε, εp 0.07%
	70	0+	-61.590	41.1 m	ε
	71	5/2-	-63.090s	4.74 m 5	ε
	72	0+	-67.897	8.40 d 8	ε
	73	9/2+	-68.215	7.15 h 8	ε
	73m	1/2-	-68.189	39.8 m 13	IT 73%, ε 27%
	74	0+	-72.215	0.9% 1	
	75	5/2+	-72.171	119.770 d 10	ε
	76	0+	-75.254	9.0% 2	
	77	1/2-	-74.602	7.6% 2	
	77m	7/2+	-74.440	17.45 s 10	IT
	78	0+	-77.028	23.6% 6	
	79	7/2+	-75.919	≤65000 y	β-
	79m	1/2-	-75.823	3.91 m 5	IT
	80	0+	-77.762	49.7% 7	
	81	(1/2)-	-76.392	18.5 m 1	β-
	81m	(7/2)+	-76.289	57.25 m 9	IT, β- 0.07%
	82	0+	-77.596	1.4×10^20 y 9.2% 5	2β-
	83	(9/2)+	-75.343	22.5 m 2	β-
	83m	(1/2)-	-75.123	70.4 s 3	β-
	84	0+	-75.952	3.2 m 2	β-
	85	(5/2+)	-72.420	31.7 s 9	β-
	86	0+	-70.540	15.3 s 9	β-
	87		-66.710s	5.55 s 20	β-, β-n 0.16%
	88	0+		1.53 s 6	β-, β-n 0.8%
	89			0.41 s 4	β-, β-n 5%
	91			0.27 s 5	β-, β-n≈21%
35 Br	69		-46.790s		
	70		-51.190s	80 ms	ε
	71		-56.590s		
	72	(3)	-59.030s	78.6 s 24	ε
	73	3/2-	-63.640	3.4 m 3	ε
	74	(0-,1-)	-65.300	25.3 m 3	ε
	74m	4-	-65.105	41.5 m 15	ε
	75	3/2-	-69.161	97 m 2	ε
	76	1-	-70.302	16.2 h 2	ε
	76m	(4)+	-70.199	1.31 s 2	IT 99.4%, ε 0.6%
	77	3/2-	-73.237	57.036 h 6	ε
	77m	9/2+	-73.131	4.28 m 1	IT
	78	1+	-73.454	6.46 m 4	ε≥99.99%, β-≤0.01%
	79	3/2-	-76.070	50.69% 5	
	79m		-75.863	4.864 s 35	IT
	80	1+	-75.891	17.68 m 2	β- 91.7%, ε 8.3%
	80m	5-	-75.805	4.42 h 1	IT
	81	3/2-	-77.977	49.31% 5	
	82	5-	-77.499	35.30 h 3	β-
	82m	2-	-77.453	6.13 m 8	IT 97.6%, β- 2.4%
	83	(3/2)-	-79.011	2.39 h 2	β-
	84	2-	-77.778	31.80 m 8	β-

11

Bottom left

Z El A		Jπ	Δ (MeV)	T1/2 or Abundance	Decay Mode
31 Ga	82			0.60 s	β-, β-n
	83			0.31 s	β-, β-n
32 Ge	61			40 ms 15	ε, εp
	63		-47.390s		
	64	0+	-54.430	63.7 s 25	ε
	65	3/2-,5/2-	-56.410	30.9 s 7	ε, εp 0.013%
	66	0+	-61.622	2.26 h 5	ε
	67	(1/2-)	-62.656	18.7 m 5	ε
	68	0+	-66.978	271 d	ε
	69	5/2-	-67.097	39.05 h 10	ε
	70	0+	-70.561	20.5% 5	
	71	1/2-	-69.906	11.8 d 4	ε
	72	0+	-72.584	27.4% 6	
	73	9/2+	-71.295	7.8% 2	
	73m	1/2-	-71.228	0.499 s 11	IT
	74	0+	-73.423	36.5% 7	
	75	1/2-	-71.858	82.78 m 4	β-
	75m	7/2+	-71.717	47.7 s 7	IT 99.97%, β- 0.03%
	76	0+	-73.215	7.8% 2	
	77	7/2+	-71.215	11.30 h 1	β-
	77m	1/2-	-71.055	52.9 s 6	β- 79%, IT 21%
	78	0+	-71.863	88 m 1	β-
	79	(1/2-)	-69.530	19.1 s 3	β-
	79m	(7/2+)	-69.344	39.0 s 10	β- 96%, IT 4%
	80	0+	-69.380	29.5 s 4	β-
	81	1/2+	-66.310	7.6 s	β-, β-
	82	0+	-65.380	4.6 s 4	β-
	83		-61.240s	1.9 s 4	β-
	84	0+		1.2 s 3	β-
33 As	65		-47.310s		
	66		-52.070	95.8 ms 4	ε
	67	(3/2,5/2)	-56.650	42.5 s 12	ε
	68	3(+)	-58.880	2.530 m 17	ε
	69	5/2-	-63.080	15.2 m 2	ε
	70	4(+)	-64.340	52.6 m 3	ε
	71	5/2-	-67.893	62 h	ε
	72		-68.228	26.0 h 1	ε
	73	3/2-	-70.955	80.30 d 6	ε
	74	2-	-70.861	17.78 d 3	ε 65.8%, β- 34.2%
	75	3/2-	-73.035	100%	
	76	2-	-72.291	26.32 h 7	β-
	77	3/2-	-73.918	38.83 h 5	β-
	78	(2-)	-72.816	90.7 m 2	β-
	79	3/2-	-73.639	9.01 m 15	β-
	80	1+	-72.165	15.2 s 2	β-
	81	(3/2-)	-72.535	33 s 2	β-
	82	(1+)	-70.078	19 s	β-
	82	(5-)	-70.078	14 s	β-
	83		-69.880	13 s	β-, β-n
	84		-66.080s	5.5 s 3	β-, β-n 0.1%
	84m		-66.080s	0.65 s	β-
	85	(3/2-)	-63.510s	2.028 s 12	β-, β-n 23%
	86			0.9 s 2	β-, β-n≈4%

10

Bottom right

Z El A		Jπ	Δ (MeV)	T1/2 or Abundance	Decay Mode
35 Br	84m	(5,6)	-77.458	6.0 m 2	β-
	85	(3/2-)	-78.607	2.90 m 6	β-
	86	(2-)	-75.640	55.0 s 8	β-
	87	(3/2-)	-73.880	55.69 s 13	β-, β-n 2.3%
	88	(1-)	-70.720	16.3 s 3	β-, β-n 6%
	89	(3/2-)	-68.420s	4.53 s 10	β-, β-n 13%
	90	(2-)	-64.260s	1.71 s 14	β-, β-n 25%
	91			0.541 s 5	β-, β-n 9%
	92			0.365 s 7	β-, β-n 21%
	94				β-, β-n>0%
36 Kr	71		-46.490s	0.1 s	ε
	72	0+	-53.970s	17.2 s 3	ε, εp 0.68%
	73		-56.890	27.0 s 12	ε
	74	0+	-62.140	11.50 m 11	ε
	75	+	-64.246	4.3 m 1	ε
	76	0+	-68.969	14.8 h 1	ε
	77	5/2+	-70.227	74.4 m 6	ε
	78	0+	-74.151	0.35% 2	
	79	1/2-	-74.442	35.04 h 10	ε
	79m	7/2+	-74.312	50 s 3	IT
	80	0+	-77.892	2.25% 2	
	81	7/2+	-77.697	2.1×10^5 y 2	ε
	81m	1/2-	-77.507	13.3 s	IT
	82	0+	-80.591	11.6% 1	
	83	9/2+	-79.983	11.5% 1	
	83m	1/2-	-79.941	1.83 h 2	IT
	84	0+	-82.431	57.0% 3	
	85	9/2+	-81.477	10.72 y 2	β-
	85m	1/2-	-81.172	4.480 h 8	β- 79%, IT 21%
	86	0+	-83.262	17.3% 2	
	87	5/2+	-80.706	76.3 m 5	β-
	88	0+	-79.687	2.84 h 2	β-
	89	(5/2+)	-76.720	3.07 m 9	β-
	90	0+	-74.960	32.32 s 9	β-
	91	(5/2+)	-71.370	8.57 s 5	β-
	92	0+	-68.680	1.85 s 1	β-, β-n 0.03%
	93	(1/2+)	-64.150	1.289 s 12	β-, β-n 1.9%
	94	0+		0.20 s 1	β-, β-n 5.7%
	95			0.78 s 3	β-, β-n
	96?			<0.1 s	β-
37 Rb	73		-46.590s		
	74	(0+)	-51.750	65 ms	ε
	75		-57.280	17.2 s 3	ε
	76	1	-60.580	39.1 s 6	ε
	77	3/2-	-64.950	3.70 m 15	ε
	78	0(+)	-66.980	17.66 m 8	ε
	78m	4(-)	-66.877	5.74 m 6	ε 90%, IT 8%
	79	5/2+	-70.837	22.9 m 5	ε
	80	1+	-72.173	34 s 4	ε
	81	3/2-	-75.461	4.58 h 1	ε, IT
	81m	9/2+	-75.376	32 m	ε, IT
	82	1+	-76.202	1.25 m 3	ε
	82m	5-	-76.102	6.2 h 5	ε
	83	5/2-	-79.044	86.2 d 1	ε

12

510

Isotope Z El A	Jπ	Δ (MeV)	T1/2 or Abundance	Decay Mode
37 Rb 84	2-	-79.746	32.87 d 11	ε 96%, β- 4%
84m	6-	-79.282	20.49 m 17	IT
85	5/2-	-82.164	72.165% 13	
86	2-	-82.743	18.66 d 2	β-, ε 0.005%
86m	(6-)	-82.187	1.017 m 3	IT>99.7%, β-<0.3%
87	3/2-	-84.592	4.80·10^10 y 13 27.835% 13	β-
88	2-	-82.600	17.8 m 1	β-
89	(3/2-)	-81.713	15.2 m 1	β-
90	(1-)	-79.353	153 s 3	β-
90m	(4-)	-79.246	258 s 5	β- 95.7%, IT 4.3%
91	3/2(-)	-77.794	58.4 s 4	β-
92	0(-)	-74.836	4.50 s 2	β-, β-n 0.012%
93	(5/2-)	-72.679	5.85 s 1	β-, β-n 1.3%
94	3(-)	-68.529	2.702 s 5	β-, β-n 10.4%
95	5/2	-65.808	384 ms 6	β-, β-n 9.1%
96	2+	-61.140	0.199 s 3	β-, β-n 13%
97	3/2(-)	-58.280	171.8 ms 16	β-, β-n 24.6%
98	(0)	-54.060	0.114 s 5	β-, β-n 15.9%
99	(3/2-)	-50.860	59 ms	β-, β-n
100			50 ms	β-, β-n
102?			90 ms 20	β-
38 Sr 77	(≥ 5/2)	-57.890	9.0 s 10	ε, εp<0.25%
78	0+	-63.650s	30.6? m 23	ε
79	(3/2-)	-65.340s	2.25 m 10	ε
80	0+	-70.190	106.3 m 15	ε
81	(1/2-)	-71.470	22.2 m 1	ε
82	0+	-75.997	25.6 d	ε
83	7/2+	-76.788	32.4 h 2	ε
83m	1/2-	-76.529	4.95 s 12	IT
84	0+	-80.640	0.56% 1	
85	9/2+	-81.099	64.84 d 3	ε
85m	1/2-	-80.860	67.66 m 7	IT 87.3%, ε 12.7%
86	0+	-84.518	9.86% 1	
87	9/2+	-84.875	7.00% 1	
87m	1/2-	-84.487	2.81 h 1	IT 99.7%, ε 0.3%
88	0+	-87.916	82.58% 1	
89	5/2+	-86.210	50.55 d 9	β-
90	0+	-85.942	28.6 y 3	β-
91	(5/2)+	-83.661	9.52 h 5	β-
92	0+	-82.956	2.71 h 1	β-
93	(7/2+)	-80.121	7.6 m 2	β-
94	0+	-78.836	75.1 s 7	β-
95	(1/2+)	-75.090	25.1 s 2	β-
96	0+	-72.890	1.06 s 4	β-
97		-68.800	0.42 s 3	β-, β-n 0.27%
98	0+	-66.490	0.65 s 3	β-, β-n 0.8%
99		-62.180	0.29 s	β-, β-n
100	0+	-60.020s	0.2 s	β-
101			121 ms 6	β-
102	0+		355 ms 50	β-
39 Y 79		-58.240s		
80	(>2)	-61.190s	33.8 s 6	ε

13

Isotope Z El A	Jπ	Δ (MeV)	T1/2 or Abundance	Decay Mode
40 Zr 92	0+	-88.457	17.15% 1	
93	5/2+	-87.119	1.53·10^6 y 10	β-
94	0+	-87.268	17.38% 2	
95	5/2+	-85.659	64.02 d 4	β-
96	0+	-85.442	>3.56·10^17 y 2.80% 1	
97	1/2+	-82.950	16.90 h 5	β-
98	0+	-81.288	30.7 s 4	β-
99	1/2+	-77.740	2.1 s	β-
100	0+	-76.620	7.1 s 4	β-
101	(3/2)	-73.100	2.1 s 3	β-
102	0+	-71.760s	2.9 s 2	β-
103		-67.610s		
104	0+		1.2 s 1	β-
41 Nb 84	(3+)	-66.740s	12 s 3	ε
85		-66.740s		
86	≈8	-69.580s	1.45 m	ε
87	(9/2+)	-74.180	2.60 m 7	ε
87m	(1/2-)	-74.180	3.82 m 9	ε
88	(8+)	-76.430s	14.3 m	ε
88m	(4-)	-76.430s	7.8 m	ε
89	(1/2)-	-80.622	66 m 2	ε
89	(9/2+)	-80.622	122 m 4	ε
90	8+	-82.659	14.60 h 5	ε
90m	4-	-82.534	18.8 s 1	IT
91	(9/2)+	-86.638	7·10^2 y	ε
91m	(1/2)-	-86.534	62 d	IT 95%, ε 5%
92	(7)+	-86.451	3.5·10^7 y 3	ε
92m	(2)+	-86.315	10.15 d 2	ε
93	9/2+	-87.210	100%	
93m	1/2-	-87.180	13.6 y 3	IT
94	6+	-86.368	2.03·10^4 y 16	β-
94m	3+	-86.327	6.26 m 1	IT 99.5%, β- 0.5%
95	9/2+	-86.784	34.97 d 3	β-
95m	1/2-	-86.548	3.61 d 3	IT 94.4%, β- 5.6%
96	6+	-85.605	23.35 h 5	β-
97	9/2+	-85.608	72.1 m 7	β-
97m	1/2-	-84.865	60 s 8	IT
98	1+	-83.528	2.86 s 6	β-
98m	(5)+	-83.444	51.3 m 4	β-
99	9/2+	-82.327	15.0 s	β-
99m	(1/2)+	-81.958	2.6 m 2	β-, ITw
100		-79.950	3.1 s 3	β-
100		-79.950	1.5 s 3	β-
101		-78.880	7.1 s 3	β-
102	low	-76.350	1.3 s 2	β-
102	high	-76.350	4.3 s 4	β-
103	(5/2+)	-75.110s	1.5 s 2	β-
104		-71.780s	0.91 s 10	β-
104		-71.780s	4.8 s 4	β-
105		-70.140s	1.8 s 8	β-
106			1.1 s 1	β-
42 Mo 87		-67.440	14.6 s 15	ε
88	0+	-72.830s	8.2 m 5	ε

15

Isotope Z El A	Jπ	Δ (MeV)	T1/2 or Abundance	Decay Mode
39 Y 81		-65.950	72 s	ε
82	1+	-68.180	9.5 s	ε
83	(9/2+)	-72.380	7.06 m 8	ε
83	(1/2)-	-72.380	2.85 m 2	ε
84	(5-)	-74.230	40 m 1	ε
84m	1+	-74.230	4.6 s 2	ε
85	(1/2-)	-77.839	2.68 h 5	ε
85m	(9/2+)	-77.819	4.86 h 13	ε
86	4-	-79.278	14.74 h 2	ε
86m	(8)+	-79.060	48 m 1	IT 99.31%, ε 0.7%
87	1/2-	-83.014	80.3 h 3	ε
87m	9/2+	-82.633	12.9 h 4	IT 98.43%, ε 1.57%
88	4-	-84.294	106.64 d 8	ε
89	1/2-	-87.702	100%	
89m	9/2+	-86.793	16.06 s 4	IT
90	2-	-86.488	64.1 h 1	β-
90m	7+	-85.806	3.19 h 1	IT, β- 0.002%
91	1/2-	-86.347	58.51 d 6	β-
91m	(9/2)+	-85.791	49.71 m 4	IT
92	2-	-84.844	3.54 h 1	β-
93	1/2-	-84.235	10.1 h 2	β-
93m	9/2+	-83.476	0.82 s 1	IT
94	2-	-82.348	18.7 m 1	β-
95	1/2-	-81.214	10.3 m 2	β-
96	0-	-78.300	6.2 s 2	β-
96m	(3+)	-78.200	9.6 s 2	β-
96m		-78.200	2.3 m 1	β-
97	(1/2-)	-76.270	3.5 s 2	β-, β-n 0.06%
97m	(9/2+)	-75.603	1.23 s 2	β-, β-n 0.3%
98	1+	-72.370	0.64 s 3	β-, β-n 0.3%
98m	(4-)	-72.370	2.0 s 2	β-
99	(1/2-)	-70.130	1.5 s	β-, β-n 1%
100	(3+)	-66.720s	0.94 s	β-
100m		-66.720s	0.5 s	β-
101	(5/2)	-64.380s	0.50 s 5	β-
102			0.27 s 7	β-
40 Zr 81		-58.790	11 m	ε
82	0+	-64.180	2.5 m	ε
83		-66.360	44 s	ε
83			8 s	ε
84	0+	-71.430s	28 m	ε
85	(7/2+)	-73.150	7.86 m 4	ε
85m	(1/2-)	-72.858	10.9 s 3	IT
86	0+	-77.980s	16.5 h 1	ε
87	(9/2+)	-79.349	1.73 h 8	ε
87m	(1/2-)	-79.012	14.0 s 2	IT
88	0+	-83.626	83.4 d 3	ε
89	9/2+	-84.869	78.43 h 8	ε
89m	1/2-	-84.281	4.18 m 1	IT 93.76%, ε 6.24%
90	0+	-88.770	51.45% 2	
90m	5-	-86.451	809.2 ms 20	IT
91	5/2+	-87.893	11.22% 2	

14

Isotope Z El A	Jπ	Δ (MeV)	T1/2 or Abundance	Decay Mode
42 Mo 89	9/2+	-75.004	2.2 m	ε
89m	1/2-	-75.004	0.19 s	IT
90	0+	-80.172	5.67 h 5	ε
91	9/2+	-82.200	15.49 m 1	ε
91m	1/2-	-81.547	65.2 s 8	IT 50.1%, ε 49.9%
92	0+	-86.808	14.84% 4	
93	5/2+	-86.804	3.5·10^3 y 7	ε
93m	(21/2)+	-84.379	6.85 h 7	IT 99.88%, ε 0.12%
94	0+	-88.413	9.25% 2	
95	5/2+	-87.709	15.92% 4	
96	0+	-88.792	16.68% 4	
97	5/2+	-87.542	9.55% 2	
98	0+	-88.113	24.13% 6	
99	1/2+	-85.967	66.0 h 2	β-
100	0+	-86.186	9.63% 2	
101	1/2+	-83.513	14.6 m 1	β-
102	0+	-83.559	11.3 m 2	β-
103	(3/2+)	-80.610s	67.5 s 15	β-
104	0+	-80.480s	60 s 2	β-
105		-77.140s	50 s	β-
105		-77.140s	30 s 1	β-
106	0+	-76.430s	8.4 s 5	β-
107		-72.510s	3.5 s 5	β-
108	0+		1.5 s 4	β-
43 Tc 89		-68.000s		
90	1+	-70.970s	8.3 s	ε
90	high	-70.970s	49.2 s	ε
91	(9/2+)	-75.980	3.14 m 2	ε
91m	(1/2-)	-75.630	3.3 m 1	ε
92	(8+)	-78.938	4.4 m 3	ε
93	9/2+	-83.606	2.75 h 5	ε
93m	1/2-	-83.216	43.5 m 10	IT 80%, ε 20%
94	7+	-84.157	4.88 h 2	ε
94m	(2)+	-84.084	52 m 1	ε, IT<0.1%
95	9/2+	-86.018	20.0 h 1	ε
95m	1/2-	-85.979	61 d 2	ε 96%, IT 4%
96	7+	-85.819	4.28 d 6	ε
96m	4+	-85.785	51.5 m 10	IT 98%, ε 2%
97	9/2+	-87.222	2.6·10^6 y 4	ε
97m	1/2-	-87.126	90.5 d 10	IT
98	(6)+	-86.429	4.2·10^6 y 3	β-
99	9/2+	-87.324	2.13·10^5 y 5	β-
99m	1/2-	-87.181	6.02 h 3	IT, β-w
100	1+	-86.017	15.8 s 1	β-
101	9/2+	-86.325	14.2 m 1	β-
102	1+	-84.573	5.28 s 15	β-
102m	(4,5)	-84.273	4.35 m 7	β-≈98%, IT≈2%
103	5/2+	-84.606	54.2 s 2	β-
104	(3+)	-82.480	18.3 m 3	β-
105		-82.140s	7.7 m 2	β-
106	(1,2)	-79.630s	36 s 1	β-
107		-78.960s	21.2 s 2	β-
108	(3)	-75.990s	5.17 s 7	β-

16

Isotope Z El A	Jπ	Δ (MeV)	T1/2 or Abundance	Decay Mode
43 Tc 109		-74.910s	1.4 s 4	β-
110			0.83 s 4	β-
44 Ru 91		-68.400s		
92	0+	-74.410s	3.65 m 5	ε
93	(9/2)+	-77.270	60 s	ε
93m	(1/2-)	-76.536-	10.8 s	ε 79%, IT 21%
94	0+	-82.567	51.8 m 6	ε
95	5/2+	-83.449	1.64 h 1	ε
96	0+	-86.071	5.52% 5	
97	5/2+	-86.111	2.9 d 1	ε
98	0+	-88.225	1.88% 5	
99	5/2+	-87.618	12.7% 1	
100	0+	-89.220	12.6% 1	
101	5/2+	-87.951	17.0% 1	
102	0+	-89.099	31.6% 2	
103	(3/2)+	-87.261	39.26 d 2	β-
104	0+	-88.098	18.7% 2	
105	3/2+	-85.937	4.44 h 2	β-
106	0+	-86.330	371.63 d 17	β-
107	(5/2+)	-83.710	3.75 m 5	β-
108	0+	-83.700	4.55 m 5	β-
109		-80.810s	34.5 s 10	β-
109?		-80.810s	12.9 s	β-, IT?
110	0+	-80.340s	14.6 s 10	β-
111		-76.920s	1.5 s	β-
112	0+		4.65 s 14	β-
113			2.69 s 10	β-
45 Rh 93		-69.110s		
94	(3+)	-72.970s	70.6 s 6	ε
94	(8+)	-72.970s	25.8 s 2	ε
95	(9/2+)	-78.340	5.02 m 10	ε
95m	(1/2-)	-77.797	1.96 m 4	IT 88%, ε 12%
96	5(+)	-79.630	9.6 m	ε
96m	2(+)	-79.578	1.51 m 2	IT 60%, ε 40%
97	(9/2+)	-82.600	31.1 m 8	ε
97m	(1/2-)	-82.342	44.3 m 8	ε 95.1%, IT 4.9%
98	(2)+	-83.168	8.7 m 2	ε
98m	(5+)	-83.118	3.5 m 3	ε
99	(1/2-)	-85.519	16.1 d	ε
99m	9/2+	-85.454	4.7 h	ε
100	1-	-85.590	20.8 h 1	ε
100m	(5+)	-85.250	4.7 m	IT 93%, ε 7%
101	1/2-	-87.413	3.3 y 3	ε
101m	9/2+	-87.256	4.34 d 1	ε 92.3%, IT 7.7%
102	6(+)	-86.803	≈2.9 y	ε 75%, β- 20%, IT 5%
102m	(1-,2-)	-86.733	207 d 3	
103	1/2-	-88.027	100%	
103m	7/2+	-87.987	56.12 m 1	IT
104	1+	-86.954	42.3 s 4	β- 99.55%, ε 0.45%
104m	5+	-86.825	4.34 m 5	IT 99.87%, β- 0.13%
105	7/2+	-87.853	35.36 h 6	β-

Isotope Z El A	Jπ	Δ (MeV)	T1/2 or Abundance	Decay Mode
47 Ag 103	7/2+	-84.780	65.7 m 7	ε
103m	1/2-	-84.646	5.7 s 3	IT
104	5+	-85.118	69.2 m 10	ε
104m	2+	-85.111	33.5 m 20	ε 67%, IT 33%
105	1/2-	-87.076	41.29 d 7	ε
105m	7/2+	-87.051	7.23 m 16	IT 98.12%, ε 1.88%
106	1+	-86.944	24.0 m 1	ε≥99%, β-≤1%
106m	6+	-86.854	8.46 d 10	ε
107	1/2-	-88.407	51.839% 5	
107m	7/2+	-88.314	44.3 s 2	IT
108	1+	-87.605	2.37 m 1	β- 97.15%, ε 2.85%
108m	6+	-87.496	127 y 21	ε 91.3%, IT 8.7%
109	1/2-	-88.720	48.161% 5	
109m	7/2+	-88.632	39.6 s 2	IT
110	1+	-87.458	24.6 s 2	β- 99.7%, ε 0.3%
110m	6+	-87.340	249.76 d 4	β- 98.64%, IT 1.36%
111	1/2-	-88.218	7.45 d 1	β-
111m	7/2+	-88.158	64.8 s 8	IT 99.3%, β- 0.7%
112	2-	-86.623	3.14 h 2	β-
113	1/2-	-87.041	5.37 h 5	β-
113m	(7/2)+	-86.998	68.7 s 50	IT 80%, β- 20%
114	1+	-84.990	4.6 s 2	β-
115	(1/2-)	-84.950	20.0 m 5	β-
115	(7/2+)	-84.950	18.0 s 7	β-
116		-82.720	2.68 m 1	β-
116m		-82.639	10.4 s 8	β- 98%, IT 2%
117	(1/2-)	-82.250	72.8 s +20-7	β-
117	(7/2+)	-82.250	5.34 s 5	β-
118		-79.580	4.0 s	β-
118m		-79.452	2.8 s 3	β- 59%, IT 41%
119	(7/2+)	-78.590	2.1 s 1	β-
120		-75.770	1.17 s 5	β-
120m		-75.567	0.32 s 4	β- 63%, IT 37%
121		-74.550	0.8 s 1	β-
122			1.5 s 5	β-, β-n
123			0.39 s 3	β-, β-n
48 Cd 97			3 s	ε, εp
98	0+		≦8 s	ε, εp?
99		-69.990s	16 s	ε, εp
100	0+	-74.310s	1.1 m 3	ε
101	(5/2+)	-75.690	1.2 m 2	ε
102	0+	-79.700s	5.5 m 5	ε
103	(5/2+)	-80.620	7.3 m 1	ε
104	0+	-83.974	57.7 m 10	ε
105	5/2+	-84.339	55.5 m 4	ε
106	0+	-87.132	1.25% 3	
107	5/2+	-86.990	6.50 h 2	ε
108	0+	-89.260	0.89% 1	
109	5/2+	-88.536	462.9 d 20	ε
110	0+	-90.351	12.49% 9	
111	1/2+	-89.255	12.80% 6	

Isotope Z El A	Jπ	Δ (MeV)	T1/2 or Abundance	Decay Mode
45 Rh 105m	1/2-	-87.723	45 s	IT
106	1+	-86.370	29.80 s 8	β-
106m	(6)+	-86.230	130 m 2	β-
107	(7/2+)	-86.862	21.7 m 4	β-
108		-85.090	6.0 m 3	β-
108m	1+	-85.020	16.8 s 5	β-
109	7/2+	-85.014	80 s 2	β-
110	1+	-82.940	3.2 s 2	β-
110	(≥ 2)	-82.940	28.5 s 15	β-
111		-82.320s	11 s 1	β-
112		-79.730s	0.8 s 1	β-
113?			0.91 s 8	β-
114?		-78.840s	1.68 s 7	β-
46 Pd 94	0+		9.0 s 5	ε
95		-70.150s		
95m	(21/2+)	-68.150s	13.3 s 3	ε, εp 0.93%
96	0+	-76.370s	2.0 m	ε
97	(5/2+)	-77.800	3.1 m 1	ε
98	0+	-81.299	17.7 m 3	ε
99	(5/2+)	-82.192	21.4 m	ε
100	0+	-85.207	3.63 d 9	ε
101	5/2+	-85.431	8.47 h 6	ε
102	0+	-87.902	1.020% 12	
103	5/2+	-87.455	16.991 d 19	ε
104	0+	-89.397	11.14% 8	
105	5/2+	-88.419	22.33% 8	
106	0+	-89.910	27.33% 5	
107	5/2+	-88.374	6.5·10^6 y 3	β-
107m	11/2-	-88.159	21.3 s 5	IT
108	0+	-89.522	26.46% 9	
109	5/2+	-87.604	13.7 h 1	β-
109m	11/2-	-87.415	4.69 m 1	IT
110	0+	-88.337	11.72% 9	
111		-86.020	23.4 m 2	β-
111m	11/2-	-85.848	5.5 h 1	IT 73%, β- 27%
112	0+	-86.329	21.05 h 1	β-
113		-83.680	98 s	β-
113		-83.680	89 s	β-
114	0+	-83.540	2.4 m 1	β-
115		-80.490s	47 s	β-
116	0+	-80.110	12.7 s 4	β-
117			5.0 s 6	β-
118	0+		3.1 s 3	β-
47 Ag 96			5.1 s	ε, εp
97		-70.900s	21 s 3	ε
98	(7+)	-73.090s	44.5 s 12	ε, εp>0%
99	(9/2+)	-76.760	2.07 m	ε
99m	(1/2-)	-76.760	11 s	IT
100	(5+)	-78.120	2.0 m	ε
100		-78.120	2.3 m	ε
101	9/2+	-81.220	11.1 m 3	ε
101m	1/2-	-80.947	3.10 s 10	IT
102	5+	-82.020	12.9 m 3	ε
102m	2+	-82.011	7.7 m 5	ε 51%, IT 49%

Isotope Z El A	Jπ	Δ (MeV)	T1/2 or Abundance	Decay Mode
48 Cd 111m	11/2-	-88.859	48.6 m 3	IT
112	0+	-90.582	24.13% 11	
113	1/2+	-89.051	9.3·10^15 y 19	β- 12.22% 6
113m	11/2-	-88.787	13.7 y	β- 99.86%, IT 0.14%
114	0+	-90.023	28.73% 21	
115	1/2+	-88.092	53.46 h 10	β-
115m	11/2-	-87.911	44.6 d 3	β-
116	0+	-88.721	7.49% 9	
117	1/2+	-86.417	2.49 h 4	β-
117m	(11/2)-	-86.281	3.36 h 5	β-
118	0+	-86.710	50.3 m 2	β-
119	(1/2+)	-83.940	2.69 m 2	β-
119m	(11/2-)	-83.793	2.20 m 2	β-
120	0+	-83.973	50.80 s 21	β-
121		-80.950	13.5 s 3	β-
121m		-80.950	8 s	β-
122	0+	-80.580s	5.5 s 1	β-
123		-77.320s		β-
124	0+		0.9 s 2	β-
126	0+		0.506 s 15	β-
128	0+		0.94 s 5	β-
49 In 100				ε, εp
101		-68.410s		ε
102	(5)	-70.420s	23 s 4	ε
103	(9/2+)	-74.420s	65 s 7	ε
104	5+	-75.970s	1.7 m 2	ε
105	9/2+	-79.589	4.9 m 3	ε
105m	(1/2-)	-78.915	43 s	IT
106	(7)+	-80.590	6.2 m 1	ε
106m	(3)+	-80.190	5.2 m 1	ε
107	9/2+	-83.570	32.4 m 3	ε
107m	1/2-	-82.892	50.4 s 6	IT
108	3+	-84.135	39.6 m 7	ε
108	(6+)	-84.135	58.0 m 12	ε
109	9/2+	-86.505	4.2 h 1	ε
109m	1/2-	-85.855	1.34 m 7	IT
109m	(19/2+)	-84.395	0.21 s 1	IT
110	2+	-86.410	69.1 m 5	ε
110	7+	-86.410	4.9 h 1	ε
111	9/2+	-88.391	2.83 d 1	ε
111m	1/2-	-87.854	7.7 m 2	IT
112	1+	-87.994	14.4 m 2	ε 56%, β- 44%
112m	4+	-87.837	20.9 m 2	IT
113	9/2+	-89.367	4.3% 2	
113m	1/2-	-88.975	1.658 h 1	IT
114	1+	-88.570	71.9 s 1	β- 99.5%, ε 0.5%
114m	5+	-88.380	49.51 d 1	IT 95.7%, ε 4.3%
115	9/2+	-89.534	4.41·10^14 y 25	β- 95.7% 2
115m	1/2-	-89.198	4.486 h 4	IT 95%, β- 5%
116	1+	-88.247	14.10 s 3	β- >99.94%, ε <0.06%

Isotope Z El A	Jπ	Δ (MeV)	T1/2 or Abundance	Decay Mode
49 In 116m	5+	-88.120	54.15 m 6	β-
116m	8-	-87.957	2.18 s 4	IT
117	9/2+	-88.943	43.8 m 7	β-
117m	1/2-	-88.628	116.5 m 7	β- 52.9%, IT 47.1%
118	1+	-87.450	5.0 s 3	β-
118m	(5+)	-87.390	4.45 m 5	β-
118m	(8-)	-87.250	8.5 s 3	IT 98.5%, β- 1.5%
119	9/2+	-87.730	2.4 m 1	β-
119m	1/2-	-87.419	18.0 m 3	β- 97.5%, IT 2.5%
120	(5)+	-85.800	44.4 s -10	β-
120	1+	-85.800	3.08 s 8	β-
121	9/2+	-85.840	23.1 s 6	β-
121m	1/2-	-85.526	3.88 m 10	β- 98.8%, IT 1.2%
122		-83.580	10.0 s 5	β-
122	(1+)	-83.580	1.5 s 3	β-
123	(9/2)+	-83.420	5.98 s 6	β-
123m	(1/2-)	-83.100	47.8 s 5	β-
124	3+	-81.060	3.17 s 5	β-
124m	(5 to 8)	-80.870	2.4 s 4	β-
125	(9/2+)	-80.420	2.33 s 4	β-
125m	(1/2-)	-80.240	12.2 s 1	β-
126	(6,7,8)	-77.810	1.45 s 22	β-
126m	3+	-77.660	1.5 s 2	β-
127	(9/2+)	-77.010	1.15 s 5	β-
127m	(1/2-)	-76.850	3.76 s 3	β-, β-n
128	(2,3)+	-74.000	0.9 s 1	β-, β-n
129	(9/2+)	-73.030	0.59 s 2	β-, β-n
129m	(1/2-)	-72.830	1.26 s 2	β-, β-n
130	10-	-69.990s	0.51 s	β-, β-n
130		-69.990s	0.53 s 5	β-, β-n
131	(9/2+)	-68.550	0.27 s 2	β-, β-n
132			0.22 s	β-, β-n
50 Sn 102	0+	-64.800s		
103		-66.920s	7 s 3	ε, εp
104	0+	-71.470s		ε
105		-73.270	31 s	ε, εp
106	0+	-77.290s	2.10 m 15	ε
107	5/2+,7/2+	-78.370s	2.90 m 5	ε
108	0+	-82.090	10.30 m 8	ε
109	7/2(+)	-82.630	18.0 m 2	ε
110	0+	-85.830	4.11 h 10	ε
111	7/2+	-85.939	35.3 m 8	ε
112	0+	-88.654	0.97% 1	
113	1/2+	-88.328	115.09 d 4	ε
113m	1/2+	-88.251	21.4 m 4	IT 91.1%, ε 8.9%
114	0+	-90.557	0.65% 1	
115	1/2+	-90.032	0.36% 1	
116	0+	-91.523	14.53% 11	
117	1/2+	-90.396	7.68% 7	
117m	11/2-	-90.081	13.61 d 4	IT
118	0+	-91.652	24.22% 11	
119	1/2+	-90.066	8.58% 4	
119m	11/2-	-89.976	293.0 d 13	IT

Isotope Z El A	Jπ	Δ (MeV)	T1/2 or Abundance	Decay Mode
51 Sb 126m	(5+)	-86.382	19.0 m 3	β- 86%, IT 14%
126m	(3-)	-86.360	≈11 s	IT
127	7/2+	-86.705	3.85 d 5	β-
128	8-	-84.600	9.01 h 3	β-
128m	5+	-84.580	10.4 m 2	β- 96.4%, IT 3.6%
129	7/2+	-84.631	4.40 h 1	β-
129		-84.631	17.7 m	β-
130	(8-)	-82.360	38.4 m	β-
130m	(4,5)+	-82.330	6.3 m 2	β-
131	(7/2+)	-82.020	23 m 2	β-
132	(8-)	-79.730	4.1 m	β-
132	(4+)	-79.730	3.07 m 2	β-
133		-74.000	2.5 m	β-
134	(0-)	-74.000	0.85 s 10	β-
134	(7-)	-74.000	10.43 s 14	β-, β-n 0.1%
135		-70.310s	1.71 s	β-, β-n 20%
136		-70.310s	0.82 s 2	β-, β-n 32%
52 Te 106	0+	-58.050s	0.06 ms	α
107		-60.510s	3.6 ms +6-4	α 70%, ε 30%
108	0+	-65.620s	2.1 s 1	α 68%, ε 32%, εp
109		-67.650	4.6 s 3	ε 96%, εp, α 4%
110	0+	-72.140s	18.6 s 8	ε, α
111		-73.470	19.3 s 4	ε, εp, α
112	0+	-77.300	2.0 m 2	ε
113	(7/2+)	-78.320s	1.7 m 2	ε
114	0+	-81.760s	15.2 m 7	ε
115	7/2+	-82.250	5.8 m 2	ε
115m	(1/2)+	-82.230	6.7 m 4	ε
116	0+	-85.280	2.49 h 4	ε
117	1/2+	-85.110	62 m 2	ε
118	0+	-87.647	6.00 d 2	ε
119	1/2+	-87.178	16.05 h 5	ε
119m	11/2-	-86.878	4.69 d 4	ε
120	0+	-89.380	0.096% 2	
121	1/2+	-88.542	16.78 d 35	ε
121m	11/2-	-88.248	154 d 7	IT 88.6%, ε 11.4%
122	0+	-90.309	2.60% 1	
123	1/2+	-89.172	1.3·10^13 y 0.908% 3	ε
123m	11/2-	-88.925	119.7 d 1	IT
124	0+	-90.525	4.816% 8	
125	1/2+	-89.025	7.14% 1	
125m	11/2-	-88.880	58 d 1	IT
126	0+	-90.067	18.95% 1	
127	3/2+	-88.286	9.35 h 7	β-
127m	11/2-	-88.198	109 d 2	IT 97.6%, β- 2.4%
128	0+	-88.993	>8.·10^24 y 2 31.69% 2	2β-
129	3/2+	-87.008	69.6 m 2	β-
129m	11/2-	-86.903	33.6 d 1	IT 64%, β- 36%
130	0+	-87.348	2.51·10^21 y 27 33.80% 2	2β-
131	3/2+	-85.206	25.0 m 1	β-
131m	11/2-	-85.024	30 h 2	β- 77.8%, IT 22.2%

Isotope Z El A	Jπ	Δ (MeV)	T1/2 or Abundance	Decay Mode
50 Sn 120	0+	-91.102	32.59% 10	
121	3/2+	-89.202	27.06 h 4	β-
121m	11/2-	-89.196	55 y 5	IT 77.6%, β- 22.4%
122	0+	-89.945	4.63% 3	
123	11/2-	-87.820	129.2 d 4	β-
123m	3/2+	-87.795	40.08 m 7	β-
124	0+	-88.237	5.79% 5	
125	11/2-	-85.898	9.64 d 3	β-
125m	3/2+	-85.870	9.52 m 5	β-
126	0+	-86.021	≈1.0·10^5 y	β-
127	(11/2-)	-83.504	2.10 h 4	β-
127m	(3/2+)	-83.499	4.13 m 3	β-
128	0+	-83.310	59.1 m 5	β-
128m	(7-)	-81.219	6.5 s 5	IT
129	(3/2+)	-80.630	2.16 m 4	β-
129m	(11/2-)	-80.595	6.7 m 4	β-, IT 0.0002%
130	0+	-80.190	3.72 m 11	β-
130m	(7-)	-78.390	1.7 m 1	β-
131	(3/2+)	-77.370	61 s 3	β-
131m		-77.370	39 s	β-
132	0+	-76.610	40 s 1	β-
133		-69.990s	1.47 s	β-, β-n
134	0+		1.04 s 2	β-, β-n 17%
51 Sb 104		-59.270s		
105		-64.090s		
106		-66.490s		
107		-70.670s		
108		-72.530s	7.0 s 5	ε, εp
109	(5/2+)	-76.250	17.0 s 7	ε
110	3+	-77.530s	23.0 s 4	ε
111	(5/2+)	-80.840s	75 s 1	ε
112	3+	-81.589	51.4 s 10	ε
113	5/2+	-84.421	6.67 m 7	ε
114	3+	-84.670	3.49 m 3	ε
115	5/2+	-87.002	32.1 m 3	ε
116	3+	-86.816	15.8 m 8	ε
116m	8-	-86.326	60.3 m 6	ε
117	5/2+	-88.641	2.80 h 1	ε
118	1+	-87.995	3.6 m	ε
118m	8-	-87.783	5.00 h 1	ε
119	5/2+	-89.472	38.1 h 2	ε
120	1+	-88.421	15.89 m 4	ε
120	8-	-88.421	5.76 d 2	ε
121	5/2+	-89.591	57.3% 9	
122	2-	-88.326	2.70 d 1	β- 97.62%, ε 2.38%
122m	(8-)	-88.163	4.2 m 2	IT
123	7/2+	-89.223	42.7% 9	
124	3-	-87.619	60.20 d 3	β-
124m	5+	-87.608	93 s 5	IT 75%, β- 25%
124m	8-	-87.582	20.2 m 2	IT
125	7/2+	-88.258	2.73 y 3	β-
126	(8-)	-86.400	12.4 d 5	β-

Isotope Z El A	Jπ	Δ (MeV)	T1/2 or Abundance	Decay Mode
52 Te 132	0+	-85.217	78.2 h 8	β-
133	(3/2+)	-82.990	12.45 m 28	β-
133m	(11/2-)	-82.656	55.4 m 4	β- 83%, IT 17%
134	0+	-82.410	41.8 m 8	β-
135		-77.850	19.2 s	β-
136	0+	-74.410	18 s	β-, β-n 0.7%
137	(7/2-)	-69.480s	2.5 s	β-, β-n 2%
138	0+		1.4 s 4	β-, β-n 6%
53 I 108		-52.550s		
109		-57.760s		
110		-60.490s	0.65 s 2	ε 83%, α 17%, εp, εα
111		-64.970s	7.5 s	ε 99.9%, α 0.1%
112		-67.120s	3.42 s 11	ε, εp, εα, α
113		-71.120	5.9 s 5	ε, εα, α
114		-72.860s	2.1 s 2	ε
115	(5/2+)	-76.300s	28 s	ε
116	1+	-77.520	2.91 s 15	ε
117	(5/2+)	-80.610s	2.3 m 1	ε
118	(2-)	-81.250s	14.3 m	ε
118m	8-	-81.146s	8.5 m 5	ε, IT
119	(5/2+)	-83.810	19.1 m 4	ε
120	2-	-83.980	81.0 m 6	ε
120m		-83.080	53 m 4	ε
121	5/2+	-86.263	2.12 h 1	ε
122	1+	-86.075	3.62 m 6	ε
123	5/2+	-87.939	13.2 h 1	ε
124	2-	-87.368	4.18 d 2	ε
125	5/2+	-88.847	60.14 d 11	ε
126	2-	-87.912	13.02 d 7	ε 56.3%, β- 43.7%
127	5/2+	-88.984	100%	β-
128	1+	-87.738	24.99 m 2	β- 93.1%, ε 6.9%
129	7/2+	-88.506	1.57·10^7 y 4	β-
130	5+	-86.897	12.36 h 1	β-
130m	2+	-86.849	9.0 m 1	IT 83%, β- 17%
131	7/2+	-87.455	8.04 d 1	β-
132	4+	-85.710	2.30 h 3	β-
132m	(8-)	-85.590	83.6 m 17	IT 86%, β- 14%
133	7/2+	-85.910	20.8 h 1	β-
133m	19/2-	-84.276	9 s	IT
134	4+	-83.970	52.6 m 4	β-
134m	8-	-83.654	3.69 m 7	IT 97.7%, β- 2.3%
135	7/2+	-83.813	6.61 h 1	β-
136	(2-)	-79.510	84 s 1	β-
136m	(.6-)	-79.360	45 s 1	β-
137	(7/2+)	-76.500	24.5 s 2	β-, β-n 6.4%
138	(2-)	-72.310	6.41 s 5	β-, β-n 5%
139		-68.880	2.30 s 5	β-, β-n 10%
140			0.86 s 4	β-, β-n 14%
141			0.43 s 2	β-, β-n 21.2%
142			0.2 s	β-
54 Xe 110	0+	-51.750s	0.2 s	ε, α
111		-54.380s	0.7 s	ε
111		-54.380s	0.9 s 2	ε, α

Isotope Z El A	Jπ	Δ (MeV)	T1/2 or Abundance	Decay Mode
54 Xe 112	0+	−59.880s	2.8 s 2	ε 99.16%, α 0.84%
113		−62.130	2.8 s 2	ε, εp, εα, α
114	0+	−66.910s	10.0 s 4	ε
115		−68.670s	18 s 4	ε, εp 0.3%
116	0+	−73.020s	56 s 2	ε
117	(1/2+)	−74.290s	61 s 2	ε, εp 0.003%
118	0+	−78.050s	4 m	ε
119	(7/2+)	−78.820	5.8 m 3	ε
120	0+	−82.030	40 m 1	ε
121	(5/2+)	−82.490	40.1 m 2	ε
122	0+	−85.540	20.1 h 1	ε
123	(1/2)+	−85.261	2.08 h 2	ε
124	0+	−87.660	0.10% 1	
125	(1/2)+	−87.191	16.9 h 2	ε
125m	(9/2)−	−86.939	57 s 1	IT
126	0+	−89.162	0.09% 1	
127	(1/2+)	−88.323	36.4 d 1	ε
127m	(9/2−)	−88.026	69.2 s 9	IT
128	0+	−89.861	1.91% 3	
129	1/2+	−88.697	26.4% 6	
129m	11/2−	−88.461	8.89 d 2	IT
130	0+	−89.881	4.1% 1	
131	3/2+	−88.426	21.2% 4	
131m	11/2−	−88.262	11.9 d 1	IT
132	0+	−89.290	26.9% 5	
133	3/2+	−87.665	5.245 d 6	β−
133m	11/2−	−87.431	2.188 d 8	IT
134	0+	−88.125	10.4% 2	
134m	7−	−86.166	290 ms 17	IT
135	3/2+	−86.509	9.09 h 1	β−
135m	11/2−	−85.982	15.29 m 3	IT, β− 0.004%
136	0+	−86.431	8.9% 1	
137	7/2−	−82.385	3.818 m 13	β−
138	0+	−80.130	14.08 m 8	β−
139		−75.700	39.68 s 14	β−
140	0+	−73.020	13.60 s 10	β−
141		−68.360	1.73 s 1	β−, β−n 0.044%
142	0+	−65.550	1.22 s 2	β−, β−n 0.41%
143			0.30 s 3	β−
143			0.96 s 2	β−
144	0+		1.15 s 20	β−
145			0.9 s 3	β−, β−n
55 Cs 113		−51.610s		
114	(1+)	−54.710s	0.57 s 2	ε, εp 7%, εα 0.16%, α 0.02%
115		−59.550s	1.4 s 8	ε, εp 0.3%
116	≥ 4+	−62.300	3.81 s 16	ε, εp, εα
116	(1+,2+,3+)	−62.300	0.71 s 8	ε, εp, εα
117		−66.230	6.5 s	ε
118		−68.240	16.4 s 12	ε, εp 0.04%, εα 0.0024%
119	9/2(+)	−72.200	37.7 s 10	ε, εα?
119	3/2	−72.200	28 s 1	ε
120	2	−73.770	64 s	ε, εp, εα

Isotope Z El A	Jπ	Δ (MeV)	T1/2 or Abundance	Decay Mode
56 Ba 131	1/2+	−86.721	11.8 d 2	ε
131m	9/2−	−86.533	14.6 m 2	IT
132	0+	−88.453	0.101% 2	
133	1/2+	−87.572	10.74 y 5	ε
133m	11/2−	−87.284	38.9 h 1	IT 99.99%, ε 0.01%
134	0+	−88.972	2.417% 27	
135	3/2+	−87.873	6.592% 18	
135m	11/2−	−87.605	28.7 h 2	IT
136	0+	−88.909	7.854% 39	
136m	7−	−86.879	0.306 s	IT
137	3/2+	−87.736	11.23% 4	
137m	11/2−	−87.074	2.5513 m 7	IT
138	0+	−88.276	71.70% 7	
139	7/2−	−84.928	84.63 m 34	β−
140	0+	−83.294	12.746 d 10	β−
141	3/2−	−79.771	18.27 m 7	β−
142	0+	−77.910	10.6 m 2	β−
143	5/2+	−74.070	14.5 s 5	β−
144	0+	−71.870	11.4 s 5	β−
145	5/2+	−68.040	4.0 s	β−
146	0+	−65.100	2.20 s 3	β−
147		−61.260s	0.72 s 7	β−, β−n
148	0+	−58.510s	0.607 s 25	β−, β−n
149		−53.390s		
57 La 123			17 s 3	ε ---
124			29 s 3	ε
125	(11/2−)		76 s 6	ε
126			1.0 m 3	ε
127	(3/2+)	−77.980s	3.8 m 5	ε
127m	(11/2−)	−77.980s	5.0 m ?	ε
128	4−,5−	−78.880	5.0 m 3	ε
129	(3/2+)	−81.380	11.6 m 2	ε
129m	11/2−	−81.208	0.56 s 5	IT
130	(3+)	−81.600s	8.7 m 1	ε
131	3/2+	−83.760	59 m 2	ε
132	2−	−83.740	4.8 h 2	ε
132m	6−	−83.551	24.3 m 5	IT 76%, ε 24%
133	5/2(+)	−85.570s	3.912 h 8	ε
134	1+	−85.270	6.45 m 16	ε
135	5/2+	−86.673	19.5 h	ε
136	1+	−86.040	9.87 m 3	ε
137	7/2+	−87.130	$6 \cdot 10^4$ y 2	ε
138	5+	−86.531	$1.28 \cdot 10^{11}$ y 12 ε 66.7%, β− 33.3% 0.09% 1	
139	7/2+	−87.238	99.91% 1	
140	3−	−84.328	40.272 h 7	β−
141	7/2(+)	−83.000	3.92 h 3	β−
142	2−	−80.025	91.1 m 5	β−
143	7/2+	−78.320	14.23 m 14	β−
144		−74.850	40.9 s 4	β−
145		−72.990	24.8 s 20	β−
146	(2−)	−69.370	6.27 s 10	β−
146m	(6)	−69.090	10.0 s 1	β−

Isotope Z El A	Jπ	Δ (MeV)	T1/2 or Abundance	Decay Mode
55 Cs 120		−73.770	60.2 s 15	ε
121	3/2+	−77.090	136 s	ε
121m	9/2(+)	−77.090	121 s	IT, ε
122	8	−78.160	4.5 m 2	ε
122	1(+)	−78.160	21.0 s 7	ε
123	1/2+	−81.050	5.87 m 5	ε
123m	(11/2−)	−80.891	1.60 s 15	IT
124	1+	−81.720	30.8 s 5	ε
124m	(7)+	−81.257	6.3 s 2	IT
125	1/2+	−84.092	45 m 1	ε
126	1+	−84.334	1.64 m 2	ε
127	1/2+	−86.231	6.25 h 10	ε
128	1+	−85.926	3.62 m 2	ε
129	1/2+	−87.536	32.06 h 6	ε
130	1+	−86.859	29.2 m	ε 98.4%, β− 1.6%
131	5/2+	−88.079	9.69 d 1	ε
132	2(−)	−87.160	6.475 d 10	ε 98%, β− 2%
133	7/2+	−88.093	100%	
134	4+	−86.913	2.062 y 5	β−, ε 0.0003%
134m	8−	−86.774	2.91 h 1	IT
135	7/2+	−87.668	$3 \cdot 10^6$ y	β−
135m	(19/2−)	−86.041	53 m 2	IT
136	5+	−86.361	13.16 d 3	β−
136	8	−86.361	19 s 2	β−
137	7/2+	−86.561	30.17 y	β−
138	3(−)	−82.900	32.2 m 1	β−
138m	(6−)	−82.820	2.90 m 10	IT 81%, β− 19%
139	7/2	−80.715	9.27 m 5	β−
140	1−	−77.076	63.7 s 3	β−
141	(7/2+)	−74.515	24.94 s 6	β−, β−n 0.029%
142		−70.590	1.8 s	β−, β−n 0.28%
143	3/2+	−67.790	1.78 s 1	β−, β−n 1.7%
144	1	−60.940	1.02 s 3	β−, β−n 3%
144?	>3	−63.410	<1 s	β−
145	3/2+	−60.240	0.59 s 1	β−, β−n 12%
146	(2−)	−55.690	0.343 s 7	β−, β−n 14%
147		−52.380	0.22 s 1	β−, β−n
148		−47.590	170 ms 7	β−
56 Ba 117		−56.930s	1.9 s 2	ε, εp
118	0+	−61.950s		
119		−63.960s	5.35 s 30	ε, εp
120	0+	−68.470s	32 s	ε
121	0+	−70.140s	29.7 s 15	ε, εp 0.02%
122	0+	−74.360s	2.0 m	ε
123		−75.260s	2.7 m 4	ε
124	0+	−78.820s	11.9 m 10	ε
125	(1/2+)	−79.510	3.5 m 4	ε
125		−79.510	8 m	ε
126	0+	−82.660s	100 m 2	ε
127	(1/2+)	−82.780	12.7 m 4	ε
128	0+	−85.478	2.43 d 5	ε
129	1/2+	−85.100	2.23 h 11	ε
129m	(7/2)+	−85.092	2.17 h 4	ε
130	0+	−87.299	0.106% 2	

Isotope Z El A	Jπ	Δ (MeV)	T1/2 or Abundance	Decay Mode
57 La 147		−66.970s	4.4 s 5	β−, β−n
148	(2)−	−63.910s	1.05 s 1	β−
149		−61.190s	1.2 s 4	β−
150		−57.890s		
58 Ce 124	0+		6 s 2	ε
125			9 s	ε, εp
126	0+		50 s 6	ε
127			32 s 4	ε
128	0+		6 m	ε
129			3.5 m 5	ε
130	0+	−79.400s	25 m 2	ε
131		−79.860s	10 m 1	ε
131		−79.860s	5 m 1	ε
132	0+	−82.440s	3.5 h	ε
133	9/2(−)	−82.570s	5.40 h 5	ε
133	1/2(+)	−82.570s	97 m	ε
134	0+	−84.870s	75.9 h 9	ε
135	1/2(+)	−84.657	17.6 h	ε
135m	(11/2−)	−84.212	20 s	IT
136	0+	−86.500	0.19% 1	
137	3/2+	−85.910	9.0 h 3	ε
137m	11/2−	−85.656	34.4 h 3	IT 99.22%, ε 0.78%
138	0+	−87.575	0.25% 1	
139	3/2+	−86.973	137.66 d 13	ε
139m	11/2−	−86.219	56.4 s 5	IT
140	0+	−88.089	88.48% 10	
141	7/2−	−85.446	32.501 d 5	β−
142	0+	−84.542	$>5 \cdot 10^{16}$ y 11.08% 10	β−
143	3/2−	−81.615	33.0 h 2	β−
144	0+	−80.442	284.4 d	β−
145	5/2+	−77.100	2.98 m 15	β−
146	0+	−75.760	13.52 m 13	β−
147		−72.160	56.4 s 12	β−
148	0+	−70.410	56 s 1	β−
149		−67.290s	5.2 s 3	β−
150	0+	−65.510s	4.0 s 6	β−
151		−62.260s	1.02 s 6	β−
59 Pr 121			1.4 s 8	ε, εp
129			24 s 5	ε
130			28 s	ε
132		−75.340s	1.6 m 3	ε
133	5/2(+)	−78.070s	6.5 m 3	ε
134	2−	−78.770s	17 m 2	ε
134m	(5−)	−78.770s	11 m	ε
135	3/2(+)	−80.910	25 m	ε
136	2+	−81.380	13.1 m 1	ε
137	5/2+	−83.200	1.28 h 2	ε
138	1+	−83.138	1.45 m 5	ε
138m	7−	−82.774	2.1 h 1	ε
139	5/2+	−84.844	4.41 h 4	ε
140	1+	−84.701	3.39 m 1	ε
141	5/2+	−86.027	100%	

Isotope Z El A	Jπ	Δ (MeV)	T1/2 or Abundance	Decay Mode
59 Pr 142	2-	-83.799	19.12 h 4	β- 99.98%, ε 0.02%
142m	5-	-83.795	14.6 m 5	IT
143	7/2+	-83.077	13.58 d 3	β-
144	0-	-80.760	17.28 m 5	β-
144m	3-	-80.701	7.2 m 2	IT 99.96%, β- 0.04%
145	(7/2)+	-79.636	5.98 h 2	β-
146	(2)-	-76.780	24.15 m 18	β-
147	(5/2+)	-75.470	13.6 m 5	β-
148	1-	-72.460	2.27 m 4	β-
148m	(4)	-72.370	2.0 m 1	β-
149	(5/2+)	-70.988	2.26 m 7	β-
150	(1)	-68.590s	6.19 s 16	β-
151		-67.160s	4.0 s 7	β-
152		-64.560s	3.2 s	β-
60 Nd 129			5.9 s 6	ε, εp
130	0+		28 s	ε
132	0+		1.8 m	ε
133			1.2 m	ε
134	0+	-75.570s	8.5 m 15	ε
135	9/2(-)	-76.210s	12 m	ε
135m		-75.910s	≈5.5 m	ε
136	0+	-79.170	50.65 m 33	ε
137	1/2+	-79.400	38.5 m 15	ε
137m	11/2-	-78.880	1.60 s 15	IT
138	0+	-82.140s	5.04 h 9	ε
139	3/2+	-82.040	29.7 m 5	ε
139m	11/2-	-81.809	5.5 h 2	ε 88%, IT 12%
140	0+	-84.481	3.37 d 2	ε
141	3/2+	-84.213	2.49 h 3	ε
141m	11/2-	-83.456	62.4 s 9	IT 99.97%, ε 0.03%
142	0+	-85.959	27.13% 10	
143	7/2-	-84.012	12.18% 5	
144	0+	-83.757	$2.1 \cdot 10^{15}$ y 4	α 23.80% 10
145	7/2-	-81.441	$>6 \cdot 10^{16}$ y	α 8.30% 5
146	0+	-80.935	17.19% 8	
147	5/2-	-78.156	10.98 d 1	β-
148	0+	-77.418	5.76% 3	
149	5/2-	-74.385	1.725 h 7	β-
150	0+	-73.694	$>1 \cdot 10^{18}$ y 5.64% 3	2β-
151	(3/2+)	-70.957	12.44 m 7	β-
152	0+	-70.160	11.4 m 2	β-
153		-67.370s		
154	0+	-65.770s	40 s 10	β-
61 Pm 132			4 s	ε
133			12 s	ε
134			24 s 2	ε
135	(11/2-)		0.8 m	ε
136	(5+)	-71.280s	107 s 6	ε

29

Isotope Z El A	Jπ	Δ (MeV)	T1/2 or Abundance	Decay Mode
61 Pm 137	(11/2)-	-74.100s	2.4 m 1	ε
138		-75.140s	3.24 m 5	ε
139	(5/2)+	-77.520	4.15 m 5	ε
140	1+	-78.410	9.2 s 2	ε
140m	(7-)	-78.010	5.95 m 5	ε
141	5/2+	-80.480	20.90 m 5	ε
142	1+	-81.070	40.5 s 5	ε
143	5/2+	-82.969	265 d 7	ε
144	5-	-81.424	363 d 14	ε
145	5/2+	-81.280	17.7 y 4	ε, αw
146	3-	-79.450	5.53 y 5	ε 66.1%, β- 33.9%
147	7/2+	-79.052	2.6234 y 2	β-
148	1-	-76.874	5.370 d 9	β-
148m	6-	-76.737	41.29 d 11	β- 95.4%, IT 4.6%
149	7/2+	-76.074	53.08 h 5	β-
150	(1-)	-73.607	2.68 h 2	β-
151	5/2+	-73.400	28.40 h 4	β-
152	1+	-71.270	4.1 m 1	β-
152m	(4-)	-71.150	7.52 m 8	β-
152m	≥6	-71.150	15 m 1	β-, IT
153	5/2-	-70.669	5.4 m 2	β-
154	(0,1)	-68.470	1.7 m 2	β-
154	(3,4)	-68.470	2.7 m 1	β-
155	(5/2)	-67.100s	48 s 4	β-
156		-64.480s		
62 Sm 133			32.0 s	ε, εp
134	0+		12 s 3	ε
135			10 s	ε, εp
136	0+		42 s	ε
137			44 s 8	ε
138	0+	-71.340s	3.0 m 3	ε
139	(1/2+)	-72.090	2.57 m 10	ε
139m	(11/2-)	-71.632	9.5 s 10	IT 93.7%, ε 6.3%
140	0+	-75.410s	14.82 m 10	ε
141	5/2+	-75.942	10.2 m 2	ε
141m	11/2-	-75.766	22.6 m 2	ε 99.69%, IT 0.31%
142	0+	-78.986	72.49 m 5	ε
143	3/2+	-79.526	8.83 m 1	ε
143m	11/2-	-78.772	66 s 2	IT 99.76%, ε 0.24%
144	0+	-81.974	3.1% 1	
145	7/2-	-80.660	340 d 3	ε
146	0+	-80.992	$10.3 \cdot 10^7$ y 5	α
147	7/2-	-79.276	$1.06 \cdot 10^{11}$ y 2 15.0% 2	α
148	0+	-79.346	$7 \cdot 10^{15}$ y 3 11.3% 2	α
149	7/2-	-77.147	$>2 \cdot 10^{15}$ y 13.8% 1	α?
150	0+	-77.061	7.4% 1	
151	5/2-	-74.587	90 y 6	β-
152	0+	-74.773	26.7% 2	
153	3/2+	-72.566	46.7 h 1	β-

30

Isotope Z El A	Jπ	Δ (MeV)	T1/2 or Abundance	Decay Mode
62 Sm 154	0+	-72.466	22.7% 2	
155	3/2-	-70.202	22.1 m 2	β-
156	0+	-69.380	9.4 h 2	β-
157		-66.870	8.0 m 5	β-
158	0+	-65.200s	5.51 m 9	β-
63 Eu 138			35 s 6	ε
139			1.5 s 4	ε
139			22 s 3	ε
140		-67.210s	20 s +15-1	ε
140		-67.210s	1.3 s 2	ε
141	5/2+	-69.980	40.0 s 7	ε
141m	11/2-	-69.884	3.3 s 3	ε 67%, IT 33%
142	1+	-71.590	2.4 s 2	ε
142m	8-	-71.410	1.22 m 2	ε
143	5/2+	-74.380	2.63 m 5	ε
144	1+	-75.645	10.2 s 1	ε
145	5/2+	-77.998	5.93 d 4	ε
146	4-	-77.114	4.59 d 3	ε
147	5/2+	-77.555	24 d 1	ε, αw
148	5-	-76.266	54.5 d 5	ε, αw
149	5/2+	-76.452	93.1 d 4	ε
150	0(-)	-74.798	12.62 h 10	β- 89%, ε 11%
150	(4-,5-)	-74.798	35.8 y 10	ε
151	5/2+	-74.663	47.8% 5	
152	3-	-72.897	13.33 y 4	ε 72.08%, β- 27.92%
152m	0-	-72.851	9.32 h 1	β- 72%, ε 28%
152m	(8)-	-72.749	96 m 1	IT
153	5/2+	-73.379	52.2% 5	
154	3-	-71.749	8.8 y 1	β- 99.98%, ε 0.02%
154m	(8-)	-71.592	46.0 m 3	IT
155	5/2+	-71.829	4.96 y 1	β-
156	0+	-70.094	15.19 d 6	β-
157	5/2+	-69.473	15.15 h 4	β-
158	(1-)	-67.250	45.9 m 2	β-
159	(5/2+)	-66.059	18.1 m 1	β-
160	(0-)	-63.450s	44 s 4	β-
64 Gd 142	0+	-67.190s	1.5 m 3	ε
143	(1/2+)	-68.480s	39 s 2	ε
143m	11/2-	-68.380s	1.83 m 1	ε, IT?
144	0+	-71.940s	4.5 m 1	ε
145	1/2+	-72.950	23.9 m 1	ε
145m	11/2-	-72.201	85 s 3	IT 95.3%, ε 4.7%
146	0+	-76.100	48.27 d 10	ε
147	7/2-	-75.505	38.1 h 1	ε
148	0+	-76.278	74.6 y 30	α
149	7/2-	-75.131	9.4 d 3	ε, αw
150	0+	-75.766	$1.79 \cdot 10^6$ y 8	α
151	7/2-	-74.198	120 d 20	ε, αw
152	0+	-74.719	$1.08 \cdot 10^{14}$ y 8 0.20% 1	α
153	3/2-	-72.895	241.6 d 2	ε
154	0+	-73.718	2.18% 3	

31

Isotope Z El A	Jπ	Δ (MeV)	T1/2 or Abundance	Decay Mode
64 Gd 155	3/2-	-72.082	14.80% 5	
156	0+	-72.547	20.47% 4	
157	3/2-	-70.835	15.65% 3	
158	0+	-70.702	24.84% 12	
159	3/2-	-68.573	18.56 h 8	β-
160	0+	-67.954	21.86% 4	
161	5/2-	-65.518	3.66 m 5	β-
162	0+	-64.260	8.4 m 2	β-
163			68 s 3	β-
65 Tb 144	(1)	-62.940s	5 s	ε
145		-66.200s	30 s	ε
146	1+	-67.860	8 s 4	ε
146m	5-	-67.860	23 s 2	ε
147	(5/2+)	-70.960	1.65 h 10	ε
147	(11/2-)	-70.960	1.83 m 6	ε
148	2-	-70.670	60 m 1	ε
148m	(9)+	-70.670	2.20 m 5	ε
149	(1/2+)	-71.495	4.13 h 2	ε 82.8%, α 17.2%
149m	(11/2-)	-71.455	4.16 m 4	ε, α 0.022%
150	(2-)	-71.102	3.27 h 10	ε, α≤0.05%
150	(9)+	-71.102	5.8 m 2	ε
151	1/2(+)	-71.632	17.6 h 1	ε, α 0.01%
151m	11/2-	-71.632	50 s 1	IT
152	(2)-	-70.869	17.5 h 3	ε
152m	(8)-	-70.367	4.3 m 2	IT 78.9%, ε 21.1%
153	5/2+	-71.316	2.34 d 1	ε
154	0(-)	-70.160	21.4 h 5	ε
154m	3(-)	-70.160	9.0 h 5	ε 78.2%, IT 21.8%
154m	(7-,8-)	-70.160	22.6 h 6	ε 98.2%, IT 1.8%
155	3/2+	-71.260	5.32 d 6	ε
156	3(-)	-70.103	5.34 d 9	ε
156m	(0+)	-70.015	5.0 h 1	IT, ε, β-w
156m	(4+)	-70.015	24.4 h 10	IT
157	3/2+	-70.773	150 y 30	ε
158	3-	-69.480	150 y 30	ε 82%, β- 18%
158m	0-	-69.370	10.5 s 2	IT
159	3/2+	-69.544	100%	
160	3-	-67.848	72.3 d 2	β-
161	3/2+	-67.473	6.90 d 2	β-
162	1-	-65.660	7.76 m 10	β-
163	3/2+	-64.690	19.5 m 3	β-
164	(5+)	-62.120	3.0 m 1	β-
165	3/2+		2.11 m 10	β-
66 Dy 145			18 s	ε
146	0+	-62.860s	29 s 3	ε
147	1/2+	-64.570	80 s	ε
147m	11/2-	-63.819	58 s	IT, ε, εp
148	0+	-67.980	3.1 m 1	ε
149	(7/2-)	-67.890s	4.23 m 18	ε
150	0+	-69.325	7.17 m 2	ε 64%, α 36%
151	7/2-	-68.902	16.9 m 5	ε 94.4%, α 5.6%
152	0+	-70.127	2.38 h 2	ε 99.9%, α 0.1%
153	7/2(-)	-69.146	6.4 h 1	ε 99.99%, α 0.01%
154	0+	-70.393	$3 \cdot 10^6$ y	α

32

Top-left table:

Isotope Z El A	Jπ	Δ (MeV)	T1/2 or Abundance	Decay Mode
66 Dy 155	3/2-	-69.166	10.0 h 3	ε
156	0+	-70.536	>1.0·10^18 y	
			0.06% 1	
157	3/2-	-69.434	8.1 h 1	ε
158	0+	-70.419	0.10% 1	
159	3/2-	-69.178	144.4 d 2	ε
160	0+	-69.683	2.34% 5	
161	5/2+	-68.065	18.9% 1	
162	0+	-68.190	25.5% 2	
163	5/2-	-66.390	24.9% 2	
164	0+	-65.977	28.2% 2	
165	7/2+	-63.622	2.334 h 6	β-
165m	1/2-	-63.514	1.258 m 6	IT 97.76%, β- 2.24%
166	0+	-62.594	81.6 h 1	β-
167	(1/2-)	-59.940	6.2 m	β-
168	0+		8.5 m 5	β-
67 Ho 146	(10+)		3.9 s 8	ε
147			?	ε, εp
148	1+	-58.370s	2.2 s 11	ε
148m	(4-)	-58.370s	9 s 1	ε
149	(11/2-)	-61.850s	21.4 s 18	ε
150	(2-)	-62.220s	72 s 10	ε
150	9+	-62.220s	24 s	ε
151	(11/2-)	-63.803	35.6 s 4	ε 90%, α 10%
151	(5/2+)	-63.803	47 s 2	ε 80%, α 20%
152	(3+)	-63.740	2.35 m 1	ε 88%, α 12%
152	(9)	-63.740	52.3 s 5	ε 89.5%, α 10.5%
153	(11/2-)	-65.023	2.0 m 1	ε 99.95%, α 0.05%
153m	(5/2)	-64.963	9.3 m 5	ε 99.82%, α 0.18%
154	1	-64.637	11.8 m 5	ε, α 0.017%
154	(8+)	-64.637	3.2 m 1	ε, α<0.002%
155	5/2(+)	-66.064	48 m 1	ε, α
156	(5+)	-65.540s	2 m	ε
156m	1	-65.540s	55.6 m 6	IT, ε
156?		-65.540s	7.4 m	
157	7/2-	-66.890	12.6 m 2	ε
158	5+	-66.200	11.3 m 4	ε
158m	(9+)	-66.200	21.3 m 23	ε
158m	2-	-66.133	27 m 2	IT 65%, ε 35%
159	7/2-	-67.342	33 m 1	ε
159m	1/2+	-67.136	8.30 s 8	IT
160	5+	-66.397	25.6 m 3	ε
160m	2-	-66.337	5.02 h 5	IT 65%, ε 35%
160m	(1+)	-66.325	<2 m	?
161	7/2-	-67.208	2.48 h 5	ε
161m	1/2+	-66.997	6.73 s 10	IT
162	1+	-66.051	15 m 1	ε
162m	6-	-65.945	67 m 1	IT 63%, ε 37%
163	(7/2-)	-66.387	>10 y	ε
163m	(1/2+)	-66.089	1.09 s 3	IT
164	1+	-64.939	29 m 1	ε 58%, β- 42%
164m	(6)-	-64.799	37.5 m +15-5	IT
165	7/2-	-64.908	100%	

Top-right table:

Isotope Z El A	Jπ	Δ (MeV)	T1/2 or Abundance	Decay Mode
69 Tm 163	1/2+	-62.738	1.81 h 6	ε
164	1+	-61.990	2.0 m 1	ε
164m	6(-)	-61.990	5.1 m 1	IT 80%, ε 20%
165	1/2+	-62.939	30.06 h 11	ε
166	2+	-61.888	7.70 h 3	ε
167	1/2+	-62.552	9.24 d 2	ε
168	3(+)	-61.321	93.1 d 1	ε, β-?
169	1/2+	-61.282	100%	
170	1-	-59.804	128.6 d 3	β- 99.85%, ε 0.15%
171	1/2+	-59.219	1.92 y 1	β-
172	2-	-57.383	63.6 h 2	β-
173	(1/2+)	-56.267	8.24 h 8	β-
174	(4-)	-53.860	5.4 m 1	β-
175	(1/2+)	-52.300	15.2 m 5	β-
176	(4+)	-49.600s	1.9 m 1	β-
70 Yb 152	0+	-46.320s		
153		-47.270s	4.0 s 5	α
154	0+	-50.120s	0.42 s 2	α
155		-50.740s	1.65 s 15	α≈90%, ε
156	0+	-53.380	24 s 1	ε, α 21%
157		-53.620s	38.6 s 10	ε 99.5%, α 0.5%
158	0+	-56.023	1.38 m 14	ε, α 0.003%
159		-55.930s	12 s	ε, α
160	0+	-58.060s	4.8 m 2	ε
161	(3/2-)	-57.810s	4.2 m 2	ε
162	0+	-59.750s	18.87 m 19	ε
163	(3/2-)	-59.370	11.05 m 25	ε
164	0+	-60.990s	75.8 m 17	ε
165	(5/2)-	-60.176	9.9 m	ε
166	0+	-61.595	56.7 h 1	ε
167	5/2-	-60.598	17.5 m 2	ε
168	0+	-61.578	0.13% 1	
169	7/2+	-60.374	32.022 d 8	ε
169m	1/2-	-60.350	46 s 2	IT
170	0+	-60.772	3.05% 5	
171	1/2-	-59.315	14.3% 2	
172	0+	-59.264	21.9% 3	
173	5/2-	-57.560	16.12% 18	
174	0+	-56.953	31.8% 4	
175	7/2-	-54.704	4.19 d 1	β-
176	0+	-53.502	12.7% 1	
176m	(8)-	-52.451	11.4 s 5	IT
177	9/2+	-50.997	1.9 h 1	β-
177m	1/2-	-50.665	6.41 s 2	IT
178	0+	-49.706	74 m 3	β-
179			8 m	β-
71 Lu 151		-38.380s	0.09 s	p
153		-39.630s	1.0 s	ε
154		-42.770s	0.07 s 2	α, ε
155	high	-43.720s	≈0.5 s	α, ε
156	low	-43.720s	0.23 s 3	α
157		-46.630s	5.5 s 3	ε 94%, α 6%

Bottom-left table:

Isotope Z El A	Jπ	Δ (MeV)	T1/2 or Abundance	Decay Mode
67 Ho 166	0-	-63.081	26.80 h 2	β-
166m	(7-)	-63.076	1.20·10^3 y 18	β-
167	(7/2-)	-62.292	3.1 h 1	β-
168	3+	-60.280	3.0 m 1	β-
169	7/2-	-58.806	4.7 m 1	β-
170	(6+)	-56.250	2.8 m	β-
170	1(+)	-56.250	43 s	β-
68 Er 147			2.5 s	ε, εp
148	0+		4.5 s 4	ε
149			9 s 2	ε, εp
150	0+	-58.020s	18.5 s 7	ε
151		-58.500s	23 s 2	ε
152	0+	-60.620	10.3 s	α 90%, ε 10%
153		-60.670s	37.1 s	α 53%, ε 47%
154	0+	-62.623	3.75 m 12	ε, α 0.5%
155	(7/2-)	-62.360	5.3 m 3	ε ≥99.98%, α≥0.02%
156	0+	-64.000s	20 m	ε
157	3/2(-)	-63.420	25 m 3	ε
158	0+	-65.200s	2.25 h 7	ε
159	3/2-	-64.573	36 m	ε
160	0+	-66.063	28.59 h 9	ε
161	3/2-	-65.209	3.21 h 3	ε
162	0+	-66.347	0.14% 1	
163	5/2-	-65.177	75.0 m 4	ε
164	0+	-65.952	1.61% 1	
165	5/2-	-64.531	10.36 h 4	ε
166	0+	-64.935	33.6% 2	
167	7/2+	-63.299	22.95% 13	
167m	1/2-	-63.091	2.28 s 3	IT
168	0+	-62.999	26.8% 2	
169	1/2-	-60.931	9.40 d 2	β-
170	0+	-60.118	14.9% 1	
171	5/2-	-57.728	7.52 h 3	β-
172	0+	-56.493	49.3 h 5	β-
173	(7/2-)	-53.770	1.4 m 1	β-
69 Tm 147			0.5 s	p
148	(10+)		0.7 s 2	ε
150	(4-,5-,6-)		3.5 s 6	ε
151		-51.000s		ε
152		-51.740s	5 s	ε
153	low	-54.180s	1.59 s 8	α≥95%, ε≤5%
154		-54.630s	8.3 s	ε, α
154	(9+)	-54.630s	3.4 s	α, ε
155		-56.810	25 s	α, ε
156	low	-56.970s	80 s 3	ε, α
156	high	-56.970	19 s 3	α
157		-58.790s	3.5 m 2	ε, αw
158	(2-)	-58.700s	4.02 m 10	ε
159	5/2+	-60.570s	9 m	ε
160	1-	-60.460s	9.2 m 4	ε
161	7/2+	-62.010s	38 m 4	ε
162	1-	-61.560	22.0 m 7	ε
162m	5+	-61.368	24.3 s 17	IT 90%, ε 10%

Bottom-right table:

Isotope Z El A	Jπ	Δ (MeV)	T1/2 or Abundance	Decay Mode
71 Lu 158		-47.240	10.4 s 1	ε >98.5%, α<1.5%
159		-49.850	12 s	ε, α
160		-50.260s	35.5 s 8	ε
161		-52.510s	72 s 6	ε
162	(1-)	-52.660s	1.37 m 2	ε
162m	(4-)	-52.660s	≈1.5 m	IT
162m		-52.660s	≈1.5 m	IT
163		-54.740s	4.1 m	ε
164		-54.690s	3.17 m	ε
164?		-54.690s	2 m	
165	1/2	-56.380s	12 m 1	ε
166	(6-)	-56.120	2.8 m	ε
166m	(3-)	-56.086	1.41 m 10	ε 58%, IT 42%
166m	(0-)	-56.077	2.12 m 10	ε >80%, IT<20%
167	7/2+	-57.470	51.5 m 10	ε
168	(6)-	-57.110	5.5 m 1	ε
168m	(3)+	-56.890	6.7 m 4	ε, IT?
169	7/2+	-58.081	34.06 h 5	ε
169m	1/2-	-58.052	160 s 10	IT
170	0+	-57.332	2.00 d 3	ε
170m	4-	-57.239	0.67 s 10	IT
171	7/2+	-57.836	8.24 d 3	ε
171m	1/2-	-57.765	79 s 2	IT
172	(4-)	-56.743	6.70 d 3	ε
172m	(1-)	-56.701	3.7 m 5	IT
173	7/2+	-56.887	1.37 y 1	ε
174	(1-)	-55.577	3.31 y 5	ε
174m	(6-)	-55.406	142 d 2	IT 99.35%, ε 0.65%
175	7/2+	-55.173	97.41% 2	
176	7-	-53.395	3.60·10^10 y 16	β- 2.59% 2
176m	1-	-53.269	3.68 h 1	β-
177	7/2+	-52.395	6.71 d 1	β-
177m	23/2-	-51.425	160.9 d 3	β- 79%, IT 21%
178	1(+)	-50.336	28.4 m 2	β-
178m	(9)-	-50.036	23 m	β-
179	(7/2+)	-49.130	4.59 h 6	β-
180	(3-,4-)	-46.690	5.7 m 1	β-
181	(7/2)+		3.5 m 3	β-
182			2.0 m	β-
183			58 s	β-
72 Hf 154	0+	-32.820s	2 s	ε
155		-34.440s	0.9 s	ε
156	0+	-37.860s	25 ms 4	α
157		-38.960s	110 ms 6	α 91%, ε 9%
158	0+	-42.300s	2.9 s 2	ε 54%, α 46%
159		-43.090s	5.6 s	ε, α
160	0+	-46.060	≈12 s	ε, α
161		-46.480s	17 s 2	α
162	0+	-49.179	37.6 s 8	ε
163		-49.390s	40 s	ε
164	0+	-51.790s	2.8 m 2	ε
165		-51.650s		

Top-left table (page 37):

Isotope Z El A	Jπ	Δ (MeV)	T1/2 or Abundance	Decay Mode
72 Hf 166	0+	-53.790s	6.77 m 30	ε
167	(5/2-)	-53.470s	2.05 m 5	ε
168	0+	-55.210s	25.95 m 2	ε
169	(5/2-)	-54.730	3.24 m 4	ε
170	0+	-56.130s	16.01 h 13	ε
171	(7/2+)	-55.440s	12.1 h 4	ε
172	0+	-56.390	1.87 y 3	ε
173	1/2-	-55.290s	23.6 h	ε
174	0+	-55.849	$2.0 \cdot 10^{15}$ y 4	α
			0.162% 2	
175	5/2-	-54.486	70 d 2	ε
176	0+	-54.581	5.206% 4	
177	7/2-	-52.893	18.606% 3	
177m	23/2+	-51.578	1.08 s 6	IT
177m	37/2-	-50.153	51.4 m 5	IT
178	0+	-52.447	27.297% 3	
178m	8-	-51.300	4.0 s 2	IT
178m	(16,17+)	-49.947	31 y 1	IT
179	9/2+	-50.476	13.629% 5	
179m	(1/2-)	-50.101	18.68 s 6	IT
179m	25/2-	-49.370	25.1 d	IT
180	0+	-49.793	35.100% 6	
180m	8-	-48.651	5.5 h 1	IT
181	1/2-	-47.417	42.39 d 6	β-
182	0+	-46.063	$9 \cdot 10^6$ y 3	β-
182m	(8-)	-44.890	61.5 m 15	β- 54%, IT 46%
183	(3/2-)	-43.290	64 m 1	β-
184	0+	-41.500	4.12 h 5	β-
73 Ta 156		-25.920s		
157		-29.580s	5.3 ms 18	α
158		-31.000s	36.8 ms 16	α 93%, ε 7%
159		-34.600s	0.6 s	α, ε
160		-35.740s	?	α
161		-38.920s	?	α
162		-39.910s		
163		-42.690s		
164		-43.440s	13.6 s 2	ε, α 0.02%
165		-45.880s		
166		-46.310s	32 s	ε
167		-48.370s	2.9 m 15	ε
168		-48.610s	2.5 m 12	ε
169		-50.280s	4.9 m 4	ε
170	(3+)	-50.330s	6.76 m 6	ε
171	(5/2-)	-51.540s	23.3 m 3	ε
172	(3-)	-51.470	36.8 m 3	ε
173	(5/2-)	-52.490s	3.65 h 5	ε
174	3(-)	-51.850s	1.18 h 5	ε
175	7/2+	-52.490s	10.5 h 2	ε
176	1-	-51.480	8.08 h 7	ε
177	7/2+	-51.735	56.6 h 1	ε
178	1+	-50.540	9.31 m 3	ε
178	(7)-	-50.540	2.4 h	ε
179	(7/2+)	-50.366	664.9 d 42	ε
180	1+	-48.940	8.1 h 1	ε 87%, β- 13%

37

Top-right table (page 39):

Isotope Z El A	Jπ	Δ (MeV)	T1/2 or Abundance	Decay Mode
75 Re 167		-34.850s	2.0 s	α
168		-35.820s	2.9 s	ε, α
169		-38.450s	short	ε
170		-39.090s	8 s	ε
171		-41.340s		
172		-41.680s	55 s	ε
172		-41.680s	15 s	ε
173?		-43.560s	2 m	α
174		-43.610s	2.3 m 1	ε
175		-45.240s	4.6 m	ε
176	(3+)	-45.180s	5.3 m	ε
177	(5/2-)	-46.230s	14.0 m 10	ε
178	(3)	-45.790	13.2 m 2	ε
179	(5/2+)	-46.620s	19.7 m 5	ε
180	(1)-	-45.850	2.43 m 6	ε
181	5/2+	-46.560s	19.9 h 7	ε
182	2+	-45.450	12.7 h 2	ε
182m	(7+)	-45.410	64.0 h 5	ε
183	(5/2)+	-45.814	70.0 d 11	ε
184	3-	-44.218	38.0 d 5	ε
184m	8+	-44.030	165 d 5	IT 74.7%, ε 25.3%
185	5/2+	-43.826	37.40% 2	
186	1(-)	-41.933	90.64 h 9	β- 93.5%, ε 6.5%
186m	(8+)	-41.783	$2.0 \cdot 10^5$ y	IT
187	5/2+	-41.224	$5 \cdot 10^{10}$ y 2	β-
			62.60% 2	
188	1-	-39.025	16.98 h 2	β-
188m	(6)-	-38.853	18.6 m 1	IT
189	(5/2+)	-37.987	24.3 h	β-
190	(2)-	-35.540	3.1 m 3	β-
190m	(6-)	-35.367	3.2 h 2	β- 54.5%, IT 45.5%
191	(3/2+,1/2+)	-34.361	9.8 m 5	β-
192		-31.790s	16 s 1	β-
76 Os 163		-16.450s	?	α
164	0+	-20.460s	41 ms 20	α
165		-21.870s	65 ms +70-30	α
166	0+	-25.640s	0.18 s	α, ε
167		-26.740s	0.7 s	α, ε
168	0+	-30.110	2.1 s 2	α, ε
169		-30.880s	3.2 s 2	ε 84%, α 16%
170	0+	-33.934	7.1 s 5	ε, α
171		-34.570s	8.0 s 7	ε 98.3%, α 1.7%
172	0+	-37.260s	19 s 2	ε, α≤0.3%
173		-37.540s	16 s 5	ε 99.98%, α 0.02%
174	0+	-39.950s	44 s 4	ε 99.98%, α 0.02%
175		-40.070s	1.4 m 1	ε
176	0+	-42.030s	3.6 m	ε
177	1/2-	-41.930s	2.8 m	ε
178	0+	-43.550s	5.0 m 4	ε
179		-43.010s	7 m	ε
180	0+	-44.350s	22 m 3	ε
181	(7/2)-	-43.530s	2.7 m 1	ε
181	1/2-	-43.530s	105 m 3	ε

39

Bottom-left table (page 38):

Isotope Z El A	Jπ	Δ (MeV)	T1/2 or Abundance	Decay Mode
73 Ta 180m	9-	-48.908	$>1.2 \cdot 10^{15}$ y	ε, β-
			0.012% 2	
181	7/2+	-48.445	99.988% 2	
182	3-	-46.437	114.5 d	β-
182m	5+	-45.917	0.283 s	IT
182m	10-	-45.917	15.84 m 10	IT
183	7/2+	-45.299	5.1 d 1	β-
184	(5-)	-42.844	8.7 h 1	β-
185	(7/2+)	-41.403	49 m 2	β-
186	(3-)	-38.620	10.5 m 5	β-
74 W 158	0+	-23.780s	?	α
159		-25.550s	7 ms	α
160	0+	-29.360s	41 ms 20	α
161		-30.610s	410 ms 40	α 82%, ε
162	0+	-34.200s	1.39 s 4	ε 54%, α 46%
163		-35.150s	2.8 s	α,ε≈50%
164	0+	-38.360s	6.4 s 8	ε 97.4%, α 2.6%
165		-39.020s	5.1 s 5	ε, α 0.15%
166	0+	-41.899	16 s	α
167		-42.370s		
168	0+	-44.910s		
169		-45.020s		
170?	0+	-47.240s	4 m 1	ε
171?		-47.240s	9.0 m 15	ε
172	0+	-48.970s	6.7 m 10	ε
173		-48.710s	16.5 m 5	ε
174	0+	-50.150s	29 m 1	ε
175	(1/2-)	-49.590s	34 m 1	ε
176	0+	-50.680s	2.5 h	ε
177	(1/2-)	-49.730s	135 m 3	ε
178	0+	-50.450	21.7 d 3	ε
179	(7/2-)	-49.307	37.5 m 5	ε
179m	(1/2-)	-49.085	6.4 m	IT 99.8%, ε 0.2%
180	0+	-49.648	$>1.1 \cdot 10^{15}$ y	
			0.13% 3	
181	9/2+	-48.259	120.98 d 12	ε
182	0+	-48.250	26.3% 2	
183	1/2-	-46.370	14.3% 1	
183m	(11/2)+	-46.061	5.15 s 3	IT
184	0+	-45.710	$>3 \cdot 10^{17}$ y	
			30.67% 15	
185	3/2-	-43.393	75.1 d 3	β-
185m	11/2+	-43.196	1.67 m 3	IT
186	0+	-42.517	28.6% 2	
187	3/2-	-39.912	23.9 h 1	β-
188	0+	-38.676	69.4 d 5	β-
189	(3/2-)	-35.490	11.5 m 3	β-
190	0+	-34.270	30.0 m 15	β-
75 Re 161		-20.710s	10 ms +15-5	α
162		-22.300s	100 ms 30	α>3%
163		-26.110s	0.26 s	α, ε
164		-27.390s	0.88 s 24	α 58%, ε 42%
165		-30.840s	2.4 s 6	ε, α 13%
166		-31.910s	2.2 s	α

38

Bottom-right table (page 40):

Isotope Z El A	Jπ	Δ (MeV)	T1/2 or Abundance	Decay Mode
76 Os 182	0+	-44.600	21.5 h	ε
183	(9/2+)	-43.510s	13.0 h 5	ε
183m	(1/2-)	-43.339s	9.9 h 3	ε 89%, IT 11%
184	0+	-44.257	$>1 \cdot 10^{17}$ y	
			0.02% 1	
185	1/2-	-42.811	93.6 d 5	ε
186	0+	-43.007	$2.0 \cdot 10^{15}$ y 11	α
			1.58% 10	
187	1/2-	-41.227	1.6% 1	
188	0+	-41.145	13.3% 2	
189	3/2-	-38.995	16.1% 3	
189m	9/2-	-38.964	5.8 h	IT
190	0+	-38.717	26.4% 4	
190m	10-	-37.012	9.9 m 1	IT
191	9/2-	-36.403	15.4 d 1	β-
191m	3/2-	-36.329	13.10 h 5	IT
192	0+	-35.893	41.0% 3	
192m	(10-)	-33.878	6.1 s	IT
193	3/2-	-33.406	30.5 h 4	β-
194	0+	-32.441	6.0 y 2	β-
195		-29.700	6.5 m	β-
196	0+	-28.300	34.9 m 2	β-
77 Ir 166		-13.170s	>5 ms	α
167		-17.140s	>5 ms	α
168		-18.560s		α
169		-22.140s	0.4 s 1	α
170		-23.320s	1.0 s	α
171		-26.360s	1.5 s 1	α
172		-27.430s	2.1 s	α
173		-30.220s	3.0 s 1	α
174		-31.060s	4 s 1	α, ε
175		-33.400s	4.5 s 10	α
176		-34.020s	8 s 1	α
177		-36.000s	21 s 2	α
178		-36.290s	12 s	ε
179		-38.020s	4 m	ε
180		-37.950s	1.5 m 1	ε
181	(7/2+)	-39.460s	4.92 m 13	ε
182	(3)	-39.150s	15 m 1	ε
183	(1/2+,3/2+)	-40.320s	57 m 4	ε
184	5	-39.540	3.02 h 6	ε
185	5/2(-)	-40.310s	14.0 h 9	ε
186	(5)	-39.176	15.8 h 3	ε
186	(2-)	-39.176	1.75 h 15	ε
187	3/2(+)	-39.730s	10.5 h 3	ε
188	(2-)	-38.350	41.5 h 5	ε
189	3/2+	-38.460	13.2 d	ε
190	(4)+	-36.720	11.78 d 10	ε
190m	(7)+	-36.694	1.2 h	IT
190m	(11)-	-36.545	3.2 h 2	ε 94.4%, IT 5.6%
191	3/2+	-36.716	37.3% 5	
191m	11/2-	-36.545	4.94 s 3	IT
191m		-34.669	5.5 s 7	IT
192	4(-)	-34.857	73.831 d 8	β- 95.4%, ε 4.6%

40

517

Top-left table (p. 41)

Z El A	Jπ	Δ (MeV)	T1/2 or Abundance	Decay Mode
77 Ir 192m	1(+)	-34.799	1.45 m 5	IT 99.98%, β- 0.02%
192m	(9+)	-34.702	241 y 9	IT
193	3/2+	-34.543	62.7% 5	
193m	11/2-	-34.463	10.60 d 11	IT
194	1-	-32.538	19.15 h 3	β-
194m	(11)	-32.098	171 d 11	β-
195	(3/2+)	-31.702	2.8 h	β-
195m	(11/2-)	-31.582	3.8 h 2	β- 96%, IT 4%
196	(0-)	-29.460	52 s 2	β-
196m	(10,11-)	-29.050	1.40 h 2	β-
197	(3/2+)	-28.290	5.8 m 5	β-
197m	(11/2-)	-28.290	8.9 m 3	β-, IT
198		-25.930	8 s 1	β-
78 Pt 168	0+	-11.150s	?	α
169		-12.600s	2.5 ms	α
170	0+	-16.510s	6 ms	α
171		-17.710s	25 ms 9	α
172	0+	-21.220	0.10 s	α
173		-22.100s	0.34 s	α 84%, ε
174	0+	-25.326	0.90 s 1	α 83%, ε 17%
175		-25.960s	2.52 s 8	α 64%, ε
176	0+	-28.940s	6.33 s 15	ε 58%, α 42%
177		-29.470s	11 s 2	ε 91%, α 9%
178	0+	-31.950s	21.0 s 7	ε, α
179		-32.350s	43 s	ε, α 0.27%
180	0+	-34.350s	52 s 3	ε, α 0.3%
181	1/2-	-34.380s	51 s 5	ε, α≈0.06%
182	0+	-36.180s	2.6 m 1	ε≈99.98%, α≈0.02%
183		-35.740s	6.6 m 9	ε, α 0.0013%
183m	(7/2-)	-35.740s	43 s	ε
184	0+	-37.330s	17.3 m 2	ε≈0.001%
185	(9/2+)	-36.610s	70.9 m 24	ε
185m	(1/2-)	-36.481s	33.0 m 8	ε
186	0+	-37.850	2.0 h 1	ε, α≈0.0001%
187	3/2(-)	-36.830s	2.35 h 3	ε
188	0+	-37.832	10.2 d 3	ε>99.99%, α<0.01%
189	3/2-	-36.499	10.89 h 1	ε
190	0+	-37.338	$6 \cdot 10^{11}$ y 1 0.01% 1	α
191	3/2-	-35.710	2.9 d 1	ε
192	0+	-36.311	0.79% 5	
193	(1/2)-	-34.487	50 y 9	ε
193m	(13/2)+	-34.337	4.33 d 3	IT
194	0+	-34.787	32.9% 5	
195	1/2-	-32.821	33.8% 5	
195m	13/2+	-32.562	4.02 d 1	IT
196	0+	-32.671	25.3% 5	
197	1/2-	-30.446	18.3 h 3	β-
197m	13/2+	-30.046	94.4 m 8	IT 96.7%, β- 3.3%
198	0+	-29.930	7.2% 2	
199	(5/2-)	-27.430	30.8 m 4	β-

41

Bottom-left table (p. 42)

Z El A	Jπ	Δ (MeV)	T1/2 or Abundance	Decay Mode
78 Pt 199m	(13/2+)	-27.006	13.6 s 4	IT
200	0+	-26.625	12.5 h 3	β-
201	(5/2-)	-23.750	2.5 m 1	β-
79 Au 173		-12.840s		
174		-14.210s	120 ms 20	α
175		-17.340s	≈0.1 s	α
176		-18.570s	1.25 s 30	α
177		-21.500s	1.3 s 4	α
178		-22.580s	2.6 s 5	α
179		-24.990s	7.5 s	α
180		-25.800s	8.1 s	ε
181		-27.830s	11.4 s 5	ε 98.9%, α 1.1%
182		-28.330s	21 s 2	ε≈99.96%, α≈0.04%
183	(3/2+)	-30.130s	42 s	ε, α 0.3%
184		-30.240s	53.0 s 14	ε 99.98%, α 0.02%
185	5/2(-)	-31.850s	4.3 m 8	ε≤99.9%, α≥0.093%
185m		-31.850s	6.8 m 3	ε, IT
186	3	-31.580s	10.7 m 5	ε
186		-31.580s	<2 m	ε
187	1/2+	-33.110s	8.0 m 4	ε, α?
188	1(-)	-32.530s	8.84 m 6	ε
189	1/2+	-33.800s	28.7 m	ε
189m	11/2-	-33.553s	4.59 m 1	ε, IT>0%
190	1-	-32.896	42.8 m 10	ε
191	3/2+	-33.880	3.18 h 8	ε
191m	(11/2-)	-33.613	0.92 s 11	IT
192	1-	-32.796	4.94 h 9	ε
193	3/2+	-33.490s	17.65 h 15	ε
193m	(11/2)-	-33.200s	3.9 s 3	IT≈99.97%, ε≈0.03%
194	1-	-32.278	39.5 h 5	ε
195	3/2+	-32.591	186.1 d	ε
195m	11/2-	-32.272	30.5 s 2	IT
196	2-	-31.165	6.183 d 10	ε 92.5%, β- 7.5%
196m	5+	-31.080	8.1 s 2	IT
196m	12-	-30.569	9.7 h 1	IT
197	3/2+	-31.165	100%	ε
197m	11/2-	-30.756	7.8 s 1	IT
198	2-	-29.606	2.696 d 2	β-
198m	(12-)	-28.794	2.30 d 4	IT
199	3/2+	-29.119	3.139 d 7	β-
200	1(-)	-27.320	48.4 m 3	β-
200m	12-	-26.330	18.7 h 5	β- 82%, IT 18%
201	(3/2+)	-26.413	26 m 1	β-
202	1-,0-	-24.370s	28 s 2	β-
203	(3/2+)	-23.153	53 s 2	β-
204	(2-)	-20.220	40 s 3	β-
80 Hg 175		-8.270s		
176	0+	-11.890		
177		-12.940s	0.17 s	α, ε
178	0+	-16.323	0.47 s 14	α≈84%, ε≈16%
179		-17.110s	1.09 s	α 53%, ε 47%, εpv

42

Top-right table (p. 43)

Z El A	Jπ	Δ (MeV)	T1/2 or Abundance	Decay Mode
80 Hg 180	0+	-20.260s	2.9 s 3	ε, α
181	1/2(-)	-20.760s	3.6 s 3	ε>87%, α<13%, εpv, εαv
182	0+	-23.520s	11.2 s 10	ε 91%, α 9%
183	1/2(-)	-23.890s	8.8 s 5	ε 88%, α 12%, εpv
184	0+	-26.260s	30.6 s 3	ε 98.75%, α 1.25%
185	1/2-	-26.170s	50 s 2	ε≤95%, α≥5%
185m	13/2+	-26.146s	20 s 2	ε, IT, α
186	0+	-28.550s	1.38 m 10	ε, α 0.02%
187	3/2-	-28.170s	2.4 m 3	ε, α>0.0001%
187	13/2+	-28.170s	1.9 m 3	ε, α>0.0002%
188	0+	-30.200s	3.25 m 15	ε>99.99%, α<0.01%
189	3/2-	-29.600s	7.6 m	ε
189	13/2+	-29.600s	8.6 m	ε
190	0+	-31.300s	20.0 m 5	ε
191	(3/2-)	-30.540	49 m 10	ε
191m	(13/2+)	-30.400	50.8 m 15	ε
192	0+	-32.000s	4.85 h 20	ε
193	3/2-	-31.150s	3.80 h 15	ε
193m	13/2+	-31.009s	11.8 h 2	ε 92%, IT 8%
194	0+	-32.238	520 y	ε
195	1/2-	-31.000s	9.5 h	ε
195m	13/2+	-30.894	40.0 h	IT 54.2%, ε 45.8%
196	0+	-31.851	0.14% 10	
197	1/2-	-30.565	64.1 h 1	ε
197m	13/2+	-30.266	23.8 h 1	IT 93%, ε 7%
198	0+	-30.979	10.02% 7	
199	1/2-	-29.571	16.84% 11	
199m	13/2+	-29.039	42.6 m 2	IT
200	0+	-29.529	23.13% 11	
201	3/2-	-27.687	13.22% 11	
202	0+	-27.370	29.80% 14	
203	5/2-	-25.292	46.60 d 2	β-
204	0+	-24.716	6.85% 5	
205	1/2-	-22.312	5.2 m 1	β-
206	0+	-20.969	8.15 m 10	β-
207	(9/2+)	-16.270	2.9 m 2	β-
81 Tl 179		-7.920s		
180		-9.350s		
181		-12.320s		
182		-13.550s		
183		-16.170s		
184		-17.070s	11 s 1	ε 98%, α 2%
185	(1/2+)	-19.400s		
185m	(9/2-)	-18.947s	1.8 s 2	IT, α
186		-20.020s	28 s	ε, αv
186m		-20.020s	4 s	IT
187	(1/2+)	-22.130s	50 s	ε
187m	(9/2-)	-21.800s	15.60 s 12	IT, ε, α
188	(2-)	-22.470s	71 s 10	ε
188m	(7+)	-22.470s	71 s 1	ε
189	(1/2+)	-24.400s	2.3 m	ε
189m	(9/2-)	-24.400s	1.4 m	ε

43

Bottom-right table (p. 44)

Z El A	Jπ	Δ (MeV)	T1/2 or Abundance	Decay Mode
81 Tl 190	(2-)	-24.700s	2.6 m 3	ε
190m	(7+)	-24.700s	3.7 m 3	ε
191		-26.240s		
191m	(9/2-)	-26.240s	5.22 m 16	ε
192	(2-)	-25.970s	9.6 m 4	ε
192m	(7+)	-25.970s	10.8 m 2	ε
193	1/2(+)	-27.460	21.6 m 8	ε
193m	(9/2-)	-27.095	2.11 m 15	IT 75%, ε 25%
194	2-	-27.090s	33.0 m 5	ε
194m	(7+)	-26.790s	32.8 m 2	ε
195	1/2+	-28.290	1.16 h 5	ε
195m	9/2-	-27.807	3.6 s 4	IT
196	2-	-27.520s	1.84 h 3	ε
196m	(7+)	-27.125s	1.41 h 2	ε 95.5%, IT 4.5%
197	1/2+	-28.420	2.84 h 4	ε
197m	9/2-	-27.812	0.54 s 1	IT
198	2-	-27.520	5.3 h 5	ε
198m	7+	-26.976	1.87 h 3	ε 54%, IT 46%
199	1/2+	-28.070	7.42 h 8	ε
200	2-	-27.075	26.1 h 1	ε
201	1/2+	-27.205	73.1 h 2	ε
202	2-	-26.003	12.23 d 2	ε
203	1/2+	-25.784	29.524% 9	
204	2-	-24.369	3.78 y 2	β- 97.45%, ε 2.55%
205	1/2+	-23.846	70.476% 9	
206	0-	-22.278	4.20 m 2	β-
206m	(12-)	-19.635	3.76 m 4	IT
207	1/2+	-21.048	4.77 m 2	β-
207m	11/2-	-19.700	1.33 s 11	IT
208	(5+)	-16.778	3.053 m 7	β-
209	(1/2+)	-13.662	2.20 m 7	β-
210	(5+)	-9.263	1.30 m 3	β-, β-n 0.007%
82 Pb 183		-7.730s		α
184	0+	-11.060s	0.6 s	α
185		-11.650s	4.1 s 3	α
186	0+	-14.630s	7.9 s 16	ε, α≈2.4%
187	(13/2+)	-15.060s	18.3 s 3	ε 98%, α 2%
187		-15.060s	15.2 s 3	α, ε
188	0+	-17.720s	24.5 s 15	ε 78%, α 22%
189		-17.900s	51 s	ε>99%, α≈0.4%
190	0+	-20.430s	1.2 m 1	ε 99.1%, α 0.9%
191		-20.340s	1.33 m 8	ε 99.99%, α 0.01%
191m	(13/2+)	-20.340s	2.2 m	ε
192	0+	-22.550s	3.5 m 1	ε 99.99%, α 0.01%
193	(13/2+)	-22.250s	5.8 m 2	ε
194	0+	-24.240s	10 m	ε
195	(3/2-)	-23.770s	15 m	ε
195m	(13/2+)	-23.570s	15.8 m 2	ε
196	0+	-25.450s	37 m 3	ε, α<0.0001%
197	3/2-	-24.820s	8 m	ε
197m	13/2+	-24.501s	43 m	ε 81%, IT 19%, α
198	0+	-26.120s	2.4 h 1	ε
199	5/2-	-25.270	90 m 10	ε

44

Isotope Z El A	Jπ	Δ (MeV)	T1/2 or Abundance	Decay Mode
82 Pb 199m	13/2+	-24.846	12.2 m 3	IT 90%, ε
200	0+	-26.270s	21.5 h 4	ε
201	5/2-	-25.300	9.33 h	ε
201m	13/2+	-24.671	61 s 2	IT
202	0+	-25.957	$5.3 \cdot 10^4$ y	ε
202m	9-	-23.787	3.53 h	IT 90.5%, ε 9.5%
203	5/2-	-24.811	51.88 h	ε
203m	29/2-	-21.861	0.48 s 2	IT
203m	13/2+	-23.986	6.3 s 2	IT
204	0+	-25.132	$\geq 1.4 \cdot 10^{17}$ y 1.4% 1	
204m	9-	-22.946	66.9 m	ε
205	5/2-	-23.792	$1.52 \cdot 10^7$ y 7	ε
206	0+	-23.809	24.1% 1	
207	1/2-	-22.476	22.1% 1	
207m	13/2+	-20.843	0.796 s	IT
208	0+	-21.772	52.4% 1	
209	9/2+	-17.638	3.253 h 14	β-
210	0+	-14.752	22.3 y 2	β-, α▼
211	(9/2+)	-10.493	36.1 m 2	β-
212	0+	-7.572	10.64 h 1	β-
213		-3.250s	10.2 m 3	β-
214	0+	-0.188	26.8 m	β-
83 Bi 188		-7.380s		α
189		-9.710s	≤1.5 s	α
190		-10.730s	5.4 s 5	α≈90%, ε
191		-12.940s	13 s 1	ε, α≈40%
191m		-12.940s	20 s 15	ε, α
192		-13.600s	42 s 5	ε 80%, α 20%
193		-15.660s	64 s 4	α 60%, ε 40%
193m		-15.660s	3.5 s 2	ε, α≈25%
194	(10-)	-16.260s	105 s 15	ε>99.8%, α<0.2%
195		-17.980s	170 s 20	ε>99.8%, α<0.2%
195m		-17.980s	90 s 5	ε, α 4%
196	(10-)	-17.990	4.6 m 5	ε
197	(9/2-)	-19.670	?	ε
197m	(1/2+)	-19.170	9.5 m 10	ε≈99.89%, α≈0.11%
198	(7+)	-19.570	11.85 m 18	ε
198m	(10-)	-19.321	7.7 s 5	IT
199	9/2-	-20.940	27 m 1	ε
199m	(1/2+)	-20.140	24.70 m 5	ε, α
200	7+	-20.410	36.4 m 5	ε
200m	(2+)	-20.210	31 m 2	ε
201	9/2-	-21.490	108 m 3	ε
201m	(1/2+)	-20.644	59.1 m 6	ε, IT 10%, α▼
202	5+	-20.810	1.72 h 5	ε
203	9/2-	-21.590	11.76 h 5	ε, α▼
204	6+	-20.740	11.22 h 10	ε
205	9/2-	-21.085	15.31 d 4	ε
206	6+	-20.048	6.243 d 3	ε
207	9/2-	-20.078	32.2 y 9	ε
208	(5+)	-18.894	$3.68 \cdot 10^5$ y 4	ε

Isotope Z El A	Jπ	Δ (MeV)	T1/2 or Abundance	Decay Mode
85 At 194		-0.810s		α
195		-3.110s		α
196		-3.970s	0.3 s 1	α
197	(9/2-)	-6.140s	0.4 s 1	α≈96%, ε≈4%
198		-6.940s	4.9 s	α
198m		-6.840s	1.5 s 3	α, IT?
199	(9/2-)	-8.770s	7.0 s 1	α 53%, ε 47%
200	(5+)	-8.970	43 s 2	ε 65%, α 35%
200m	(10-)	-8.830	4.3 s 3	α, IT?
201	(9/2-)	-10.770	89 s 3	α 71%, ε 29%
202	(5+)	-10.790	181 s 3	ε 88%, α 12%
202m		-10.790	1.1 s 3	IT
203	9/2-	-12.310	7.37 m 20	ε 69%, α 31%
204	(5+)	-11.920	9.2 m 2	ε 95.6%, α 4.4%
205	9/2-	-13.050	26.2 m 5	ε 90%, α 10%
206	5+	-12.500	29.4 m 3	ε 99.04%, α 0.96%
207	9/2-	-13.290	1.80 h 4	ε 91.3%, α 8.7%
208	6+	-12.560	1.63 h 3	ε 99.45%, α 0.55%
209	9/2-	-12.902	5.41 h 5	ε 95.9%, α 4.1%
210	5+	-11.992	8.1 h 4	ε 99.82%, α 0.18%
211	9/2-	-11.672	7.214 h 7	ε 58.3%, α 41.7%
212	(1-)	-8.640	0.314 s 2	α
212m	(9-)	-8.415	0.119 s 3	α
213	9/2-	-6.603	0.11 μs 2	α
214	(1-)	-3.403	2 μs	α
215	(9/2-)	-1.269	0.10 ms 2	α
216	1(-)	2.226	0.30 ms 3	α
217	9/2(-)	4.373	32.3 ms 4	α 99.99%, β- 0.01%
218		8.089	≈2 s	α 99.9%, β- 0.1%
219		10.520	0.9 m 1	α 97%, β- 3%
220		14.290s		
86 Rn 199		-1.610s	0.5 s	α
199m		-1.610s	0.29 s	α
200	0+	-4.000s	1.0 s 2	α≈98%, ε≈2%
201	(3/2-)	-4.120s	7.0 s 4	α≈80%, ε≈20%
201m	(13/2+)	-3.920s	3.8 s 4	α, ε
202	0+	-6.310s	9.85 s 20	α≈85%, ε≈15%
203	(5/2-)	-6.220s	45 s 3	α 66%, ε 34%
203m	(13/2+)	-5.910s	28 s 2	α
204	0+	-8.070s	1.24 m 3	α 68%, ε 32%
205	(5/2-)	-7.780s	2.83 m 7	ε 77%, α 23%
206	0+	-9.180s	5.67 m 17	α 68%, ε 32%
207	5/2-	-8.670	9.3 m 2	ε 77%, α 23%
208	0+	-9.680s	24.35 m 14	α 52%, ε 48%
209	5/2-	-8.970	28.5 m 10	ε 83%, α 17%
210	0+	-9.623	2.4 h 1	α 96%, ε 4%
211	1/2-	-8.779	14.6 h 2	ε 74%, α 26%
212	0+	-8.682	24 m 2	α
213	(9/2+)	-5.723	25.0 ms 2	α
214	0+	-4.342	0.27 μs 2	α
215	(9/2+)	-1.192	2.30 μs 10	α
216	0+	0.232	45 μs 5	α
217	9/2+	3.634	0.54 ms 5	α

Isotope Z El A	Jπ	Δ (MeV)	T1/2 or Abundance	Decay Mode
83 Bi 209	9/2-	-18.282	100%	
210	1-	-14.815	5.013 d 5	β-, α 0.0001%
210m	9-	-14.544	$3.0 \cdot 10^6$ y 1	α
211	9/2-	-11.872	2.14 m 2	α 99.72%, β- 0.28%
212	1(-)	-8.146	60.55 m 6	β- 64.06%, α 35.94%, β-α 0.014%
212m	(9-)	-7.896	25 m	α≈93%, β-≥7%
212m	(15-)	-7.446	9 m	β-
213	9/2(-)	-5.254	45.59 m 6	β- 97.84%, α 2.16%
214	(1-)	-1.219	19.9 m 4	β- 99.98%, α 0.02%
215	(9/2-)	1.710	7.4 m 6	β-
216		5.960s		
84 Po 192	0+	-7.980s	0.034 s 3	α
193		-8.370s	450 ms 150	α
193m		-8.370s	0.42 s	α
194	0+	-11.010s	0.7 s	α
195		-11.170s	4.5 s 5	α
195m		-11.170s	2.0 s 2	α
196	0+	-13.470s	5.5 s 5	α, ε
197	(3/2-)	-13.410s	56 s 3	ε 56%, α 44%
197m	(13/2+)	-13.206s	26 s 2	α 84%, ε≤16%, IT?
198	0+	-15.500s	1.76 m 3	α 70%, ε 30%
199	(3/2-)	-15.270s	5.2 m 1	ε 88%, α 12%
199m	(13/2+)	-14.960s	4.2 m 1	ε 61%, α 39%
200	0+	-17.040s	11.5 m 1	ε 85%, α 15%
201	3/2-	-16.590s	15.3 m 2	ε 98.4%, α 1.6%
201m	13/2+	-16.166s	8.9 m 2	ε 57%, IT 40%, α≈3%
202	0+	-17.990s	44.7 m 5	ε 98%, α 2%
203	5/2-	-17.320s	35 m	ε 99.89%, α 0.11%
203m	13/2+	-16.709s	1.2 m 2	IT 95.5%, ε 4.5%
204	0+	-18.360s	3.53 h 2	ε 99.34%, α 0.66%
205	5/2-	-17.560s	1.66 h 2	ε 99.96%, α 0.04%
206	0+	-18.206s	8.8 d 1	ε 94.55%, α 5.45%
207	5/2-	-17.168s	5.80 h 2	ε 99.98%, α 0.02%
207m	19/2-	-15.785s	2.8 s 2	IT
208	0+	-17.492	2.898 y 2	α, ε 0.0018%
209	1/2-	-16.391	102 y 5	α 99.74%, ε 0.26%
210	0+	-15.977	138.376 d 2	α
211	(25/2+)	-12.457	0.516 s 3	α
211m	(25/2+)	-10.994	25.2 s 6	α
212	0+	-10.394	0.298 μs 3	α
212m	(16+)	-7.473	45.1 s 6	α
213	9/2+	-6.676	4.2 μs 8	α
214	0+	-4.494	164.3 μs 18	α
215	(9/2+)	-0.542	1.780 ms 4	α, β- 0.0002%
216	0+	1.759	0.15 s	α
217		5.830s	<10 s	α 95%, β- <5%
218	0+	8.352	3.11 m	α 99.98%, β- 0.02%

Isotope Z El A	Jπ	Δ (MeV)	T1/2 or Abundance	Decay Mode
86 Rn 218	0+	5.198	35 ms 5	α
219	(5/2+)	8.829	3.96 s 1	α
220	0+	10.589	55.6 s 1	α
221	(7/2+,9/2+)	14.410s	25 m 2	β- 78%, α 22%
222	0+	16.367	3.8235 d 3	α
223			43 m 3	β-
224	0+		107 m 3	β-
225			4.5 m 3	β-
226	0+		6.0 m 5	β-
87 Fr 201		3.830s	48 ms 15	α
202		3.080s	0.34 s 4	α
203		0.930s	0.55 s	α, α?
204		0.630	2.1 s 2	α≈80%, ε≈20%
205	(9/2-)	-1.290	3.85 s 10	α, ε<1%
206		-1.440	16.0 s 2	α 85%, ε 15%
206m		-1.440	0.7 s	IT, α
207	9/2-	-2.980	14.8 s 1	α 95%, ε 5%
208	7+	-2.720	59.0 s 20	α 77%, ε 23%
209	9/2-	-3.840	50.0 s 3	α 89%, ε 11%
210	6+	-3.410	3.18 m 6	α 60%, ε 40%
211	9/2-	-4.200	3.10 m 2	α>70%, ε<30%
212	5(+)	-3.610	20.0 m 1	ε 57%, α 43%
213	9/2-	-3.573	34.6 s 3	α 99.45%, ε 0.55%
214	(1-)	-0.980	5.0 ms 2	α
214m	(9-)	-0.858	3.35 ms 5	α
215	9/2-	0.289	0.12 μs	α
216	(1-)	2.960	0.70 μs 2	α
217	9/2-	4.293	22 μs 5	α
218		7.036	0.7 ms 6	α
219	(9/2-)	8.609	21 ms 1	α
220	1	11.451	27.4 s 3	α≈99.65%, β-≈0.35%
221	5/2(-)	13.255	4.9 m 2	α, β?
222	(2)	16.360	14.4 m 4	β-, α?
223	(3/2)	18.381	21.8 m 4	β- 99.99%, α 0.01%
224	(1)	21.630	2.67 m 20	β-
225		23.840	3.9 m 2	β-
226		27.200	48 s 1	β-
227	1/2+	29.590	2.4 m 2	β-
228		33.140s	39 s 1	β-
229			50 s 20	β-
88 Ra 204	0+	6.020s		α
205		5.800s		α, ε?
206	0+	3.540s	0.4 s 2	α, ε?
207	(5/2,3/2-)	3.480s	1.3 s 2	α, ε?
208	0+	1.630s	1.4 s 4	α
209		1.790s	4.6 s 2	α
210	0+	0.400s	3.7 s 2	α 96%, ε 4%
211	(5/2-)	0.800	13 s 2	α>93%, ε<7%
212	0+	-0.220s	13.0 s 2	α 94%, ε≈6%
213	(1/2-)	0.310	2.74 m 6	α 80%, ε 20%
213m		2.080	2.1 ms 1	IT 99%, α 1%
214	0+	0.074	2.46 s 3	α>99.9%, ε<0.1%

Isotope Z El A	Jπ	Δ (MeV)	T1/2 or Abundance	Decay Mode
88 Ra 215	(9/2+)	2.510	1.59 ms 9	α
216	0+	3.269	182 ns 10	α
217	(9/2+)	5.863	1.6 μs 2	α
218	0+	6.630	14 μs 2	α
219		9.365	10 ms 3	α
220	0+	10.250	23 ms 5	α
221		12.938	29 s	α
222	0+	14.301	38.0 s 5	α
223	1/2(+)	17.234	11.434 d 4	α, ^{14}Cw
224	0+	18.803	3.66 d 4	α
225	(3/2)+	21.987	14.8 d 2	β-
226	0+	23.663	1600 y 7	α
227	(3/2+)	27.173	42.2 m 5	β-
228	0+	28.936	5.75 y 3	β-
229		32.480	4.0 m 2	β-
230	0+	34.460s	93 m 2	β-
89 Ac 209		8.870	0.10 s 5	α
210		8.590	0.35 s 5	α 96%, ε 4%
211	(9/2-)	7.070	0.25 s 5	α>99.8%, ε<0.2%
212		7.230	0.93 s 5	α≈98%, ε≈2%
213	(9/2-)	6.090	0.80 s 5	α
214	(5+)	6.370	8.2 s 2	α≥89%, ε≤11%
215	(9/2-)	5.970	0.17 s 1	α 99.91%, ε 0.09%
216	(1-)	8.060	≈0.33 ms	α, α
217	9/2-	8.684	111 ns 7	α
218		10.820	0.27 μs 4	α
219	(9/2-)	11.540	7 μs 2	α
220		13.730	26.1 ms 5	α
221		14.500	52 ms 2	α
222		16.603	4.2 s 5	α
222m		16.603	66 s 3	α, IT?, ε?
223	(5/2-)	17.818	2.2 m 1	α 99%, ε 1%
224	0(-),1	20.200	2.9 h 2	ε≈90%, α≈10%
225	(3/2-)	21.615	10.0 d 1	α
226	(1-)	24.298	29 h	β- 82.8%, ε 17.2%, αw
227	3/2(-)	25.849	21.773 y 3	β-98.62%, α 1.38%
228	(3+)	28.890	6.13 h	β-
229	(3/2+)	30.720	62.7 m 5	β-
230	(1+)	33.760s	122 s 3	β-
231	(1/2+)	35.910	7.5 m 1	β-
232	(1,0)	39.240s	35 s 5	β-
90 Th 212	0+	12.000s	30 ms	α
213		12.060s	150 ms 25	α
214	0+	10.650s	0.09 s	α
215	(1/2-)	10.890	1.2 s 2	α
216	0+	10.270s	0.028 s 2	α
217		12.160	0.252 ms 7	α
218	0+	12.346	109 ns 13	α
219		14.450	1.05 μs 3	α
220	0+	14.646	9.7 μs 6	α
221		16.916	1.68 ms 6	α
222	0+	17.183	2.8 ms 3	α

Isotope Z El A	Jπ	Δ (MeV)	T1/2 or Abundance	Decay Mode
92 U 234	0+	38.142	2.45·10⁵ y 2 / 0.0055% 5	SFw
235	7/2-	40.916	703.8·10⁶ y 5 / 0.7200% 12	α, SFw
235m	1/2+	40.917	≈25 m	IT
236	0+	42.442	2.3415·10⁷ y 14	α, SFw
237	1/2+	45.387	6.75 d 1	β-
238	0+	47.306	4.468·10⁹ y 3 / 99.2745% 15	α, SFw
239	5/2+	50.571	23.50 m 5	β-
240	0+	52.711	14.1 h 1	β-
242	0+		16.8 m 5	β-
93 Np 227?			60 s 5	SF
229		33.740	4.0 m 2	α>50%, ε<50%
230		35.220	4.6 m 3	ε<97%, α≥3%
231	(5/2)	35.620	48.8 m 2	ε 98%, α 2%
232	(4+)	37.280s	14.7 m 3	ε, α≈0.003%
233	(5/2+)	38.000s	36.2 m 1	ε, α≤0.001%
234	(0+)	39.950	4.4 d 1	ε
235	5/2+	41.039	396.2 d 12	ε, α 0.0014%
236	(6-)	43.370	115·10³ y 12	ε 91%, β- 8.9%, α
236m	1(-)	43.420	22.5 h 4	ε 52%, β- 48%
237	5/2+	44.868	2.14·10⁶ y 1	α
238	2+	47.452	2.117 d 2	β-
239	5/2+	49.307	2.355 d 4	β-
240	(5+)	52.210	61.9 m 2	β-
240m	1(-)	52.210	7.22 m 2	β- 99.88%, IT 0.12%
241	(5/2+)	54.260	13.9 m 2	β-
242	1(+)	57.410	2.2 m 2	β-
242	(6+)	57.410	5.5 m 1	β-
243		59.921		
94 Pu 231		38.390s		
232	0+	38.349	34.1 m 7	ε≈80%, α≈20%
233		40.043	20.9 m 4	ε 99.88%, α 0.12%
234	0+	40.333	8.8 h 1	ε 94%, α 6%
235	(5/2+)	42.160	25.3 m 10	ε, α 0.0027%
236	0+	42.879	2.851 y 8	α, SFw
237	7/2-	45.086	45.3 d 2	ε, α 0.005%
237m	1/2+	45.231	0.18 s 2	IT
238	0+	46.160	87.74 y 4	α, SFw
239	1/2+	48.585	24119 y 26	α, SF
240	0+	50.122	6570 y 6	α, SFw
241	5/2+	52.952	14.35 y 10	β-, α 0.0025%
242	0+	54.714	3.763·10⁵ y 12	α, SFw
243	7/2+	57.751	4.956 h 3	β-
244	0+	59.801	8.08·10⁷ y 10	α 99.88%, SF 0.12%
245	(9/2-)	63.174	10.5 h 1	β-
246	0+	65.365	10.85 d 2	β-
95 Am 232			55 s 7	ε≈98%, α≈2%, εSF
233		43.170s		
234		44.340s	2.6 m 2	ε, α?
235		44.640s	?	

Isotope Z El A	Jπ	Δ (MeV)	T1/2 or Abundance	Decay Mode
90 Th 223		19.243	0.66 s 1	α
224	0+	19.980	1.04 s 5	α
225	(3/2+)	22.283	8.0 m 5	α≈90%, ε≈10%
226	0+	23.180	31 m	α
227	(3/2+)	25.805	18.718 d 5	α
228	0+	26.748	1.9131 y 9	α
229	5/2+	29.580	7340 y 160	α
230	0+	30.859	7.538·10⁴ y 30	α, SF?
231	5/2(+)	33.812	25.52 h 1	β-, αw
232	0+	35.444	1.405·10¹⁰ y 6 / 100%	α, SF?
233	(1/2+)	38.729	22.3 m 1	β-
234	0+	40.607	24.10 d 3	β-
235	(1/2+)	44.250	6.9 m 2	β-
236	0+		37.1 m 15	β-
91 Pa 215		17.660	14 ms	α
216		17.660	0.20 s 4	α
217		17.000	4.9 ms 6	α
217m		17.000	1.6 ms	α
218		18.590	0.12 ms	α
219		18.490s		
220		20.180s		
221		20.310s	6 μs	α
222		21.940	4.3 ms	α
223		22.310	6.5 ms 10	α
224		23.780	0.95 s 15	α
225		24.310	1.8 s 3	α
226		26.015	1.8 m 2	α 74%, ε 26%
227	(5/2-)	26.825	38.3 m 3	α≈85%, ε≈15%
228	(3+)	28.852	22 h 1	ε≈98%, α≈2%
229	(5/2+)	29.876	1.4 d 4	ε 99.75%, α 0.25%
230	(2-)	32.162	17.4 d 5	ε 91.6%, β- 8.4%, αw
231	3/2-	33.422	3.276·10⁴ y 11	α, SF?
232	(2-)	35.923	1.31 d 2	β-, ε≈0.2%
233	3/2-	37.486	27.0 d 1	β-
234	4(+)	40.337	6.70 h 5	β-
234m	(0-)	40.411	1.17 m 3	β- 99.87%, IT 0.13%
235	(3/2-)	42.320	24.1 m 2	β-
236	(1-)	45.540	9.1 m 2	β-, SFw
237	(1/2+)	47.640	8.7 m 2	β-
238	(3-)	51.270	2.3 m 1	β-
92 U 222	0+		1 μs	α
226	0+	27.170	0.5 s 2	α
227		28.870s	1.1 m 3	α
228	0+	29.208	9.1 m 2	α≥95%, ε≤5%
229	(3/2+)	31.181	58 m 3	ε≈80%, α≈20%
230	0+	31.598	20.8 d	α
231	(5/2-)	33.780	4.2 d 1	ε, α 0.006%
232	0+	34.586	68.9 y 4	α, SFw
233	5/2+	36.914	1.592·10⁵ y 2	α, SFw
234	0+	38.142	2.45·10⁵ y 2	α

Isotope Z El A	Jπ	Δ (MeV)	T1/2 or Abundance	Decay Mode
95 Am 236		46.000s		
237	(5/2-)	46.630s	73.0 m 10	ε 99.97%, α 0.03%
238	1+	48.420	98 m 2	ε, α 0.0001%
239	(5/2-)	49.385	11.9 h 1	ε 99.99%, α 0.01%
240	(3-)	51.491	50.9 h 2	ε, αw
241	5/2-	52.931	432.2 y 5	α, SF
242	1-	53.465	16.02 h 2	β- 82.7%, ε 17.3%
242m	5-	55.512	141 y 2	IT 99.5%, α 0.5%, SFw
243	5/2-	57.171	7380 y 40	α, SFw
244	(6-)	59.876	10.1 h 1	β-
244m		59.945	≈26 m	β-, ε w
245	(5/2)+	61.893	2.05 h 1	β-
246	(7-)	64.991	39 m 3	β-
246m	2(-)	64.991	25.0 m 2	β-
247	(5/2)	67.230s	22 m 3	β-
248		70.590s		
96 Cm 235		48.020s		
236	0+	47.870s		
237		49.150s		
238	0+	49.390	2.4 h 1	ε≥90%, α≤10%
239	(7/2-)	51.090s	≈2.9 h	ε, α<0.1%
240	0+	51.701	27 d	α>99.5%, ε<0.5%, SFw
241	1/2+	53.696	32.8 d 2	ε 99%, α 1%
242	0+	54.801	162.8 d 2	α, SFw
243	5/2+	57.177	28.5 y 2	α 99.76%, ε 0.24%
244	0+	58.449	18.10 y 2	α, SFw
245	7/2+	60.998	8500 y 100	α
246	0+	62.613	4730 y 100	α 99.97%, SF 0.03%
247	9/2-	65.528	1.56·10⁷ y 5	α
248	0+	67.388	3.40·10⁵ y 3	α 91.74%, SF 8.26%
249	1/2(+)	70.746	64.15 m 3	β-
250	0+	72.985	≈7400 y	SF 65%, α≈28%, β-≈7%
251	(1/2+)	76.650s	16.8 m 2	β-
252	0+		<2 d	β-
97 Bk 237		53.190s		
238		54.170s		
239		54.270s		
240		55.600s	5 m 2	ε, εSFw
241		56.100s		
242		57.700s	7.0 m 13	ε, α<1%, SF<0.03%
243	(3/2-)	58.682	4.5 h 2	ε 99.85%, α 0.15%
244	(4-)	60.690	4.35 h 15	ε, α 0.006%
245	3/2-	61.812	4.94 d 3	ε 99.88%, α 0.12%
246	2(-)	64.010s	1.80 d 2	ε, α<0.2%
247	(3/2-)	65.485	1380 y 250	α
248	(6+,8-)	68.099	>9 y	α>70%, β-<30%, ε?
248m	1(-)	68.099	23.7 h 2	β- 70%, ε 30%

Isotope Z El A	Jπ	Δ (MeV)	T1/2 or Abundance	Decay Mode
97 Bk 249	7/2+	69.844	320 d 6	β−, α 0.0015%, SFw
250	2−	72.948	3.22 h 1	β−
251	(3/2−)	75.230s	56 m 2	β−, α≈0.0001%
252		78.530s	?	
98 Cf 239		58.240s	39 s +37−1	α
240	0+	58.020s	1.06 m 15	α
241		59.170s	3.8 m 7	ε≈90%, α≈10%
242	0+	59.330	3.49 m 12	α, ε?
243	(1/2+)	60.910s	10.7 m 5	ε≈86%, α≈14%
244	0+	61.459	19.4 m 6	α, ε?
245		63.377	43.6 m 8	ε≈70%, α≈30%
246	0+	64.087	35.7 h 5	α, SF 0.0002%, ε<0.0001%
247	(7/2+)	66.150s	3.11 h 3	ε 99.96%, α 0.03%
248	0+	67.239	333.5 d 28	α, SFw
249	9/2−	69.718	350.6 y 21	α, SFw
250	0+	71.167	13.08 y 9	α 99.92%, SF 0.08%
251	1/2+	74.128	898 y 44	α, SFw
252	0+	76.030	2.638 y 10	α 96.91%, SF 3.09%
253	(7/2+)	79.296	17.81 d 8	β− 99.69%, α 0.31%
254	0+	81.337	60.5 d 2	SF 99.69%, α 0.31%
255			1.4 h 3	β−
256	0+		12.3 m 12	SF
99 Es 241		63.820s		
242		64.690s		
243		64.720s	21 s 2	ε≤70%, α≥30%
244		65.960s	37 s 4	ε 96%, α 4%
245		66.380s	1.33 m 15	ε 60%, α 40%
246		67.930s	7.7 m 5	ε 90.1%, α 9.9%, εSF
247		68.550	4.7 m 3	ε≈93%, α≈7%
248	(2−,0+)	70.270s	27 m 3	ε, α≈0.25%
249	7/2(+)	71.110	1.70 h 1	ε 99.43%, α 0.57%
250	(6+)	73.270s	8.6 h 1	ε≥97%, α≤3%
250m	1(−)	73.270s	2.22 h 5	ε≥99%, α≤1%
251	(3/2−)	74.507	33 h 1	ε 99.5%, α 0.5%
252	(5−,4+)	77.263	471.7 d 19	α 76%, ε 24%, β−≈0.01%
253	7/2+	79.008	20.47 d 3	α, SFw
254	(7+)	81.990	275.5 d 5	α
254m	2+	82.068	39.3 h 2	β− 99.59%, α 0.33%, ε 0.08%
255	(7/2+)	84.090s	39.8 d 12	β− 92%, α 8%, SF
256	(1+)	87.160s	25 m 4	β−
256m	(8+)	87.160s	≈7.6 h	β−
257			?	
100 Fm 242	0+		0.8 ms 2	SF, α?
243		69.360s	0.18 s +8−4	α
244	0+	69.040s	3.7 ms 4	SF≈99%, α≈1%

53

Isotope Z El A	Jπ	Δ (MeV)	T1/2 or Abundance	Decay Mode
103 Lr 252			≈1 s	α≈90%, ε≈10%, SF<1%
253		88.640s	1.4 s	α, SF<2%, ε 1%
254		89.730s	≈20 s	α, ε, SF<0.1%
255		90.050s	22 s 5	α>70%, ε<30%, SF<1%
256		91.740s	28 s 3	α>80%, ε<20%, SF<0.03
257		92.670s	0.646 s 25	α>85%, ε<15%
258		94.750s	4.3 s 5	α>95%, ε<5%
259		95.850	5.4 s 8	α>90%, SF<10%, ε<1%
260		98.100	180 s 30	α, ε?
104 Rf 253?			≈1.8 s	SF≈50%, α≈50%
254?	0+		0.5 ms 2	SF, α≈0.3%
255?	0+	94.310s	≈2 s	α≈50%, SF≈50%
256?	0+	94.280s	≈5 ms	SF
257		95.890s	4.8 s 3	α≈82%, ε≈18%
258	0+	96.350s	11 ms 2	SF≈90%, α≈10%
259		98.300s	3.1 s 7	α≈93%, SF≈6%, ε≈0.3%
260?	0+	99.020s	20 ms	SF, α<10%
261		101.240s	65 s 10	α>80%, SF<10%, ε<10%
262?	0+		63 ms	SF
105 Ha 255?			1.5 s	SF
257		100.390s	1 s	α≈80%, SF≈20%, ε?
258		101.550s	4 s	α, ε
259?		102.070s	1.2 s	SF
260		103.440s	1.52 s 13	α≈90%, SF≤9.6%, ε<2.5%
261		104.160s	1.8 s 4	α>50%, SF<50%
262		105.970s	34 s 4	SF 71%, α 27%, ε<5%
106 259?			7 ms 3	SF≈70%, α≈30%
260	0+	106.910s		
261		108.220s		
262	0+	108.470s		
263		110.120s	0.8 s 2	SF≈70%, α≈30%
107 261?			1.5 ms 5	α≈80%, SF≈20%
262		114.510s	115 ms	α
262m		114.510s	4.7 ms	α
263		114.800s		
264		115.960s		
108 264	0+	120.130s		
265		121.240s		
109 266?		128.210s	5 ms	α

55

Isotope Z El A	Jπ	Δ (MeV)	T1/2 or Abundance	Decay Mode
100 Fm 245		70.100s	4.2 s 13	α
246	0+	70.130	1.1 s 2	α 92%, SF 8%
247		71.540s	35 s 4	α 50%, ε 50%
247		71.540s	9.2 s 23	α
248	0+	71.885	36 s 3	α 99%, ε≈1%, SF≈0.05%
249	(7/2+)	73.500s	2.6 m 7	ε≈85%, α≈15%
250	0+	74.063	30 m 3	α>90%, SF≈0.0006%, ε<10%
250m	0+	74.063	1.8 s	IT
251	(9/2−)	76.000s	5.30 h 8	ε 98.2%, α 1.8%
252	0+	76.817	25.39 h 5	α, SFw
253	1/2+	79.339	3.00 d 12	ε 88%, α 12%
254	0+	80.897	3.240 h 2	α 99.94%, SF 0.06%
255	7/2+	83.787	20.07 h 7	α, SFw
256	0+	85.482	2.63 h 2	SF 91.9%, α 8.1%
257	(9/2+)	88.585	100.5 d 2	α 99.79%, SF 0.21%
258	0+		380 μs 60	SF
259			1.5 s 3	SF
101 Md 247		76.040s	3 s	α
248		77.080s	7 s 3	ε 80%, α 20%
249		77.260s	24 s 4	α≈70%, ε≈30%
250		78.600s	52 s 6	ε 94%, α 6%
251		79.020s	4.0 m 5	ε≥94%, α≤6%
252		80.540s	2.3 m 8	ε>50%, α<50%
253		81.240s		
254		83.490s	10 m 3	ε
254		83.490s	28 m 8	ε
255	(7/2−)	84.842	27 m 2	ε 92%, α 8%
256		87.522	76 m 4	ε 90.7%, α 9.3%
257	(7/2−)	89.030s	5.2 h 5	ε 90%, α 10%
258	(8−)	91.820s	55 d 4	α
258m	(1−)	91.820s	43 m 4	ε
259	(7/2−)		103 m	SF>95%, α<5%, ε?
102 No 250	0+		0.25 ms 5	SF, α≈0.1%
251		82.780s	0.8 s 3	α, ε≈1%
252	0+	82.856	2.30 s 22	α 73.1%, SF 26.9%
253	(9/2−)	84.330s	1.7 m 3	α≈80%, ε≈20%, SF≈0.001%
254	0+	84.723	55 s 5	α>99%, ε<1%, SF<0.06%
254m		85.223	0.28 s 4	IT
255	(1/2+)	86.870s	3.1 m 2	α 61.4%, ε 38.6%
256	0+	87.796	3.3 s 2	α 99.7%, SF≈0.25%
257	(7/2+)	90.220	25 s 2	α
258	0+	91.420s	1.2 ms	SF
259	(9/2+)	94.018	60 m 5	α 78%, ε 22%, SF<2%
260	0+		?	SF

54

521

Answers to Even-Numbered Problems

Chapter 1

2. 2.1×10^7 Hz

4. 2.7×10^{-17} kg·m/s, 3.7×10^5 Hz

6. 3.3 T

8. (a) 8.421×10^{-14} kg; (b) 29.821, 28.192, 29.875, 28.244, 34.911, 26.613, 41.679, and 33.295×10^{-19} C; (c) 18.0, 16.0, 17.0, 17.0, 21.0, 16.0, 25.1, and $20.1 \times (1.66 \times 10^{-19}$ C)

10. 0.72 T

12. ^{44}Ca

14. 0.683

18. $\rho = \rho_0 e^{-mgz/kT}$

20. 4.0×10^{-21} J; 1.0×10^{-10} m/s, 2.9×10^{-11} radian

22. 1.6×10^{-12} m/s, 1.6×10^{-13} m

24. 9.0×10^{-15} C, 1.3×10^{-9} A

26. 3.4 cm, 3.4 cm

30. $v_p = 2.0$ m/s, $v_g = 3$ m/s

32. $v_g = \sqrt{gh}$

34. (a) $v_g = (3/2)v_p$; (b) $v_p = 2.1$ m/s, $v_g = 3.1$ m/s

36. 3.7×10^{-18} N, 6.2×10^{-18}, 2.4×10^{-3} m/s^2 toward Sun

38. 1.00867 u

40. 1.7×10^{11} J, 1.9×10^{-6} kg, 1.9×10^{-4}%

42. 0.3651 Å

Chapter 2

2. (a) $\sqrt{1 - v^2/c^2}$; (b) for unequal arms, such a contraction would *not* make the phase difference zero

6. $\theta' = 43.9°$

10. 6.82×10^{-8}%

12. 3.61 light-years, 7.62 years

14. (a) 25.9 years; (b) 23.7 years

18. 1.63×10^{-18} m

20. $v_y = v'_y \sqrt{1 - V^2/c^2}/(1 + Vv_x/c^2)$, $v_z = v'_z \sqrt{1 - V^2/c^2}/(1 + Vv_x/c^2)$

22. (a) $\theta' = \tan^{-1}[v/(c\sqrt{1-V^2/c^2})] = 0.0057 = 20''$;
(b) $\theta' = \tan^{-1}(v/c)$; difference is not detectable
24. $a'_x = a_x[(1-V^2/c^2)^{3/2}/(1-Vv_x/c^2)^3]$
$a'_y = a_y[(1-V^2/c^2)/(1-Vv_x/c^2)^2]$
$a'_z = a_z[(1-V^2/c^2)/(1-Vv_x/c^2)^2]$
26. 0.0040%   **28.** 132 m/s, 5.33×10^{-16} kg·m/s
30. $V = 0.905c$
32. $E_\nu = 29.9$ MeV, $p_\nu = 29.9$ MeV/c; $E_\mu = 110$ MeV, $p_\mu = 29.9$ MeV/c in opposite direction

Chapter 3

2. 75 W   **4.** (a) 1.7×10^{11} Hz; (b) 2.0×10^9 W
6. 279 K or 6°C   **14.** no
16. 1.5×10^6/s   **18.** (a) 3.2×10^{20}/s; (b) 1.3×10^{16}/s
20. (a) all photoelectrons reach anode (saturation); (b) current is twice as large; (c) stopping potential is twice as large
22. 1.02×10^{15} Hz   **26.** 0.713 Å, 0.696 Å, 0.679 Å
28. 22 eV   **30.** 1.32×10^{-5} Å
34. 31 mm   **36.** 0.048 cm
38. 0.12   **40.** diamond, lead

Chapter 4

2. yellow, He   **4.** fifth line ($n = 9$)
6. (a) 4102.9 Å and 4341.7 Å in Balmer series, lengthened by a factor of 1.0035; (b) 1.0×10^6 m/s
8. (a) $p = -0.0407$, $d = -0.0016$; (b) Principal corresponds to Lyman, Diffuse to Balmer; (c) $1/\lambda = R[1/(3+d)^2 - 1/(n+f)^2]$
10. (a) 0.25 Å; (b) 1.75×10^{16} Hz   **12.** 33 MeV
14. (a) 542, 458; (b) 212, 115
18. 1.24×10^{-26} m² for 20°, 5.38×10^{-27} m² for 30°, 7.04×10^{-27} m² for 20° to 30°
20. $c/137$, $c/274$ **22.** 1215.7 Å
24. 1.79 Å, 1.32 Å; larger for first line   **26.** 0.18 Å, -122 eV
28. 2.85×10^{-3} Å, $-(2.53 \times 10^3$ eV$)/n^2$, 6.541 Å
30. $n\hbar^2/(GMm^2)$; (b) 2.54×10^{74}; (c) 1.18×10^{-63} m
34. 0.130 Å, cobalt
36. tungsten   **38.** 7.00

Chapter 5

2. 0.087 Å; yes   **4.** 0.061 Å
6. 6.2×10^{-9} m, 1.9×10^{-15} m, 2.4×10^{-23} m, 1.5×10^{-25} m
8. (a) 1.19×10^{-20} J, 3.76×10^3 m/s; (b) 1.05×10^{-10} m
10. 6.2×10^{-20} m   **12.** 6.0×10^{-6} eV
14. (a) 1.65 Å; (b) 1.67 Å   **16.** 6.02 cm, 14.8 cm, 39.8 cm
18. $\theta = 44.0°$
20. (b) equal intensities, but unequal phases, unless the phase change per reflection is (fortuitously) 0° or 180°
22. 5×10^{-6} m, 1×10^{-28} kg·m/s   **24.** 6×10^{-3} m, 0; 3×10^{-6} m, 0

26. with $\Delta p \simeq h/\Delta x \simeq h/2a_0$, the uncertainty in wavelength is of the same order of magnitude as the wavelength, i.e., the wavelength is undefined

30. (a) 2×10^{-40} kg·m/s; (b) 5×10^2 Hz

32. 3×10^{-23} s

Chapter 6

2. 3.7×10^3 MeV, 1.1×10^4 MeV, no   **4.** 2×10^{-3}, 1.5×10^{-3}

6. $\bar{x} = \frac{1}{2}L$, $\overline{x^2} = L^2[\frac{1}{3} - 1/(2n^2\pi^2)]$ **8.** $n^2\pi^2\hbar^2/(mL^3)$

12. $E_{111} = 3\hbar^2\pi^2/(2mL^2)$, $E_{211} = E_{121} = E_{112} = 6\hbar^2\pi^2/(2mL^2)$

16. $\Psi = e^{-iEt/\hbar}e^{ikx} + e^{-iEt/\hbar}e^{-ikx}(ik + \kappa)/(ik - \kappa)$ for $x < 0$

   $\Psi = e^{-iEt/\hbar}e^{-\kappa x}(2ik)/(ik - \kappa)$ for $x > 0$

   with $k = \sqrt{2mE/\hbar^2}$ and $\kappa = \sqrt{2m(U_0 - E)/\hbar^2}$

18. $0.74\hbar^2\pi^2/(2mL^2)$, $2.82\hbar^2\pi^2/(2mL^2)$ **22.** 2.5×10^{-7}

24. $R = -\dfrac{e^{2ika}(1 - e^{-2\kappa(b-a)})(\kappa^2 + k^2)}{(\kappa - ik)^2 - (\kappa + ik)^2 e^{-2\kappa(b-a)}}$ **26.** $1 - (k' - k)^2/(k' + k)^2$

28. 9.0×10^{-57} **30.** $P = e^{-\pi A\sqrt{B}\sqrt{2m/\hbar^2}}$

32. $\hbar\omega_0$ **38.** $(n + \frac{1}{2}) \times 0.540$ eV; 2.29×10^4 Å

40. $x^2 = \hbar/(2m\omega_0)$ **42.** $E_n = (3Anh/4)^{2/3}/(2m)^3$

44. $l = 0, 1, 2$; $m_l = 0, \pm1, \pm2$ **46.** $-m_ee^4/[2(4\pi\varepsilon_0)^2\hbar^2]$

48. 0.323

50. $U = -m_ee^4/[(4\pi\varepsilon_0)^2\hbar^2]$, $K = m_ee^4/[2(4\pi\varepsilon_0)^2\hbar^2]$

Chapter 7

2. 1, 4, 5   **4.** $\frac{1}{2}\sqrt{63}\,\hbar$, $\frac{1}{2}\sqrt{35}\,\hbar$

6. $^2S_{1/2}$, $^2S_{1/2}$, $^2P_{1/2}$, $^2P_{3/2}$, $^2P_{1/2}$, $^2P_{3/2}$, $^2D_{3/2}$, $^2D_{5/2}$

8. (a) $p = qL/2 - (8/9)(\sqrt{3}/\pi^2)qL\cos[(E_1 - E_2)t/\hbar]$

   (b) $p = qL/2$; forbidden

10. 0, $\pm e\hbar/(2m_e)$, 0, $\pm\frac{1}{3}e\hbar/(2m_e)$

12. 0.086 Å, 0.034 Å, -0.043 Å, -0.086 Å

14. 2.31×10^{-5} eV, 0.77×10^{-5} eV, -0.77×10^{-5} eV, -2.31×10^{-5} eV

16. 7 T   **18.** (b) the second

20. (a) $\psi_1(x)\psi_1(x') = (2/L)\sin(\pi x/L)\sin(\pi x'/L)$

   (b) $\psi_1(x)^2\psi_1(x)^2(dx)^2 = (2/L)^2(dx)^2$

   (c) $\psi_1(x)^2\,dx = 2\,dx/L$

22. 1s2s for $^3S_1$, 1s2s for $^1S_0$, 1s2p for $^3P_0$, 1s2p for $^1P_1$

24. $^1S_0$, same as argon

26. $^2S_{1/2}$ for IA, $^1S_0$ for IIA, various for IIIB, $^3F_2$ for IVB, $^4F_{3/2}$ for VB, $^7S_3$ for VIB, $^6S_{5/2}$ for VIIB, various for VIII, $^2S_{1/2}$ for IB, $^1S_0$ for IIB, $^2P_{1/2}$ for IIIA, $^3P_0$ for IVA, $^4S_{3/2}$ for VA, $^3P_2$ for VIA, $^2P_{3/2}$ for VIIA, $^1S_0$ for 0

28. 1.39×10^{-46} kg·m^2, 1.09 Å   **30.** 11 eV

32. 8.97×10^{13} Hz

Chapter 8

2. 2.82 Å   **4.** 8.47×10^{28}/m^3

6. 4.8×10^{-6} m/s, 4.9 days   **8.** (c) yes, reversed; (d) 1.8×10^{-4} V

10. 3.1×10^5 J   **12.** $\bar{v} = (3/4)v_F$, $\sqrt{\overline{v^2}} = \sqrt{2/5}\,v_F$

14. 19 MeV, 6.0×10^7 m/s
16. 1800 K
20. $m^* = \hbar/Ba$
22. no
24. 4.149 K, 4.138 K, 4.118 K
26. 5.7×10^{-4} T·m^2, $2.8 \times 10^7\,\Phi_0$
28. 3.61×10^{-4}, 1.66×10^{-4}, 3.31×10^{-4}, 2.87×10^{-3}, 1.03×10^{-3},
1.129×10^{-3}, 2.0×10^{-4}, 3.6×10^{-6}, 1.260×10^{-3}, and 2.18×10^{-3} eV
30. 2.2×10^{-15} T·m^2

Chapter 9

4. 5.70 MeV
6. 1.88 MeV
8. 2.3×10^{17} kg/m^3
10. 32 MeV
12. 3.4 fm, 3.7 fm, 4.2 fm; yes, $R = 1.5$ fm $\times$ A$^{1/3}$
14. $E_F = 53$ MeV $\times (Z/A)^{2/3}$, $\bar{E} = 32$ MeV $\times (Z/A)^{2/3}$; $E_F = 32$ MeV, $\bar{E} = 19$ MeV;
$E_F = 28$ MeV, $\bar{E} = 17$ MeV
16. 1.3×10^4 N
18. $0.8800 e\hbar/(2m_p)$
20. 17.0209, 18.0068, 19.0009, 19.9934, 20.9928, 21.9900, 22.9933, 23.9941,
25.0004, 26.0039, 27.0126, and 28.0183 u, in good agreement to three decimal
places
24. $7.3459 A^{2/3} - 0.8345 Z^2/A^{1/3}$, 25.5 MeV
26. 0.177 MeV, 0.149 MeV, 0.243 MeV; 0.0701 Å, 0.0832 Å, 0.0510 Å
28. 15.9996 u, noticeably larger than the actual value, because mass formula does not
take into account magic numbers
30. (a) spin of nucleus is 1/2, which gives two possible total spin states for every
electronic angular momentum; (b) 5.9×10^{-6} eV
32. 2.166, $3.249 e\hbar/(2m_p)$

Chapter 10

2. $\sqrt{(t - \tau)^2} = \tau$, $t_{\text{prob}} = 0$
4. 0.88 kg, 17.4 years
6. (a) 4.57×10^9 years; (b) native lead would have contained a mix of isotopes
8. 1.49×10^2 (MeV) from Fig. 10.5, 1.59×10^2 (MeV) from Eq. (33)
10. 10^7 m/s
12. 0.0862 MeV, 4.8707 MeV
14. 0.00119 u
16. 6.1×10^9/s
18. 11.90 MeV, 1.25 MeV
22. 0.93 kg
24. 7.7×10^{10} years
26. (a) 1.6×10^{-10}, 1.6×10^{-59}; (b) 2.2×10^9 K

Chapter 11

2. 1.3 T
4. 2.70×10^{-8} m
8. 2×10^{-18} m
10. first forbidden by charge conservation, fourth by baryon number, fifth by
lepton number
12. weak, weak
14. $K^- + p \rightarrow K^0 + K^+ + \Omega^-$, strong; $\Omega^- \rightarrow \Xi^0 + \pi^-$, weak; $\Xi^0 \rightarrow (\gamma + \gamma) + \Lambda^0$,
weak; $\Lambda^0 \rightarrow \pi^- + p$, weak; $\gamma \rightarrow e + \bar{e}$, electromagnetic
16. 1677 MeV/c^2
20. $\bar{s}d$, $\bar{s}u$, $\bar{u}d$, $u\bar{d}$, $s\bar{u}$, $s\bar{d}$
22. yes, $\mu_p/\mu_n = -1.913/2.793 \simeq -\frac{2}{3}$

Bibliography

The list of books in this bibliography is highly selective; it unabashedly reflects what the author happens to be acquainted with, and likes or finds useful.

1. Classical Mechanics and Electromagnetism

The following introductory textbooks provide the background in elementary classical mechanics and electromagnetism prerequisite for the study of modern physics:

Halliday, D., and Resnick, R., *Fundamentals of Physics*, (Wiley, New York, 1981)

Ohanian, H. C., *Physics* (Norton, New York, 1985)

Tipler, P. A., *Physics* (Worth, New York, 1976)

2. Modern Physics

There are many books that deal with modern physics at roughly the same level as this book. Some of these are:

Acosta, V., Cowan, C. L., and Graham, B. J., *Essentials of Modern Physics* (Harper and Row, New York, 1973)

Anderson, E. E., *Introduction to Modern Physics* (Saunders, Philadelphia, 1982)

Beiser, A., *Concepts of Modern Physics* (McGraw-Hill, New York, 1981)

Bitter, F., and Medicus, H. A., *Fields and Particles* (Elsevier, New York, 1973)

Enge, H. A., Wehr, M. R., and Richards, J. A., *Introduction to Atomic Physics* (Addison-Wesley, Reading, Massachussetts, 1972)

French, A. P., and Taylor, E., *An Introduction to Quantum Physics* (Norton, New York, 1978)

Leighton, R. B., *Principles of Modern Physics* (McGraw-Hill, New York, 1959)

Richtmeyer, F. K., Kennard, E. H., and Cooper, J. N., *Introduction to Modern Physics* (McGraw-Hill, New York, 1969)

Semat, H., *Introduction to Atomic and Nuclear Physics* (Rinehart, New York, 1972)

Sproull, R. L., and Phillips, W. A., *Modern Physics* (Wiley, New York, 1980)

Tipler, P. A., *Modern Physics* (Worth, New York, 1978)

Weidner, R. T., and Sells, R. L., *Elementary Modern Physics* (Allyn and Bacon, Boston, 1980)

Wichmann, E. H., *Quantum Physics* (McGraw-Hill, New York, 1971)

Two older books, first published in the 30's, have become classics and still remain valuable:

Born, M., *Atomic Physics*, translated by J. Dougall (Hafner, New York, 1962)

Herzberg, G., *Atomic Spectra and Atomic Structure*, translated by J. W. T. Spinks (Prentice-Hall, New York, 1937; reprinted by Dover, New York, 1944)

3. Relativity

The two best elementary introductions to the theory of Special Relativity are:

French, A. P., *Special Relativity* (Norton, New York, 1968)

Taylor, E. F., and Wheeler, J. A., *Spacetime Physics* (Freeman, San Francisco, 1966)

More advanced treatments will be found in:

Jackson, J. D., *Classical Electrodynamics* (Wiley, New York, 1975), Chapter 11

Møller, C., *The Theory of Relativity* (Oxford University Press, London, 1952), Chapters I–VII

Pauli, W., *Theory of Relativity*, (Pergamon Press, London, 1958)

Rindler, W., *Essential Relativity*, (Springer, New York, 1977), Chapters 1–6

Synge, J. L., *Relativity: The Special Theory* (Interscience, New York, 1956)

4. Quantum Theory and Its Applications

The following textbooks are written at a higher level than those listed in Section 2; most of them presume some acquaintance with the basic concepts of modern physics.

Beard, D. B., and Beard, G. B., *Quantum Mechanics with Applications* (Allyn and Bacon, Boston, 1970)

Bohm, D., *Quantum Theory* (Prentice-Hall, Englewood Cliffs, New Jersey, 1951)

Dicke, R. H., and Wittke, J. P., *Introduction to Quantum Mechanics* (Addison-Wesley, Reading, Massachusetts, 1960)

Eisberg, R. M., and Resnick, R., *Quantum Physics of Atoms, Molecules, Solids, Nuclei and Particles* (Wiley, New York, 1974)

Feynman, R. P., Leighton, R. B., and Sands, M., *The Feynman Lectures on Physics*, Vol. III (Addison-Wesley, Reading, Massachusetts, 1965)

Frenkel, J., *Wave Mechanics—Elementary Theory* (Dover, New York, 1950)

Matthews, P. T., *Introduction to Quantum Mechanics* (McGraw-Hill, Maidenhead, Berkshire, England, 1968)

Powell, J. L., and Craseman, B., *Quantum Mechanics* (Addison-Wesley, Reading, Massachusetts, 1961)

Saxon, D. S., *Elementary Quantum Mechanics* (Holden-Day, San Francisco, 1968)

5. Solid State Physics, Nuclear Physics, and Elementary Particles

Dekker, A. J., *Solid State Physics* (Prentice-Hall, Englewood Cliffs, New Jersey, 1960)

Dodd, J. E., *The Ideas of Particle Physics* (Cambridge University Press, Cambridge, 1984)

Evans, R. D., *The Atomic Nucleus* (McGraw-Hill, New York, 1955)

Feinberg, G., *What Is the World Made Of?* (Doubleday, Garden City, New York, 1977)

Fermi, E., *Nuclear Physics* (University of Chicago Press, Chicago, 1950)

Fritzsch, H., *Quarks* (Basic Books, New York, 1983)

Halliday, D., *Introductory Nuclear Physics* (Wiley, New York, 1955)

Kittel, C., *Introduction to Solid State Physics*, 5th ed. (Wiley, New York, 1976)

Polkinghorne, J. C., *The Particle Play* (Freeman, Oxford, 1979)

Trefil, J. S., *From Atoms to Quarks* (Scribner's, New York, 1980)

6. History

Born, M., *The Born-Einstein Letters* (Walker, New York, 1971)

Boorse, H. A., and Motz, L., *The World of the Atom* (Basic Books, New York, 1966)

Gamow, G., *Thirty Years That Shook Physics* (Doubleday, Garden City, New York, 1966)

Heisenberg, W., *Physics and Beyond—Encounters and Conversations* (Harper and Row, New York, 1971)

Jammer, M., *The Conceptual Development of Quantum Mechanics* (McGraw-Hill, New York, 1966)

Schilpp, P. A., ed., *Albert Einstein, Philosopher-Scientist* (Harper, New York, 1959)

Segré, E., *From X-Rays to Quarks* (Freeman, San Francisco, 1980)

van der Waerden, B. L., *Sources of Quantum Mechanics* (Dover, New York, 1968)

Weinberg, S., *The Discovery of Subatomic Particles* (Freeman, New York, 1983)

7. General References

Brookhaven National Laboratory (M. D. Oldberg, et al.), *Neutron Cross Sections* (Upton, New York, 1966)

Condon, E. U., and Shortley, G. H., *The Theory of Atomic Spectra* (University Press, Cambridge, 1964)

Frauenfelder, H., and Henley, E. M., *Subatomic Physics* (Prentice-Hall, Englewood Cliffs, New Jersey, 1974)

Frazer, W. R., *Elementary Particles* (Prentice-Hall, Englewood Cliffs, New Jersey, 1966)

General Electric Company, *Chart of the Nuclides*, 13th edition (Knolls Atomic Power Laboratory, Schenectady, New York, 1983)

Gentner, W., Maier-Leibnitz, H., and Bothe, W., *An Atlas of Typical Expansion Chamber Photographs* (Interscience, New York, 1954)

Glasstone, S., ed., *The Effects of Nuclear Weapons* (U. S. Atomic Energy Commission, Washington, 1962)

Halzen, F., and Martin, A. D., *Quarks and Leptons* (Wiley, New York, 1984)

Herzberg, G., *Molecular Spectra and Molecular Structure* (Van Nostrand, New York, 1950)

Hughes, D. J., and Harvey, J. A., *Neutron Cross Sections* (McGraw-Hill, New York, 1955)

Hughes, I. S., *Elementary Particles* (Cambridge University Press, Cambridge, 1985)

Kuhn, H. G., *Atomic Spectra* (Academic Press, New York, 1962)

Landau, L. D., and Lifshitz, E. M., *Statistical Physics* (Addison-Wesley, Reading, Massachusetts, 1958)

Leon, M., *Particle Physics: An Introduction* (Academic Press, New York, 1973)

Millikan, *The Electron*, 2nd ed. (University of Chicago Press, Chicago, 1924)

Morgan, W. W., Abt, H. A., Tapscott, J. W., *Revised MK Spectral Atlas for Stars Earlier Than the Sun* (Yerkes Observatory and Kitt Peak National Observatory, 1978)

Particle Data Group (Aguilar-Benitez, M., et al.), *Review of Particle Properties* (Berkeley, California, April 1986)

Preston, M. A., *Physics of the Nucleus* (Addison-Wesley, Reading, Massachusetts, 1962)

Schiff, L. I., *Quantum Mechanics* (McGraw-Hill, New York, 1968)

Segré, E., ed., *Experimental Nuclear Physics* (Wiley, New York, 1953)

Shutt, R. P., ed., *Bubble and Spark Chambers* (Academic Press, New York, 1967)

Sommerfeld, A., *Atombau und Spektrallinien* (Vieweg, Braunschweig, 1931)

Tuli, J. K., *Nuclear Wallet Cards* (National Nuclear Data Center, Brookhaven National Laboratory, Upton, New York, 1985)

Williams, W. S. C., *An Introduction to Elementary Particles* (Academic Press, New York, 1961)

Index

FUNDAMENTAL CONSTANTS

Speed of light	$c = 3.00 \times 10^8 \, \text{m/s}$
Planck's constant	$h = 6.63 \times 10^{-34} \, \text{J} \cdot \text{s} = 4.14 \times 10^{-21} \, \text{MeV} \cdot \text{s}$
	$\hbar = \dfrac{h}{2\pi} = 1.05 \times 10^{-34} \, \text{J} \cdot \text{s} = 6.58 \times 10^{-22} \, \text{MeV} \cdot \text{s}$
Gravitational constant	$G = 6.67 \times 10^{-11} \, \text{N} \cdot \text{m}^2/\text{kg}^2$
Permittivity constant	$\varepsilon_0 = 8.85 \times 10^{-12} \, \text{F/m} \qquad \dfrac{1}{4\pi\varepsilon_0} = 8.99 \times 10^9 \, \text{F/m}$
Permeability constant	$\mu_0 = 1.26 \times 10^{-6} \, \text{H/m} \qquad \dfrac{\mu_0}{4\pi} = 10^{-7} \, \text{H/m}$
Avogadro's number	$N_A = 6.02 \times 10^{23}/\text{mole}$
Boltzmann's constant	$k = 1.38 \times 10^{-23} \, \text{J/K} = 8.62 \times 10^{-5} \, \text{eV/K}$
Electron charge	$-e = -1.6 \times 10^{-19} \, \text{C}$
Electron mass	$m_e = 9.11 \times 10^{-31} \, \text{kg}$
Proton mass	$m_p = 1.673 \times 10^{-27} \, \text{kg}$
Neutron mass	$m_n = 1.675 \times 10^{-27} \, \text{kg}$
Proton-electron mass ratio	$m_p/m_c = 1836$
Stefan-Boltzmann constant	$\sigma = \dfrac{\pi^2 k^4}{60\hbar^3 c^2} = 5.67 \times 10^{-8} \, \text{W}/(\text{m}^2 \cdot \text{K}^4)$
Rydberg constant	$R_H = \dfrac{e^4 m_e}{4\pi(4\pi\varepsilon_0)^2 \hbar^3 c} = 1.097 \times 10^7/\text{m}$
Bohr radius	$a_0 = \dfrac{4\pi\varepsilon_0 \hbar^2}{e^2 m_e} = 0.529 \, \text{Å}$
Fine-structure constant	$\alpha = \dfrac{e^2}{4\pi\varepsilon_0 \hbar c} = \dfrac{1}{137}$
Compton wavelength	$\dfrac{h}{m_e c} = 2.43 \times 10^{-12} \, \text{m}$
Electron magnetic moment	$\mu_{e,z} = -\dfrac{1.001 e\hbar}{2m_e} = 9.28 \times 10^{-24} \, \text{J/T}$
Proton magnetic moment	$\mu_{p,z} = \dfrac{2.79 e\hbar}{2m_p} = 1.41 \times 10^{-26} \, \text{J/T}$
Neutron magnetic moment	$\mu_{n,z} = -\dfrac{1.91 e\hbar}{2m_n} = -9.66 \times 10^{-27} \, \text{J/T}$
Flux quantum	$\Phi = \dfrac{h}{2e} = 2.07 \times 10^{-15} \, \text{T} \cdot \text{m}^2$